AF362046

Future Advances in Basin Modeling

Future Advances in Basin Modeling

Suggestions from Current Observations, Analyses, and Simulations

Editors

Willy Fjeldskaar
Lawrence Cathles

MDPI • Basel • Beijing • Wuhan • Barcelona • Belgrade • Manchester • Tokyo • Cluj • Tianjin

Editors

Willy Fjeldskaar   Lawrence Cathles
Tectonor AS   Cornell University
Norway   USA

Editorial Office
MDPI
St. Alban-Anlage 66
4052 Basel, Switzerland

This is a reprint of articles from the Special Issue published online in the open access journal *Geosciences* (ISSN 2076-3263) (available at: https://www.mdpi.com/journal/geosciences/special_issues/basin_modelling).

For citation purposes, cite each article independently as indicated on the article page online and as indicated below:

LastName, A.A.; LastName, B.B.; LastName, C.C. Article Title. *Journal Name* **Year**, *Volume Number*, Page Range.

ISBN 978-3-0365-0276-2 (Hbk)
ISBN 978-3-0365-0277-9 (PDF)

Cover image courtesy of Ivar Grunnaleite, Tectonor AS.

Contents

About the Editors vii

Preface to "Future Advances in Basin Modeling" ix

Lawrence Cathles and Willy Fjeldskaar
A Summary of "Future Advances in Basin Modeling: Suggestions from Current Observations,
Analyses and Simulations"
Reprinted from: *Geosciences* **2020**, *10*, 506, doi:10.3390/geosciences10120506 1

Lawrence Cathles
On the Processes that Produce Hydrocarbon and Mineral Resources in Sedimentary Basins
Reprinted from: *Geosciences* **2019**, *9*, 520, doi:10.3390/geosciences9120520 9

Christopher C. Barton and Jacques Angelier
Direct Inversion Method of Fault Slip Analysis to Determine the Orientation of Principal
Stresses and Relative Chronology for Tectonic Events in Southwestern White Mountain Region
of New Hampshire, USA
Reprinted from: *Geosciences* **2020**, *10*, 464, doi:10.3390/geosciences10110464 35

**Antoine Bouziat, Nicolas Guy, Jérémy Frey, Daniele Colombo, Priscille Colin,
Marie-Christine Cacas-Stentz and Tristan Cornu**
An Assessment of Stress States in Passive Margin Sediments: Iterative Hydro-Mechanical
Simulations on Basin Models and Implications for Rock Failure Predictions
Reprinted from: *Geosciences* **2019**, *9*, 469, doi:10.3390/geosciences9110469 57

Ingrid F. Løtveit, Willy Fjeldskaar and Magnhild Sydnes
Tilting and Flexural Stresses in Basins Due to Glaciations—An Example from the Barents Sea
Reprinted from: *Geosciences* **2019**, *9*, 474, doi:10.3390/geosciences9110474 79

Daniele Cerroni, Mattia Penati, Giovanni Porta, Edie Miglio and Paolo Ruffo
Multiscale Modeling of Glacial Loading by a 3D Thermo-Hydro-Mechanical Approach
Including Erosion and Isostasy
Reprinted from: *Geosciences* **2019**, *9*, 465, doi:10.3390/geosciences9110465 95

Andrés Cedeño, Luis Alberto Rojo, Néstor Cardozo, Luis Centeno and Alejandro Escalona
The Impact of Salt Tectonics on the Thermal Evolution and the Petroleum System of Confined
Rift Basins: Insights from Basin Modeling of the Nordkapp Basin, Norwegian Barents Sea
Reprinted from: *Geosciences* **2019**, *9*, 316, doi:10.3390/geosciences9070316 119

Ivar Grunnaleite and Arve Mosbron
On the Significance of Salt Modelling—Example from Modelling of Salt Tectonics, Temperature
and Maturity Around Salt Structures in Southern North Sea
Reprinted from: *Geosciences* **2019**, *9*, 363, doi:10.3390/geosciences9090363 145

Larry D. Brown and Doyeon Kim
Extensive Sills in the Continental Basement from Deep Seismic Reflection Profiling
Reprinted from: *Geosciences* **2020**, *10*, 449, doi:10.3390/geosciences10110449 171

Magnhild Sydnes, Willy Fjeldskaar, Ivar Grunnaleite, Ingrid Fjeldskaar Løtveit and Rolf Mjelde
Transient Thermal Effects in Sedimentary Basins with Normal Faults and Magmatic Sill Intrusions—A Sensitivity Study
Reprinted from: *Geosciences* **2019**, *9*, 160, doi:10.3390/geosciences9040160 . . . . **191**

Magnhild Sydnes, Willy Fjeldskaar, Ivar Grunnaleite, Ingrid Fjeldskaar Løtveit and Rolf Mjelde
The Influence of Magmatic Intrusions on Diagenetic Processes and Stress Accumulation
Reprinted from: *Geosciences* **2019**, *9*, 477, doi:10.3390/geosciences9110477 . . . . **223**

Peter E. Malin, Peter C. Leary, Lawrence M. Cathles and Christopher C. Barton
Observational and Critical State Physics Descriptions of Long-Range Flow Structures
Reprinted from: *Geosciences* **2020**, *10*, 50, doi:10.3390/geosciences10020050 . . . . **251**

Charles Sicking and Peter Malin
Fracture Seismic: Mapping Subsurface Connectivity
Reprinted from: *Geosciences* **2019**, *9*, 508, doi:10.3390/geosciences9120508 . . . . **267**

Mona Wetrhus Minde and Aksel Hiorth
Compaction and Fluid—Rock Interaction in Chalk Insight from Modelling and Data at Pore-, Core-, and Field-Scale
Reprinted from: *Geosciences* **2020**, *10*, 6, doi:10.3390/geosciences10010006 . . . . **301**

Lawrence Cathles and Alain Prinzhofer
What Pulsating H_2 Emissions Suggest about the H_2 Resource in the Sao Francisco Basin of Brazil
Reprinted from: *Geosciences* **2020**, *10*, 149, doi:10.3390/geosciences10040149 . . . . **317**

Frédéric-Victor Donzé, Laurent Truche, Parisa Shekari Namin, Nicolas Lefeuvre and Elena F. Bazarkina
Migration of Natural Hydrogen from Deep-Seated Sources in the São Francisco Basin, Brazil
Reprinted from: *Geosciences* **2020**, *10*, 346, doi:10.3390/geosciences10090346 . . . . **335**

Jacob Simon, Patrick Fulton, Alain Prinzhofer and Lawrence Cathles
Earth Tides and H_2 Venting in the Sao Francisco Basin, Brazil
Reprinted from: *Geosciences* **2020**, *10*, 414, doi:10.3390/geosciences10100414 . . . . **351**

About the Editors

Willy Fjeldskaar received his Dr. Scient. in geophysics from the University of Bergen (Norway) in 1981. He worked for more than 20 years in the research foundation Rogaland Research/IRIS (now Force). In addition to his management position in geology, his own research was focused on basin modeling and glacial isostatic adjustment. For 10 years, he was Adjunct Professor in Reservoir Geology at the University of Stavanger (Norway). He now carries out applied research at Tectonor AS. He has published many peer-reviewed papers in Earth science, mostly related to sedimentary basins and modeling of glaciation isostasy.

Lawrence Cathles received his Ph.D. in geophysics from Princeton University in 1971. He has researched mineral and industrial processes at the Kennecott Copper Corporation's Ledgemont Laboratory in Massachusetts, massive sulfide deposits at the Pennsylvania State University, gold, sulfide, and CO2 at the Chevron California Oil Field Research Laboratory, and, since 1987, multiphase flow dynamics, chemical alteration, nanoparticles, and global resources at Cornell. He is a fellow of the American Association for the Advancement of Science, was the 2011 Distinguished Lecturer of the Society of Economic Geologists, and is the 2021 recipient of the Penrose Gold Medal of the SEG. He has published over 140 peer-reviewed publications and a book titled "The Viscosity of the Earth's Mantle".

Preface to "Future Advances in Basin Modeling"

This volume describes the nature, causes, and consequences of the diverse fluid movements that produce energy and mineral resources in sedimentary basins. The contained papers point to new capabilities in basin analysis methods and models. The processes that operate in the resource-producing thermo-chemical-structural reactors we call sedimentary basins are reviewed. Efficient ways to infer the tectonic history of basins are described. Impacts on hydrocarbon maturation and migration of glacial tilting, magmatic intrusion, salt migration, and fracturing are illustrated. The conditions under which subsurface flow will channel with distance traveled are identified. Seismic methods that can image and map subsurface permeability channels are described. The surface maturation, surface charge, and chemical reaction foundations of creep subsidence are set forth. Dynamic aspects of the hydrogen resource in basins are analyzed. There is much that is new that is presented in these papers with the intent of stimulating thinking and enthusiasm for the advances that will be made in future decades.

Willy Fjeldskaar, Lawrence Cathles
Editors

Editorial

A Summary of "Future Advances in Basin Modeling: Suggestions from Current Observations, Analyses and Simulations"

Lawrence Cathles [1,*] and Willy Fjeldskaar [2,*]

1   Department of Earth and Atmospheric Sciences, Cornell University, Ithaca, NY 14853, USA
2   Tectonor AS, P.O. Box 8034, NO-4068 Stavanger, Norway
*   Correspondence: lmc19@cornell.edu (L.C.); wf@tectonor.com (W.F.)

Received: 9 December 2020; Accepted: 18 December 2020; Published: 21 December 2020

1. Introduction

The objective of this volume differs from that of the usual review of current advances. While the state of the art remains the basis for departure, our main objective is to identify areas where advances in understanding and modeling could significantly impact exploration effectiveness. Our criterion is not what is the most exciting and important current science, but what could be the most important in the future. We encouraged our authors to avail themselves of the increased latitude for speculation that this future focus affords, and solicited the papers contained in this volume with this perspective.

A first step in any discussion of basin processes must be an observational appreciation of their scale and diversity. The scale of operation is huge, and the driving forces planetary. The first paper is an overview that reminds us of this. Subsequent papers address more specific basin phenomena:

- Two papers illustrate how changes in the stress tensor can be inferred and used.
- Two papers show how oil migration is impacted by glacial changes in strata tilt.
- Two papers calculate the delay in hydrocarbon maturation caused by salt diapirism.
- Three papers assess the impact of magmatic intrusion on the timing of hydrocarbon maturation and sediment alteration.
- Two papers address the progressive focusing of flow with distance traveled and how flow pathways can be detected seismically.
- One paper investigates the nature of compaction and alteration that is related to water flooding.
- Three papers describe and analyze basin hydrogen seeps and the possibility of a "new" nonhydrocarbon (H_2) basin energy resource.

2. Review of Volume Papers

2.1. Overview

In the first overview paper, Cathles [1] reminds us that basin formation is driven by global plate tectonics and that the basin fluids that accumulate resources can move over many hundreds of kilometers. Brine and petroleum fluid movements produce alteration patterns that can be vectors to resource accumulation. Flow can be steady or episodic. Different styles of flow produce different types of mineral and hydrocarbon resources. Permeability is not always an intrinsic property of a strata but is often dynamically controlled either by pressure or the presence of nonaqueous fluids. Shosa seals (the kind that can trap variably pressured gas for hundreds of millions of years) can massively fail and then reheal, repeatedly propelling large volumes of gas into subsurface aquifers. Seals can migrate through the stratigraphy, and porosity profiles record this migration and the causative pore pressure changes. Departures from the expected path of paleomagnetic pole migration may indicate the style of resource that should be sought in a basin. Basin processes are dynamic, diverse, and so large-scale that they are easy to overlook.

2.2. Stress

Who could guess that two weeks of outcrop mapping could reveal all the tectonic events that impacted the White Mountains over the last 410 Ma? Barton and Angelier [2] show how slip indicators on oriented outcrop faults can be inverted to identify the changes in the stress tensor that mark these events. The fresh, glacially exhumed New Hampshire outcrops are ideal for this purpose, but the dramatic success of the inversion suggests efforts to collect similar data (from drill holes, seismics, etc.) could quickly yield very valuable tectonic information.

The paper by Bouziat et al. [3] suggests one way this might be done. They determine the stress tensor by coupling a basin simulator with a finite element mechanical solver. Evaluating this method on a set of synthetic passive margin siliciclastic sediment accumulation histories, they predict the spatial–temporal pattern of stress tensor rotation and zones of weakness in the section from the growing sediment load and the changing basement tilt. For common basin parameters, the distal part of the sediment wedge is at all times compressive, the proximal part of the wedge is always extensional (but more so during lowstands), and areas of weakness develop under the continental slope late in the sedimentation history.

2.3. Reservoir Tilting

Many of the Earth's sedimentary basins are affected by glaciation. Repeated glaciations over the last millions of years have had great influence on the physical conditions in sedimentary basins and basin structure. Sedimentary basins near the former ice margin can be tilted enough to significantly alter the pathways of hydrocarbon migration. Løtveit et al. [4] present some of the major effects that ice sheets have on sedimentary basins by modeling data from the Norwegian part of the Barents Sea. Among the most important effects are movements of the solid Earth caused by glacial loading and unloading and related flexural stresses. Future basin models should include glacial loading/unloading when dealing with petroleum potential in former glaciated areas.

Cerroni et al. [5] describe a new model of the hydromechanical changes induced by a glacial cycle. They address the generation of the computational grid and the algorithm for the numerical solution. They present a multiscale approach that accounts for the global deformation of the lithosphere and couple it with the thermo-hydro-mechanical feedback of the ice load on a representative domain of smaller scale.

2.4. Salt Diapirs

Salt diapirs act as heat pipes and can depress temperatures in the nearby oil window by nearly 100 °C, delaying hydrocarbon maturation enough to affect exploration strategies. Two studies of the specific analysis of the impact of salt on basin temperature illustrate how the modeling of salt diapirs can be integrated into future exploration models. Cedeño et al. [6] used Slumberger software to analyze the impact of salt diapirism on the subsurface temperature and the timing of maturation in a confined salt-bearing basin in the Norwegian Barents Sea, showing that the densely packed diapirs depress temperatures by 50–70 °C and delay maturation. With examples from the Nordkapp Basin, they show that the temperatures along the diapir flanks are 70 °C cooler and are exceptionally low (~150 °C) at depths of ~9 km beneath the salt.

Grunnaleite and Mosbron [7] show that salt structures on the Eastern flank of the Central Graben of the Norwegian North Sea depress temperatures by 85 °C and vitrinite R_o by up to 1.0%. They use the BMT (Basin Modeling Toolbox) software to accurately track the changing shape of the diapirs (especially the retraction of the root), and present what may be the most realistic salt model of a specific site yet published. They show the timing and geometrical evolution of salt structures depends critically on correctly defining the geometry of salt volumes and having a good geohistory model.

2.5. Magmatic Sills

Magmatic sills can increase the temperature of organic-rich strata and cause them to mature earlier than otherwise expected. Brown and Kim [8] set the stage by reviewing crustal reflection profiling seismic data to show how common sills are in the crystalline crust that underlies all basins. These sills can transfer heat from the mantle, change crustal rheology, and potentially affect overlying basin evolution in a fashion that impacts hydrocarbon and mineral resource potential.

Sydnes et al. [9] report the results of a sensitivity study of the impact of sills on temperature and maturation when attendant faulting is taken into account. They show that omitting structural changes related to magmatic intrusion may lead to over- or underestimation of the thermal effects of magmatic intrusions and the timing of maturation.

Sydnes et al. [10] evaluate the impact of sill emplacement on diagenetic processes and stress accumulations. Based on data from the Vøring Basin (Norwegian Sea), the modeling shows that basins with magmatic intrusions have thermal histories that enhance diagenetic processes during and after sill emplacement. Areas located between clusters of sills are particularly prone to diagenetic changes. The chemical alteration changes the stress pattern.

2.6. Fluid Flow

Fluids transport and concentrate all resources. Understanding and identifying the pathways of flow is perhaps the most important challenge for the explorationist. Two fundamental questions arise: First: how are the flow pathways likely to change with the distance of flow? Second: particularly if the pathways are increasingly concentrated, where are they located in a specific subsurface volume? Malin et al. [11] address the first question by showing that two factors control whether flow will concentrate with distance traveled: the spread (standard deviation) of log permeability about its mean, and whether permeability is distributed in a scale-invariant fashion. If the spread is substantial and the distribution scale invariant, flow will become increasingly concentrated with distance traveled. The spread of permeability is notoriously large, and fractures are scale-invariant in their distribution. This is a matter of observation. The reason is probably that the Earth's crust is in a state of near failure and scale-invariant systems are commonly observed near the critical point of failure. Knowing that flow is likely concentrated in any particular volume is a good perspective, but of little use if the actual flow channels in the volume cannot be located.

Sicking and Malin [12] describe how subsurface flow channels can be located from the seismic energy emitted by Krauklis waves trapped on water-filled fractures. The contrasting seismic wave velocity between water in the fractures and the surrounding rock traps seismic energy, and seismic waves bounce back and forth from the ends of the fracture. Episodes of harmonic humming, often minutes or more in duration, can be extracted from what is usually considered seismic noise using specialized processing techniques. Permeable channels are presumed to be where fluid-filled fractures are most abundant. A substantial number of field studies support this hypothesis. Necessary analysis can be carried out by processing existing 3D seismic data.

2.7. Chemical Alteration

Compaction is important in basin models; decompaction is an indispensable step in their construction. Compaction is of practical importance in Norwegian North Sea petroleum production. The Ekofisk platform has subsided 9 m over 40 years. Half of this subsidence is due to production-related decreases in pore pressure, which can be arrested by maintaining reservoir pressure; but half is due to the injection of water for secondary hydrocarbon recovery, which results in a slow plastic creep that declines with time and is independent of the elastic compaction that immediately attends changes in pore pressure (effective stress). Minde and Hiorth [13] analyze this second kind of creep compaction, first in terms of sliding block observations, and then in terms of water chemistry and chemical alteration. Sliding blocks slip at a declining rate after the initiation of slipping because the contact surface area between them increases with time. Creep experiments can be interpreted similarly for the slip between grains. The activity of water affects this creep.

When SO_4 is present in the pore fluid, creep is faster than expected from the sliding block model. This is because sulfate ions make the surface charge of calcite much more negative and produce a disjoining (osmotic) pressure, which pushes the grains apart and weakens the chalk. Flooding with $MgCl_2$ causes still more rapid compaction: three times more rapid than flooding with sulfate. This is because during the core flooding calcite is replaced volumetrically with magnesite, which is 10% more dense. Chalk creep compaction is thus due to at least three factors: changes in grain contact area, changes in the surface charge of calcite, and chemical replacement reactions. Translating from the laboratory to the reservoir requires taking into account the movement of the thermal front associated with the injection of cold seawater and chemical reactions tied to this thermal front, as well as the changes in water salinity (displacement of connate brine with injected seawater) that move more rapidly through the reservoir. Minde and Hiorth [13] show how careful interpretations of reservoir production phenomena identify concepts that might be transferred to more sophisticated future basin models.

2.8. H_2 Basin Resources

The last three papers discuss a completely new and surprising kind of nonhydrocarbon basin energy resource. Hydrogen gas has been observed venting from circular depressions (fairy circles) in many basins. As described by Cathles and Prinzhofer [14], H_2 concentrations at 1 m depth in the Sao Francisco Basin in Brazil occur mainly at the margins of a circular depression ~550 m in diameter, and are nonzero for about 6 h a day with peak concentrations occurring at ~1:00 pm. The periodic venting could be caused by atmospheric pressure tides, which have a very regular diurnal cycle at this location. The maximum rate of atmospheric pressure decrease occurs at 1:00 pm. The changing atmospheric pressure pushes air into and pulls it out of the subsurface in an accordion-like fashion. The volume of the gas pulled in and out of the shallow subsurface vents must be 1000 times less than the volume of gas in the subsurface reservoir that is compressed and decompressed. H_2 losses to the atmosphere can be supplied by a H_2 flux of ~0.1 m^3 m^{-2} d^{-1}.

Donzé et al. [15] address the critical question of whether the H_2 venting observed in many basins could constitute a significant energy resource by placing the venting rate analyzed by Cathles and Prinzhofer in contest with the local geology and global venting rate estimates. The H_2 generation rate expected from the 10^8 km^2 of Precambrian lithosphere, scaled to the 300,000 km^2 area of the Sao Francisco basin, suggests the entire basin could generate 90 to 266 tons H_2/y by radiolysis and/or 113 to 1018 tons H_2/y by serpentinization. Both generation mechanisms are possible in and under the Sao Francisco Basin: faults cut deeply into the crust, and storage reservoirs are present. However, the H_2 venting rate estimated by Cathles and Prinzhofer of 200 to 5400 tons H_2 per year from a single vent is as high as that expected from the entire Sao Francisco Basin, which suggests either the global estimates are too low or the venting at the study site is presently unusually strong.

Finally, Simon et al. [16] discount the possibility that solid Earth tides could be the cause of the variable venting in the Sao Francisco Basin by showing that solid Earth tides at the site have two co-equal peaks per day, but the H_2 venting has only one. We are just beginning to understand the H_2 system. We are at a stage similar to when we knew of a few hydrocarbon seepages, but had no concept of the magnitude or importance of the petroleum system.

3. Discussion

Perhaps what stands out most from the papers in this volume is the magnitude of the challenge of properly incorporating the diverse basin processes into models that can be usefully deployed to analyze basins in a resource context. The application papers in this volume indicate the potential of this. Capturing the stratigraphy, faulting, and salt diapirism realistically impacts the timing of hydrocarbon maturation and suggests the pattern of metamorphic alteration. Calculating stress changes from sediment loading predicts the locations of rock failure (fracturing). Strata tilting, sediment compaction, and fluid flow driven by glaciers can be analyzed with important resource implications. However, the effects of magmatic sill intrusion (even in the basement) can change everything, and assuming pressure depends on regular compaction is probably inappropriate. The movement of Shosa-type capillary seals can shift the pattern of overpressuring and fluid flow, and nonlinear stress feedbacks can be expected from alteration, faulting, and slumping. Identifying and properly incorporating all the process interactions will be very challenging.

The flip side of this incorporation challenge is the insights that can be obtained from diverse observations. Compaction is a process that changes physically with grain comminution and with the chemical activity of water, water chemistry, and rock alteration. Insights from investigations of the subsidence of oil platforms can be transferred to future basin models. Chemical alteration tied to water flooding can be transferred to mineral exploration basin models.

Basins reflect physical fundamentals, and the fundamentals pose powerful constraints. The scale invariance of strata and fractures, a consequence of the state of incipient failure of the atmosphere and lithosphere, indicates that the progressive channeling of fluid flow with distance traveled must be expected. It is thus of the greatest significance that subsurface permeability might be mapped by the intensity of seismic energy trapped on fluid-filled fractures. The ability to define the permeability structure at a specific site could change resource exploration in the most dramatic and exciting fashions.

Basins are giant thermo-chemical-structural reactors that produce mineral and diverse hydrocarbon resources. Hydrogen, seeping from Proterozoic basins worldwide, is a basin energy resource that combusts only to water vapor and generates no CO_2. The size and significance of this resource is yet to be determined, but the periodicity of venting reveals much about near-surface permeability and reservoirs of H_2.

New ground is cut by nearly all the papers in the volume. Shosa seals have many important implications. The pulses of flow they allow can reset the paleomagnetic pole on a subcontinental scale. Fault slip and other stress change-related observations are most powerful if interpreted as temporal changes in the stress tensor. Glaciation can tilt strata sufficiently to change the directions of petroleum migration. Sill intrusions are common in basins and the underlying lithosphere, and can change temperature and stress. There is a fundamental tendency for flow to become increasingly channelized with distance traveled. Seismic waves trapped in fluid-filled fractures indicate flow channels. Grain comminution, changes in water activity, aqueous chemistry-related changes in surface change, and replacement reactions all contribute to reservoir creep compaction. The H_2 system is a new and exciting kind of basin energy resource whose significance is yet to be defined.

The papers in this volume by no means cover all the topics that could be of interest and significance in basin modeling. We would, for example, have liked to include papers on gas adsorption and desorption. Gas seepages in glaciated areas may reflect the release of absorbed gas following the very recent glacial unloading (Jay Leonard, pc 2018). Hydrocarbon alteration during migration is touched upon in the overview, but gas washing is only one part of this important phenomenon, and it would have been nice to have a good overview paper summarizing what is known and what is not about organic chemical changes related to phase fractionation (the separation of supercritical gas–oil into distinct gas and oil phases), gas condensation, mixing, and bacterial and thermal degradation. Secondary hydrocarbon migration is also not treated at all in this volume. Crustal flexure is important in basin margins, and there is much yet to be learned from observed deformation patterns.

There is surely also much that is important that we do not currently perceive. As a community, we have stumbled over some things that could have been obvious much earlier (such as the potential significance of glacial tilting). We have addressed H_2 venting enough to recognize that we know almost nothing about the importance of this noncarbon energy system. We know a bit about the CO_2 generation–migration–trapping system in basins, but there are probably other chemical systems that are of significance. It would be wonderful to know more about the general chemical chromatography of basins.

Basins host mineral and nonhydrocarbon energy resources that will be important in the future. Knowledge of basins, in part an important heritage of the hydrocarbon era, provides a strong foundation, yet there is a lot that remains to be learned about the thermo-chemical-structural reactors we call sedimentary basins. The future is very bright for the next generation of researchers that will seek to address basin processes and resources. We hope that this volume will stimulate enthusiasm and encourage the research into basin resources that is necessary to meet the mineral and energy needs of the future.

Author Contributions: The authors contributed equally to the preparation of this summary. Both authors have read and agreed to the published version of the manuscript.

Funding: This research received no external funding.

Data Availability Statement: All data in this review is contained in the volume papers reviewed.

Acknowledgments: The authors thank all the scientists who contributed their high-quality papers to this volume, and the regular editors of *Geosciences* who made the submission and review process for the guest editors and authors very easy.

Conflicts of Interest: The authors declare no conflict of interest.

References

1. Cathles, L. On the Processes that Produce Hydrocarbon and Mineral Resources in Sedimentary Basins. *Geosciences* **2019**, *9*, 520. [CrossRef]
2. Barton, C.C.; Angelier, J. Direct Inversion Method of Fault Slip Analysis to Determine the Orientation of Principal Stresses and Relative Chronology for Tectonic Events in Southwestern White Mountain Region of New Hampshire, USA. *Geosciences* **2020**, *10*, 464. [CrossRef]
3. Bouziat, A.; Guy, N.; Frey, J.; Colombo, D.; Colin, P.; Cacas-Stentz, M.-C.; Cornu, T. An Assessment of Stress States in Passive Margin Sediments: Iterative Hydro-Mechanical Simulations on Basin Models and Implications for Rock Failure Predictions. *Geosciences* **2019**, *9*, 469. [CrossRef]
4. Løtveit, I.F.; Fjeldskaar, W.; Sydnes, M. Tilting and Flexural Stresses in Basins Due to Glaciations—An Example from the Barents Sea. *Geosciences* **2019**, *9*, 474. [CrossRef]
5. Cerroni, D.; Penati, M.; Porta, G.; Miglio, E.; Zunino, P.; Ruffo, P. Multiscale Modeling of Glacial Loading by a 3D Thermo-Hydro-Mechanical Approach Including Erosion and Isostasy. *Geosciences* **2019**, *9*, 465. [CrossRef]
6. Cedeño, A.; Rojo, L.A.; Cardozo, N.; Centeno, L.; Escalona, A. The Impact of Salt Tectonics on the Thermal Evolution and the Petroleum System of Confined Rift Basins: Insights from Basin Modeling of the Nordkapp Basin, Norwegian Barents Sea. *Geosciences* **2019**, *9*, 316. [CrossRef]
7. Grunnaleite, I.; Mosbron, A. On the Significance of Salt Modelling—Example from Modelling of Salt Tectonics, Temperature and Maturity Around Salt Structures in Southern North Sea. *Geosciences* **2019**, *9*, 363. [CrossRef]
8. Brown, L.D.; Kim, D. Extensive Sills in the Continental Basement from Deep Seismic Reflection Profiling. *Geosciences* **2020**, *10*, 449. [CrossRef]
9. Sydnes, M.; Fjeldskaar, W.; Grunnaleite, I.; Løtveit, I.F.; Mjelde, R. Transient Thermal Effects in Sedimentary Basins with Normal Faults and Magmatic Sill Intrusions—A Sensitivity Study. *Geosciences* **2019**, *9*, 160. [CrossRef]
10. Sydnes, M.; Fjeldskaar, W.; Grunnaleite, I.; Løtveit, I.F.; Mjelde, R. The Influence of Magmatic Intrusions on Diagenetic Processes and Stress Accumulation. *Geosciences* **2019**, *9*, 477. [CrossRef]
11. Malin, P.E.; Leary, P.C.; Cathles, L.M.; Barton, C.C. Observational and critical state physics descriptions of long-range flow structures. *Geosciences* **2020**, *10*, 50. [CrossRef]
12. Sicking, C.; Malin, P. Fracture Seismic: Mapping Subsurface Connectivity. *Geosciences* **2019**, *9*, 508. [CrossRef]
13. Minde, M.W.; Hiorth, A. Compaction and Fluid—Rock Interaction in Chalk Insight from Modelling and Data at Pore-, Core-, and Field-Scale. *Geosciences* **2019**, *10*, 6. [CrossRef]
14. Cathles, L.; Prinzhofer, A. What Pulsating H_2 Emissions Suggest about the H_2 Resource in the Sao Francisco Basin of Brazil. *Geosciences* **2020**, *10*, 149. [CrossRef]

15. Donzé, F.-V.; Truche, L.; Shekari Namin, P.; Lefeuvre, N.; Bazarkina, E.F. Migration of Natural Hydrogen from Deep-Seated Sources in the São Francisco Basin, Brazil. *Geosciences* **2020**, *10*, 346. [CrossRef]
16. Simon, J.; Fulton, P.; Prinzhofer, A.; Cathles, L. Earth Tides and H_2 Venting in the Sao Francisco Basin, Brazil. *Geosciences* **2020**, *10*, 414. [CrossRef]

Publisher's Note: MDPI stays neutral with regard to jurisdictional claims in published maps and institutional affiliations.

 *geosciences*

Article

On the Processes that Produce Hydrocarbon and Mineral Resources in Sedimentary Basins

Lawrence Cathles

Department of Earth and Atmospheric Sciences, Cornell University, Ithaca, NY 14853, USA; lmc19@cornell.edu

Received: 8 November 2019; Accepted: 13 December 2019; Published: 17 December 2019

Abstract: Sedimentary basins are near-planetary scale stratigraphic-structural-thermochemical reactors that produce a cornucopia of organic and inorganic resources. The scale over which fluid movements coordinate in basins and the broad mix of processes involved is remarkable. Easily observed characteristics indicate the style of flow that has operated and suggest what kind of resources the basin has likely produced. The case for this proposition is built by reviewing and interpreting observations. Features that future basin models might include to become more effective exploration and development tools are suggested.

Keywords: sedimentary basins; fluid flow; capillary seals; chemical alteration; resources; basin modeling

---

1. Introduction

Sedimentary basins might seem dead and uninteresting. They are, after all, simply the places where the Earth deposits its debris. In fact, they are remarkably dynamic and host a wide diversity of processes that have broad implications. We know a lot about basins because the energy and mineral resources they host have inspired the collection of massive amounts of data and funded extensive research. This paper reviews the mantle, fluid, capillary, and gas-generating processes that observations indicate operate in basins. The review is brief and intended to convey the scale and diversity of the processes involved, indicate why they matter for resources, and suggest how identification of the particular processes operating in a particular basin might be used in resource exploration.

2. Basin Processes

2.1. Mantle Dynamics

2.1.1. McKenzie's Stretching Model for Basin Formation

McKenzie's [1] lithosphere-stretching model was the first to quantitatively tie basin development to mantle dynamics. In this model the crust and its underlying lithosphere were assumed sutured together and stretched by tensional tectonic forces such that their original thickness (individually and in sum) was reduced by a factor β^{-1}, and the width of any portion of the crust/lithosphere was increased by β. The temperature at the base of the lithosphere/top of the asthenosphere was considered to be 1350 °C, and the lithosphere and any ocean water load on it was considered to be in isostatic equilibrium (e.g., the crust/lithosphere floated on the underlying asthenospheric mantle). The consequence of the stretching was an immediate increase in heat flow, an immediate subsidence or uplift depending on the thickness of the crust, and an ensuing slow subsidence as the thinned lithosphere lost heat and grew back to its original thickness. For typical crust and mantle density and thermal expansion, ocean water loading, and complete oceanization (β large) of a portion of continent, the initial subsidence is ~4.2 km if the crust is 35 km thick, 0 km if it is 15 km thick, and −3.2 km (3.2 km of uplift) if the crust is initially 0 km thick. The thermal subsidence over the next ~60 million years is 3.2 km. The heat flow as a function of time can be calculated from the stretching factor, and the maturation of organic material in the sediments can be computed from this time-temperature history. Sediment loading causes additional isostatic subsidence, but this can also be taken into account.

The McKenzie stretching model proved a very useful baseline for analyzing when buried organic material in a basin might generate hydrocarbons and fill structural traps. It spawned a vast literature investigating the cooling effects of pore fluid movement, modeling the thermal conductivity of sediments, taking into account the impact of differential stretching of the crust and underlying lithosphere, accounting for the effects of gradual stretching and crustal flexure, the cooling effects of rapid sediment deposition, heating by sill intrusion, etc. There proved to be so many factors that needed to be taken into account that most exploration companies chose to use maturity indicators to calibrate the heat flow history in their basins rather than try to predict it from the ground up. However, where from-first-principle predictions can be made they are very informative. An example is given in what follows.

2.1.2. Rifting and Base Metal Deposition

Focused rifting can produce another kind of basin resource. Japan provides a good example [2]. Japan has been subject to multiple episodes of focused rifting. As illustrated in Figure 1A, Japan detached from China ~60 million years ago by the rifting and spreading that opened the Japan Basin. Between 38 and 20 Ma Japan rifted again, and the Yamato Basin opened. The Yamato Rise is a fragment of Japan that was left behind. Thirteen million years ago Japan rifted again, splitting its volcanic chain in half. This rift failed, although the volcanoes on either side of the segmented basin network remain active.

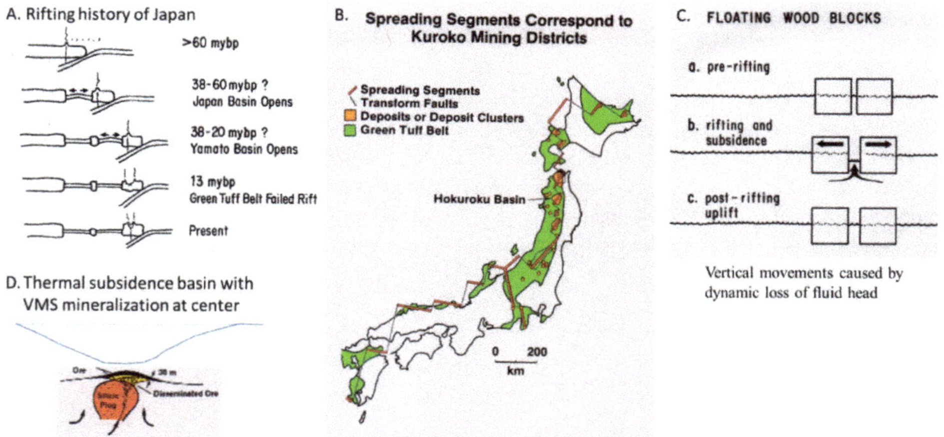

Figure 1. Formation of Kuroko-type volcanogenic massive sulfide (VMS) deposits in failed rifts, an example from Japan. (**A**) The rifting history of Japan. (**B**) The pattern of districts in Japan superimposed on hypothetical rift segments. (**C**) A floating wood block illustration of how dynamic loss of asthenosphere head can explain the inferred vertical movements associated with VMS formation. (**D**) VMS deposits may lie beneath rift basin centers. Figure panels A–C are from [2].

Just as suggested by McKenzie's stretching model, the rifting produced basins. One is the Hokuroku Basin indicated in Figure 1B. Sediments flowed into the Hokuroku Basin from all sides. A good analogy is the Guaymas Basin in the Gulf of California, which overlies a seafloor spreading segment of the East Pacific Rise and hosts hydrothermal circulation today. The Hokuroku basin contains a collection of volcanogenic massive sulfide (VMS) deposits called the Hokuroku mining district. VMS deposits are formed by the same processes that are producing seafloor massive sulfide (SMS) deposits at ocean ridges today. Hundreds of these SMS deposits have been documented [3].

The SMS deposits form when non-boiling 350 °C fluids are thermally quenched near the seafloor, producing black smokers and massive sulfide deposits containing copper, zinc, and gold. Boiling would produce a vein deposit, so the Kuroku VMS deposits, which are mined today at ~500 m depth, must have formed at depths >1.5 km below sea level. Thus, the Hokuroku basin must have subsided more than a kilometer before the VMS deposits formed, and then uplifted after the rifting aborted. This kind of vertical tectonic behavior is expected. There is a dynamic loss of asthenospheric hydraulic head as the asthenosphere seeks to fill the void opened by the extension of the lithosphere and crust. As illustrated in Figure 1C, what happened in Japan is similar to what happens when two wood blocks in a bathtub are pulled apart. When the blocks are moving apart viscous resistance to the upwelling of water between the blocks causes the water level to be depressed. When one stops pulling the blocks apart, the fluid level returns to the bathtub level. If one thinks of the bathtub water being the asthenosphere, and the blocks the rifted Japanese lithosphere, the vertical motions of the Hokuroku and other rift basins in Japan can be understood. Basins such as the Hokuroku can have quite a dynamic history, can be expected to be bounded by significant faults, and may be underlain by hydrothermally altered rock and VMS mineralization (see [2] for more discussion).

2.1.3. Juxtaposition of Sediments of Contrasting Oxidation State

Brines form where surface conditions favor net evaporation, and being denser than fresh water they sink into the stratigraphy. This tends to happen nearly everywhere. Water with salinity low enough to be potable is usually confined to within 300 m of the surface.

Where there is net evaporation from isolated seas, the seas often become saline enough to stratify. This is the case in the Black Sea today, for example. The stratified waters tend to be stagnant and any flux of organic material into them will cause them to become anoxic. Thereafter, any organic material that settles to the floor of the basin will not be consumed by biological organisms or oxidized. Therefore, this is an excellent setting for the production of hydrocarbon source rock, and hydrocarbons can be expected to have associated pore waters that are highly saline. Metal solubility increases very strongly with salinity and so a relationship between basin oil field brines and base metal resources might be anticipated.

There are tectonic connections that are important. Eugster [4] noted that Red Beds tend to be capped by shales and evaporates. Oxidized sediments will be deposited in arid, rifted continental settings. As the lithosphere cools and subsides, marine waters will incur, restricted access to the ocean is likely, stratified brines pools with unusually high the primary productivity are likely to form, and organic rich shales will be deposited and covered by evaporates. Saline lakes have high primary productivity, and shales tend to be the first member of an evaporate sequence. Physical and chemical processes can in such circumstances be linked in what Eugster called an orderly (by which he meant expected) succession.

2.2. Fluid Dynamics

2.2.1. Dynamic Permeability

Fluids extract resources from one location and deposit them in a location where they accumulate. One might think that fluid movement requires basin sediments be permeable, and that intrinsic sediment properties are in control. In some cases, this is certainly true. But in others fluids make their own permeability.

Consider mud depositing in a basin and converting first to mudstone and then to shale as it is buried. Regardless of induration, the sediments compact as they are buried. Porosity, near 90% at the sea or lake bottom, is ultimately reduced to a few percent at depth. The reduction in porosity pressurizes the pore fluids up to ~0.85 of the total weight of sediments and water above (e.g., up to

~0.85 of lithostatic). The fluid pressure does not rise further because at ~0.85 lithostatic the fluids hydro-fracture the rock and create the permeability needed for their escape to the surface.

The rate of decrease in porosity caused by a specified rate of sedimentation can be calculated for both a pore pressure gradient 0.85 of lithostatic, and 10% greater than hydrostatic. The permeability required for water expelled from the pores to escape to the surface can then be calculated.

Figure 2 shows that the dynamic permeability calculated in this fashion depends surprisingly weakly on sedimentation rate, and is similar to that measured for shales. Pore fluid expulsion and the associated hydro-fracturing seem to set the permeability of low permeability rocks such as shales in sedimentary basins. Malin et al. [5] suggest, for quite different reasons, that generally pore fluids conspire with mechanical processes to control not only the permeability but the distribution of permeability in the crust.(Calculations and additional discussion can be found in [5,6]).

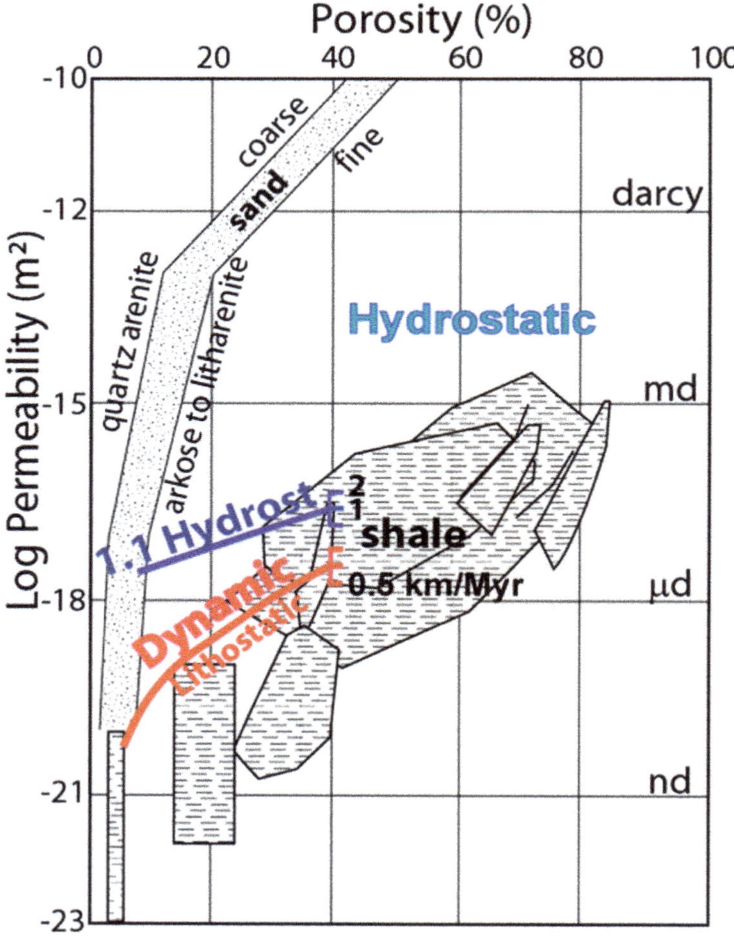

Figure 2. Shales have the permeability needed for pore fluids to escape as compaction occurs. The horizontal hachure shows the measured permeability of shale as a function of porosity. The red curve shows the permeability required for pore fluids to escape under overpressured conditions, and that the required permeability depends only very weakly on the sedimentation rate (scale at end shows how the curve would shift if the sedimentation rate was 2 km/ma or 0.5 km/ma instead of 1 km/ma). Sandstone permeability is from [7,8], shale permeability is from [9]. Figure is from [6].

2.2.2. Hydrocarbon Migration

Basin sediments are heated as they are buried and move downward through thermal gradients typically 20 to 30 °C km^{-1}. Organic material is thermally cracked, first to oil (between 90–110 °C), and then to progressively dry (pure CH$_4$) gas (between 110–130 °C). Oil and gas under basin conditions are lighter than water, and both will thus rise buoyantly through the sediments. If their rate of generation is slow compared to the rate at which they rise, they will rise as isolated rivulets of oil or in gas chimneys. If their rate of generation is fast compared to the rate at which they can rise, they will displace the pore water and completely fill the pores with oil or gas. The former is the conventional oil and gas scenario where the rising hydrocarbons are ponded beneath low permeability strata forming oil and gas reservoirs. The latter is the unconventional oil and gas scenario that has been of much interest recently. We'll return to this latter case later. First consider conventional hydrocarbon migration.

If hydrocarbons mature in a basin with extensive sand layers that are more permeable than ~1 millidarcy (10^{-15} m^2), vertically migrating hydrocarbons can be intercepted and diverted laterally over hundreds of kilometers. The oils migrate along only the upper centimeters of the carrier beds and fill reservoirs where the carrier beds are folded or offset by faults. The North Sea presents many examples of this kind of lateral migration and trapping, and the process is well described in [10]. Even a slight change in the tilt of the North Sea carrier sands can be important, and Løtviet et al. [11] show how tilting by glacial isostatic adjustment can redistribute oil in the North Sea.

The Gulf of Mexico is a currently active hydrocarbon generating area where there are no strata permeable enough to laterally divert the rising hydrocarbons in a very substantial fashion. In the 1990s we selected a portion of the Gulf of Mexico basin we thought would be large enough to capture the processes operating there. We called this ~200 × 100 km area of offshore Louisiana the "GRI Corridor" after our funding source, the Gas Research Institute. The corridor tells quite a remarkable story [12–17]. Briefly (elaboration follows): the oils in reservoirs in the northern half of this Corridor were sourced by Jurassic carbonate and Eocene shale strata, but in the south only by Jurassic carbonates. The deeper Jurassic oils matured and migrated first. The Eocene oils matured and migrated later, mixing with the earlier Jurassic oils in the north but not in the south. The Jurassic carbonates then generated gas which altered the oils by "gas washing" in a fashion that allows both the depth of washing and the amount of washing gas to be determined. This small area of one basin generated more hydrocarbon resources than have been extracted and consumed by humans over the entire petroleum era. Ninety percent of the generated oil was either expelled into the ocean or retained in the source strata. Discovered reservoirs in the Corridor constitute less than a fifth of the hydrocarbons currently migrating within it. The current reservoirs were filled recently (all perhaps in the last 100,000 years). In this flow-through hydrocarbon system "the present is the key to the present" (statement by Glen Gatenby, 2001).

Perhaps the most immediately obvious and remarkable feature of the GRI Corridor is the regular N-S change in the chemistry of its reservoir oils (Figure 3) [14]. Over 90 wt% of the +10 n-alkane component of the oils in the Tiger Shoals field at the north end of the Corridor have been removed by gas washing. At the South Marsh Island Block 9 field (SMI 9) slightly to the south, 50% have been removed. At South Eugene Island Block 330 15% have been removed. At the Jolliet field at the south end of the transect, the oils have not been washed at all, and 0 Wt% of the n-alkanes have been removed.

Figure 3B illustrates how the n-alkane depletion is measured. Unaltered oils have a linear logarithmic decrease in the n-alkane mole fraction with carbon number, called the Kissen slope. This unaltered trend is shown by the black slanted line that melds to the measured n-alkane mole fractions of the un-depleted oil (red squares) at carbon numbers greater than 24. The mass depletion (in this case 90 wt%) is the purple shaded area. The depletion results from dry gas (methane) interacting with the oils. At basin pressures, methane can dissolve a lot of oil, and the lower carbon numbers are more soluble than the higher ones, which accounts for the greater depletion of the low carbon numbers (the rollover of the mole fraction curve). The depletion caused by gas washing can be distinguished from other kinds of alteration such as bacterial degradation. Modeling [18–21] shows that the break number at which the oil composition departs from the Kissen trend measures the depth at which the oil was washed, and the degree of n-alkane depletion measures the amount of gas that washed the oil. The reservoir oils in the transect appear to have been gas washed in the first sand layer encountered by the hydrocarbons rising to fill the reservoirs.

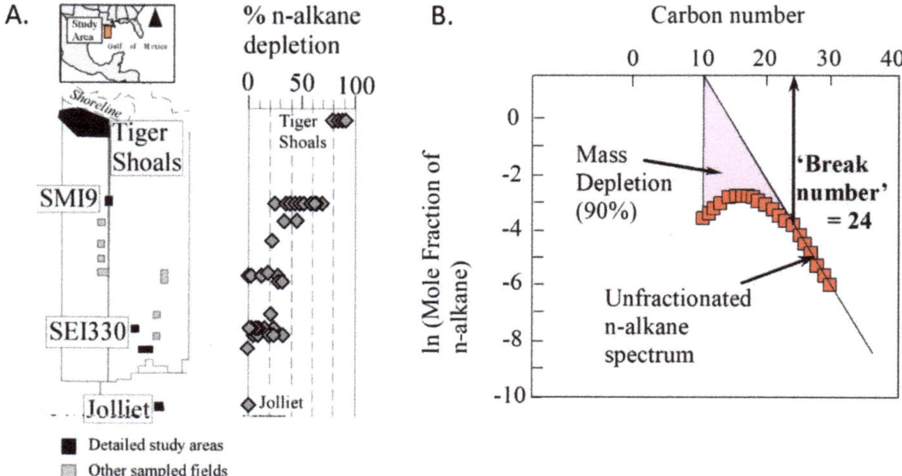

Figure 3. Systematic changes in the n-alkane chemistry of 138 oils from N to S across a 202 km (N-S) by 125 km (E–W) transect in the offshore Louisiana Gulf of Mexico. (**A**) The transect is red rectangle in the insert. Below the insert the sample locations are shown on the left, and the percent n-alkane depletion on the right. (**B**) Illustration of how the n-alkane depletion is calculated (purple area) with reference to the un-washed n-alkane distribution (bounding straight line melding onto the measured n-alkane mole fractions indicted by red squares). The illustration is for Tiger Shoals. Figure simplified from [18].

Extensive modeling was carried out to determine if the washing could have been produced by the oil and gas generated in the Corridor [15]. First the McKenzie stretching factor was determined from the sediment thickness and water depth along a 1050 km section near the Corridor that was contributed by Exxon. The section ran from the Lousiana border with Mississippi to the Sigsbee Knolls in the Gulf of Mexico. The evolution of the basin was reconstructed by back-stripping and decompaction. Heat flow, temperature, and vitrinite reflectance were calculated using a finite element grid tied to deposited strata and extending to the 150 km depth. The mantle heat flux determined from the stretching factor was applied there. Radiogenic heat production in the crust and sediments, changes in the surface temperature with water depth, the cooling due to sediment deposition, and the effect of compaction on lithology-specific thermal conductivity were taken into account. This modeling matched Exxon's measured heat flow data as well as temperature and vitrinite reflectance depth profiles using unadjusted literature parameters. Heat flow is about half normal (30 mW m^{-2}) near the shelf edge due to cooling by rapid sedimentation there, and about half this low heat flow is due to radiogenic heat production in the basin sediments. Despite the low heat flow, the temperature gradient is ~20 °C km^{-1} because of the low thermal conductivity of shales in the vertical direction. Agreement between the model and observations is excellent, and identification of the sources of heat is instructive, but great care had to be taken to include and properly specify all the important parameters in the model.

Heat flow at the base of the sediment section was then extracted from the portion of the Exxon line corresponding to the Corridor, and temperature calculated for the evolution of a much more realistic stratigraphy. Radiogenic heat production in the sediments was included, and the stratigraphic evolution included inversion of the Louann salt to a surface sill which was then buried. Salt diapirism and mini-basin formation was simulated. Maturation of a 100 m thick 5 wt% TOC (total organic carbon) carbonate Jurassic source strata (Type II kerogen) across the full section, and a 30 m thick 4 wt% TOC Type III Eocene coal across the northern half of the section was then calculated. The matured hydrocarbons were moved vertically out of the source strata once the saturation in the source strata exceeded 20 vol%, and moved upward thereafter when the pores of an element were filled to a specified migration saturation.

Gas is venting actively at hundreds of locations along the Corridor and there are many discovered hydrocarbon reservoirs. The modeling indicates that for gas to vent, the pore saturation of migrating hydrocarbons must be much less than 0.5%. The oils in the northern half of the section contain oleanane, a biomarker from plants that evolved in mid-Cretaceous time. The northern oils are also low in sulfur. Both indicate these oils came from the silicate Eocene source rock. For the Eocene oils to be dominant in the northern half of the Corridor, the migrating hydrocarbon pore saturation must be <0.05%. For a migration pore saturation of 0.025%, the oils are 85% Eocene at Tiger Shoals and 50% Eocene at the middle of the Corridor close to the end of the Eocene section, and there is sufficient late-generated gas to wash the oils as observed. Overall, the Corridor generated 184 Btoe (billion tons of oil equivalent) of hydrocarbons, mainly from Jurassic source beds, 37 Btoe were retained in the source rocks, 15 Btoe are in migration pathways between the source strata and the surface, and 131 Btoe (~1000 billion barrels- about 20% more than humans have consumed across the entire petroleum era) have been expelled into the ocean. Most of the oil has either been expelled (71%) or retained in the source strata (20%). The hydrocarbons discovered in the Corridor (1.4 Btoe) constitute ~9% of the hydrocarbons expelled from the source but not yet vented from the Corridor. The migration and filling is ongoing and the reservoirs were all filled recently, some very recently. The Jolliet reservoirs are hosted in 0.6 to 1.8 Ma strata. The model hydrocarbon flux across the 0.95 Ma horizon at the Jolliet location, assuming draw from a 40 km diameter mini-basin size area, is 190 and 110 tons of oil and gas per year respectively. Thus the Jolliet reservoirs (4400 kt oil and 3100 kt gas) could have been filled in 23 and 29 ka, respectively. The gas venting rate at Bush Hill near the Joliet field is estimated to be 900 t per year, which suggests the filling rate could be even more rapid [22].

2.2.3. Nature of Fluid Flow in Basins

Steady Expulsion: The Kupferschiefer Deposits in Germany and Poland

Base metal enrichment of the Kupferschiefer shale in Germany and Poland provides one of the most spectacular and best-documented examples of basin base metal mineralization.

The Kupferschiefer is the lowermost unit of the Zechstein evaporate sequence that extends from the England across the North Sea to southeastern Poland. It is a thin (usually 30 to 60 cm, but sometimes up to 1m thick), 258 Ma old pyrite and organic rich (~6% carbon) black shale, enriched in Cu, Pb, Zn, Au and other metals, and mined since at least 1199 AD. The technical challenges of mining made the Germans early leaders in technology. The deposits in Poland were discovered in 1957.

In the early Carboniferous (~350 ma) the 1700 × 500 km area later covered by the Zechstein sediments was the site of Hercynian clastic foreland basin sedimentation. In the Autunian (295–285 Ma), wrench tectonics extended the area and heated the lithosphere. Oxidized Rotliegende sediments accumulated in a basin and range topography with sill injection and some mafic volcanism. Erosion in a semi-desert setting of seasonal rivers and playa lakes leveled the topography. The peneplane was then partly covered by white aeolian dune sands (the Weissliegendes) which were later partially reworked into beach sands. As the lithosphere cooled, the area that would become the Permian Basin subsided in a broad down warp to 100 s of meters below sea level, but remained subaerial until ~259 Ma when it was suddenly flooded [23]. The laminated shale/shaly limestone or dolostone Kupferschiefer could thus be deposited throughout the basin below wave level in calm, shallow waters. The waters in this restricted basin were saline, the organic productivity high, and the bottom anaerobic [24]. The reduced, ubiquitous Kupferschiefer starkly contrasted with the oxidized sediments below. Four to five carbonate-anhydrite-salt cycles were then deposited in a shallow marine setting on top of the Kupferschefer. By the end of the Cretaceous up to 8 km of sediments had accumulated over the Kupferschiefer [6,23–29].

The mineralization occurred as post-Kupferschiefer sediment deposition loaded and compacted the underlying sediments. Oxidized brines, expelled from the compacting pores, forced their way through the Kupferscheifer and immediately adjacent strata, and pushed through in greater quantities where these strata were less resistive to flow. The metal-bearing oxidized brines were reduced as they moved through the reducing strata, and base metals were deposited. Where there was sufficient brine throughput, the Kupferschiefer was completely oxidized (called Rote Fäule or red fooling rock because it is barren of mineralization and red because the iron-bearing sulfides have been replaced by hematite) [24,27,30]. This process of metal deposition is not unusual. It also occurred in the midcontinent Proterozoic rift in the upper Peninsula of Michigan, for example, where reduced siltstones and shales of the Nonesuch Formation were oxidized by brines expelled from the underlying Copper Harbor conglomerate [6,31–33]. What is remarkable about the Polish Kupferschiefer mineralzation is the documentation of its metallization by 774 drillholes (50,000 analyzed samples) across all of Poland [34] (now there are 1700 drill holes [26]). The metallization maps allow us to estimate the volume of brine required to produce the metal enrichment. The estimated volume is so large that a large fraction of the brine expelled from the basin must have participated in the mineralization.

Figure 4 shows the copper surface density in kg/m^2 from one of the maps in the Metallogenic Atlas of the Zechstein Copper-bearing Series in Poland [34]. Similar maps in the Atlas show the zinc and lead metal surface density. The metal surface density shown in Figure 4D is the kilograms of metal in the Kupferschiefer and adjacent strata under each m^2 of plan area. The entire mineralized portion of the Lower Zechstein was analyzed down to a cutoff grade of 0.1%. The maximum thickness of the mineralized interval was 123 m, but typically the thickness analyzed was between 10 and 60 m. The metal density contours can be integrated to obtain the total metal added. I did this by tracing and summing the area of each metal density interval (e.g., 1–5, 5–10, 10–50 kg m^{-2}, etc.) in the Atlas maps, and multiplying by the log average metal density added in each interval, as shown in Table 1. For example, the metal added between the 1 and 5 kg m^{-2} contours equals the area of this contour interval times 2.24 (= $10^{0.5(\log 1 + \log 5)}$). The metal added between the 5 and 10 kg m^{-2} contours equals the area of this contour interval times 4.83. The 4.83 kg m^{-2} metal added equals the metal under this contour (7.07 = $10^{0.5(\log 5 + \log 10)}$) minus the 2.24 kg m^{-2} log average surface density of the first contour interval. Table 1 shows that 824, 927, and 1523 million tons of Cu, Pb and Zn, respectively, were added to the Kupferschiefer and immediately adjacent sediments in the area covered by the Atlas maps. The reserves of the Lubin, Polkowice-Sieroszosice and Rudna mining districts (near the black mining symbol at the north border of the red-outlined Fore-Sudetic Block in Figure 4D) are 30.4 Mt Cu [26]. The mine reserves thus constitute only 3.6% of the metallization.

The 9th row of Table 1 (labeled Brine in column 1) indicates the concentration of metals that would have to precipitate from the brine to account for the metallization in the 8th row of the table, if the brine volume expelled through the Kupferschiefer were 200,000 km^3. For this volume of brine, 4.1, 4.6 and 7.6 ppm of Cu, Pb, and Zn respectively would have to precipitate for the observed metal tonnages to be deposited. Metal concentrations more than those required have been measured in oil field brines (last 3 rows in Table 1). The 200,000 km^3 volume of brine is very large. It could be supplied if compaction reduced the porosity of a 4 km thick strata covering a 500 × 500 km area (the area of Poland) by 20%. Such a reduction of porosity is possible (see discussion below), so the hypothesis of mineralization-by-brine-expulsion is plausible. Much could be discussed (e.g., the magnitude of the porosity change early and later in burial, how much basement rocks might compact, how much brine was expelled unrecorded through the Rote Fäule vents, etc.). The important point made by this rough calculation is that the brine expulsion recorded by the metallization mapped in Poland constitutes a significant portion of the brine that could be expelled by compaction from the sub-Kupferschiefer eastern portion of the Southern Permian Basin.

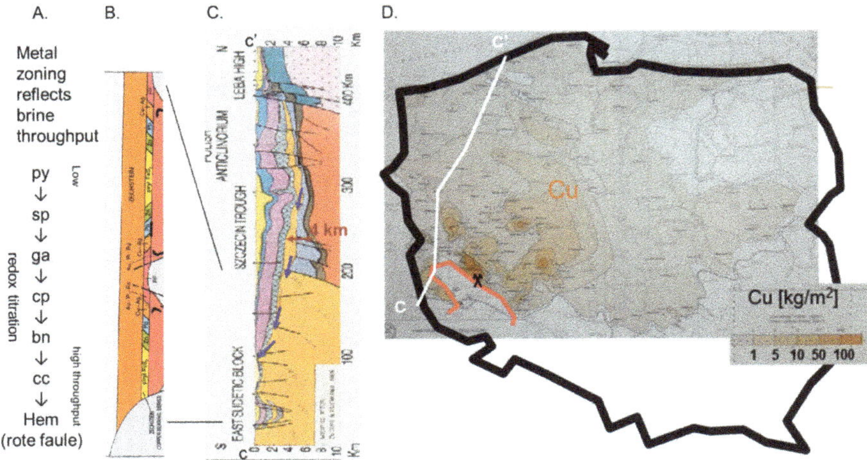

Figure 4. Key aspects of the Kupferschiefer mineralization in Poland. (**A**) Pyrite in the Kupfereschiefer is replaced first by Zn, then Pb, then increasingly copper rich sulfides and finally flushed completely of metals (Rote Fäule). (**B**) The metallization and Cu enrichment and depletion is zoned around sites of more intense brine discharge near the Fore-Sudetic Block (red outline in D). Figure from [34]. (**C**) The Zechstein (blue cross hatched) and its basal Kupferschifer shale are underlain by the Rotliegend sediments (orange), Carboniferous sediments (blue), and Variscan (light brown), Caledonian (red), and pre-Cambrian (pink) basement rocks. Section from [28]. (**D**) Copper surface density determined by 774 drill holes through the Kupferschiefer [34]. Red outlines the Fore-Sudetic Block. The Lubin, Polkowice-Sieroszosice and Rudna mines are all adjacent to this block at the location marked by the black crossed rock picks.

Table 1. Integration of metal surface densities in Metallogenic Atlas of Poland [34]. The first column indicates the metal surface density interval in the Atlas maps, and the second the log average metal content added by the interval, as discussed in the text. The next 3 columns show the area covered by each surface density interval, and the last 3 columns the additional metal introduced in each surface density interval, (e.g., the metallization) in millions of tons of Cu, Pb, and Zn. The 7th and 8th rows (labeled Sum and Brine in columns 1 and 2) show the total Cu, Pb, and Zn introduced in the area covered by the map, and the ppm that must have been extracted from a brine volume of 200,000 km^3 to account for the introduced metals. The last 3 rows show metal concentrations observed in oil field brines.

Surface Density Interval	Log Average Metal Added	A_{Cu}	A_{Pb}	A_{zn}	M_{Cu}	M_{Pb}	M_{Zn}
(kg m^{-2})		1000 km^2			10^6 tons		
1 to 5	2.24	60.0	70.0	162.0	134	157	363
5 to 10	4.83	32.0	42.9	82.4	155	207	398
10 to 50	13.05	20.0	27.0	58.0	261	352	757
50 to 100	39.04	3.3	5.4	0.1	129	211	5
100 to 500	121.23	1.2			145	0	0
	Sum	116.5	145.3	302.5	824	927	1523
Brine (km^3)	200,000			ppm	4.1	4.6	7.6
Rotliegende [35], ppm					1	50	50
Akkrum field [36], ppm					<0.5	50	60
Chelekin, 50–80 °C, [37], ppm					0.9 to 15		

Another remarkable aspect of the Kupferschiefer mineralization, illustrated schematically in Figure 4A,B, is a regular regional scale metal zonation. As shown by a color-coded metal dominance map in the Atlas (but not reproduced here), the northeast of the map area in Figure 4D is largely unenriched in base metals and the original Kupferschiefer pyrite dominates. To the west of this pyrite zone lies an NE-SW trending band of about equal area where Zn is dominant. To the west of this lies a smaller band where Pb is dominant, to farther still to the lies an area where Cu is dominant, and still further west is an area of barren Rote Fäule.

What this suggests is that the oxidized brines have forced their way through the Kupferschiefer. Where only a little brine was forced through, sphalerite is the dominant mineral and Zn dominates. With more brine throughput Zn is flushed out, and Pb is the dominant metal. With still more throughput, successively more Cu-rich sulfides (chalcopyrite, then bornite, then chalcocite) dominate. With still more brine throughput, all the sulfides are oxidized, all the base metals are flushed out, and only hematite remains (Rote Fäule). Gold and platinum-group metals precipitated at the base of the zone of copper enrichment and the top of the encroaching Rote Fäule [38]. The metal enrichment process involves oxidative titration of the reduced Kupferschiefer shale by brine throughput. This story is a common one and applies to other types of base metal deposits, for example Kuroko-type volcanogenic massive sulfide deposits [39].

The progressive westerly increase of brine movement through the Kupferschiefer in Poland suggests the brines below the Kupferschiefer moved to the west. Near the Fore-Sudedetic Block the leakage through the Kupferschiefer was optimal for metal enrichment and the mines are located there, as indicated in Figure 4. Why was leakage easier through the Kupferschiefer in western Poland, and why does the brine throughput per unit area change so regularly? The brines must have had remarkably equal access to the base of the Kupferschiefer. Perhaps this is not surprising. The first Zechstein deposition was a thin basal limestone which could be very permeable and the Variscan sediments could be permeable. If the leakage was slow, uniform access to the base of the Kupferschiefer could be expected provided only that the underlying rocks were dramatically more permeable, as seems very likely. What is important here is that the metal zoning is remarkably coherent, so brine access must have been relatively uniform, and, in this case at least, the regular increase in leakage can be followed westward to the locations where it produced mineral deposits.

A couple of brief comments: First, transgressive reduced black shales lying between evaporates and red beds of continental origin are commonly enough mineralized that a "Kupferschiefer" sub-type of the sediment-hosted statiform copper deposit class has been distinguished [40]. Second, the estimate of the volume of brine expelled is for the eastern portion of the Southern Permian Basin. An equally large area of the western part of the basin would be required to produce the German Kupferschiefer deposits. Third, the few mm thick horizontal chalcocite and other veins in the Kupferschiefer indicate the brines forcing their way through were nearly lithostatically over-pressured, and that flow persisted through the Kupferschiefer for protracted periods of time. Near lithostatic pressures were required to jack the horizontal veins open, and time was required to fill them with mineralization. Over 80% of the mineralization is disseminated (replacement of framboids) which appears to have occurred early in the mineralization history, but the veins indicate overpressure. Fourth, convective flow is not responsible for the metal zoning. The Kupferschiefer is underlain by very permeable strata (the Weissliegend sands and conglomerates) which preclude the possibility of horizontal pressure gradient even vaguely large enough for brine convection to drive flow horizontally through the Kupfersciefer. Flow was driven by fluid overpressures (pressures greater than hydrostatic) produced by compaction, or possibly positive volume change hydrocarbon maturation reactions. Overpressured fluids were driven vertically (with perhaps slight, few meter scale, lateral diversions) through the Kupferschiefer. The metal zoning is due to the amount of brine throughput, as discussed above. Finally, the Kupferschiefer metallization records regional scale brine flow and documents leakage through a sealing shale capped by evaporites.

Episodic Expulsion: Mississippi Valley Type Pb-Zn Deposits

The sediment-hosted base metal deposits, of which the Kupferschiefer is a sub-type, seem to have formed by quiet, steady expulsion of over-pressured brine. Mississippi Valley-Type (MVT) Pb-Zn deposits formed by short, sudden pulses of brine expulsion. Ore deposition is by cooling rather than reduction.

From the homogenization temperature of sphalerite fluid inclusions we know that ore deposition in MVT deposits occurred at T > 80 °C (Figure 5A), but the low maturation of conodonts indicates the sites could have been heated, cumulatively, for only a short period of time (<200,000 yrs; [41]). The deposits formed within a kilometer of the surface. Figure 5B shows that for the near subsurface to be even slightly warmed, a million years of compactive brine expulsion must occur in a few years. Temperature constraints thus suggest that the ore deposition occurred in short pulses. Other observations support this conclusion. For example, 8 pulses of mineralization have been documented in the Buick mine in the Viburnum Trend (carbonate reef escape hatch; [42]). Eight episodes of chalcopyrite deposition, 6 of sphalerite, 5 of galena and quartz have been documented in the Tri-state district [43]. Thermal pulses are indicated by cathodoluminescent banding in hydrothermal dolomite that is coherent over 275 km south of the Viburnum Trend [44]). The ore minerals show corrosion between pulses of deposition, and the cathodoluminescent bands show unconformities, as would be expected if the hydrothermal discharge were pulsed, and between pulses, cool meteoric water incurred and partly dissolved the mineralization.

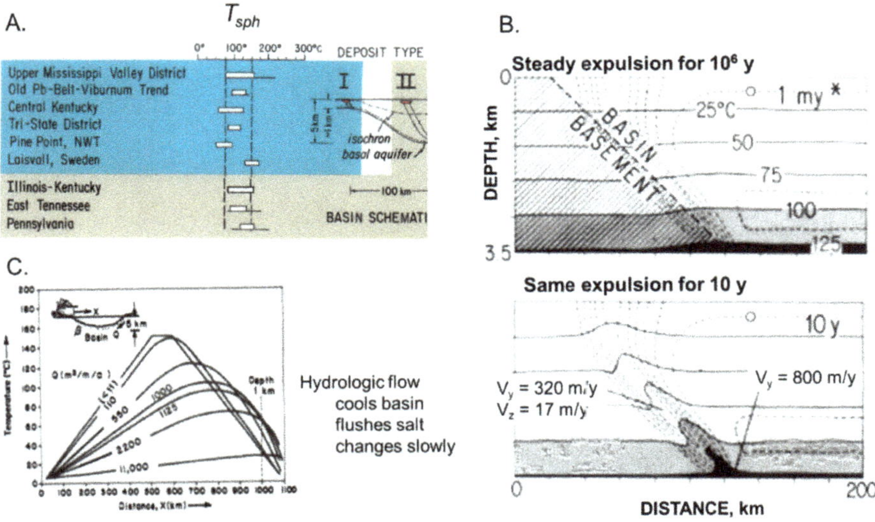

Figure 5. (**A**) Mississippi Valley-Type (MVT) lead-zinc deposits formed at shallow depths and temperatures >80 °C. From [45]. (**B**) To warm the near-surface significantly requires a million years of compactive expulsion to occur in <10 years. (**C**) Cross basin hydrologic flow warms the discharge margin but also cools the basin. Temperatures at 1 km depth can be raised only to half the maximum in the basin. From [46].

Hydrologic flow across the basin is too steady to easily explain pulses in mineralization (topography changes slowly), cross basin flow would flush the brines before the margins are warmed, and cross basin flow at the rates required to warm the margins will cool the basin as shown in Figure 5C (see reviews in [6,41]). What caused the pulses of brine expulsion? For over 30 years I could find no good explanation, but now there is a mechanism that might just work: gas suddenly introduced to the brine-filled aquifer following the failure of a capillary seal, as discussed below.

2.3. Capillary Dynamics

2.3.1. Basin Pressure Compartments

On 30 July 1987 Dave Powley (Amoco Production Company) made a presentation to the Gas Research Institute (GRI) in Chicago entitled "Subsurface Fluid Compartments" [47]. His presentation described how, over the previous two decades, Amoco studies by John S. Bradley and himself showed that basins are commonly over-pressured (but sometimes under-pressured) and compartmented at depths greater than 3 and sometimes just 1 km. The pressure is different between compartments, but the gradient is hydrostatic within compartments. He described the compartments as "huge [beer] bottles [which have been variably shaken]. Each one has a thin, essentially impermeable, outer seal and an internal volume which exhibits effective internal hydraulic communication." Figure 6A reproduces his presentation figure. He stated: " … the compartments have an amazing longevity" and can " … cut indiscriminately across structures, facies, formations, and geologic time horizons … ". He gave examples from Romania, Norway, Burma, and Alaska. He urged GRI to investigate the causes of basin pressure compartmentation.

Over the next 10 years, the Gas Research Institute funded research on basin compartmentation and sealing. The first phase independently confirmed pressure compartmentation. AAPG Memoir 61 [48] documented that basins are commonly pressure-compartmented. Its poster child example is the Anadarko [49]. A highly over-pressured, gas-rich (20 tcf), ~240 × 113 × 5 km thick portion of this basin is overlain and underlain by normal hydrostatic pressures, as illustrated in Figure 6B. This zone has been overpressured for >250 Ma. It is so internally compartmented that Zuhair Al-Shaieb termed it a megacompartment complex. Compartments are nested within compartments in an almost fractal fashion.

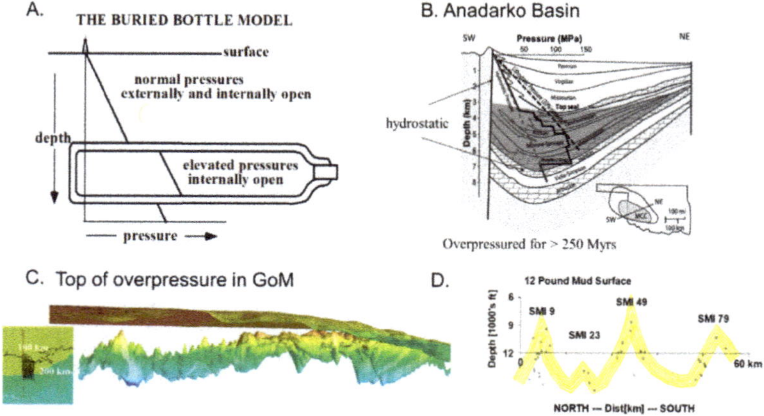

Figure 6. (**A**) Buried bottle illustration of basin pressure compartmentation from Powley's 1987 address to the Gas Research Institute (GRI) [47]. (**B**) Overpressure in the Anadarko Basin (gray) from [49]. (**C**) Top of overpressure (12 pound per gallon mudweight) from 2131 wells in the GRI Corridor. Deepest portion of the surface lies about 3.8 km under central Louisiana. (**D**) topographic highs in the top of overpressure tend to underlie oil reservoirs. For relation in full Corridor see [12].

That overpressure compartments can cross cut stratigraphy and have irregular surfaces is illustrated by Figure 6C, which shows the top of overpressure in the offshore Louisiana GRI Corridor (located in the insert in Figure 6C). The top of overpressure (TOOP) is defined in this figure by the 12 lb/gal mud weight surface (lithostatic is ~22.7 lb/gal) interpolated from mud weights in the header logs of 2131 wells and Krieged to produce the surface shown. The TOOP transgresses from 112 Ma Cretaceous to 2.4 Ma Quaternary strata as it shallows toward the continental slope from ~3.8 km under central Louisiana to <1 km depth on the slope. The TOOP surface is highly irregular, with topographic highs rising ~1km from the baseline surface. The topographic highs are spatially associated with discovered oil fields (Figure 6D; [12]). Twenty of 29 hydrocarbon fields in the Corridor are near topographic highs in the TOOP.

2.3.2. Capillary Seals

The second phase of GRI funding addressed the origin of the seals that bound the pressure compartments. How capillary forces impede the migration of hydrocarbons was understood: Water typically wets silicate surfaces and surface tension pulls water strongly into a shale. Oil, on the other hand, must be compressed to move from the large pores of a sand into the finer openings of a shale, and, unless the required entry pressure is exceeded, it will pond below the shale [50]. How a capillary seal of this nature could impede the movement of both brine and water was not clear, and Bradley and Powley [51] were careful to distinguish the compartment "pressure seal" that impedes both brine and hydrocarbons, from a capillary seal which impedes just hydrocarbons.

As illustrated in Figure 7, experiments carried out by Jennifer Shosa [52] show that, under the right circumstances, capillary seals can impede both brine and hydrocarbons. The right circumstance is that sufficient quantities of both wetting and non-wetting fluid phases be present in the pore space. Shosa passed CO_2-laden water through a 0.5 m long 12.7 mm inner diameter sand-packed tube that contained 2 to 8 fine-grained impeding layers. The impeding layers consisted of 13 to 15 mm thick intervals of 2 μm diameter sand held in place by adjacent 45 μm diameter sand layers ~11 mm thick, as illustrated in the insert. Flow was driven through the tube and through these fine impeding layers by a high-performance liquid chromatography (HPLC) piston pump. Pressure was controlled with a backpressure regulator at the discharge end, and reduced slowly with time. Until the pressure dropped below 290 psi and the CO_2 began to exsolve, flow through the tube was single phase and the pressure difference between the entry and exit ends was very small (first $5\frac{1}{4}$ h in Figure 7). When the CO_2 began to exsolve, flow through the tube was terminated and the pressure at the discharge and was reduced to atmospheric. The pressure at the entry end dropped to 163 psi over a short transition period (cross hatched band), and then remained unchanged for 3 weeks. The pore fraction of CO_2 gas after this decompression was 54%.

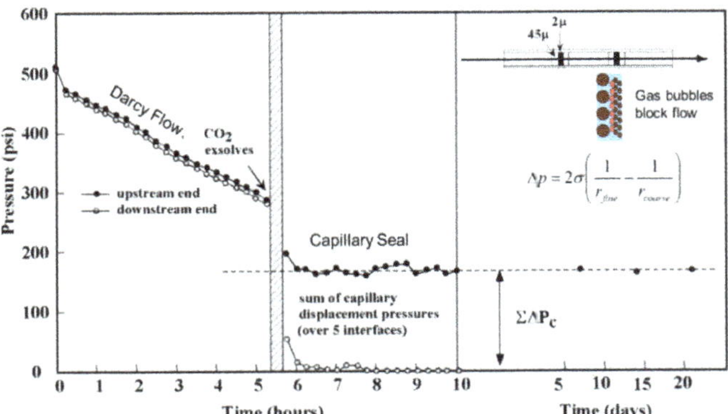

Figure 7. Pressure changes as CO_2-laden water is driven across a sand-packed tube with 5 fine-sand barriers as pressure is reduced. Until the pressure is reduced to the point that CO_2 exsolves, the pressure difference between in inlet and outlet is small and predicted by the darcy flow equation. After CO_2 exsolves, the flow of both gas and water stops despite a 163 psi pressure drop. The fine-sand barriers are illustrated in the insert, as is the nature of the sealing (red gas bubbles blocking flow into all the pores of the fine-grained layer). Figure modified from [52].

The total pressure drop is linearly related to the number of barriers. Six barriers have twice the pressure drop of 3 barriers and six times the drop of one barrier. Re-pressuring past 290 psi restores single phase flow and the single phase permeability is unchanged and predicted well by the Kozeny-Carmen relationship. The capillary pressure drop is predicted well by the Laplace relationship (insert) which equates the pressure drop across a fine-grained layer to the product of twice the CO_2-water interfacial tension, σ, and the difference of the inverse radii of the fine and coarse pores on either side of the interface. The Laplace relationship does not contain permeability, and the temperature dependence of the capillary pressure drop, measured between 20 and 100 °C, is that of CO_2-water interfacial tension. The two-phase barrier to flow in the experiment is thus clearly of a capillary nature. The capillary seals remained intact when 0.3 tube pore volumes of CO_2-saturated fluid were passed through them. The flow blockage can be visualized as gas bubbles preventing flow into all the pores in the fine sand like toilet plungers (see insert). Capillary seals are remarkably easy to form in the laboratory, and remarkably durable.

The kind of seal formed in Shosa's experiments has many new properties and implications. It is not a lithologic seal. Although it may seem tied to lithology because it forms at a particular fine-grained layer or sequence of fine-grained layers, it can shift to other fine-grained layers. The top of overpressure can migrate upwards. Since sealing depends on both wetting (brine) and non-wetting (usually gas) phases being present, where either is dominant the seal does not exist. In a gas chimney, for example, gas is free to move inside the pipe and water is free to move outside the pipe. Only in the transition zone between gas and water are both fluids present in sufficient quantities to be immobilized. A proper-mixture-proportion impermeable sheath confines the pipe. A consequence is that a Shosa seal can fail completely, and release a great volume of gas from a pressurized compartment, and then re-heal. Finally, Shosa seals should form naturally and spontaneously in basins, and produce just the kind of pressure compartments observed. While recognizing that there is much we don't know about the kind of seals formed in Shosa's experiments, the following paragraphs discuss our current perceptions.

Pressure Compartmentation is Spontaneous and Natural

Shosa seals should form spontaneously and pressure compartment basins just as observed. Hydrocarbon maturation will saturate brines in a basin, and compaction and reactions with positive volume change will oblige those brines to move through the stratigraphy. As the brine moves upward and decompresses, a bit of gas will exsolve and plug some of the pores in a shale parting or other comparatively fine-grained barrier, reducing its permeability. The decrease in permeability will divert the brine to an adjacent location, and exsolved gas will reduce its permeability. Eventually, enough partings will be sealed that the flow is stopped. In this fashion local pressure compartments will form. The overpressures in the compartments will be variable, and they will exchange brine as they are buried, further compacted, and further pressured. The details will be complicated and probably unpredictable, but pressure compartments in basins can be expected to form naturally and spontaneously.

Incarcerated Gas

Gas flowing upward in a gas chimney is in a sense incarcerated by its bounding Shosa seals. Gas chimneys are common in hydrocarbon basins and are generally cylindrical, with diameters up to 3 km or more. But the best example of incarcerated gas is basin center gas (see [6] for a more extensive summary than offered here). Huge volumes of gas are incarcerated in the centers of basins where the permeability is <0.1 md ($<10^{-16}$ m^2), and the gas is typically under-pressured with respect to hydrostatic. Examples are the Appalachian basin from New York to Tennessee documented with over 76,500 wells [53,54], and the Western Canada Sedimentary basin, well documented by [55]. In the 600 × ~100 km portion of the Western Canada basin it is not possible to drill a well that does not hit gas; the issue is only to find zones permeable enough to produce it. This is also the situation in the Arkoma Basin which lies just east of the Anadarko Basin discussed above. The water saturation is so low in the Arkoma that it is impossible to produce water from it. The pressure gradient is gas-static (~1 MPa km^{-1}), 1/10th the usual hydrostatic gradient. The Marcellus Shale is dry as a bone and contains over-pressured gas [56].

In all these cases, except the gas chimney, the gas has been incarcerated for geologic periods of time (hundreds of millions of years). The incarceration is by Shosa-type capillary seals. The basins are rich enough in organic material that enough gas has been generated to not only expel all the water but also, in some cases, blow dry the pore space to desiccation. Normally, capillary forces would draw water into the formations, but Shosa-type capillary seals formed in the transition zone to water saturated sediments prevent the imbibition of water.

Seal Migration

Shosa seals can migrate, and porosity profiles record the history of their formation and movement. Under hydrostatic conditions porosity is reduced almost linearly with depth. But because compaction depends on effective vertical stress (the weight of the overlying sediments and water minus the pore pressure), over-pressuring will arrest compaction.

Figure 8 illustrates how porosity changes with depth can be interpreted in terms of seal formation and migration. The black points indicate the porosity determined by density log. The yellow band is the model interpretation. The width of the band reflects the range in initial sediment porosities indicated by the measurements.

The pore pressure is hydrostatic and the compaction normal above 1430 m. If the blue-shaded H. selli shale became impermeable when its base was at 550 m water depth, the fixed seal compartment beneath it would have steadily increased in pressure (with no compaction because the pressure increase would keep the effective stress constant) until it began to hydrofrac and leak at a burial depth of 1430 m. As the seal top was buried from 1430 and the 2020 m depth, leakage caused the top of overpressure to migrate upward with sedimentation, capturing porosity at a constant depth and leading to a constant porosity-depth profile. The theoretical background and analysis of porosity depth profiles for 40 Gulf of Mexico wells can be found in [57]. The point here is that a porosity profile can tell us about when seals formed and how they migrated.

Fluids Migrate Together

One implication of a compartmented basin is that brines and hydrocarbons will tend to migrate together more than they otherwise might. Both will pass through the compartment seals only where and when they leak. Because the effective stress is lowest there, seals will tend to leak at topographic highs in the top of overpressure. Thus it is no surprise that hydrocarbon reservoirs are located preferentially near these highs. Brines will also flow preferentially through the topographic highs, and flow down the pressure gradient of the seal (and the higher temperature gradient it tends to host) will alter the seal. In the final volume of the GRI report, Jennifer Shosa lays the foundation for using the alteration as a flow meter for expelled brine [16].

Fluid Release Valve

Shosa seals can open and shut like valves. When open, very large volumes of gas can be released very quickly. Figure 9 gives a calculated example for a series of three 1cm-thick partings with 0.1 md permeability. The caption explains the valve-like operation.

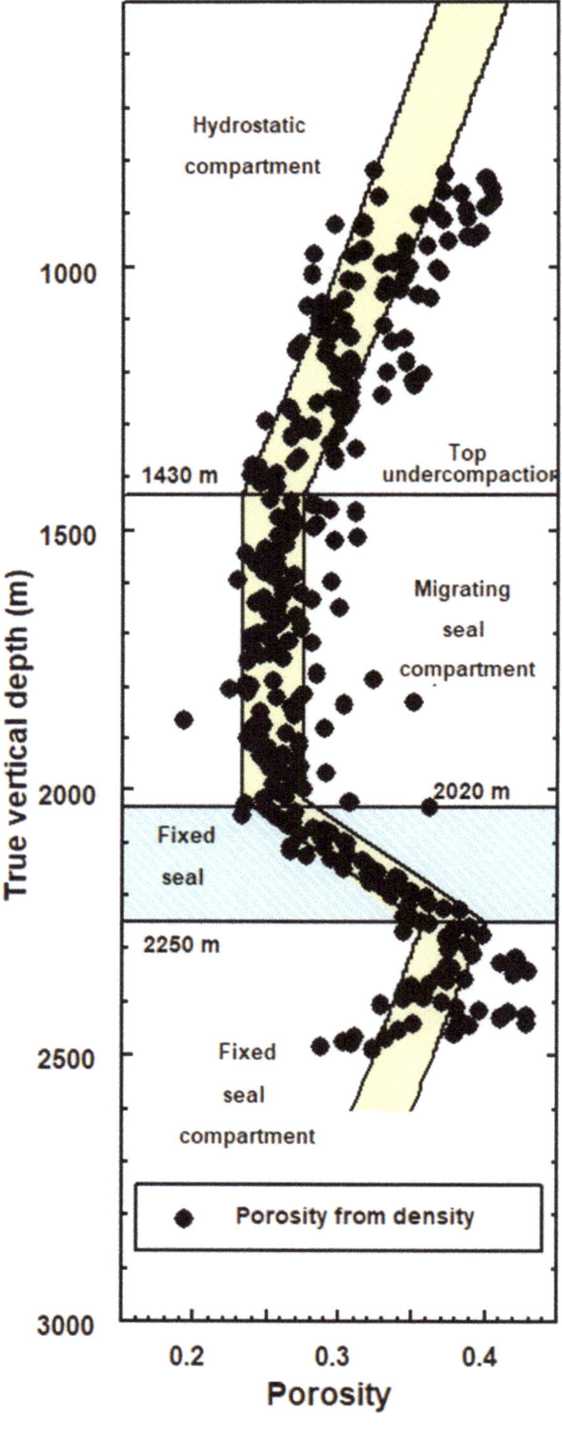

Figure 8. Observed (black points) and modeled (yellow band) porosity-depth profile from Gulf of Mexico. Profile records initiation and migration of a seal. Figure from [57].

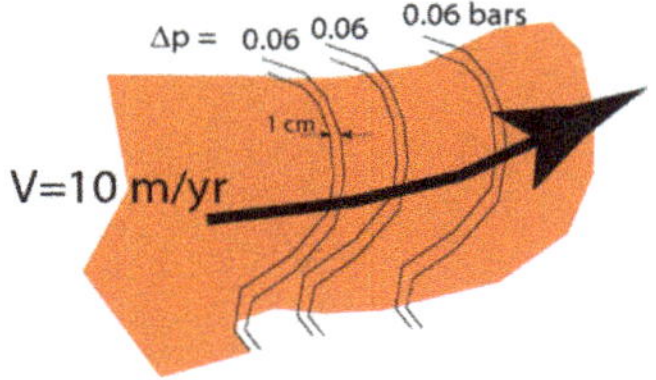

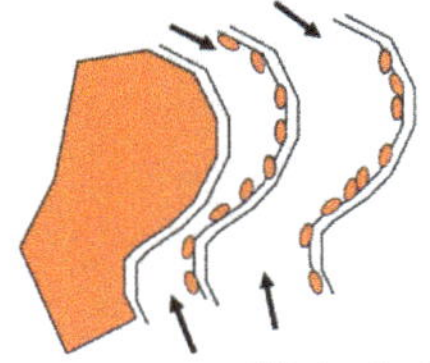

Figure 9. Illustration of Shosa seal operating as a pressure release valve. (**Top**) The capillary-blocked pressure drop across the 3-parting seal is 22.5 bars. (**Middle**) If gas were to penetrate the seal, as shown in the second panel, resistance to flow would be only the permeability of the partings. An expulsion Darcy flux of 10 m y^{-1} would produce a pressure drop across all three partings of less than 0.2 bars. Gas in the compartment could be expelled rapidly and completely. (**Bottom**) Once pressure in the compartment had been dissipated, water would be imbibed into the seal strata until the seal was restored, as shown in the bottom panel. Pressure in the compartment might then build up again if gas generation continued, until a second rupture and discharge occurred. Figure modified from [41].

Gas Pulsar Formation of Mississippi Valley-Type (MVT) Deposits

Figure 10 illustrates how failure of the seal between the gas-filled Arkoma and the darcy permeability aquifer system underlying the interior of the North American continent could have delivered pulses of hot brine to the sites of metal deposition. Weathering of Precambrian rocks in North America produced the extensive mid-continent darcy permeability Lamotte and Mt. Simon sandstones. These were subsequently overlain by karstic (and therefore very permeable) carbonates. The bines that produced the mid-continent MVT deposits discharged from these aquifers where the basal sands or carbonates approached the surface or were intersected by numerous or major faults. The pattern of hematite reduction in the Lamotte and the decrease northward from the Hicks Dome source of anomalous flourine in the St. Peter sandstone, suggest the mineralizing brines came from the Arkoma (Figure 10).

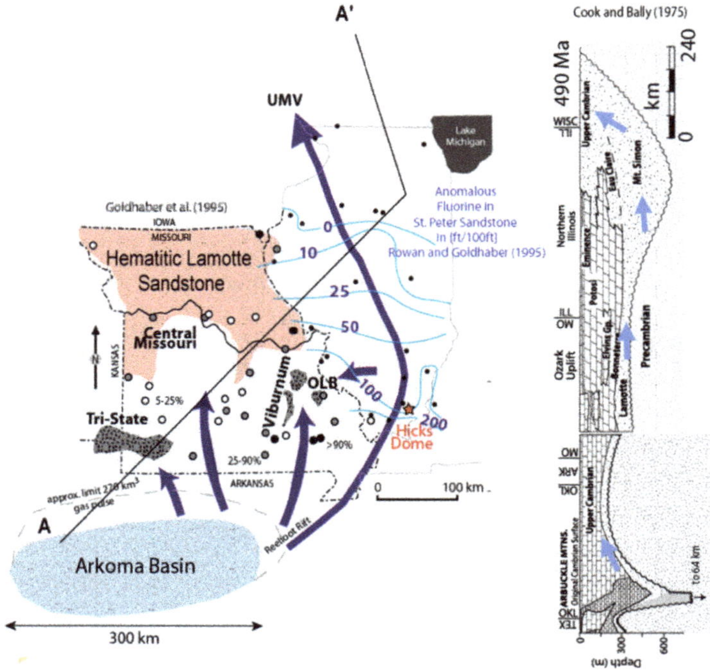

Figure 10. Periodic massive expulsion of gas from the Arkoma into the Lamotte and karstic Cambrian carbonates could displace sufficient volumes of brine rapidly enough to form the MVT deposits found in the North American interior (grey stippled areas). Brine flow from the Arkoma is suggested by the pattern of hematite reduction in the Lamotte, and the pattern of F carried north from the Hicks Dome and associated intrusions. Figure from [41] is a composite of figures from [58–60].

How could the presently gas-filled and under-pressured Arkoma have propelled the brines? The hypothesis is that back in the Permian when the Arkoma was actively generating gas it was overpressured. Periodically the capillary barrier between the Arkoma and the Lamotte and overlying carbonate was invaded by gas, rendered permeable, and perhaps as much as 3240 km^3 of gas was rapidly injected into the aquifers. The brine displaced by this gas exited the aquifer system rapidly at all the most permeable excape locations, e.g., at exactly the sites of MVT deposition. After decompression, imbibition of water resealed the barrier seal, the Arkoma repressured and then delivered a second pulse of gas which caused a second pulse of brine and mineralization, etc. The expulsion could be rapid enough to warm the sites of MVT deposition as observed [41]. The underpressured state of the gas in the Arkoma indicates that the gas pressure after the last expulsive pulse was, and is presently, controlled by capillary seals at the top of the gas zone (e.g., [55]).

Late Paleozoic Remagnetization of the North American Mid-Continent

A plethora of paleomagnetic studies have confirmed that the entire mid-continent of the U.S. was re-magnetized in the Late Paleozoic. The area re-magnetized, by magnetite deposition, is that shown in Figure 10, but extends into Indiana, Kentucky and Ohio as well. It is considered a consequence of tectonically-driven brine migration [61], and occurred at the time the MVT deposits formed [62–64]. It is logical that the pulses of massive brine expulsion discussed in the previous section were also responsible for this continent-scale re-magnitization.

Impact on Oil Production

If gas exsolves from oil during production, capillary seals may reduce production. Capillary barriers can be accounted for by changing Darcy's law to a plastic flow law wherein a small pressure gradient must be exceeded for any flow to occur. For a typical reservoir system gas exsolution and capillary barriers might reduce hydrocarbon flow to a well by 20% in 183 days. The damage will be fastest and greatest for the best producing wells because the drawdown cone for such wells extends further and encounters more capillary barriers. If production is stopped, the pressure will recover, the exsolved gas will dissolve, and the well will produce as it did initially [65]. Shosa seals have production implications.

2.3.3. Mud Volcanoes

Mud volcanism, such as is occurring in Azerbaijan, results when sediments lack sufficient induration to allow maturing gas and oil to escape. Hydrocarbons maturing at ~14 km depth in Azerbaijan produce overpressures that make the sediments quick (as in quick sand). The result is oil- and gas-spewing mud volcanoes 1 to 3.5 km in diameter. As illustrated in a spectacular geologic atlas [66], the stratigraphy is literally turning itself inside out, and there is an observed triad of oil, gas, and diapiric structures (Figure 11).

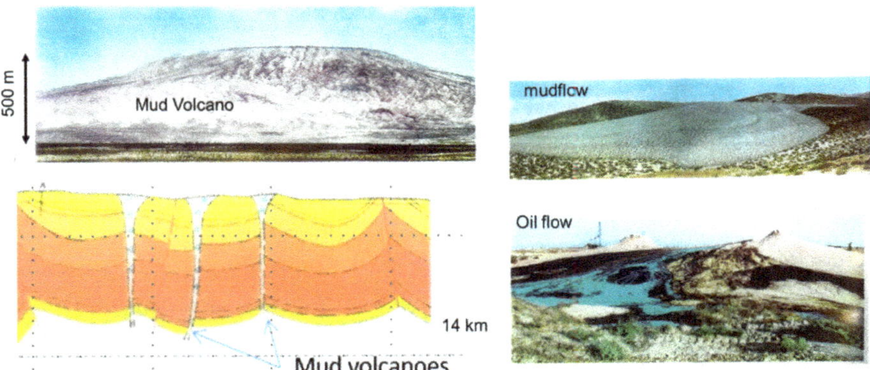

Figure 11. Mud volcanoes result when hydrocarbons are generated in un-indurated sediments. Mud Volcanoes of the Azerbaijan Atlas [66].

2.4. Non-Hydrocarbon Gas Dynamics

Non-hydrocarbon gases generated in basins contribute information on how basins operate. Consider two: CO_2 and H_2.

2.4.1. CO_2 Generation and Titration

Some reservoirs trap almost pure CO_2, and it is not uncommon for gas in reservoirs to contain a few percent CO_2. In the latter case, the partial pressure of CO_2 depends systematically on temperature, which indicates chemical equilibrium with siderite or magnesite [67], as shown in Figure 12A. Most reservoirs contain almost no CO_2. These observations can be understood and modeled as indicated in Figure 12.

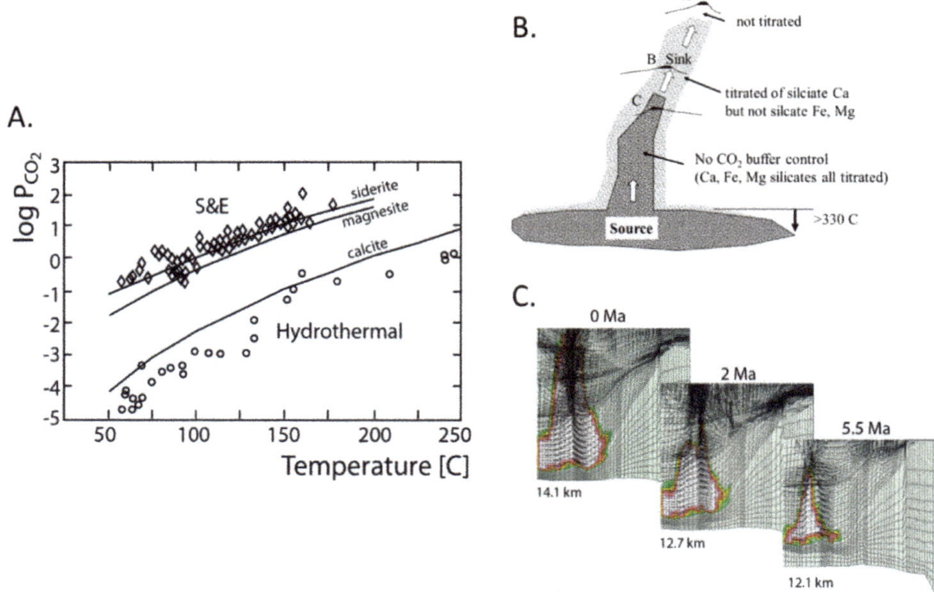

Figure 12. (**A**) Log partial pressure of CO_2 in equilibrium with aluminosilicates and various carbonate minerals. From [67]. (**B**) Whether basin reservoirs can contain almost no, a few percent, or nearly 100% CO_2 depends on whether the sediments along the pathways between the CO_2 source at >330 C and the reservoir contain Ca aluminosilicates (almost no CO_2 at location A), just Fe or Mg aluminosilicates (a few percent CO_2 at location), or no aluminosilicates (~100% CO_2 at location C). From [68]. (**C**) shows how the titration process has been implemented in basin models. From [68].

At 350 °C the partial pressure of water vapor and CO_2 exceeds the fluid pressure in over-pressured basins, and a separate gas phase is formed. But dissolved CO_2 is highly reactive, and as it migrates upward it will react with any silicate minerals that contain Ca, Mg, or Fe to form carbonates. The CO_2 in the rising gas will be consumed. So long as the stratigraphy contains calcium aluminosilicates, the partial pressure in the gas and in reservoirs will be very low. If the Ca aluminosilicates have been fully reacted but Mg or Fe aluminosilicates remain, the partial pressure of CO_2 in the reservoirs will depend on temperature and lie on the Smith and Ehrenberg (S&E) [67] buffer trend shown in Figure 12A. When the Mg or Fe aluminosilicates are titrated, there is no buffer control, and the reservoirs can be filled with pure CO_2. The titration is illustrated in Figure 12B. Reservoirs at location C can contain 100% CO_2, reservoirs at location B a few percent CO_2, and reservoirs at location A will contain no CO_2. Figure 12C shows how this titration process has been implemented in basin models (see [68]). Documentation of titrated migration pathways could reveal a lot about how gas migrates over substantial intervals of time in basins.

2.4.2. H$_2$ Generation

Prinzhofer and Cathles [69] describe how H_2, generated at 250 °C from ammonium breakdown, migrates to the surface to vent in pulses in Brazil. The daily pulses of H_2 appear to be related to atmospheric tidal pressure changes. Modeling the pulses could help constrain the economic potential of the H_2 resource.

3. Summary and Discussion

The above discussion by no means reviews all the interesting phenomena in basins, but hopefully has reviewed enough to ground a discussion of what might be beneficially incorporated into future basin models.

Space for sediment deposition is produced by rifting and spreading, processes governed by plate tectonics. On occasion models can be constructed that successfully predict the warming of sediments from first principles, and when this is done large scale basin properties are constrained, such as the degree of cooling by sediment deposition and warming by sediment radiogenicity. When first-principle models deviate from measured temperature or maturity profiles, processes such as cooling or warming by ground water circulation, or warming by sill intrusion may be inferred and perhaps quantified. Thus, ground-up models could produce insights useful enough to justify the effort required to construct them. Papers in this volume that assess the impact of sill emplacement are a step in this direction and illustrate the practical value that could be realized.

Defining permeability is perhaps the most important step in exploring for and producing basin resources. But what if permeability is set by the rate of fluid escape required by the sedimentation rate as suggested above? What if the distribution of permeability must be scale invariant as suggested by Malin et al. [5]. What if the most permeable flow pathways in this distribution could be imaged seismically as suggested by Sicking and Malin [70]. Incorporating such considerations and methods could produce far more accurate basin and production models.

Fluid flow is what we want to know because it is what redistributes and accumulates basin resources. The scale of flow in basins is stunning. Half the eastern portion of the Southern Permian Basin contributed the brine that mineralized the Kupferschiefer and produced the deposits in Poland. Pulses of brine flow remagnetized a substantial portion of the North American mid-continent while producing the MVT deposits found there. Knowing that a basin does not follow its paleomagnetic polar wander path, and therefore that it has been remagnetized, might be a clue to look for MVT deposits in that basin. Knowing that a basin has experienced pulses of rapid fluid flow (the kind that can remagnetize large areas) suggests the basin did not sustain the overpressured conditions for long periods as is required to produce Kupferschiefer style mineralization. Pulses of brine expulsion may not be optimum for hydrocarbon accumulation, or may displace accumulations from their expected positions. But pulses can produce MVT deposits.

Chemical alteration tells a lot. It is striking that the Cu enrichment of the Kupferschiefer and immediately surrounding strata is 28 times larger than the Cu in the discovered Polish deposits. The ubiquity of the Kupferschiefer metallization is remarkable. Large areas in Poland show metal enrichment (Zn, Pb, or Cu) well above background. The relative enrichment of Zn, Pb, and Cu shows a regular westward increase in leakage through the Kupferschiefer that ultimately produces ore deposits near the Fore-Sudedetic-Block. The Polish example suggests that if enriched metal content in a shale overlying red beds is noted, particularly if the shale shows indications of fluid overpressuring at the time of mineralization, one should look for base metal zoning in the shale and follow the gradient toward more intense leakage and mineralization. Similarly, measured reservoir CO_2 concentrations on the Smith and Ehrenberg concentration-temperature trend indicate CO_2 reservoirs might be found down section. A lot of useful information might be extracted from the patterns of chemical alteration. Patterns of chemical alteration could be used to train more capable basin models.

Capillary seal dynamics of the Shosa type deserves a lot of attention. Porosity profiles that tell when seals formed and how they migrated constrain the timing of hydrocarbon maturation and overpressuring. Since Shosa seals require a non-aqueous (usually hydrocarbon) fluid phase, basin flow was completely different before organic material was buried in sufficient quantity to produce hydrocarbon fluids. Paleozoic MVT deposits should not exist, and this seems to be the case [71].

It is hoped that this short paper has convinced the reader that sedimentary basins are indeed giant stratigraphic-structural-thermochemical reactors with surprising and fascinating characteristics and useful mineral and hydrocarbon products. Thanks to the focused exploration and research over the last 100 years we know enough about basins to begin to appreciate how they operate and how their processes interact. But, I suspect our understanding is just beginning. In the future we will understand a great deal more about their large scale interactions, and models that incorporate this understanding will be more effective exploration and extraction tools.

Funding: No funding supported the preparation of this paper. Past funding from the Gas Research Institute (GRI Grants 5093-260-2689 and 5097-260-3787), the DoE (DE-AC26-99BC15217), The Petroleum Research Fund (PRF 19767-AC2), Chevron Petroleum Technology Co. (CPTC 4505567 and 0070) and support (financial and data) from the Global Basin Research Network corporate affiliates program is gratefully acknowledged.

Acknowledgments: The author thanks three anonymous reviewers for excellent suggestions, and Richard Chuchla for inviting a presentation that formed the initial basis of this paper at the annual 2018 meeting of the Society of Economic Geologists in Keystone, Colorado.

Conflicts of Interest: The author declares no conflict of interest. The funders had no role in the design of the study; in the collection, analyses, or interpretation of data; in the writing of the manuscript, or in the decision to publish the results.

References

1. McKenzie, D. Some remarks on the development of sedimentary basins. *Earth Planet. Sci. Lett.* **1978**, *40*, 25–32. [CrossRef]
2. Cathles, L.M.; Guber, A.L.; Lenagh, T.C.; Dudás, F.Ö. Kuroko-Type Massive Sulfide Deposits of Japan: Products of an Aborted Island-Arc Rift. In *The Kuroko and Related Volcanogenic Massive Sulfide Deposits*; Society of Economic Geologists: Littleton, CO, USA, 1983.
3. German, C.R.; Petersen, S.; Hannington, M.D. Hydrothermal exploration of mid-ocean ridges: Where might the largest sulfide deposits be forming? *Chem. Geol.* **2016**, *420*, 114–126. [CrossRef]
4. Eugster, H.P. Oil shales, evaporites and ore deposits. *Geochim. Cosmochim. Acta* **1985**, *49*, 619–635. [CrossRef]
5. Malin, P.; Leary, P.; Cathles, L.M.; Barton, C.C. Observational and critical state physics descriptions of long-range flow structures. *Geosciences* **2019**. submitted.
6. Cathles, L.M.; Adams, J.J. Fluid Flow and Petroleum and Mineral Resources in the Upper (<20-km) Continental Crust. In *One Hundredth Anniversary Volume*; Society of Economic Geologists: Littleton, CO, USA, 2005; pp. 77–110.
7. Nelson, P.H. Permeability-porosity Relationships in Sedimentary Rocks. *Soc. Petrophys. Well-Log Anal.* **1994**, *35*, 38–62.
8. Nelson, P.H. Evolution of Permeability-Porosity Trends in Sandstones. In Proceedings of the SPWLA 41st Annual Logging Symposium, Dallas, TX, USA, 4–7 June 2000; p. 14.
9. Neuzil, C.E. How permeable are clays and shales? *Water Resour. Res.* **1994**, *30*, 145–150. [CrossRef]
10. England, W.A.; Mackenzie, A.S.; Mann, D.M.; Quigley, T.M. The movement and entrapment of petroleum fluids in the subsurface. *J. Geol. Soc.* **1987**, *144*, 327–347. [CrossRef]
11. Løtveit, I.F.; Fjeldskaar, W.; Sydnes, M. Tilting and Flexural Stresses in Basins Due to Glaciations—An Example from the Barents Sea. *Geosciences* **2019**, *9*, 474. [CrossRef]
12. Cathles, L.M.; Wizevich, M.; Losh, S. Volume II: Geology, geophysics, geochemistry, and GoCAD database. In *Seal Control of Hydrocarbon Migration and Its Physical and Chemical Consequences*; GRI-03/0065; Cathles, L.M., Ed.; Gas Research Institute: Chicago, IL, USA, 2002; p. 51.
13. Whelan, J.K.; Eglinton, L. Volume III: Organic geochemical consequences in a North South transect in the northern Gulf of Mexico. In *Seal Control of Hydrocarbon Migration and Its Physical and Chemical Consequences*; GRI-03/0065; Cathles, L.M., Ed.; Gas Research Institute: Chicago, IL, USA, 2002; p. 85.
14. Losh, S.; Cathles, L.M. Volume IV: Gas washing of oil and its implications. In *Seal Control of Hydrocarbon Migration and Its Physical and Chemical Consequences*; GRI-03/0065; Cathles, L.M., Ed.; Gas Research Institute: Chicago, IL, USA, 2002; p. 74.
15. Cathles, L.M.; Losh, S. Volume V: A modeling analysis of the hydrocarbon chemistry and gas washing, hydrocarbon fluxes, and reservoir filling. In *Seal Control of Hydrocarbon Migration and Its Physical and Chemical Consequences*; GRI-03/0065; Cathles, L.M., Ed.; Gas Research Institute: Chicago, IL, USA, 2002; p. 63.

16. Cathles, L.M.; Shosa, J.D. Volume VI: A theoretical analysis of the inorganic alteration by flow of brines through seals. In *Seal Control of Hydrocarbon Migration and Its Physical and Chemical Consequences*; GRI-03/0065; Cathles, L.M., Ed.; Gas Research Institute: Chicago, IL, USA, 2002; p. 70.

17. Cathles, L.M. Hydrocarbon generation, migration, and venting in a portion of the offshore Louisiana Gulf of Mexico basin. *Lead. Edge* **2004**, *23*, 760–770. [CrossRef]

18. Losh, S.; Cathles, L.; Meulbroek, P. Gas washing of oil along a regional transect, offshore Louisiana. *Org. Geochem.* **2002**, *33*, 655–663. [CrossRef]

19. Meulbroek, P. Equations of state in exploration. *Org. Geochem.* **2002**, *33*, 613–634. [CrossRef]

20. Meulbroek, P.; Cathles, L.; Whelan, J. Phase fractionation at South Eugene Island Block 330. *Org. Geochem.* **1998**, *29*, 223–239. [CrossRef]

21. Meulbroek, P.; Cathles, L.; Goddard, W.A. HCToolkit/EOS interface: An open source, multi-platform phase equilibria framework for exploring phase behaviour of complex mixtures. *Geol. Soc. Lond. Spec. Publ.* **2004**, *237*, 89–98. [CrossRef]

22. Chen, D.F.; Cathles, L.M. A kinetic model for the pattern and amounts of hydrate precipitated from a gas steam: Application to the Bush Hill vent site, Green Canyon Block 185, Gulf of Mexico: A Kinetic Model for Hydrate Precipitation. *J. Geophys. Res.* **2003**, *108*. [CrossRef]

23. Van Wees, J.-D.; Stephenson, R.A.; Ziegler, P.A.; Bayer, U.; McCann, T.; Dadlez, R.; Gaupp, R.; Narkiewicz, M.; Bitzer, F.; Scheck, M. On the origin of the Southern Permian Basin, Central Europe. *Mar. Pet. Geol.* **2000**, *17*, 43–59. [CrossRef]

24. Oszczepalski, S. Kupferschiefer in Southwestern Poland: Sedimentary environments, metal zoning, and ore controls. In *Sediment-Hosted Stratiform Copper Deposits*; Special Paper; Geological Association of Canada: St. John's, NL, Canada, 1989; pp. 571–600.

25. Cathles, L.M.; Oszczepalski, S.; Jowett, E.C. Mass balance evaluation of the late diagenetic hypothesis for Kupferschiefer Cu mineralization in the Lubin Basin of southwestern Poland. *Econ. Geol.* **1993**, *88*, 948–956. [CrossRef]

26. Oszczepalski, S.; Speczik, S.; Zieliński, K.; Chmielewski, A. The Kupferschiefer Deposits and Prospects in SW Poland: Past, Present and Future. *Minerals* **2019**, *9*, 592. [CrossRef]

27. Jowett, E.C.; Rydzewsk, A.; Jowett, R.J. The Kupferschiefer Cu–Ag ore deposits in Poland: A re-appraisal of the evidence of their origin and presentation of a new genetic model. *Can. J. Earth Sci.* **1987**, *24*, 2016–2037. [CrossRef]

28. Ziegler, P.A. *Evolution of the Arctic-North Atlantic and the Western Tethys/Book and Map*; American Association of Petroleum Geologists: Tulsa, OK, USA, 1988.

29. Schito, A.; Corrado, S.; Trolese, M.; Aldega, L.; Caricchi, C.; Cirilli, S.; Grigo, D.; Guedes, A.; Romano, C.; Spina, A.; et al. Assessment of thermal evolution of Paleozoic successions of the Holy Cross Mountains (Poland). *Mar. Pet. Geol.* **2017**, *80*, 112–132. [CrossRef]

30. Oszsczepalski, S.; Rydzewski, A.; Speczik, S. Rote Fäule-related Au–Pt–Pd mineralization in SW Poland: New data. In Proceeding of the Fifth Biennial SGA Meeting, London, UK, 22–25 August 1999.

31. White, W.S. A paleohydrologic model for mineralization of the White Pine copper deposit, northern Michigan. *Econ. Geol.* **1971**, *66*, 1–13. [CrossRef]

32. Brown, A.C. Sediment-hosted Stratiform Copper Deposits. *Geosci. Canada* **1992**, *19*, 125–141.

33. Mauk, J.L.; Kelly, W.C.; van der Pluijm, B.A.; Seasor, R.W. Relations between deformation and sediment-hosted copper mineralization: Evidence from the White Pine part of the Midcontinent rift system. *Geology* **1992**, *20*, 427–430. [CrossRef]

34. Oszczepalski, S.; Rydzewski, A.; Geologiczny, P.I. *Metallogenic Atlas of the Zechstein Copper-Bearing Series in Poland*; Państwowy Instytut Geologiczny: Warsaw, Poland, 1997.

35. Downorowicz, S. Geothermics of the copper ore deposit of the Fore-Sudetic monocline: Prace Inst. *Ge-. ol. (Poland)* **1983**, *106*, 88.

36. Hallager, W.S.; Carpenter, A.B.; Campbell, W.L. Smectite clays in red beds as a source of base metals. In *Geological Society America Abstracts Programs*; Geological Society America: Boulder, CO, USA, 1991; pp. 31–32.

37. Tooms, J.S. *Review of Knowledge of Metalliferous Brines and Related Deposits*; Institution of Mining & Metallurgy: London, UK, 1970.

38. Piestrzyński, A.; Pieczonka, J.; Głuszek, A. Redbed-type gold mineralisation, Kupferschiefer, south-west Poland. *Min. Dep.* **2002**, *37*, 512–528. [CrossRef]

39. Eldridge, C.S.; Barton, P.; Ohmoto, H. Mineral textures and their bearing on formation of Kuroko orebodies. In *The Kuroko and Related Volcanogenic Massive Sulfide Deposits*; Economic Geology Monograph; Economic Geology Publishing Co.: Littleton, CO, USA, 1983; pp. 241–281.

40. Cox, P.; Lindsey, D.A.; Singer, D.A.; Diggles, M.F. Sediment-Hosted Copper Deposits of the World: Deposit Models and Database. *US Geol. Surv. Open-File Rep.* **2003**, *3*, 50.

41. Cathles, L.M. Changes in sub-water table fluid flow at the end of the Proterozoic and its implications for gas pulsars and MVT lead–zinc deposits. *Geofluids* **2007**, *7*, 209–226. [CrossRef]

42. Sverjensky, D.A. The origin of a mississippi valley-type deposit in the Viburnum Trend, Southeast Missouri. *Econ. Geol.* **1981**, *76*, 1848–1872. [CrossRef]

43. Hagni, R.D. Tri-State Ore Deposits: The Character of Their Host Rocks and Theis Genesis. In *Cu, Zn, Pb and Ag Deposits*; en Wolf, K.H., Ed.; Elsevier: Amsterdam, The Netherlands, 1976; Volume 6.

44. Cathles, L.M. *A Discussion of Flow Mechanisms Responsible for Alteration and Mineralization in the Cambrian Aquifers of the Ouachita-Arkoma Basin-Ozark System: Chapter 8: Diagenesis and Basin Hydrodynamics*; American Association of Petroleum Geologists: Tulsa, OK, USA, 1993.

45. Cathles, L.M.; Smith, A.T. Thermal constraints on the formation of mississippi valley-type lead-zinc deposits and their implications for episodic basin dewatering and deposit genesis. *Econ. Geol.* **1983**, *78*, 983–1002. [CrossRef]

46. Cathles, L.M. A simple analytical method for calculating temperature perturbations in a basin caused by the flow of water through thin, shallow-dipping aquifers. *Appl. Geochem.* **1987**, *2*, 649–655. [CrossRef]

47. Powley, D.E. Subsurface Fluid Compartments. In *Gas Research Institute Deep Gas Sands Workshop*; Gas Research Institute: Chicago, IL, USA, 1987.

48. Ortoleva, P.J. *Basin Compartments and Seals*; Memoir 61; American Association of Petroleum Geologists: Tulsa, OK, USA, 1994; p. 477.

49. Al-Shaieb, Z.; Puckette, J.O.; Abdalla, A.A.; Ely, P.B. Megacompartment Complex in the Anadarko Basin: A Completely Sealed Overpressured Phenomenon. In *Basin Compartments and Seals*; Orteleva, P.J., Ed.; American Association of Petroleum Geologists: Tulsa, OK, USA, 1994; pp. 55–68.

50. Schowalter, T.T. Mechanics of Secondary Hydrocarbon Migration and Entrapment. *AAPG Bull.* **1979**, *63*, 723–760.

51. Bradley, J.S.; Powley, D.E. Pressure Compartments in Sedimentary Basins: A Review. In *Basin Compartments and Seals*; Orteleva, P.J., Ed.; American Association of Petroleum Geologists: Tulsa, OK, USA, 1994; pp. 3–26.

52. Shosa, J.D.; Cathles, L.M. Experimental investigation of capillary blockage of two phase flow in layered porous media. *Pet. Syst. Deep-Water Basins Glob. Gulf Mexico Exp. Houston Texas GCSSEPM GCS* **2001**, *21*, 725–739.

53. Law, B.E. Basin-Centered Gas Systems. *AAPG Bull.* **2002**, *86*, 1891–1919.

54. Ryder, R.T.; Zagorski, W.A. Nature, origin, and production characteristics of the Lower Silurian regional oil and gas accumulation, central Appalachian basin, United States. *AAPG Bull.* **2003**, *87*, 847–872. [CrossRef]

55. Masters, J.A. *Lower Cretaceous Oil and Gas in Western Canada. Elmworth: Case Study of a Deep Basin Gas Field*; AAPG Memoir 38; American Association of Petroleum Geologists: Tulsa, OK, USA, 1984; pp. 1–33.

56. Engelder, T.; Cathles, M.L.; Bryndzia, T.L. The fate of residual treatment water in gas shale. *J. Unconv. Oil Gas Resour.* **2014**, *7*, 33–48. [CrossRef]

57. Revil, A.; Cathles, L.M. The porosity-depth pattern defined by 40 wells in Eugene Island South Addition, Block 330 Area, and its relation to pore pressure, fluid leakage, and seal migration. *Pet. Syst. Deep-Water Basins Glob. Gulf Mexico Exp. Houston Texas GCSSEPM GCS* **2001**, *21*, 687–712.

58. Goldhaber, M.B.; Church, S.E.; Doe, B.R.; Aleinikoff, J.N.; Podosek, F.A.; Brannon, J.C.; Mosier, E.L.; Taylor, C.D.; Gent, C.A. Lead and sulfur isotope investigation of Paleozoic sedimentary rocks from the southern Midcontinent of the United States; implications for paleohydrology and ore genesis of the Southeast Missouri lead belts. *Econ. Geol.* **1995**, *90*, 1875–1910. [CrossRef]

59. Rowan, E.L.; Goldhaber, M.B. Duration of mineralization and fluid-flow history of the Upper Mississippi Valley zinc-lead district. *Geology* **1995**, *23*, 609–612. [CrossRef]

60. Cook, T.D.; Bally, A.W. (Eds.) *Stratigraphic Atlas of North and Central America [cartographic Material]*; Princeton University Press: Princeton, NJ, USA, 1975; p. 271.

61. Lu, G.; Marshak, S.; Kent, D.V. Characteristics of magnetic carriers responsible for Late Paleozoic remagnetization in carbonate strata of the mid-continent, U.S.A. *Earth Planet. Sci. Lett.* **1990**, *99*, 351–361.

62. Voo, R.V.D.; Torsvik, T.H. The history of remagnetization of sedimentary rocks: Deceptions, developments and discoveries. *Geol. Soc. Lond. Spec. Publ.* **2012**, *371*, 23–53.
63. Oliver, J. Fluids expelled tectonically from orogenic belts: Their role in hydrocarbon migration and other geologic phenomena. *Geology* **1986**, *14*, 99–102. [CrossRef]
64. McCabe, C.; Elmore, R.D. The occurrence and origin of Late Paleozoic remagnetization in the sedimentary rocks of North America. *Rev. Geophys.* **1989**, *27*, 471–494. [CrossRef]
65. Erendi, A.; Cathles, L.M. Gas capillary inhibition to oil production. *Pet. Syst. Deep-Water Basins Glob. Gulf Mexico Exp. Houston Texas GCSSEPM GCS* **2001**, *21*, 607–618.
66. Jakubov, A.A.; Ali-Zade, A.A.; Zeinalov, M.M. *Mud Volcanoes of the Azerbaijan SSR. Atlas*; The Academy of Sciences of the Azerbaijan SSR: Baku, Azerbaijan, 1971.
67. Smith, J.T.; Ehrenberg, S.N. Correlation of carbon dioxide abundance with temperature in clastic hydrocarbon reservoirs: Relationship to inorganic chemical equilibrium. *Mar. Pet. Geol.* **1989**, *6*, 129–135. [CrossRef]
68. Cathles, L.M.; Schoell, M. Modeling CO_2 generation, migration, and titration in sedimentary basins. *Geofluids* **2007**, *7*, 441–450. [CrossRef]
69. Prinzhofer, A.; Cathles, L.M. Explaining the pulsating emission of H_2 from a sedimentary basin in Brazil. *Geosciences.* (in preparation).
70. Sicking, C.; Malin, P. Permeability structure mapping using fracture seismics. *Geosciences* **2019**, *9*, 34.
71. Leach, D.L.; Bradley, D.; Lewchuk, M.T.; Symons, D.T.; Marsily, G.; Brannon, J. Mississippi Valley-type lead–zinc deposits through geological time: implications from recent age-dating research. *Miner. Depos.* **2001**, *36*, 711–740. [CrossRef]

Article

Direct Inversion Method of Fault Slip Analysis to Determine the Orientation of Principal Stresses and Relative Chronology for Tectonic Events in Southwestern White Mountain Region of New Hampshire, USA

Christopher C. Barton [1,*] and Jacques Angelier [2]

[1] Department of Earth and Environmental Sciences, Wright State University, Dayton, OH 45435, USA
[2] Tectonique Quantitative, Universite Pierre et Marie Curie, 4 Place Jussieu, 75252 Paris, France; angelier@geoazur.obs-vlfr.fr
*   Correspondence: chris.barton@wright.edu

Received: 10 September 2020; Accepted: 26 October 2020; Published: 16 November 2020

Abstract: The orientation and relative magnitudes of paleo tectonic stresses in the western central region of the White Mountains of New Hampshire is reconstructed using the direct inversion method of fault slip analysis on 1–10-m long fractures exposed on a series of road cuts along Interstate 93, just east of the Hubbard Brook Experimental Forest in North Woodstock, NH, USA. The inversion yields nine stress regimes which identify five tectonic events that impacted the White Mountain region over the last 410 Ma. The inversion method has potential application in basin analysis.

Keywords: direct inversion method of fault slip analysis; paleo tectonic principal stress orientations; west-central New Hampshire

---

1. Introduction

Previous studies have shown fault slip analysis at the outcrop scale provides a means to deduce the orientation of the principal stress fields and their evolution through successive tectonic events [1–7]. Additional information obtained from other structures, such as joints [8] tension gashes, and stylolites [9], is also important but will not be presented here. In this paper, we define a fault as simply a parting in rock with no claim whether it formed as a Mode 1 (opening), Mode 2 (shearing), or Mode 3 (tearing) [10]. If a fault shows shear offset (Mode 2). The input for fault slip analysis is field data collected on the surfaces of individual faults which includes the orientation of the, slip direction, and sense of slip. The latter two are determined by one or more of the following displacement indicators visible on the fault surface: slickensides, asperity ploughing, slickolite spikes, crescent marks, the growth of mineral patches on the lee side of hills on a rough fault surface, mineral fibers and steps, and Reidel shears [11].

The basic assumptions behind fault slip analysis are that: (1). conjugate fault sets result from a single brittle deformation event, and (2). slip on a fracture surface occurs in the direction of maximum resolved shear stress. The first step in the analysis consists of reconstructing the "reduced stress tensor". The reduced stress tensor differs from the actual stress tensor only in that the absolute magnitudes of the principal stresses: σ_1 (maximum compressional stress), σ_2 (intermediate stress), and σ_3 (minimum stress) are not determined, only their relative magnitudes. However, the relative magnitude, order, and orientation of the three principal stresses are the same as for the actual stress tensor and enable one to define the directions of compression and extension which prevailed during tectonic events. Knowing the stress state, one determines the shear stress and hence the slip orientation expected on any plane. The first attempt at formulating and solving the inverse problem was [12]. Numerical methods have since been developed for reconstructing paleo-stress orientations from fault slip data. In the general case illustrated in this paper, any planar discontinuity in a rock may be activated as a fault. The discontinuity may be either a pre-existing fracture activated or reactivated (inherited fault) by the tectonic stress. The basic properties of the reduced stress tensor and its determination is summarized below. The indicators of the direction and sense of shear on a discontinuity reactivated in shear is to collect and analyze fault slip data. The method of direct inversion used in this paper can be found in [1–6,13,14] and in [15].

2. Geologic and Tectonic History of the Southwestern White Mountain Region

The bedrock geology of the study site and Hubbard Brook valley has been mapped (sheet 1) at a scale of 1:12,000 [16] and at a scale of 1:10,000 [17]. The study site is also included in earlier bedrock maps [18,19] and at a scale of 1:250,000 [20]. The bedrock underlying the study site (Figure 1) consists of metamorphic rocks intruded by igneous rocks and belongs to the Central Maine Trough [20]. The metamorphic rocks of the Rangeley and Perry formations were deposited as sandy, clay-rich marine sediments [21] on a continental shelf, rise, and abyssal plain of the Rheic Ocean [22] over a time spanning the Silurian (approximately 443–428 Ma). These sediments were then buried and multiply folded by at least two deformation episodes [17] in the Acadian orogeny (early Devonian, 410–390 Ma), as the Rheic Ocean closed and Avalon collided with eastern North America. At this time, the rocks were metamorphosed to the lower sillimanite grade (approximately 600 °C and 4 kb pressure, equivalent to a burial depth of approximately 15 km), which resulted in local melting (migmatization). At approximately 410 Ma, these rocks were at or near the conditions of maximum pressure and temperature and were intruded locally by the southern portion of a large pluton of the Kinsman granodiorite of the New Hampshire plutonic series. The Kinsman granodiorite underlies most of the western half of the Hubbard Brook Valley immediately to the west of the study site [16] see Figure 1. Near the bottom of Figure 1 is a low angle thrust called the Thornton Fault on [20]. This fault thrusts older Silurian rocks over younger Devonian rocks. The fault is cutoff by and therefore must be older than the intrusion of the Kinsman Granodiorite, but younger than the Littleton Formation. This fault extends below the study site and below the Rangeley Formation at the study site. This fault may have formed during Tectonic Event 1 in Tables 1 and 2. During the late Devonian (370–365 Ma) the metamorphic rocks and the Kinsman intrusion were multiply intruded by small tabular dikes and small discordant bodies of Concord granite, also of the New Hampshire plutonic series. The Concord granite is shown on the map at a scale of 1:200 [16] and in well logs [23], but is not shown in Figure 1 which is modified from the map of [20] (scale of 1:500,000).

The Alleghenian orogeny (325–260 Ma) created the Appalachian Mountains principally by collision with North Africa. While it may not have resulted in large scale deformations at the study site, it was strong enough to create or reactivate fractures.

From early to late Jurassic (194 to 155 Ma) the are immediately to the east and north of the study site was a region of extensive granitic intrusion expressed by the huge batholiths and ring dikes of the White Mountain plutonic/volcanic series [24]. During that time or possibly later (130 to 100 Ma), the metamorphic and igneous intrusive rocks at the study site were intruded by tabular diabase dikes, emplaced as part of continental rifting associated with the opening of the present Atlantic Ocean basin.

From the time of the Acadian Orogeny to the present, erosion and uplift have brought the bedrock from a depth of approximately 15 km to at or near the Earth's surface. The last episode of deformation was the loading and unloading of the bedrock by the advance and retreat of multiple glacial ice sheets over this region in the past 100,000 years [25]. Reconstructions of the thickness of the Laurentide ice sheet yield a glacial ice loading and unloading of three kilometers for New England [26].

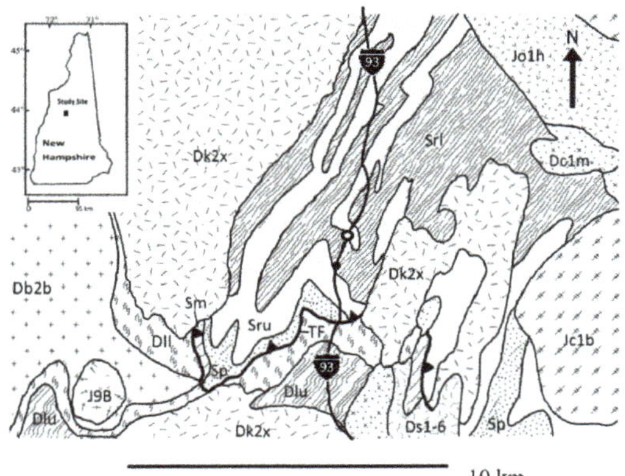

Explanation

Jurassic

Jc1b - Conway Granite (Late? and Middle Jurassic)

Jo1h - Mount Osceola Granite (Early and Middle Jurassic)

J9B - Gabbro

Devonian

Dc1m - Concord Granite (Late Devonian)

DS1-6 - Spaulding Tonalite (Early Devonian)

Db2b - Bethlehem Granodiorite (Early Devonian)

Dk2x - Kinsman Granodiorite (Early Devonian)

Dlu - Littleton Formation - unnamed upper member

Dll - Littleton Formation - unnamed lower member

Silurian

Sm - Madrid Formation (upper Silurian?)

Sp - Perry Mountain Formation (Middle? to Lower? Silurian)

Sru - Upper part of Rangeley Formation

Srl - Lower part of Rangeley Formation

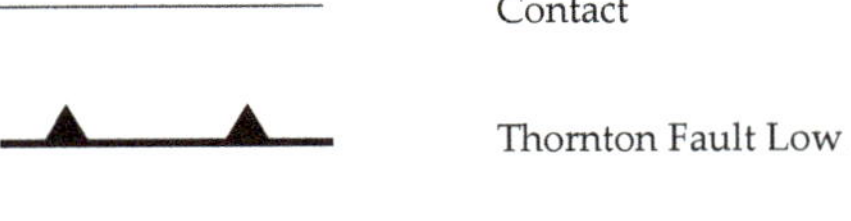

Contact

Thornton Fault Low Angle Thrust- Generally Folded (Teeth on Upper Plate)

● I-93 Roadcuts – Study Site 1

○ I-93 Roadcuts – Study Site 2

Figure 1. Bedrock geologic map for the area immediately surrounding the two study sites along the I-93 roadcuts in Woodstock, NH, USA. (After [20], sheet 1). Location map is shown in upper left. The study site is located in the Lower Rangeley Formation.

Table 1. Results of inversions for nine tectonic regimes based on the direct inversion method [4] with additional refining process [13]. Reg. = reference number of regime, also referred to in Table 2. MIFL = the control parameter, indicates the minimum individual fit level finally retained (see the term ω in [13] for detailed discussion of MIFL). N_A = number of fault planes found acceptable at this level of fit. N_R = number of fault planes rejected. The stress tensor obtained is characterized by the trends and plunges of the three principal stress axes, σ_1, σ_2, and σ_3, and by $\Phi = (\sigma_2 - \sigma_3)/(\sigma_1 - \sigma_3)$, the ratio of the principal stress differences where $0 \leq \Phi 13$ [2]. υ_m = average value of the main estimator [4], τ^{*}_m = ratio of the average shear stress to the maximum shear stress. α_m = average value of the calculated shear-actual slip angle.

Reg.	MIFL%	N_A	N_R	σ_1 degrees		σ_2 degrees		σ_3 degrees		Φ	υ_m %	τ^{*}_m%	α_m degrees
1	40	17	2	133	3	226	45	40	45	0.13	75	83	17
2	55	18	2	257	48	139	23	33	33	0.46	86	87	4
3	40	62	13	184	80	20	10	290	3	0.49	79	84	14
4	45	5	1	239	5	330	8	117	80	0.44	70	71	9
5	55	23	1	83	17	238	71	351	7	0.50	80	85	15
6–7	45	52	17	188	14	315	67	93	17	0.45	76	79	14
8	50	6	0	330	7	235	30	72	59	0.23	74	77	11
9	45	12	1	130	5	350	80	40	8	0.44	78	78	10

Table 2. Tectonic paleo-stress chronology including: tectonic events (compressional or extensional), relative chronological order, regime number, fault type, and orientation of the three principal stresses, as determined by fault-slip analysis in the present study. The time ranges are from the published literature where known tectonic events in the region have been dated as presented in Section 2 above. The orientation of the principal stresses for each tectonic event has one oriented vertical and two horizontal with azimuthal angles as shown.

Tectonic Events	Time (Ma)	Regime	Fault Type	1	2	3
1. Compression	390–375	1	Reverse	130	46	vertical
2. Extension	375–325	2	Normal	Vertical	139	33
		3	Normal	vertical	20	290
3. Compression	335–260	4	Reverse	239	330	Vertical
		5	Strike-slip	83	vertical	351
4. Compression and Extension	190–95	6–7	Strike-slip	188	Vertical	93
		8	Reverse	330	235	vertical
5. Current Compression	15–present'	9	Strike-slip	130	vertical	40

Based upon the orientation of glacial striations, the last Wisconsin Ice sheet moved over the study site from WNW to the ESE [16]. At the last glacial maximum 14,000 years ago, the minimum thickness of the ice at the study site was approximately 1.6 km [27]. The glaciers swept away the thick loess, soils and vegetation that previously covered the bedrock. In some places, the advancing glacial ice plucked automobile-sized blocks from the leeward side of the ridges and small hills in the Hubbard Brook valley. Throughout the valley, glacial ice and water carved and polished the top of the bedrock to a smooth, undulating surface. Finally, as the ice sheet melted in place, the rock debris within the glacier, worked by rivers and streams on top of, within, and beneath the melting ice, was deposited as discontinuous layers on top of the bedrock surface with thickness increasing from 0 at and near the ridge crests and stream beds to as much as 50 m in the lower part of the valley [27]. Because of the glaciation-related erosion, the present-day rock condition of bedrock exposures is extremely fresh, which makes our study of brittle structures much easier than if it were weathered rock.

The geology and faults studied in this report are exposed in roadcuts along Interstate 93 (Figure 2) which were mapped at a scale of 1:200 by [16] sheet 2. All naturally occurring fractures greater than one meter in scale are shown on the map. Fracture orientation, trace length, aperture, surface roughness, and interconnectedness were measured and analyzed [16,28]. The compositional variability in the schist persists to the millimeter scale. The schist has a well-defined foliation, which has been refolded at least twice, and the foliation can be highly variable at length scales less than a meter. At larger scales, the foliation strikes from 25 to 45 degrees east and dips steeply to the southeast, consistent with the regional tectonic fabric.

Figure 2. Photograph of a N5E striking roadcut exposure at the study site on I-93 containing NE striking, SE dipping faults included in this study. Note, most of the faults fractures exhibit dark planar surfaces. The rock type is primarily Concord granite (dark gray) with a Lower Rangeley schist block (light grey/white) exposed to the right of center above where the grass meets the bedrock and between the first and fourth drillhole from the left side of the photo. The subvertical lines are drill holes used in blasting the roadcut surface. Targets were used for rectification of photographs on which the geology and fractures were mapped by [16] (sheet 2).

The granitic rocks intrude the schist in the form of small tabular dikes and large anastomosing fingers ranging from 1000's of meters to the meter scale. The intrusion was prolific, and granitic rocks account for approximately 50% of the rock area mapped in the road cuts, Figure 2 [16] sheet 2. and in the 40 boreholes (totaling 4.6 km of wellbore) drilled in the Mirror Lake watershed, located at the eastern end of the Hubbard Brook valley [23]. Changes in lithology between granite and schist occurs every 5–9 m in the roadcuts [16] and the boreholes [23].

Bedrock fractures in the roadcuts, natural outcrops and the bedrock wells include joints (formed as Mode 1 fractures), faults (formed as Mode 2 fractures), and reactivated faults and joints. Fractures formed prior to the maximum burial and temperature (410–390 Ma) would have been destroyed by metamorphic recrystallization. We therefore assume that all the fractures that we observe in outcrop were formed after the peak metamorphic event at approximately 390 Ma. It is not possible to determine the age of fracture formation or reactivation using relative or radiometric dating. The brittle tectonic activity since 390 Ma could result from events during the Alleghenian orogeny (Permian, 299 to 251 Ma) and to younger tectonic events, such as the extension related to the opening of the northern Atlantic ocean approximately 200–175 Ma or to the glaciation-deglaciation cycles of the Quaternary (2.6 Ma to present) [27]. Little evidence for the age of brittle events can be obtained from stratigraphy or rock dating, although thin (~1 m) NE-SW striking diabase dikes occur in the study site during the early Jurassic 200–146 Ma may be a brittle episode related to the opening of the Atlantic Ocean. Large numbers of fracture surfaces display syntectonic mineral infill or fiber growth. Because syntectonic minerals like quartz could not develop during the brittle events at very shallow depth, such mineralization indicates that most of the brittle tectonic activity that produced the fault slips took place at depths up to 15 km., and hence is related to tectonic episodes that predate the glaciation-deglaciation events

3. Data Collection and Stress Inversion Method of Analysis

Two hundred and eight fault-slip data were collected at roadcuts in bedrock at two locations on Interstate-93 in Woodstock, New Hampshire as shown in Figure 1. The first location includes four sub-parallel vertical roadcut faces approximately 40 m apart, whose bedrock geology and fractures had been previously mapped [16]. The second location is a roadcut on the east side of the northbound lane of I-93, 1.3 km north of the first location. Figure 2 is a photograph of a section of a portion of roadcut at location 1 showing NE striking, SE dipping fractures in the Concord granite 370–365 Ma.

Faults were easily identifiable in the roadcuts, most of them bearing slickensides resulting from slip-parallel quartz growth. Numerous faults show minor (~1–2 mm) but clearly observable offsets of cleavage, schist-granite contacts, quartz-pegmatite veins, and along contacts of the diabase dikes and the schist and granite. Evidence of slip-parallel quartz growth was common. The strike and dip of the fault plane, the rake of the slickenside lineations, and the sense of relative offset were measured for each observable fault. The faults were numbered. All types of fault slips were found: dip-slip, strike-slip and oblique slip, with normal, reverse, right-lateral and left-lateral components of motion. This variety of fault slips indicates polyphase brittle tectonism, which was confirmed by differences in mineral fillings. The inferred tectonic regime/relative chronology (by number) and level of certainty, the roadcut location, and the fracture number on [16], were all noted. All the information recorded is listed in Appendix A Table A1.

Particular attention was paid to determination of the sense of motion on each fault. A variety of criteria were used, including: (1) offset of granite-schist and other metamorphic boundaries, (2) mineral growth along the slip direction, (3) presence of rough and polished facets along the fault surface, (4) asymmetrical striation markers, (5) striation-related micro-veins, (6) offsets of older fractures or veins, (7) presence of small Riedel's shear fractures, mainly R in type. Where possible, these criteria were cross-checked. As a result, three levels of certainty were considered concerning the senses of motion (see Appendix A, Table A1). The letter C refers to a slip sense that could be determined with certainty in the field, based on one or several unambiguous criteria. The letter P indicates that the slip sense is considered probable, which means that despite good observation some ambiguity could not be removed. The letter S refers to a poorly recorded sense of motion, in the absence of reliable criterion or with conflicting criteria. In that case an inferred "supposed" sense of motion was attributed, based on both the low-quality criterion (if any) and the behavior of the neighboring faults with well-recorded sense of motion and similar dip direction, attitude, and slip orientation.

Many faults were associated in conjugate or Riedel's type patterns with particular symmetries. Fault subsets were defined based on common geometry in terms of fault attitude, slip orientation and sense, fault dip direction, relation to other faults, and mechanical consistency. The relation between conjugate fault systems and stress has been highlighted by Daubrée's experiments [29] and Anderson's analysis [30]. In addition, Riedel's shears [11] often explain the relationships between faults at different scales.

Most fault slip data in the outcrops studied could not be interpreted in simple geometrical terms, because they resulted from reactivation of earlier faults or mechanical discontinuities (older faults, joints, veins, cleavage, contacts between rock types, etc.). Such inherited faults may have various attitudes oblique to all stress axes, contrary to the "newly formed" faults discussed above, which generally contain one principal stress axis and form symmetrical systems. For this reason, we undertook systematic inversion of the fault slip data to reconstruct the stress regimes. Such inversions are based on consideration of the stress-slip relationships proposed by [31,32], which were used by [12], who first addressed the inverse problem in their pioneering work. Later studies demonstrated that the basic assumptions underlying the method were acceptable in the first approximation and well accounted for actual slip distribution (e.g., [2,3]), and numerical modeling experiments showed that deviations from the model are significant but remain statistically minor with regard to other sources of uncertainty [33].

The direct inversion method' used here is based on a least-square minimization, with a criterion called υ (upsilon) that depends on both the angle between the calculated shear and the actual slip, and the shear stress amplitude relative to maximum shear stress. For details, the reader is referred to the paper that describes this method [4]. We also use a robust refining process that was not described in the original formulation of the method but is presented in the use of another method especially designed for the stress inversion of earthquake focal mechanisms [13]. This additional process was facilitated by the negligible runtime of the inversion method, which involves analytical means instead of numerical search. A crucial parameter is the minimum fit level required for defining acceptable data. We use a scale from −100% (total misfit) to 100% (perfect fit). The lowest bound involves maximum shear stress acting in the direction opposite to slip. At the highest bound, the shear stress is also maximum but acts in the same direction and sense as the slip. A zero value indicates that slip occurs with shear stress perpendicular to slip, as the limit between consistent and inconsistent senses of motion. Note that this minimum fit level is linearly related to the RUP % estimator defined by [4], the values −100% and +100% corresponding to the values 200% and zero (respectively) in the RUP estimator and differs from the ω estimator defined by [13].

To determine a stress regime, υ is minimized as a function of the four unknowns that describe a reduced stress tensor: the orientations of the three principal stress axes and the ratio $\Phi = (\sigma_2 - \sigma_3)/(\sigma_1 - \sigma_3)$. One obtains the smallest slip-shear angles and the largest possible shear stresses that can simultaneously exist for all the data taken together.

The real data dispersion, which depends on complex geological factors, is larger than the angular uncertainty of about 5° in our field data collection. To determine whether a stress inversion is significant or not, we use an iterative refining process that involves successive inversions with a progressively increasing demand for good individual fits. This process allows determination of the level of data rejection consistent with the data accuracy.

4. Results

Based on consideration of relationships between fault slips (crosscutting relationship, reactivation of fault surface, etc.), spatial association between faults (e.g., conjugate patterns) and syntectonic mineral growth (e.g., quartz fibers), and taking into additional account the mechanical consistency within each subset of fault slips, it was possible to separate nine data subsets (regimes), as listed in Table 1 below.

The number of Regimes is large (9). High confidence can be placed in the definition of the regimes themselves, their sequential order, and especially the directions of compression and extension of the inferred stress tensor. A second step involves grouping the Regimes into tectonic Events where the known regional tectonics coincides with the direction of the principal stresses and the time sequence of known regional tectonic events. The grouping into Events is shown in Table 2 and discussed below.

Each Regime is depicted on a lower hemisphere equal area projection below showing the great circles of each of the fault planes determined. Arrows indicate the slip on the fault planes. A separate companion plot displays the same arrows and the poles to the fault planes (open circles) and the trend and plunge of the intermediate stress (σ_2). Arrows pointing toward the center of the projection indicate compression. Arrows pointing to the perimeter of the projection indicate extension. Solid circles with two arrows pointing in opposite directions indicate the sense of strike-slip movement. Those with no arrowheads have an indeterminate sense of movement.

4.1. Event 1—Regime 1

This regime is characterized by reverse faults compatible with a NW-SE compression (Figure 3). The frequent presence of quartz veins and along-slip quartz growth indicate that slip probably occurred close to the ductile-brittle transition. The relatively deep and hot character of this tectonic deformation suggests that this event is the oldest event.

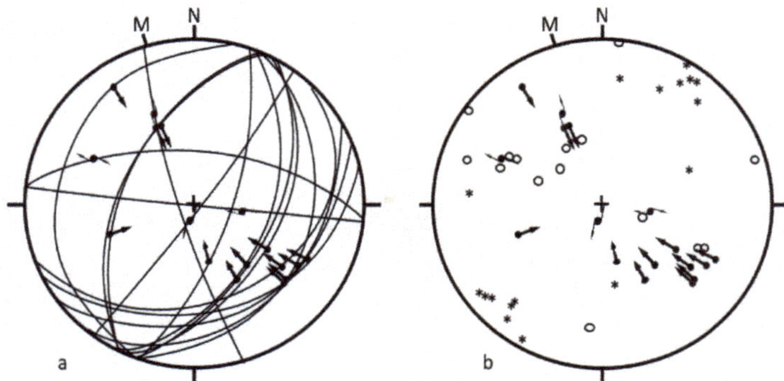

Figure 3. Regime 1. Lower Hemisphere Equal Area projection of (**a**) great circles of fault plane orientation, and (**b**) poles to fault planes (open circles) and (*) the calculated trend and plunge of σ_2. The arrows indicate the rake of slickenlines on the fault planes and point in the direction of slip. Arrows pointing toward the center of the projection indicate compression, those pointing away from the center indicate extension. Solid circles with two arrows pointing in opposite directions indicate the sense of strike-slip movement. Those with no arrowheads have an indeterminant sense of movement. M = magnetic north, N = true north.

The stress inversion of fault slip data for this data subset shows that for a reasonable threshold, MIFL = 40%, only 2 of the 19 fault slip data are considered unacceptable. Similar solutions were obtained for minimum individual fit levels of 20% (no data being eliminated) or 40% (4 data eliminated). The stress regime determined is thus stable.

The calculated stress regime indicates a nearly horizontal compression that trends 133° N. The stress axes σ_2 and σ_3 are oblique, with plunges of 45°, in agreement with the low value, 0.13, for the ratio $\Phi = (\sigma_2 - \sigma_3)/(\sigma_1 - \sigma_3)$. This low Φ reveals σ_2 and σ_3 are closer in magnitude than either is to σ_1. The solution cannot be considered very accurate because the number of acceptable fault planes inversions is small (17). The direction of compression is constrained within ±10 degrees, but the values of Φ and the attitudes of stress axes σ_2 and σ_3 may vary widely as a function of data removal within this set. In summary: the oldest brittle tectonic episode that we can recognize corresponds to a NW-SE compression that reactivated deep fractures in reverse faulting.

4.2. Event 2—Regimes 2 and 3

In contrast to Regime 1, both Regimes 2 and 3 are dominated by normal fault extension (Figure 4). Most are dip-slip, which suggests a low level of structural inheritance and reactivation of earlier structures. Most fault surfaces are planar with relatively steep dips, which suggests that they developed at shallower crustal levels than the reverse faults of Regime 1. However, the gentle dips of some of the normal faults suggests the reverse faults of Regime 1 have been reactivated as normal slips. Note that the dominate trend of normal faults in Regime 3 are the same as for the reverse faults of Regime 1 (e.g., the fault poles are similar). Syntectonic quartz is common on the surfaces of these inherited normal faults, with the quartz probably inherited from Regime 1.

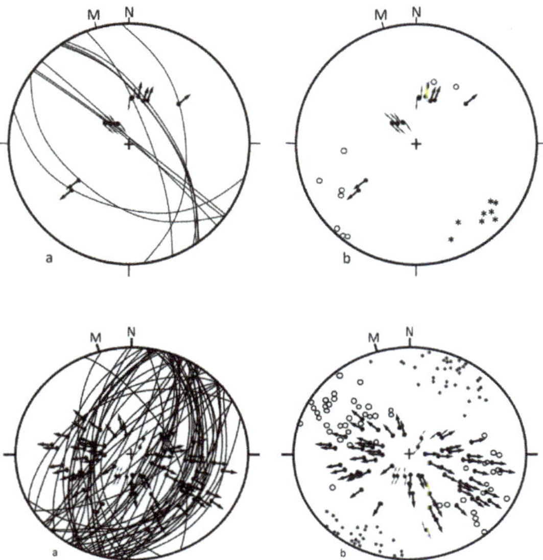

Figure 4. Regime 2 (**top**) and Regime 3 (**bottom**). Lower Hemisphere Equal Area projection of (**a**) great circles of fault planes; and (**b**) poles to fault planes (open circles) and of calculated trend and plunge (*) of σ_2. Symbols are as in Figure 3.

The two subsets of faults strike at right angles. The normal faults of Regime 2 strike NW-SE. and the normal faults of Regime 3 strike NE-SW. Regime 2 faults indicate NE-SW extension whereas Regime 3 faults indicate NW-SE extension.

The stress inversion of fault slip data for Regime 2 is stable (2 faults rejected at MIFL = 55%, no faults rejected at 20%, and only 4 of the 20 faults rejected at MIFL = 80%), but is only loosely constrained because of the limited number of data (18 acceptable faults). As with Regime 1, the numerical results are good, but the solution is highly dependent on the grouping of data. For example, removing the two nearly vertical faults results in a significantly different solution. The stress regime at MIFL = 55% indicates a 33° ± 10 degrees trending extension with oblique σ_1, σ_2 and σ_3 axes. The σ_3 axis plunges 33° NE, which is not surprising in light of the presence of nearly vertical faults with the downthrown side to the northeast. The Φ ratio of 0.46 indicates triaxial stress.

In contrast, the large number of fault slip data in Regime 3 provides a highly constrained stress tensor solution. The solution is stable (13 of 75 faults rejected at MIFL = 40%, 4 at 20% and 20 at 55%). The stress orientations and Φ are similar regardless of the MIFL, which indicates that the stress tensor is well constrained. On the other hand, the slip vectors have a large scatter (Figure 4). The number and geometrical variety of the data are more important than the average of parameter estimates and their standard deviations. Removal of fault slip data does not change the inversion results within the range of uncertainties, which confirms that geometrical constraints on the stress tensor exerted by the variety in fault slip attitudes is more important to a good interpretation. At MIFL = 40% the σ_1 axis is nearly vertical and indicates a 110° azimuth of extension (the σ_3 axis plunges only 3° to the west). The direction of extension is constrained within ±5 degrees. A Φ ratio of 0.49 indicates typical triaxial stress.

No clear chronological difference could be established between Regimes 2 and 3. The perpendicularity in fault trends and corresponding directions of extensions strongly suggest that these two regimes are linked through a permutation (relative magnitude switch) between the intermediate and minimum principal stresses, σ_2 and σ_3. These two regimes thus probably belong to a single major extensional event which we identify as Event 2. Because Regime 3 is represented by a much larger number of brittle structures than Regime 2, the dominating direction of extension is inferred to be WNW-ESE (azimuth 110°). A tensor inversion with Regimes 2–3 taken together shows that the influence of the fault slip data from Regime 3 prevails, and the combined tensor solution resembles that of Regime 3. For a MIFL = 30% the Φ ratio is lower (0.35) but the direction of extension is similar (115°). The stability of the solution is much less, which suggests the distinction between Regimes 2 and 3 is in fact significant in terms of stress states, even though both are produced by the same tectonic event. Brittle tectonic analyses have revealed significant changes in stress regimes within a single tectonic episode [4–6,13,15]. The duality of stress regimes (2 and 3) may simply result from a permutation between the σ_2 and σ_3 axes, a common phenomenon in fault tectonics.

Although there remains some indication of ductile-brittle transition for some faults with abundant quartz coating and slip-parallel quartz growth, most Event 2 faults are typically brittle, as shown by both the fault surface characteristics and their steep dips. Relative chronologies with respect to other events show that Regimes 2 (certainly) and 3 (probably) post-dated the Regime 1. Our data thus suggest that the extension of Regimes 2–3 post-dated the compression of Regime 1 and suggests that Regimes 1–3 reflect the oldest two faulting events well represented at the site.

4.3. Event 3—Regime 4

Regime 4 is poorly represented. It is characterized by only a few reverse faults that are compatible with a NE-SW compression. The stress inversion provided stable results, which has little meaning because of the very low number of faults (5). The tensor solution is very loosely constrained. Had more fault slips been identified, the result would have been subject to significant variations. For a MIFL = 45% one of the five faults is considered unacceptable, and the calculated stress regime indicates a 59° compression direction with a nearly vertical σ_3 axis and a Φ ratio of 0.44. The direction of compression may however vary within ±20°.

The reverse faults shown in Figure 5 are younger than those of Regime 1, but there is little indication of their age. Two display the same attitude, but different oblique slip vectors from the reverse faults of Regime 1 from which they are inherited.

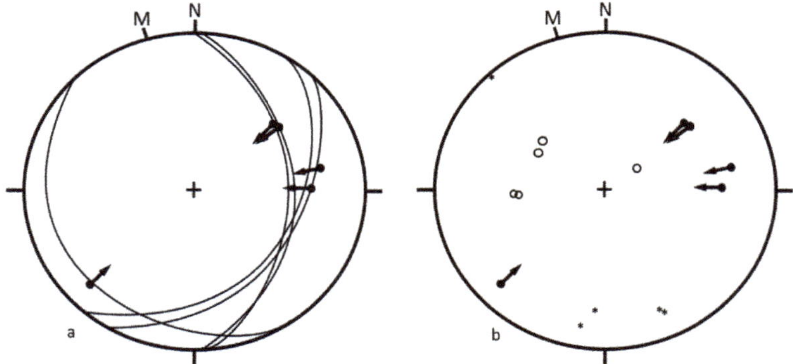

Figure 5. Regime 4. Lower Hemisphere Equal Area projection of (**a**) great circles of fault planes and (**b**) of poles to fault planes (open circles) and of calculated trend and plunge (*) of σ_2. Symbols are as in Figure 3.

4.4. Event 3—Regime 5

Regime 5 (Figure 6) is better represented than Regime 4. The stress tensor inversion indicates strike-slip faulting consistent with a nearly E-W compression and N-S extension. These strike-slip faults clearly postdate the reverse faults of Regime 1. The stress tensor inversion is stable (1 of 24 faults rejected at MIFL = 55%, none rejected at 25%, and 7 rejected at MIFL = 70%). Another indication of inversion stability is that the removal of dextral fault slips does not significantly modify the inversion results.

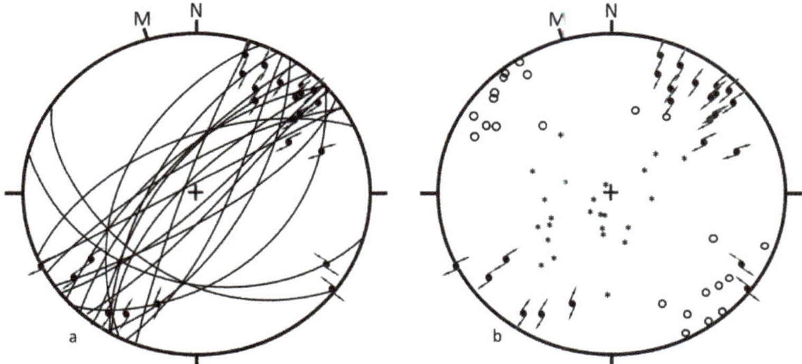

Figure 6. Regime 5. Lower Hemisphere Equal Area projection of (**a**) great circles of fault planes and (**b**) of poles to fault planes (open circles) and of calculated trend and plunge (*) of σ_2. Symbols are as in Figure 3.

The stress regime calculated at the MIFL = 55% level is characterized by a nearly horizontal σ_3 axis that trends 0° and a gently plunging σ_1 axis whose azimuth trends 83°. The direction of extension is not more tightly constrained than ±10 degrees because only 2 left-lateral faults were measured. A typical triaxial stress is indicated by the Φ ratio of 0.50.

Contrary to the typical, most strike-slip faults of Regime 5 are far from vertical. Many of the right-lateral faults dip towards the NW or SE, which suggests that they were inherited from the normal fault planes of Regime 3. Two left-lateral faults have gentle SW dips, which suggests that they were inherited from normal fault planes of Regime 2.

Although the faults of Regime 4 are reverse and the faults of Regime 5 are strike-slip, the directions of compression are similar considering the large angular uncertainty in the trend of compression of Regime 4. For this reason, we combine Regimes 4 and 5 within a single Event 3 that is dominated by a roughly E-W compression and can generate both reverse and the strike-slip faulting. The stress tensor for this combination resembles that for Regime 5 because of the larger number of faults in Regime 5. Unlike the combination of Regimes 2 and 3, the combination of Regimes 4 and 5 shows good stability (1 of 30 faults eliminated for MIFL = 20%, 9 for 55%). But the rejected faults are 2 of the 5 reverse faults of Regime 4, and the individual misfits of the three remaining Regime 4 reverse faults are large. The simplest solution suggests mixing Regimes 4 and 5 are indeed distinct.

For a reasonable fit level of 35% (2 faults rejected), the Φ ratio is 0.43, indicating triaxial stress despite the mixture of strike-slip and reverse faults. The stress regime is characterized by a gently plunging σ_3 axis with a trend azimuth of 351° and a nearly horizontal σ_1 axis with a trend of 83°. The data indicate stress regimes 4 and 5 belong to a single event dominated by WNW-ESE compression and that a permutation between σ_2 and σ_3 changes the faulting from reverse to strike slip.

4.5. Event 4—Regimes 6, 7 and 8

The strike-slip faults of Regimes 6–7 are shown and analyzed together (Figure 7). These regimes are dominated by strike-slip faulting. Sixty-nine faults are observed, the largest ones forming typical strike-slip zones composed of two walls on either side of a 1–3 m wide deformed zone with numerous smaller faults, fractures, rotated blocks, and gouge. The strike-slip faults strike approximately NNW-SSE for right-lateral faults, and NNE- SSW for left-lateral ones, indicating N-S compression.

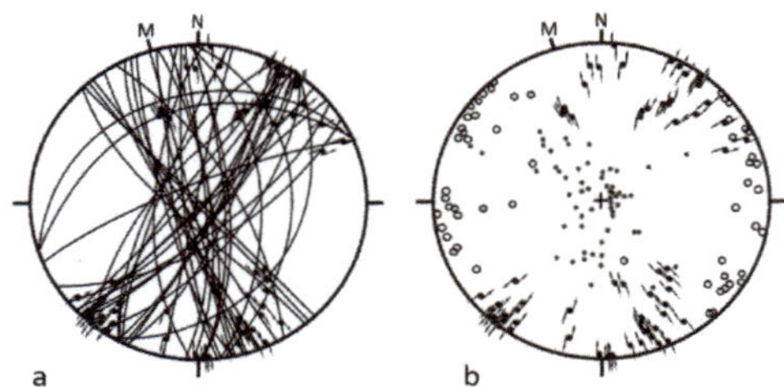

Figure 7. Regime 6–7. Lower Hemisphere Equal Area projection of great circles of fault planes (**a**); and (**b**) of poles to fault planes (open circles) and of calculated trend and plunge (*) of σ_2. Symbols are as in Figure 3.

The stress tensor solution is tightly constrained by the large number of the faults and the variety of their orientations (17 faults rejected at MIFL = 45%, 5 at 20%, and 23 at 55%). The stress orientations remain extremely stable as the MIFL increases. In addition, removal of fault slip data does not significantly affect the inversion. The geometrical constraints exerted by the variety in fault slip attitudes are strong. Because a significant overlap in stress trends is present between right-lateral and left-lateral faults, several data displayed incompatible senses of motion. This explains why faults were eliminated even for low levels of MIFL. Separation into two Regimes, 6 and 7, solved this problem and reduced the number of inconsistent senses to zero for each of the stress tensors, but was not retained because no independent qualitative evidence supported a separation of these Regimes.

The stress regime calculated at MIFL = 45% is characterized by gently plunging σ_1 and σ_3 axes (plunges of 14° and 17° degrees respectively), with a nearly N-S trending, azimuth 188°, compression. This direction of compression is constrained within less than ± 5°. The Φ ratio of 0.45 indicates typical triaxial stress.

Relative chronology data provides good evidence that this major strike-slip event postdated the normal faults of Event 3. Although some strike-slip faults of Regime 6 and 7 have relatively gentle dips suggesting that they were inherited from earlier regimes of reverse and normal faults, most of these strike-slip faults are vertical or steeply dipping, cutting through all pre-existing structures rather than reactivating them. It is likely that several NE-SW trending faults result from right-lateral reactivation of the left-lateral faults of Regime 5, but observation is speculative because of the right-lateral friction that generally destroyed the criteria supporting the evidence of an earlier left-lateral motion.

Regime 8 is represented by only a few dip-slip reverse faults (Figure 8). The relative chronology data indicate that this regime occurred before the Regimes 6 and 7. As with Regime 4, the stress inversion provides very stable solutions, but this stability is not significant because the number of faults is so small. The tensor solution is in fact poorly constrained. For MIFL = 50%, all data are acceptable, and the calculated stress regime indicates an azimuth 330° compression with a nearly horizontal σ_1 axis, a steeply plunging σ_3 axis and a Φ ratio of 0.23. The direction of compression is constrained within ±20°.

Because the direction of compression suggested by this pattern of reverse faults is not far from N–S (with an azimuth of compression approximately 160), they may be related to regimes 6–7 through a relative magnitude shift between the intermediate and minimum stress axes. If Regime 8 is added to 6 and 7 the inversion rejects all 4 faults in Regime 8. As in the case of Regimes 2–5, this suggests a common tectonic event involving a stress permutation between σ_2 axis and σ_3 axes. There is no evidence that Regime 8 resulted from a separate tectonic event.

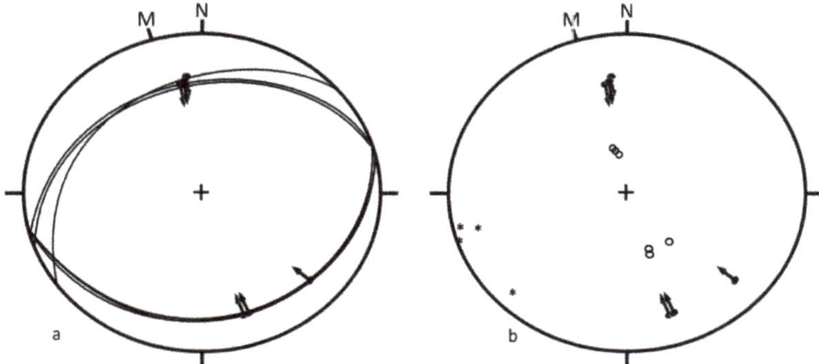

Figure 8. Regime 8. Lower Hemisphere Equal Area projection of (**a**) great circles of fault planes and (**b**) of poles to fault planes (open circles) and of calculated trend and plunge (*) of σ_2. Symbols are as in Figure 3.

Event 4 comprising Regimes 6–8 was certainly more recent than the reverse and normal faults of the ductile-brittle transition (compression of Regime 1, and the extension of Regime 2). The contacts of the diabase dikes of Jurassic age are reactivated as strike-slip faults of Regimes 6 and 7 indicating that the faulting and diabase dike intrusion in these regimes occurred 200–146 Ma or later. The NW-SE extension is compatible with the regional extension (based on local NE-SW diabase dike trends, [24]) affecting the study area during the initial opening of the north Atlantic Ocean 200–175 Ma [34], Figure 5.

4.6. Event 5—Regime 9

Regime 9 (Figure 9) corresponds to three strike-slip faults, which trend NW-SE (left-lateral) and WNW-ESE (right-lateral), and hence indicate a roughly NW-SE compression which we label as Event 5. This event is the most recent at the study site. Designation of Regime 9 and calculation of the stress tensor by three fault slips results in a high level of uncertainty.

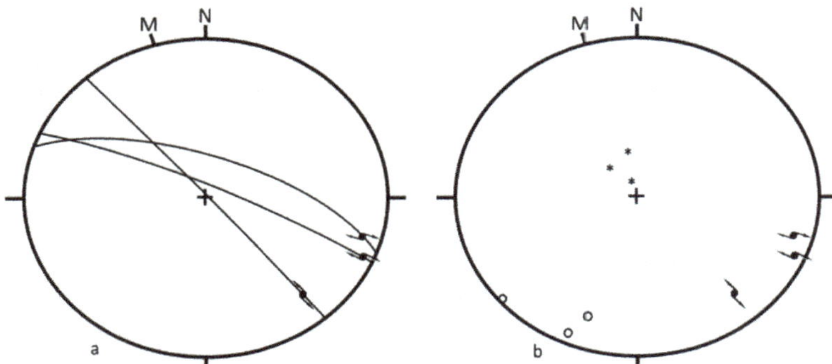

Figure 9. Regime 9. Lower Hemisphere Equal Area projection of (**a**) great circles of fault planes and (**b**) of poles to fault planes (open circles) and of calculated trend and plunge (*) of σ_2. Symbols are as in Figure 3.

5. Interpretation

We relate the relative stress tensor in our Events (Table 2) to known tectonic events summarized above in Section 2.

5.1. Event 1—Acadian Compression

This major event is the oldest episode recorded by fault slip at the study site. It was dominated by NW-SE compression (Regime 1) and took place at depth, near the ductile-brittle transition at the or near the end of the Acadian orogeny (390–375 Ma).

5.2. Event 2—Post Acadian Extension

This major event was responsible for widespread normal faulting at a depth, not far from the ductile-brittle transition. Its faults developed under a pure brittle regime at shallower depths, and the faulting was later (375–325 Ma)-than the faults in Event 1. The major NW-SE extension that occurred in this Event parallels the major NW-SE compression of Event 1. This suggests that the extension in Event 2 resulted from the exhumation of metamorphic basement that followed the major compressional phase of Event 1.

5.3. Event 3—Late Alleghenian Compression

The reverse and strike slip faulting of Regimes 4 and 5 are not numerous enough to tightly constrain the corresponding paleo-stresses, and these two Regimes are poorly located in time. They are represented by pure brittle features, and certainly postdate the brittle-ductile deformation, but may not predate Event 4. We tentatively relate these two regimes to the late tectonic evolution of the Alleghenian orogeny (325–260 Ma).

5.4. Event 4—Atlantic Extensional Opening (?)

The strike-slip faulting of Regimes 6 and 7 and the reverse faulting or Regime 8 indicate N–S compression and E–W extension. The presence of a strike-slip fault zone 1–3 m wide suggests that the offsets are relatively large (10–100's of meters) [35–37]. The contacts of the diabase dikes of Jurassic age are reactivated by the strike-slip faults of Regimes 6–7 indicating that the faulting and diabase dike intrusion in these regimes occurred 200–146 Ma or later. The NW-SE extension is compatible with the regional extension (based on local NE-SW diabase dike trends, [24]) affecting the study area during the initial opening of the Atlantic Ocean 200–175 Ma [34], Figure 5.

5.5. Event 5—Recent or Present

This Regime is represented by only a few strike-slip faults with minor offsets. The direction of compression (NW-SE) is consistent with the present state of stress in this region based on earthquake focal mechanisms [38]. We did not observe sheeting joints (sub-horizontal fractures) that can form in response to the unloading of glacial ice [39] over the past 100,000 years.

6. Discussion

This paper illustrates a method to identify distinct regional tectonic events and put them in relative chronological order from fault slip data collected on a few local roadcuts. The data was collected in this case over a period of 10 days. The volume of fault orientation and slip data is relatively small compared to regional field mapping of large scale structures field studies, but by explicitly considering the stress tensors that could have produced the observations, the fault data can be sorted and grouped into Regimes that yield compatible stress tensors that can be further grouped into distinct tectonic Events. In the southwest White Mountain region, the Events identified correspond to known tectonic events based on large scale structures over the last ~400 Ma. The orientation and relative magnitudes of principal stresses producing major regional tectonic deformations can be obtained relatively quickly from a study at one locality.

The conditions for this study were close to ideal because the roadcuts exposed unaltered rocks recently exhumed by the last glaciation. The direct inversion method could potentially be applied using oriented cores from boreholes where there are no outcrops. This paper illustrates how, if the stress tensor can be constrained, the history of overprinted deformation that could have impacted basin resources can be deduced from features on the surfaces of fractures reactivated by sequential tectonic events.

Author Contributions: C.C.B.: field collection of data 50%, writing of manuscript 90%. J.A.: field collection of data 50%, inversion of data and identification of regimes and events 95%, writing of manuscript 10%. All authors have read and agreed to the published version of the manuscript.

Funding: C.C. Barton was supported by the U.S. Geological Survey as part of his employment. J. Angelier was supported by the Universite Pierre et Marie Curie as part of his employment.

Acknowledgments: The authors wish to acknowledge support from the U.S. Geological Survey. J. Angelier was an early key developer of the Direct Inversion Method of Fault-Slip Analysis for unraveling tectonic history to deduce paleo stress orientations and relative magnitudes from surface structures on the faces rock fractures reactivated by sequential tectonic activity. Readers are directed to his many papers on this subject. He was a pleasure to work with in the field. His field notebooks were magnificent and artistically beautiful. His intelligence, enthusiasm, and good company is missed by those who knew and worked with him. Jacques Angelier died in January 2010. The authors thank Tristan Coffey who drafted the figures and tables. The authors thank C. Page Chamberlain and Gautum Mitra who provided early reviews of this manuscript and two anonymous additional reviewers. The authors also thank Lawrence Cathles who provided a thorough edit for this volume.

Conflicts of Interest: The authors declare no conflict of interest.

Data Availability: The data collected for each fault listed in tabular form in Appendix A is also available at: http://www.hubbardbrook.org/data/dataset_search.php.

Appendix A. List of Data Collected in This Study

Explanation to Columns in Table A1

Column A–Fault Reference Number

Column B–Fault Strike (azimuth)

Column C–Fault Dip and Direction

Column D–Rake of Slip Striations

Column E–Regime Number/Relative Chronology (1 = oldest, 9 = youngest)

      Certainty (C = certain, P = probable, S = inferred)

Column F–Fault Location on Interstate 93:

      SBLW = Southbound Lane, West Side of I-93, [10] sheet 2

      SBLE = Southbound Lane, East Side of I-93, [10] sheet 2.

      NBLW = Northbound Lane, East Side of I-93, [10] sheet 2

      NBLE = Northbound Lane, West Side of I-93, [10] sheet 2

Column G–Fracture Reference Numbers on [10] sheet 2

Column H–Site location (see Figure 1)

Table A1. Fault Data Collected and Analyzed in This Study.

A	B	C	D	E	F	G	H
1	173	77W	43N	6C	SBLE	1	1
2	41	40E	45N	2C 4C	SBLE	42	1
3	32	38E	62N	2C 4C	SBLE	42	1
4	95	64N	40W	1C 1C	SBLE	42A	1
5	163	83W	40N	6C	SBLE	11A	1
6	163	83W	40N	1C	SBLE	11	1
7	166	85W	44N	6C	SBLE	11A	1
8	165	83W	38N	6C	SBLE	11A	1
9	159	77W	9S	7C	SBLE	19	1
10	152	88W	20S	7C	SBLE	14B	1
11	159	78W	17S	7C	SBLE	14B'	1
12	153	89W	34S	7C	SBLE	22	1

Table A1. *Cont.*

A	B	C	D	E	F	G	H
13	137	89W	41S	7C	SBLE	22	1
14	153	77W	46S	7C	SBLE	22	1
15	165	84W	46S	7C	SBLE	22′	1
16	146	81W	25S	7C	SBLE	21	1
17	141	88W	43S	7C	SBLE	21	1
18	139	88W	36S	7C	SBLE	20	1
19	136	77W	33S	7C	SBLE	20	1
20	158	76W	24S	7C	SBLE	19′	1
21	3	69W	31S	7C	SBLE	unmapped	1
22	157	77W	23S	7C	SBLE	19′	1
23	3	69W	31S	7C	SBLE	unmapped	1
24	55	48N	52W	3C	SBLE	unmapped ou 18A	1
25	61	46N	47W	3C	SBLE	unmapped ou 18A	1
26	48	77N	76W	3C	SBLE	unmapped ou 18A	1
27	29	63W	73S	3C	SBLE	24	1
28	28	52E	68S	1C	SBLE	25	1
29	47	87S	29W	7C	SBLE	unmapped	1
30	42	88E	3S	1C 7C	SBLE	unmapped	1
31	177	42E	61N	2C 4C	SBLE	48	1
32	178	44E	21S	1C 7C	SBLE	48	1
33	178	44E	60N	2C 4C	SBLE	48	1
34	18	48E	51S	3C	SBLE	49	1
35	26	77E	25N	1C 6C	SBLE	67	1
36	26	77E	79S	2C 3C	SBLE	67	1
37	13	51E	37S	3C	SBLE	98?	1
38	12	88W	12S	7C	SBLE	unmapped	1
39	176	60E	46N	6C	SBLE	unmapped	1
40	48	87N	7E	1C 5C	SBLE	107	1
41	48	87N	75E	2C 3C	SBLE	107	1
42	30	87E	82N	3C	SBLE	106	1
43	41	78E	17N	5C	SBLE	116A proche	1
184	43	80E	15N	5C	SBLE	116A proche	1
182	28	70W	23S	7C	SBLE	116	1
183	41	75E	74S	3C	SBLE	116	1
44	54	87N	29W	5C	SBLE	116	1
45	32	66E	33S	5C	SBLE	unmapped	1
46	30	71E	81N	3C	SBLE	123	1
47	30	71E	18N	5C	SBLE	123	1
48	39	72E	83S	3C	SBLE	unmapped	1
49	51	84S	72W	1C 3C	SBLE	130	1
50	18	20W	53N	2C 1C	SBLE	138	1
51	51	84S	10E	1C 5C	SBLE	130	1
52	17	21W	53N	2C 1C	SBLE	138	1
53	50	83S	73W	3C	SBLE	130	1
54	50	83S	9E	5C	SBLE	130	1
55	73	34S	63E	1C	SBLE	unmapped	1
56	67	41S	66E	1C	SBLE	139	1
57	59	76N	28E	5C	SBLE	146	1
58	66	63N	28E	5C	SBLE	unmapped	1
59	57	76S	15W	5C	SBLE	162	1
60	24	74E	12N	1P 5C	SBLE	166	1
61	24	74E	30N	2P 7C	SBLE	166	1
62	24	74E	64S	1P 3C	SBLE	166	1
63	25	75E	30N	2P 6C	SBLE	166	1

Table A1. *Cont.*

A	B	C	D	E	F	G	H
64	47	87N	22E	6C	SBLW	66	1
65	96	89S	67E	1C	SBLW	49	1
66	57	86S	37E	5C	SBLW	48	1
67	17	88E	73S	3C	SBLE	unmapped	1
68	25	56W	61S	2C 1C	SBLW	25	1
69	25	56W	62N	1C 1C	SBLW	25	1
70	25	56W	16S	2C 5C	SBLW	25	1
71	25	54W	64N	1C 1C	SBLW	25	1
72	71	33N	82E	1C 8C	SBLW	24	1
73	71	33N	42E	2C 6C	SBLW	24	1
74	52	32N	63E	8C	SBLW	24	1
75	76	22S	58E	1C 8C	SBLW	24	1
76	71	31N	82E	2C 8C	SBLW	24	1
77	78	21S	84E	8C	SBLW	24	1
78	75	23S	88E	8C	SBLW	24	1
79	7	70E	62S	3C	SBLW	unmapped	1
80	10	43E	76S	3C	SBL	unreferenced	1
81	29	71W	24N	5C	NBLW	1	1
82	37	76W	29N	5C	NBLW	1	1
83	42	74W	35N	5C	NBLW	1	1
84	48	72N	12E	5C	NBLW	1	1
85	142	19W	85S	4C	NBLW	unmapped	1
86	68	33S	64E	3C	NBLW	9	1
87	69	35S	37E	3C	NBLW		1
88	46	52S	63E	3C	NBLW	12-11月	1
89	52	47S	63E	3C	NBLW	12-11月	1
90	51	38S	52E	3C	NBLW	12	1
91	46	42S	84E	3C	NBLW	13	1
92	45	44S	83E	3C	NBLW	13	1
93	14	20E	73S	3C	NBLW	near F29	1
94	22	25E	85S	3C	NBLW	near F29	1
95	21	33E	74S	1C	NBLW	28	1
96	21	54E	72S	1C	NBLW	28	1
97	58	33S	67E	1C	NBLW	unmapped	1
98	46	58S	66E	3C	NBLW	unmapped	1
99	41	27E	77N	1C	NBLW	35	1
100	29	50E	72N	3C	NBLW	71	1
101	31	81E	18S	5C	NBLW	73	1
102	67	81N	6E	7C	NBLW	unmapped	1
103	38	89E	9S	7C	NBLW	unmapped	1
104	24	54E	64S	3C	NBLW	79	1
105	15	30E	73S	3C	NBLW	79	1
106	10	36E	68S	3C	NBLW	79	1
107	29	46E	87S	3C	NBLW	80	1
108	29	48E	89S	3C	NBLW	80	1
135	29	48E	88S	1C	NBLW	80	1
109	110	69N	11E	9C	NBLW	unmapped	1
110	139	89E	19S	9C	NBLW	unmapped	1
111	20	85W	12N	5C	NBLW		1
112	165	57W	60S	3C	NBLW		1
113	63	86N	1W	5C	NBLW		1
114	21	65E	76N	3C	NBLW	86	1
115	160	83E	4S	7C	NBLW	122?	1
116	161	80E	14S	7C	NBLW	123?	1

Table A1. *Cont.*

A	B	C	D	E	F	G	H
117	150	75E	10S	7C	NBLW	123?	1
118	166	84E	7S	7C	NBLW	118	1
119	25	83E	2N	7C	NBLW	unmapped	1
120	177	76E	2S	7C	NBLW	unmapped	1
121	26	51E	64N	3C	NBLW	unmapped	1
122	1	82E	2S	7C	NBLW	130?	1
123	170	86E	7S	7C	NBLW	133?	1
124	4	82E	1N	7C	NBLW	131?	1
125	14	47E	79S	3C	NBLW	unmapped	1
126	47	47S	33E	5C	NBLW	135	1
127	179	62W	80N	3C	NBLW	unmapped	1
128	1	67W	84N	3C	NBLW	151	1
129	156	77W	2S	7C	NBLW	unmapped	1
130	12	69W	87S	3C	NBLW	156	1
131	33	77W	3S	7C	NBLW	156	1
132	7	57W	71S	3C	NBLW	unmapped	1
133	13	64W	77N	3C	NBLW	unmapped	1
134	8	55W	72S	3C	NBLW	unmapped	1
185	155	62E	80S	3C	NBLW	unmapped	1
186	28	82E	77S	1C 3C	NBLW	unmapped	1
187	28	82E	4N	2C 7C	NBLW	unmapped	1
136	32	79W	79N	3C	NBLE	unmapped	1
137	34	72E	74N	3C	NBLE	36	1
138	34	85E	8S	7C	NBLE	large fault	1
139	36	66E	2N	7C	NBLE	large fault	1
140	174	75E	17N	7C	NBLE	unmapped	1
141	176	83W	27N	7C	NBLE	46	1
142	175	88E	15N	7C	NBLE	46	1
143	178	79E	2S	7C	NBLE	53	1
144	30	39E	10N	2C 7C	NBLE	51	1
145	30	39E	75S	1C 3C	NBLE	51	1
146	32	88W	6N	7C	NBLE	55	1
147	41	86E	3S	7C	NBLE	64	1
148	37	82W	1S	7C	NBLE	64	1
149	17	61E	83S	3C	NBLE	58	1
150	35	73W	63S	3C	NBLE	unmapped	1
151	114	84N	5E	9C	NBLE	68	1
152	42	85E	1S	7C	NBLE	70	1
153	50	89S	7W	7C	NBLE	74	1
154	35	82W	12S	7C	NBLE	74	1
155	44	88E	6S	7C	NBLE	74	1
156	40	72W	11S	7C	NBLE	87-89	1
157	39	81W	1S	7C	NBLE	87-89	1
158	29	85W	36N	7C	NBLE	87-89	1
159	10	83W	14N	7C	NBLE	87-89	1
160	32	80E	84N	3C	NBLE	91	1
161	29	84E	28N	7C	NBLE	91	1
162	46	72N	88E	3C	NBLE	98	1
163	58	68S	76E	1C 3C	NBLE	105	1
164	58	68S	22E	2C 7C	NBLE	105	1
165	35	42E	82N	3C	NBLE	unmapped	1
166	19	73E	66S	1C	NBLE	unmapped	1
167	22	47E	80N	3C	NBLE	unmapped	1
168	52	73S	13E	7C	NBLE	111	1
169	46	89N	12E	7C	NBLE	130?	1

Table A1. *Cont.*

A	B	C	D	E	F	G	H
170	36	82E	34N	2C 7C	NBLE	136?	1
171	36	82E		1C 3C	NBLE	136?	1
172	22	49W	64S	3C	NBLE	unmapped	1
173	42	53W	53S	3C	NBLE	163?	1
174	25	67W	52S	3C	NBLE	164?	1
175	44	67W	49S	3C	NBLE	unmapped	1
176	16	49W	74N	3C	NBLE	unmapped	1
177	36	87E	82S	1C	NBLE	unmapped	1
178	23	52W	85N	3C	NBLE	unmapped	1
179	23	62W	85S	3C	NBLE	unmapped	1
180	26	65W	83S	3C	NBLE	unmapped	1
181	15	42W	89N	3C	NBLE	unmapped	1
188	35	77E	66N	3C	NBLE		2
189	32	74E	88N	3C	NBLE		2
190	33	80E	66N	3C	NBLE		2
191	36	72W	77N	3C	NBLE		2
192	26	83W	77S	3C	NBLE		2
193	146	65E	72N	2C	NBLE		2
194	106	44S	63W	1C 2C	NBLE		2
195	106	44S	19E	2C 5C	NBLE		2
196	33	52E	65S	3C	NBLE		2
197	133	87N	74W	2C	NBLE		2
198	145	65E	68N	2C	NBLE		2
199	125	48S	77W	1C 2C	NBLE		2
200	125	48S	2E	2C 5C	NBLE		2
201	174	51E	68N	2C	NBLE		2
202	6	71W	86N	3C	NBLE		2
203	159	76E	62N	2C	NBLE		2
204	24	57W	81S	3C	NBLE		2
205	11	67W	82S	3C	NBLE		2
206	148	63E	72N	2C	NBLE		2
207	127	87N	71W	2C	NBLE		2
208	128	89N	68W	2C	NBLE		2

References

1. Angelier, J. Example of informatics applied to structural analysis—Some methods for studying fault tectonics. *Revue Geographie Physique Geologie Dynamique* **1975**, *17*, 137–145.
2. Angelier, J. Tectonic analysis of fault slip data sets. *J. Geophys. Res. Solid Earth* **1984**, *89*, 5835–5848. [CrossRef]
3. Angelier, J. From orientation to magnitudes in paleostress determinations using fault slip data. *J. Struct. Geol.* **1989**, *11*, 37–50. [CrossRef]
4. Angelier, J. Inversion of field data in fault tectonics to obtain the regional stress-III. A new rapid direct inversion method by analytical means. *Geophys. J. Int.* **1990**, *103*, 363–376. [CrossRef]
5. Angelier, J. Analyse chronologique matricielle et succession régionale des événements tectoniques. *Comptes Rendus Académie Sci.* **1991**, *312*, 1633–1638.
6. Angelier, J. Palaeostress Analysis of Small-Scale Brittle Structures. In *Continental Deformation*; Hancock, P., Ed.; Pergamon Press: Oxford, UK, 1994; pp. 53–100.
7. Hu, J.C.; Angelier, J. Stress permutations: Three-dimensional distinct element analysis accounts for a common phenomenon in brittle tectonics. *J. Geophys. Res. Space Phys.* **2004**, *109*, B09403. [CrossRef]
8. Hancock, P.L. Brittle Microtectonics: Principles and Practice. *J. Struct. Geol.* **1985**, *7*, 437–457. [CrossRef]
9. Mattauer, M. *Les Déformations des Matériaux de l'Écorce Terrestre*; Herman: Paris, France, 1973; 493p.
10. Lawn, B.R. *Fracture of Brittle Solids*, 2nd ed.; Cambridge University Press: Cambridge, UK, 1993; p. 378.
11. Riedel, W. Zur mechanik geologischer brucherscheinungen. Centralblatt fur Minerologie. *Geol. Paleontol.* **1929**, 354–368.

12. Carey, E.; Brunier, B. Analyse théorique et numérique d'un modèle mécanique élémentaire appliqué à l'étude d'une population de failles. *Comptes Rendus Académie Sci.* **1974**, *279*, 891–894.

13. Angelier, J. Inversion of earthquake focal mechanisms to obtain the seismotectonic stress IV-a new method free of choice among nodal planes. *Geophys. J. Int.* **2002**, *150*, 588–609. [CrossRef]

14. Navabpour, P.; Angelier, J.; Barrier, E. Cenozoic post-collisional brittle tectonic history and stress reorientation in the High Zagros Belt (Iran, Fars Province). *Tectonophysics* **2007**, *432*, 101–131. [CrossRef]

15. Bergerat, F.; Angelier, J.; Andreasson, P.G. Evolution of paleostress fields and brittle deformation of the Tornquist Zone in Scania (Sweden) during Permo-Mesozoic and Cenozoic times. *Tectonophysics* **2007**, *444*, 93–110. [CrossRef]

16. Barton, C.C.; Camerlo, R.H.; Bailey, S.W. *Bedrock Geologic Map of Hubbard Brook Experimental Forest and Maps of Fractures and Geology in Roadcuts along Interstate I-93, Grafton County, New Hampshire*; U.S. Geological Survey Miscellaneous Investigation Series map 1-2562, 2 sheets, 1:12,000; U.S. Geological Survey: Reston, VA, USA, 1997.

17. Burton, W.C.; Walsh, G.W.; Armstrong, T.R. *Bedrock Geologic Map of the Hubbard Brook Experimental Forest, Grafton County, New Hampshire*; U.S. Geological Survey Digital Open File Report 00-45, map, text, and computer files; U.S. Geological Survey: Reston, VA, USA, 2000.

18. Hatch, N.L.; Moench, R.H. *Bedrock Geologic Map of the Wildernesses and Roadless Areas of the White Mountain National Forest, Coos, Carroll, and Grafton Counties, New Hampshire*; Miscellaneous Field Studies map—U.S. Geological Survey, Report: MF-1594-A, 1 sheet, scale 1:125,000; U.S. Geological Survey: Reston, VA, USA, 1984.

19. Moke, C.B. *The Geology of the Plymouth Quadrangle, New Hampshire*; scale 1:62,500; New Hampshire Planning and Development Commission: Concord, NH, USA, 1945; p. 21.

20. Lyons, J.B.; Bothner, W.A.; Moench, R.H.; Thompson, J.B., Jr. *Bedrock Geologic Map of New Hampshire*; U.S. Geological Survey State Geologic map, 2 sheets, scales 1:250,000 and 1:500,000; U.S. Geological Survey: Reston, VA, USA, 1997.

21. Hatch, N.L., Jr.; Moench, R.H.; Lyons, J.B. Silurian-Lower Devonian stratigraphy of eastern and south-central New Hampshire; extensions from western Maine. *Am. J. Sci.* **1983**, *283*, 739–761. [CrossRef]

22. Robinson, P.; Tucker, R.D.; Bradley, D.C.; Berry, I.V.N.H.; Osberg, P.H. Paleozoic orogens in New England, USA. *Gff* **1998**, *120*, 119–148. [CrossRef]

23. Johnson, C.D.; Dunstan, A.M. *Lithology and Fracture Characterization from Drilling Investigations in the Mirror Lake Area: From 1979 Through 1995 in Grafton County, New Hampshire*; U.S. Geological Survey Water-Resources Investigations Report 98-4183; U.S. Geological Survey: Reston, VA, USA, 1998; p. 210.

24. McHone, J.G. Tectonic and paleostress patterns of Mesozoic intrusions in eastern North America. In *Triassic-Jurassic Rifting: Continental Breakup and the Origin of the Atlantic Ocean and Passive Margins, Part B*; Manspeizer, W.R., Ed.; Elsevier: Amsterdam, The Netherlands, 1988; pp. 607–619.

25. Thompson, W.B. History of research on glaciation in the White Mountains, New Hampshire (U.S.A.). *Géographie Physique Quaternaire* **2002**, *53*, 7–24. [CrossRef]

26. Tarasov, L.; Dyke, A.S.; Neal, R.M.; Peltier, W.R. A data-calibrated distribution of deglacial chronologies for the North American ice complex from glaciological modeling. *Earth Planet. Sci. Lett.* **2012**, *315*, 30–40. [CrossRef]

27. Fowler, B.K.; Davis, P.T.; Thompson, W.B.; Eusden, J.D.; Dulin, I.T. The Alpine zone and Glacial Cirques of Mount Washington and the Northern Presidential Range, White Mountains, New Hampshire. In Proceedings of the 75th Annual Reunion of the Northeastern Friends of the Pleistocene, Pinkham Notch, NH, USA, 1–3 June 2012; p. 35.

28. Barton, C.C. Characterizing bedrock fractures in outcrops for studies of ground-water hydrology—An example from Mirror Lake, Grafton County, New Hampshire. In *U.S. Geological Survey Toxic Substances Hydrology Program, Proceedings of the Technical Meeting, Colorado Springs, CO, USA, 20–24 September 1993*; Morganwalp, D.W., Aronson, D.A., Eds.; U.S. Geological Survey Water-Resources Investigations Report 94-4015; U.S. Geological Survey: Reston, VA, USA, 1996; pp. 81–88.

29. Daubree, M. Application de la méthode expérimentale à l'étude des déformations et des cassures terrestres. *Bulletin Societe Geologique France* **1879**, *3*, 108–141.

30. Anderson, E.M. *The Dynamics of Faulting*, 2nd ed.; Oliver & Boyd: Edinburgh, UK, 1942; p. 206.

31. Bott, M.H.P. The Mechanics of Oblique Slip Faulting. *Geol. Mag.* **1959**, *96*, 109–117. [CrossRef]

32. Wallace, R.E. Geometry of Shearing Stress and Relation to Faulting. *J. Geol.* **1951**, *59*, 118–130. [CrossRef]
33. Dupin, J.M.; Sassi, W.; Angelier, J. Homogeneous stress hypothesis and actual fault slip: A distinct element analysis. *J. Struct. Geol.* **1993**, *15*, 1033–1043. [CrossRef]
34. Faure, S.; Tremblay, A.; Malo, M.; Angelier, J. Paleostress Analysis of Atlantic Coastal Extension in the Quebec Appalachians. *J. Geol.* **2006**, *114*, 435–448. [CrossRef]
35. Cowie, P.A.; Scholz, C.H. Physical explanation for the displacement length relationship of faults using a post-yield fracture- mechanics model. *J. Struct. Geol.* **1992**, *14*, 1133–1148. [CrossRef]
36. Faulkner, D.R.; Mitchell, T.M.; Jensen, E.; Cembrano, J.M. Scaling of fault damage zones with displacement and the implications for fault growth processes. *J. Geophys. Res. Space Phys.* **2011**, *116*. [CrossRef]
37. Scholz, C.H.; Dawers, N.H.; Yu, J.Z.; Anders, M.H.; Cowie, P.A. Fault growth and fault scaling laws: Preliminary results. *J. Geophys. Res. Space Phys.* **1993**, *98*, 21951–21961. [CrossRef]
38. Heidbach, O.; Tingay, M.; Barth, A.; Reinecker, J.; Kurfe, B.D.; Muller, B. *World Stress Map*, 2nd ed.; 1 sheet; Commission for the Geological map of the World: Paris, France, 2009; Available online: http://dc-app3-14.gfz-potsdam.de/pub/poster/World_Stress_map_Release_2008.pdf (accessed on 31 October 2020).
39. Jahns, R.H. Sheet Structure in Granites: Its Origin and use as a Measure of Glacial Erosion in New England. *J. Geol.* **1943**, *51*, 71–98. [CrossRef]

Publisher's Note: MDPI stays neutral with regard to jurisdictional claims in published maps and institutional affiliations.

 *geosciences*

Article

An Assessment of Stress States in Passive Margin Sediments: Iterative Hydro-Mechanical Simulations on Basin Models and Implications for Rock Failure Predictions

Antoine Bouziat [1,*], Nicolas Guy [1], Jérémy Frey [1], Daniele Colombo [1], Priscille Colin [2], Marie-Christine Cacas-Stentz [1] and Tristan Cornu [3]

[1]   IFP Energies nouvelles, 1 et 4 avenue de Bois-Préau, 92852 Rueil-Malmaison, France;
     nicolas.guy@ifpen.fr (N.G.); jeremy.frey@ifpen.fr (J.F.); daniele.colombo@ifpen.fr (D.C.);
     marie-christine.cacas-stentz@ifpen.fr (M.-C.C.-S.)
[2]   Total ABK, East Tower, Mall Street 10, Abu Dhabi 4058, UAE; priscille.colin@gmail.com
[3]   Total SA, CSTJF, Avenue Larribau, 64000 Pau, France; tristan.cornu@total.com
*   Correspondence: antoine.bouziat@ifpen.fr

Received: 21 October 2019; Accepted: 4 November 2019; Published: 6 November 2019

Abstract: Capturing the past and present hydro-mechanical behavior of passive margin sediments raises noticeable interest, notably in geo-hazard risk assessment and hydrocarbon exploration. In this work, we aim at assessing the stress states undergone by these sedimentary deposits through geological time. To do so, we use an iterative coupling between a basin simulator and a finite element mechanical solver. This method conciliates a computation of the full stress tensors with a dynamic and geologically detailed modelling of the sedimentation. It is carried out on a dedicated set of 2D synthetic basin models, designed to be representative of siliciclastic deposition in passive margins and integrating variations in their geological history. Contrary to common assumptions in operational basin modelling studies, our results imply that passive margin sedimentary wedges are multidimensional mechanical systems, which endure significant non-vertical stress without external tectonic input. Our results also highlight the variability of the stress states through space and time, with a strong control from the geometry and lithological heterogeneities of the deposits. Lastly, we used the simulation results to predict a location and timing for the development of weakness zones in the sedimentary stacks, as privileged areas for rock failure. The outcome underlines the influence of the basal tilt angle, with a slight tilt impacting the wedges stability to a similar extent as a substantial increase in sedimentation rate. Altogether, this study emphasizes the need for careful consideration of non-vertical stresses in basin simulations, including in passive tectonic contexts. It also suggests that the iterative coupling method employed is a promising way to match industrial needs in this regard.

Keywords: basin modelling; hydro-mechanical coupling; passive margins; rock failure

---

1. Introduction

Passive margins represent a cumulated length of 105,000 km of coastal areas around the world and are privileged locations for marine sediment deposition [1]. They are well-studied environments, notably in terms of morphology, geology, and underground resources [2,3]. On the latter topic, passive margins are historically key areas in the hydrocarbon industry, as they host more than 300 giant oil and gas fields [4].

Passive margin sediments are subject to a large range of hydraulic and mechanical phenomena, which creates specific interests. Remarkably, understanding their hydro-mechanical behavior is critical in civil engineering and geo-hazard prediction, as passive margin sediments are naturally likely to collapse at local or regional scales [5,6]. In petroleum exploration, analyzing slope instability and mechanical compaction through geological ages helps to accurately estimate the location of distal sand reservoirs and their volumes at the present time [7,8]. Furthermore, in all conventional petroleum systems, hydrocarbon migration is controlled by gradients of fluid overpressure and impacted by natural fracturing of cap rocks [9,10]. Lastly, assessment of present-day pore pressure appears as a crucial step to ensure drilling safety [11,12].

In the present study, our objective is to assess the stress states undergone by siliciclastic sediments in passive margins and the development of weakness zones as favorable locations for rock failure. To do so, our simulation strategy is rooted on an iterative hydro-mechanical coupling between a basin simulator, ArcTem [13], and a finite element mechanical solver, Code Aster [14]. The hydro-mechanical coupling was applied on a set of synthetic basin models, designed with the target to be representative of siliciclastic sedimentation in passive margins and of operational models in the hydrocarbon industry. This aims at depicting the stress states endured by the different parts of the sedimentary wedges through deposition, with an emphasis on non-vertical effects. Our goal is also to appraise how basal slope tilting and sedimentation rate variations affect the location and expansion of weakness areas, thus addressing the geological control on rock failure preconditioning through passive margins history.

2. Observations

Passive margins are locations of siliciclastic sediment accumulations, up to 12 km thick [15]. These accumulations usually consist of a stacking of several stratigraphic units, following well-described patterns in terms of geometry and lithology [16]. Most layers present a clinoform shape, with maximal slopes ranging from 1 to 10° in siliciclastic environments [17].

Passive margin deposits are highly compactable and experience significant volume reduction, but important under-compaction and fluid overpressure are also observed [18]. This excess of pore pressure is notably induced by high sedimentation rates and low permeability layers, keeping the hydrodynamic system out of equilibrium [19,20]. Sequence stratigraphy patterns, controlling the relative distribution of sealing shaly facies and draining sandy ones, reflect on the pore pressure profiles and cause substantial lateral flows [21,22].

Sedimentary wedges in passive margins are dynamic and unstable systems, leading to failure events of different scales and amplitudes. Locally, buried and overpressured sediments are prone to natural hydraulic fracturing, known to damage the sealing capacity of shaly layers [23,24]. At a larger scale, instability can lead to the failure of the sedimentary slope, triggering important mass transport, from superficial sliding to wider slumping and even broad landslips [25–30]. Lastly, at a regional scale, gravity collapse of passive margin sediments set common structural features, with normal faulting observed in the proximal part of the wedge and fold-and-thrust belts in the most distal areas [31,32].

Although punctual and exceptional events are often needed to trigger the amplest incidents, failure at all scales is favored by preconditioning factors resulting in weakness areas in the sedimentary stack [5,6,33]. Pore pressure rise is noted as a key factor, in relation to sealing lithology or high deposition rates [34–37]. However, the geometrical evolution of the wedge is also reported to play a major role in its destabilisation, with an influence of the clinoform slopes [38,39], progradation rates [40], and basal angles [41,42]. Lastly, mechanical heterogeneities can be decisive, with regional collapses facilitated by deep low-friction layers, for instance evaporites [43] or overpressured shales [44].

3. Simulation Strategy

3.1. Traditional Approaches

Several methods are described in the literature to model the hydro-mechanical behavior of passive margin sediments. They can be sorted into two main families.

A first approach relies on slope stability analysis. It covers a range of methods of increasing complexity, including critical Coulomb wedge theory, infinite slope and limit analyzes, their combinations, and their extensions [45–51]. These methods use slope and pore pressure estimates to compute the shear stress and the rock strength, leading to a likelihood of failure for the wedge. However, they are often based on a strong idealization of the sedimentary stack, notably in terms of geometry, lateral facies variations, and overpressured layers [6]. In addition, this kind of modelling is essentially static. Even when it is implemented in a sequential manner, each time step remains independent from the previous ones, with no consideration of fluid flows and past stress states [52].

A second approach relies on basin simulation technologies [53–59]. From a representation of the basin at the present day, the evolution of its geometry through geological ages is restored with a backward process. Then, the physical variables of interest are computed for each geological event with a forward simulation using this geometrical framework. This methodology can handle a high degree of realism in the description of the sedimentary deposits, notably in terms of slope breaks, truncated layering, lateral lithological heterogeneities, and complex overpressure distribution. The resulting models are also intrinsically dynamic: the simulation results at a given geological age are strongly dependent on the flow and stress history of the margin, and they are also significantly impacted by the evolutive geometry previously restored [60].

Nonetheless, using basin simulation methodologies, it is classical to reduce the stress state to its sole vertical component, to overcome numerical complexity or absence of calibration data [56,57,61,62]. Implicitly, the contribution of lateral efforts to porosity loss, fluid overpressure build-up and system instability is then estimated as marginal. Horizontal stresses are often considered as direct functions of the burial depth or simple ratios of the vertical stress. In the latter case, the ratios are homogenous in the sedimentary stack or solely variable with depth and lithology [57,63].

The limitations of the usual stress state simplification and their impact on hydro-mechanical simulation results have already been documented [64,65]. Notably, they can imply incorrect estimates for sediment compaction and rock failure likelihood. In a petroleum exploration context, this may produce misleading conclusions on cap rock integrity, migration flowpaths, and present-day overpressure. These observations motivated the design of several alternative methods aiming to better account for the 3D nature of the stress tensor [66,67]. Nevertheless, the focus is traditionally on basins overcoming important lateral tectonic solicitations, with large shortening or extension rates. As a result, the value of integrating non-vertical stresses in the modelling of passive geological settings appears more questionable, especially when considering the associated increase in computation costs. Basin modellers often deal with passive margin sediments as one-dimensional mechanical systems through geological ages [68–70], in line with simplistic models for horizontal stresses computation and the assumption that the maximal principal stress is perfectly aligned with the vertical axis [18].

3.2. Iterative Hydro-Mechanical Coupling

In this work, we aim at conciliating the geological realism and dynamic aspect of basin simulations with the integration of lateral mechanical effects. To do so, we use a coupling between two codes: ArcTem, a basin simulator [13], and Code Aster, a finite element mechanical solver [14]. ArcTem relies on the classical 1D stress simplification while Code Aster computes a full stress tensor.

The coupling is implemented in a prototype software named A^2 [71]. Following an iterative procedure, the two codes exchange inputs and outputs at each simulation step, until providing a consistent set of porosity, permeability, fluid pressure, and stress tensor in each cell of the basin model. This differs from the integrated numerical model of [72] but shares similar objectives.

The coupling algorithm uses an updated Lagrangian approach and is inspired from the work of [73] and [74] for reservoir simulations. The same grid is used by the two codes and the convergence criterion is based on a comparison between the porosity values they compute. A full description can be read in [71] and [75].

3.3. Rheology

The ArcTem simulator uses a direct relationship between porosity and vertical effective stress, known as the Schneider's law [76,77]. This law is widely used in the basin modelling community for its practicability, notably to reproduce porosity–depth data measured from cores or well logs [57].

In the iterative coupling, we model the rheology of passive margin sediments with a tensor extension of the Schneider's law. Concretely, the model describes an elasto-plastic behavior, with the rock stiffness varying with compaction. It accounts for the full stress tensor but is designed to trigger the same volume variations as the Schneider's law in the case of tabular sedimentation with no lateral deformation (i.e., oedometer test conditions). It is presented in detail in [71]. Other relevant models describing sediment compaction can be found in the literature [78,79]. However, we see the one used herein as a suitable compromise with several practical needs, notably applicability on geologically detailed passive margin models, easy comparison with classical basin simulators, and integration in the iterative coupling at a reasonable computation cost.

4. Set of Synthetic Basin Models

This study relies on a dedicated set of five synthetic 2D basin models, designed to be representative of actual siliciclastic sedimentation in passive margins. They share a common design, but differ in geological history, with variations in basal slope tilt and sedimentation rate.

4.1. Design

4.1.1. Geological Setting

Our modelling work focuses on a few stratigraphic sequences of siliciclastic deposition. It aims at representing the sedimentary dynamics classically observed in a passive margin, notably the morphology of stratigraphic system tracks, with gradating deposition, discordant stratification, and lateral facies' variations. These geological features are commonly integrated in operational basin models of the petroleum industry, but more rarely in published hydro-mechanical simulations, as they often involve a higher level of idealization.

Among other inspirations, the modeled sequences can be related to the sediments of Eocene age in the Brazilian oil-bearing Santos basin, as described in [80]. In this example, they were deposited in a relatively short amount of space and time (30 km and 12 My). More generally, the modeled sequences represent common sedimentation patterns, but the accumulations in passive margin basins usually consists in their repetition through longer time periods and deposited volumes. Consequently, the structural evolution of passive margins at large time or space scales (i.e., hundreds of km and My) is beyond the scope of this work.

4.1.2. Common Features

All models in the set correspond to a 33 km long section and a succession of 40 gradating sedimentary layers, distributed in stratigraphic sequences of transgressive, highstand, lowstand, and turbiditic system tracks (Figure 1). The sedimentation is globally prograding and most layers present a clinoform morphology, with a slope dipping up to 10°. This angle refers to the highest value measured in clastic deposits on passive margins around the world [17]. Maximum sediment pile thickness is 3 km. Under the sediments, 2 km of a stiff geo-material are previously set up to create a hard basis to the model.

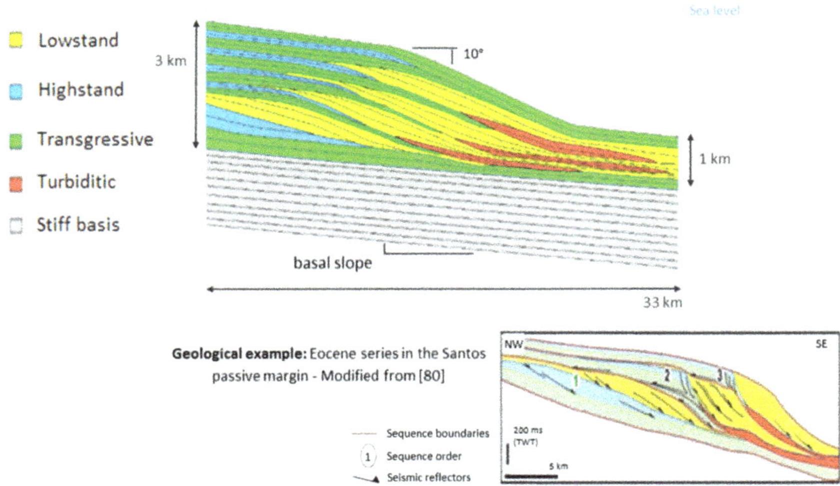

Figure 1. Geometry and stratigraphy of the study-case in the present day; vertical exaggeration × 3.

4.1.3. Variations between Models

Keeping a common design, we introduced different geological histories for the models in the set. Our objective is to appraise the geological control on failure preconditioning in passive margin sediments, and to assess the development of weakness areas in terms of timing and expansion. Among the numerous geological factors varying through a margin history, we specifically focused on the sedimentation rate and on the angle of the basal slope, which usually undergoes a progressive tilt during the subsidence of the sedimentary stack.

High sedimentation rates favor fluid overpressure build-up, and in the literature previous physical simulations illustrate how rapid deposition can create weakness zones in deeply buried parts of the sedimentary wedges [35,46]. However, weakness and rock failure are also observed in margins with slow sedimentation, and other published simulation studies showed that additional preconditioning factors than overpressure should be considered [33,49]. In this regard, the models presented in this work are a way to specifically evaluate the impact of the basal angle tilt during sedimentation.

Table 1 presents the three sedimentation rate scenarios simulated in this study. In each case, a specific duration is assigned to all sedimentation events regardless of their thickness, which makes the sedimentation rate vary with time. However, the resulting values are consistent with the ones estimated on actual passive margins around the world [26,81,82]. Table 2 presents the three tilting scenarios simulated. In every model, the bottom horizon starts horizontal and then tilts at an increment angle at each transgressive event, until reaching the target value in the present day.

Table 1. Values of sedimentation rate in the models set.

Model	Duration of Each Event (My)	Lowest Rate (m/My)	Average Rate (m/My)	Highest Rate (m/My)
Low rate	2	37	82	165
Base case	1	75	165	330
High rate	0.5	150	330	660

Table 2. Values of subsidence tilt angle in the models set.

Model	Initial Basal Slope	Tilt in Each Transgressive Event	Present-Day Basal Slope
Low tilt	0°	0.0625°	0.75°
Base case	0°	0.125°	1.5°
High tilt	0°	0.250°	3°

4.2. Present-Day Grids

For all models, a grid was built from the present-day geometry by dividing the 40 layers by 70 vertical pillars. The total number of cells is then 2760, but 47% of them are flattened because of the discordant nature of the sedimentation.

The sediments were filled with five lithological types, representative of different shale-sand proportions: "sandy" (70% sand 30% shale), "intermediate" (50% sand 50% shale), "quite shaly" (30% sand 70% shale), "shaly" (20% sand 80% shale), and "very shaly" (10% sand 90% shale). High-stand and transgressive system tracks are wholly filled with "very shaly" facies, turbiditic deposits are wholly filled with "sandy" facies, and low-stand system tracks show the full gradation from "sandy" to "very shaly" based on distality (Figure 2).

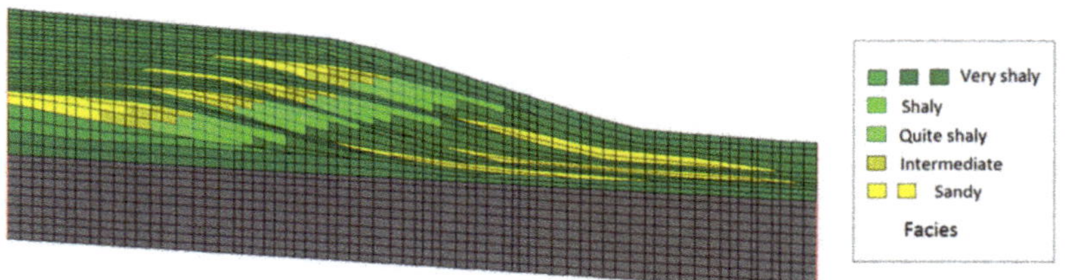

Figure 2. Mesh and lithology distribution of the study-case in the present-day; vertical exaggeration × 2.

4.3. Past Geometries

The restoration of past geometries for all models was performed with a backstripping method [83,84], implemented in the TemisFlow™ software (2016 version, Beicip-Franlab, Rueil-Malmaison, France) [85]. The uncompacting used a different Porosity vs. Depth table for each lithological type. Shalier facies have larger deposit porosities but compact more quickly with burial than the sandier ones (plots available in Figure A1 in the Appendix A).

The model bottom was used as a reference horizon. Its shape was arbitrarily input for each deposition event in order to model a progressive subsidence and tilting for the margin basement, following Table 2. The paleo-bathymetry profiles were then computed by the algorithm from the compaction tables and the basement shape. Lastly, absolute sea level variations were input by translating the full model upward or downward from the 0 mark on the vertical axis. These variations were set in accordance with the sedimentary track being deposited, and keeping the full model immersed during the whole simulation.

4.4. Simulation Parameters

Once built and restored, all models were simulated with the same parameters.

4.4.1. Mechanical Properties

The Schneider's law and its tensor extension were used with specific coefficients for each lithological type, in order to model different compaction behaviors (plots available in Figure A1 in the Appendix A). These coefficients were all calibrated to be perfectly consistent in hydrostatic conditions with the Porosity–Depth tables used in the backstripping restoration. In addition, the whole model was given a solid density of 2.7 g·cm^{-3} and a Poisson's ratio of 0.32, seen as average values for well-consolidated siliciclastic sediments [86,87].

4.4.2. Permeability

Different Permeability–Porosity tables were associated with each of the six facies (plots available in Figure A2 in the Appendix A). These tables were designed to be representative of the actual petro-physical properties of layered siliciclastic sediments at the basin scale. Insights from various references were combined, notably experimental data and empirical relationships for shale rocks [88,89] and sandstones [90,91], literature review [92], and basin-modelling guidelines [57].

For the two sandier facies, permeability decreases with compaction following a classical convex-shaped profile. Still, for the three shalier facies, a high permeability drop is modelled between 0.35 and 0.2 porosity to represent the mineralogical transformations observed during the burial of shaly sediments, with kaolinite-rich clays converting to less permeable illite-rich ones [93]. In addition, permeability anisotropy was added to the model to represent the stratification of the sedimentation. The ratios between horizontal and vertical permeability range from 1 for the sandiest facies to 18 for the shaliest one, in accordance with diverse data available in the literature [57,94,95].

4.4.3. Boundary Conditions

When running the iterative coupling, the geo-mechanical solver was strongly constrained by the backstripping restoration, and no external tectonic loading was added. Practically, the input boundary conditions were fixed nodal displacements for the bottom horizon and normal displacements blocked on lateral boundaries.

The basis was given a very stiff behavior with very little deposit porosity, similar for instance to an intact Barre granite [96]. Its permeability was defined accordingly, from several laboratory measurements on such granites [96–98].

5. Simulation Results

In this section, we analyze the simulation results in three steps. First, we observe on the base case model the classical basin modelling outputs which are porosity and pore pressure, to verify their consistency with our modelling objectives. Second, we focus on the additional output provided by the iterative hydro-mechanical coupling technology: a computation of the full stress tensor in every cell of the model at every geological age. We characterize the multidimensional nature of the simulated stress states and appraise their variability through space and time. Lastly, we use the stress tensors simulated to predict the development of weakness areas in the sedimentary stack, seen as favorable locations for rock failure. We benefit from the full set of basin models to compare the influence of two geological factors impacting the sedimentary wedge stability: deposition rate and basal angle tilting.

5.1. Pore Pressure and Porosity

Figure 3 depicts the simulation results on the base case model in terms of present-time porosity and fluid overpressure evolution through geological ages. We can assess whether compaction and overpressure levels are sufficient to be representative of actual passive margins, and whether their changes through space and time can be related to the lithology and the geometry of the modelled deposition.

Despite moderate burial, strong overpressure is simulated, with a maximum of 21.8 MPa in the present day. The highest values are located in a proximal and deep area, where the overlying sediment stack is the thickest. The simulated overpressure rises in this area during the deposition of highstand system tracks, increasing its burial. During the deposition of lowstand system tracks, the simulated overpressure rather rises beneath the continental slope, in the previously deposited lowstand wedges. The lateral facies variations then reverberate on the pressure profile: the simulated overpressure build-up concentrates in the inferior part of the wedges, presenting a shaly lithology, while the sandier superior part remains well drained.

Meanwhile, the simulated sediment compaction is significant but not extreme, with the lowest present-day porosity value at 25%, which corresponds to approximately half of the value at deposition. For a given burial, porosity estimations reflect both the lithological heterogeneities and the pore pressure profile, with higher values in the areas of sandy lithology or strong overpressures.

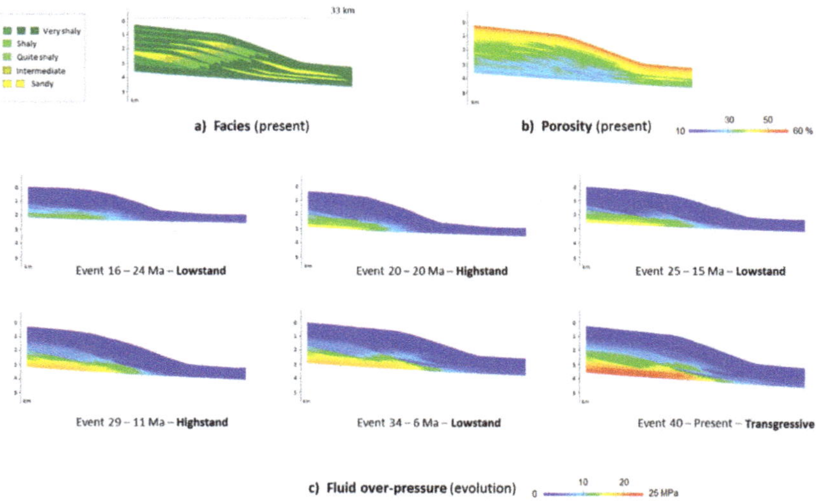

Figure 3. Porosity and overpressure simulated on the base case scenario. (**a**) reminder of the lithological facies distribution; (**b**) present-day porosity; (**c**) overview of the fluid overpressure evolution from six geological events; vertical exaggeration × 2.

At first-order and considering the strong differences in sediment stack thickness, the simulation results in the present day appear consistent with those previously predicted in real passive margins from seismic data, in situ measurements or idealised basin models, notably in New England [18,34,99], the Gulf of Mexico [36,100] or the Storegga slide area offshore Norway [46].

5.2. Stress States

With the iterative hydro-mechanical coupling, a stress tensor is computed in each cell of the basin model at each geological event. It represents the total stresses endured by the passive margin sediments during this event. In this study, we note S as this stress tensor, X the distality axis oriented from onshore to offshore, and Z the depth axis oriented downward. Consequently, S_{XX} corresponds to the horizontal stress, S_{ZZ} to the vertical stress, and S_{XZ} to the shear stress. We use the sign convention with positive compressive stresses. The stress tensor is then diagonalized to determine the principal stresses. We note S_{MAX} the maximum principal stress and S_{MIN} the minimum one. In addition, effective stresses are computed from total stresses and pore pressure values using Terzaghi's theory [101]. The resulting tensor is noted $S\prime$.

We note $p\prime$ the confinement pressure, or average effective stress, computed from $S\prime$:

$$p' = \frac{1}{3} \times (S\prime_{XX} + S\prime_{YY} + S\prime_{ZZ}) \tag{1}$$

With the confinement pressure, it is possible to isolate the deviatoric part Q of the effective stress tensor:

$$Q = S' - p' \times I \tag{2}$$

In rock mechanics, the multidimensional nature of the stress states is then commonly characterized by the equivalent deviatoric stress q, defined as follows:

$$q = \sqrt{\frac{3 \times Q \times Q}{2}} \tag{3}$$

The value of this equivalent stress is not necessarily meaningful on its own, and should be compared to the confinement pressure, notably in rock failure predictions. However, from a physical standpoint, a rise in deviatoric stress can be related to two phenomena: (1) development of shear stresses, leading to the rotation of the principal stress system, and (2) higher anisotropy in the principal stresses, leading to a more strongly extensional tectonic regime [102]. Consequently, in the following, we evaluate the significance of these two mechanical effects in our models and consider their variability through space and time, in relation to the lithology and geometry of the sediments deposited.

5.2.1. Shear Stress and Rotation of Principal Stresses

The shear stress in the present day simulated on the base case model is presented in Figure 4. The highest absolute values are concentrated in a deep area, located beneath the center of the continental slope. These values are quite moderate, as they do not exceed 3.5 MPa (Figure 4a). However, they appear much more significant when expressed as a ratio of the vertical effective stress. On the base case model, the shear stress in the present day represents up to 40% of the vertical effective stress (Figure 4b). The greatest ratios are simulated in the area of strongest shear but also in the inferior part of the overlying lowstand wedges, where the vertical effective stress is lowered by high overpressure.

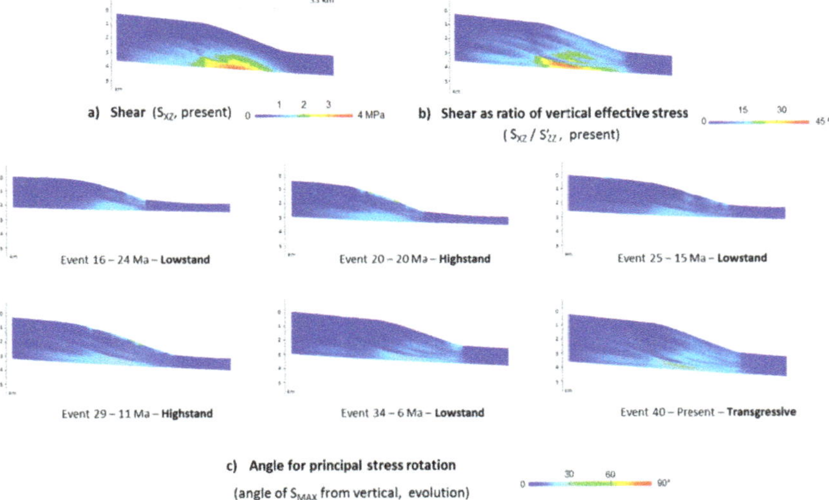

Figure 4. Shear and rotation of principal stresses simulated on the base case scenario. (**a**) present-day shear: absolute values; (**b**) present-day shear: ratios of vertical effective stress; (**c**) overview of the rotation of the principal stress system through time from six geological events; vertical exaggeration × 2.

Shear stresses reflect on the local rotation of the principal stress system, measurable by the angle from vertical for the orientation of the maximal principal stress. Figure 4c presents these rotations in the margin sediments through geological time, as simulated on the base case model. Rotations up to 32° are observed. The highest rotation angles are found in the area of highest absolute shear, deep beneath the center of the continental slope. However, strong rotations are also simulated in much shallower areas, due to low vertical stresses and sloping geometries for the deposits. These shallow zones of principal stress rotation mostly appear downslope during the deposition of the lowstand system tracks, and upslope during the deposition of highstand system tracks.

5.2.2. Extension and Compression

Even if substantial rotations of the principal stresses are simulated, the angle values displayed in Figure 4c show the normal efforts remains higher than the tangential ones in every part of the model. This means the tectonic regime of all sediments in the model remains extensional through their history. However, the ratio between minimal and maximal principal stresses varies with space and time, delineating areas and ages where the regime nears compression. These variations simulated on the base case model are represented in Figure 5.

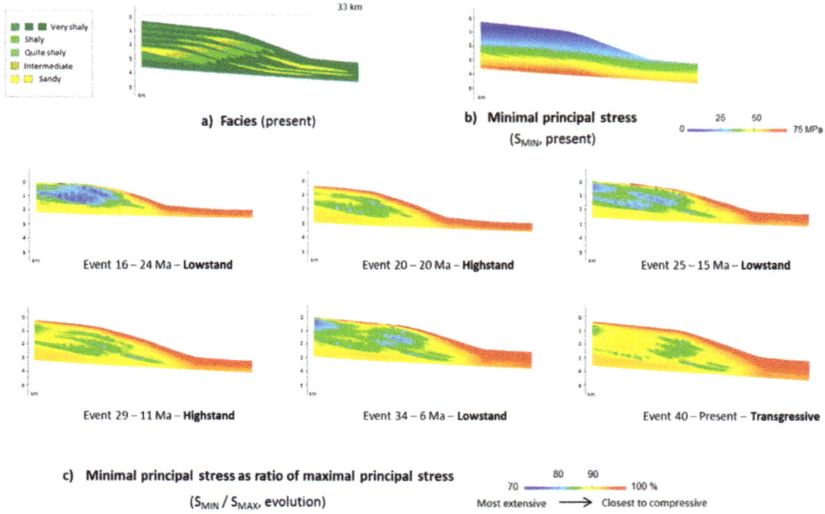

Figure 5. Minimal principal stress simulated on the base case scenario. (**a**) reminder of the lithological facies distribution; (**b**) present-day minimal principal stress: absolute values; (**c**) overview of the minimal principal stress evolution from six geological events: ratios of maximal principal stress; vertical exaggeration × 2.

A clear zoning is visible in the simulation results: the proximal part of the model hosts the lowest principal stress ratios and the most extensional regimes, while the distal part gets very close to compression with ratios above 95%. This is in line with the structural style commonly observed after the collapse of passive margin sediments. Extensional normal faults usually appear in a proximal section of the wedge and compressional fault-and-thrust belts in a distal one. However, if the principal stress ratios simulated in the distal area remain quite constant through time, those in the proximal area varies significantly with the stratigraphic units being deposited. They appear significantly lower during the lowstand periods, with large areas under 80%, than during the highstand periods, with most of the model above 85%. This can be related to a more proximal and deeply immersed sedimentation during the highstand ages.

5.3. Failure Preconditioning

5.3.1. Weakness Prediction

The simulated stress states can be used to estimate a timing and spatial distribution for the development of weakness areas in the sedimentary wedge, considered as privileged locations for local natural fracturing or wider failure initiation. In this study, we introduce a weakness criterion based on the work of [103]. As presented in [71], this criterion corresponds to the shear plasticity threshold of the rheology used in the iterative hydro-mechanical coupling and can be assimilated to a critical state line. It is defined from the confinement pressure and the equivalent deviatoric stress. Concretely, a sediment is considered weak when:

$$q > A \times p' + B \tag{4}$$

A and B are usually deduced from the internal friction angle and cohesive strength of the rock material. In this work, we used $A = 1$ (friction angle equal to 25°) and $B = 0.5$ MPa.

At each geological age in the model, the simulation results are post-processed to highlight the areas above the weakness criterion. It is worth mentioning the permeability of the sediments in these areas remain unchanged, as the criterion is considered to only represent a favorable pre-condition to failure and not directly natural fracturing. However, the corresponding cells in the model are tagged for the rest of the simulation, thus recording the progressive expansion of the weakness in the wedge. Practically, this methodology is only applied on the sediments qualified as sufficiently consolidated, which is previously defined with a porosity cut-off of 35%.

The results on the base case model are presented in Figure 6. The weakness zone in the present day corresponds to 144 cells of the base case model, covering approximatively 10% of the sedimentary wedge. Weakness appears quite late in the sedimentation history, at the 24th event out of 40. It initiates deep beneath the center of the continental slope, in a location of strong shear (visible in Figure 4). Next, the weak area grows downward and upward, forming a stripe roughly parallel to the continental slope, which seems in line with the slip plane orientations commonly observed in passive margin sediments [5,35]. Afterwards, a second weakness stripe initiates at the 33th event, also beneath the center of the continental slope, but in a shallower location. Then, both weakness areas expand through time, until joining in a single continuous weak zone. This result suggests that several levels of different shallowness can simultaneously be preconditioned to failure. This seems also in line with field observations, where superficial slides can coincide with more deeply rooted instabilities [6,26].

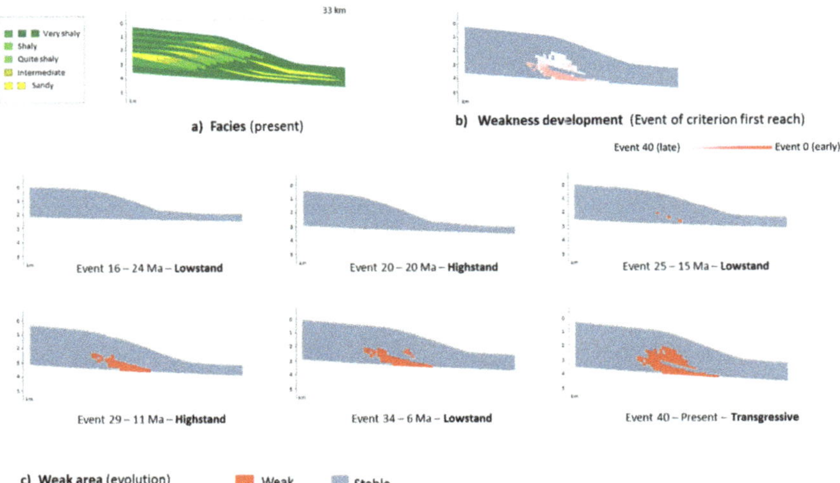

Figure 6. Weakness area simulated on the base case scenario. (**a**) reminder of the lithological facies distribution; (**b**) event of criterion first reach: the stronger the red color, the earlier the area becomes weak; (**c**) overview of the weakness expansion through time from six geological events; vertical exaggeration × 2.

5.3.2. Influence of Sedimentation Rate

The influence of sedimentation rate on failure preconditioning is addressed with the variations introduced between the different basin models of the set. The iterative hydro-mechanical coupling was run on the three models described in Table 1 and the development of weakness areas predicted from the simulated stress states. The results are presented in Figure 7.

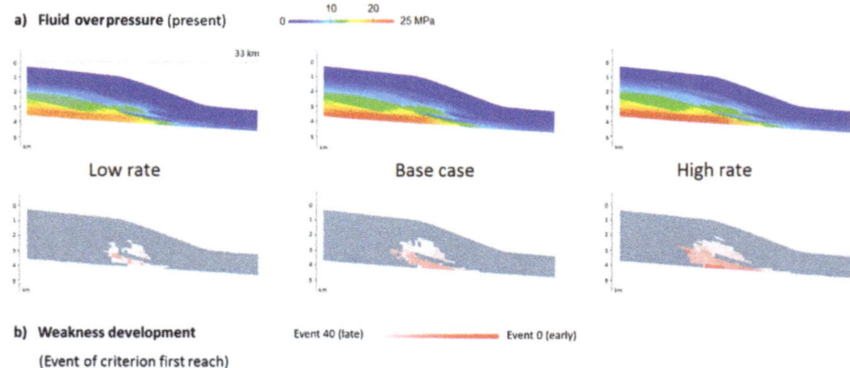

Figure 7. Influence of sedimentation rate on weakness development: globally. (**a**) present-day fluid overpressure simulated on the three scenarios; (**b**) event of criterion first reach in the three scenarios: the stronger the red color, the earlier the area becomes weak; vertical exaggeration × 2.

As expected, a sedimentation rate increase reflects on higher fluid overpressure. For instance, the strongest present-day overpressure simulated in the high rate scenario is 22.7 MPa when it only reaches 20.5 MPa in the low rate one. This leads to a weakness area of earlier initiation and wider expansion. Weakness appears at the 15th event and covers 186 cells in the high rate scenario, when it initiates 10 events later and only covers 88 cells in the low rate one. These differences between scenarios can be explained physically. With high sedimentation rates, the increase in pore pressure engenders a decrease in confinement pressure, while the shear stresses remain identical. Consequently, the ratio between confinement pressure and deviatoric stress rises compared with lower sedimentation rate scenarios, which results in the sediments getting closer to the weakness criterion.

Our physical interpretation can be confirmed by local analyses—for instance, tracking the ratio between confinement pressure and deviatoric stress through time for a given cell and comparing scenarios. Such analysis was carried out for a cell corresponding to deep shaly sediments directly overlying the deepest turbiditic deposits (Figure 8a). In a petroleum exploration context, these sediments could form a sealing cap rock above a sandy reservoir. Assessing their likelihood of failure through geological ages would then be critical to appraise the economic interest of the prospect. The output is displayed in Figure 8b. The decrease in compaction originated by the increase in pore pressure is then clearly visible, with higher sedimentation rates conducing to stress paths shifted leftward in the $p\prime$-q system. This leads the corresponding sediments to reach the weakness criterion at an earlier age in the high rate scenario than in the base case one, while the criterion is never reached in the low rate hypothesis.

5.3.3. Influence of Subsidence Tilt

Thanks to the variations in subsidence tilt angle introduced between the different basin models of the set, the influence of the basal slope on failure preconditioning can be compared with the influence of sedimentation rate. This aims at providing insights on the instability likelihood in situations of slow sedimentation, and at appraising the relative weight of geometrical factors (such as basal angle, clinoform slopes, or progradation rates) and hydro-dynamical ones (such as sedimentation rate or seal permeability) in the geological control of weakness development through the margin history. The iterative hydro-mechanical coupling was run on the three models described in Table 2, and weaker zones were delineated at each age from the simulated stress states. The results are presented in Figure 9.

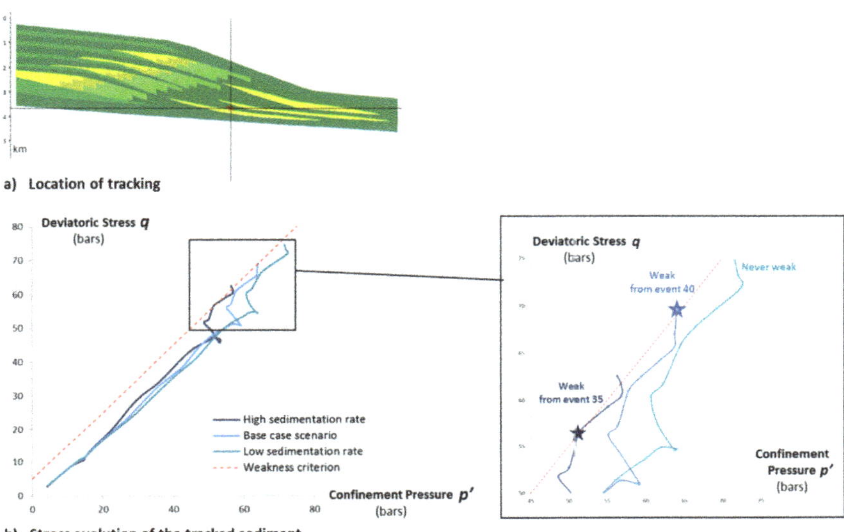

Figure 8. Influence of sedimentation rate on weakness development: locally. (**a**) location of the cell whose mechanical state was tracked through time; (**b**) evolution of the tracked cell in the p′–q system for the three scenarios. The rightmost curve corresponds to the low rate scenario and the leftmost curve to the high rate scenario. The weakness criterion is reached earlier in the high rate scenario due to higher fluid pressure resulting in lower confinement pressure.

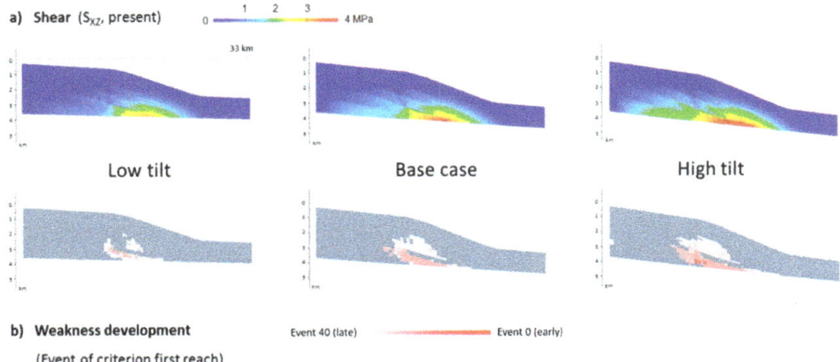

Figure 9. Influence of subsidence tilt on weakness development: globally. (**a**) present-day shear simulated on the three scenarios; (**b**) event of criterion first reach in the three scenarios: the stronger the red color, the earlier the area becomes weak; vertical exaggeration × 2.

An increase in tilt angle, contrary to an increase in sedimentation rate, barely changes the simulated pore pressure. However, it engenders stronger shear stresses and rotations of the principal stresses. In the high tilt scenario, the absolute value of the shear stress S_{XZ} reaches 3.9 MPa when it does not exceed 3.0 MPa in the low tilt scenario. This also leads to an earlier initiation and a wider expansion of the weakness area. In this regard, the differences between our tilt scenarios are of similar magnitude as the differences between our sedimentation rate scenarios (visible in Figure 7), and even slightly more. Weakness appears at the 15th event and covers 192 cells in the high tilt scenario, when it initiates 10 events later and only covers 73 cells in the low tilt one.

The physical interpretation for the impact of subsidence tilt on weakness development differs from the sedimentation rates one. With high tilt angles, the average effective stress is merely impacted, but the increase in shear stress results in an increase in deviatoric stress. Consequently, the ratio between confinement pressure and deviatoric stress rises compared with lower tilt angle scenarios, which results in the sediments getting closer to the weakness criterion. This different way of reaching failure-prone conditions is confirmed by analyzing the local stress paths in the basal tilt scenarios, with exactly the same methods as done for the sedimentation rate ones. The output of this analysis is displayed in Figure 10. The increase in deviatoric stress originated by the increase in shear stresses is then clearly visible, with higher subsidence tilts leading to stress paths shifted upward in the p'–q system. This upward shift has similar consequences as the leftward shift caused by high sedimentation rates. The studied sediments reach the weakness criterion at an earlier age in the high tilt scenario than in the base case one, while the criterion is never reached in the low tilt hypothesis.

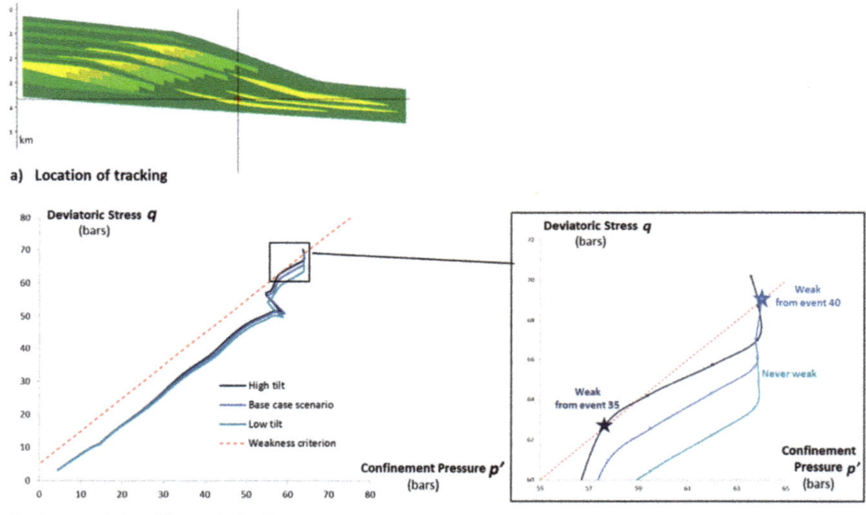

Figure 10. Influence of subsidence tilt on weakness development: locally. (**a**) location of the cell whose mechanical state was tracked through time; (**b**) evolution of the tracked cell in the p'–q system for the three scenarios. The upmost curve corresponds to the high tilt scenario and the downmost curve to the low tilt scenario. The weakness criterion is reached earlier in the high tilt scenario due to higher shear resulting in higher deviatoric stress.

The simulations presented herein confirm that geometrical factors like the evolution of the basal angle can be decisive in the destabilisation of passive margin sedimentary wedges. Although the influence of sedimentation rate is more often invoked and more thoroughly studied in the scientific literature, our study indicates a slight tilting of the sedimentary stack can be of equal significance in failure preconditioning. This advocates for not overlooking geological and geometrical uncertainties in petroleum exploration or geo-hazard risk management, as they can crucially impact predictions of local natural fracturing or wider collapse for the passive margin sediments.

6. Conclusions and Perspectives

6.1. Stress States in Passive Margin Sediments

Altogether, our simulation results reflect the complexity of stress states in siliciclastic passive margin sediments. Notably, the geo-mechanical effects simulated are far from purely vertical, with significant rotations of the principal stress system and stress regimes nearing compression in the distal part of the wedge. It is noticeable that no external tectonic solicitation is needed to trigger the simulated non-vertical effects. They are only driven by the geometric evolution of the sedimentary wedge, with a moderately sloping clinoform deposition and a slightly tilting basis as the two key features.

Another finding is the variability of the non-vertical mechanical effects through space and time. Their magnitude differs considerably depending on the zone considered in the sedimentary wedge. For instance, in our models, the proximal part of the margin hosts the most extensional regimes, whereas the tangential stresses are close to the amount of the normal ones in the abyssal plain. Meanwhile, the highest shear stresses are concentrated beneath the continental slope. These lateral variations are directed by the geometry of the sedimentary stack and reflect the lithological heterogeneities, inferring a strong geological control on the stress distribution. The magnitude of the non-vertical mechanical effects simulated also changes with the stratigraphic age considered. As an example, the highstand periods coincide with higher principal stress ratios in the proximal area, while the lowstand periods coincide with higher downslope rotation of principal stresses.

The significance and variability of the simulated non-vertical effects suggest their integration in physical simulations worth the computation cost, even when studying passive geological contexts. Our work implies that considering passive margin sedimentation as a one-dimensional mechanical system or assessing non-vertical stresses with too simplistic methods may prove misleading. For instance, simply comparing the simulated pore pressure to 85% of weight of the overlying layers, which is often the default option of industrial basin modelling scftware, would result in predicting rock failure on the most proximal and deep areas of our models, which host the strongest overpressure values, rather than under the shelf break and the continental slope. Likewise, as the magnitude of non-vertical mechanical effects in our simulations appeared strongly controlled by the changing geometry of the wedge and the represented lithological heterogereities, modellers should beware of excessive idealization of these two elements considering passive margin sediments.

6.2. Geological Control on Rock Failure Preconditioning

The importance of considering non-vertical mechanical effects in passive margin sediments was further illustrated by comparing the relative impact of sedimentation rate and subsidence tilt on our simulations results in terms of weakness development and failure preconditioning. Our models suggest that the rise in deviatoric stress caused by the tilting of the deposits can affect the stability of the wedge as significantly as the pore pressure rise caused by an increase in secimentation rate. This could explain how rock failure arises in margins of slow sedimentation with relatively well-drained accumulations. However, appraising the impact of this kind of geological and geometrical uncertainty on rock failure predictions cannot be achieved easily with traditional basin models relying on a one-dimensional mechanical scheme. It requires alternative technologies accounting for the 3D nature of the stress states, as the one used in this study.

6.3. Modelling Hypotheses and Way Forward

This study relies on several modelling choices, and we find it useful to remind the main ones here. A first series of choices relate to the synthetic basin models used for the simulation. Notably, the slope of the modelled clinoforms corresponds to the steepest siliciclastic wedges actually observed. A second series of choices relate to the physical simulation itself. Average values were picked for the solid density and the Poisson's ratio of the rock material. Thermal effects were neglected for the sake of simplicity and to focus on the hydro-mechanical phenomena. We also chose not to increase the permeability of the sediments reaching our weakness criterion, as it only represents a favorable precondition to failure. Lastly, the rheology law implemented in the coupling remains quite simple and was essentially implemented for its practicability in terms of comparison with classical basin simulations and numerical cost.

The following step in this research would be to apply a similar methodology to an actual passive margin where the results can then be related to abundant geological knowledge. Doing so, the level of geological realism of the models could be further increased by linking the iterative hydro-mechanical coupling with a forward stratigraphic simulator [104]. First attempts in a carbonate platform context are addressed in [105].

6.4. Integration of Non-Vertical Stresses in Basin Simulators

The results of this study highlight the value of integrating non-vertical mechanical effects in basin simulations. In our passive margin context, it notably appeared paramount to appraise the impact of geometrical and geological factors like basal angle tilting on the timing and location of failure preconditioning in the sedimentary wedge. In a petroleum exploration context, this would facilitate estimates and risk assessment for reservoirs' location and cap rocks' integrity. Logically, integrating non-vertical mechanical effects is even more crucial in more structurally complex areas, enduring significant lateral tectonic input. Nonetheless, before meeting industrial needs, 3D hydro-mechanical simulators must reach an appropriate compromise between detailed physical models, easy integration in existing basin modelling software, and applicability on operational basin models. This still remains a stimulating topic for research and development activities.

In this work, we appraised a method based on a sequential and iterative coupling between a classical basin simulator and a finite element mechanical solver, as described in [70] and [74]. It showed satisfactory applicability on our set of 2D synthetic basin models, designed to be representative of siliciclastic sedimentation in passive margins, including laterally gradating deposition, discordant stratification, and lateral variations of lithological and mechanical properties. It also proved to be moderately expensive in computation cost as the total simulation time for a given model did not exceed 15 minutes with a standard workstation. Consequently, we consider this method as a promising way to capture and incorporate 3D mechanical effects in basin simulations, notably in an industrial oil and gas exploration context. Further appraisals must include heavier and 3D basin models as well as simulation of hydrocarbon maturation and migration.

Author Contributions: Conceptualization, A.B.; models design and building, A.B. and P.C.; simulator implementation, N.G., J.F., and D.C.; simulations, A.B. and P.C.; results analysis, A.B., N.G., J.F., D.C., and P.C.; article writing, A.B.; article review and editing, N.G., J.F., D.C., M.C.C.S., and T.C.; project coordination and supervision, M.C.C.S. and T.C.

Funding: The study presented in this paper was carried out in the NOMBA project, a scientific partnership between IFP Energies nouvelles and Total [106].

Acknowledgments: The authors would like to thank Jean-Luc RUDKIEWICZ (IFP Energies nouvelles) and Claude GOUT (Total) for their valuable input in the design of the passive margin basin models.

Conflicts of Interest: The authors declare no conflict of interest.

Appendix A Petro-Physical Properties of the Modelled Lithological Facies

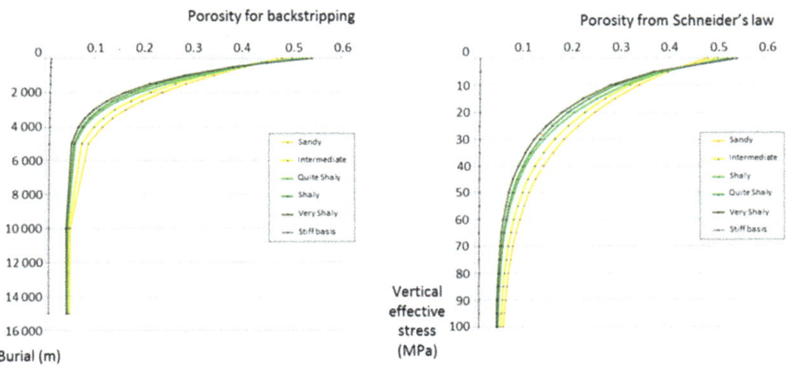

Figure A1. Porosity for the backward restoration and the forward basin simulator. At significant burial depths (e.g., 1 to 8 km), the shalier the facies, the less porous the sediment. However, this ranking is inverted at deposition.

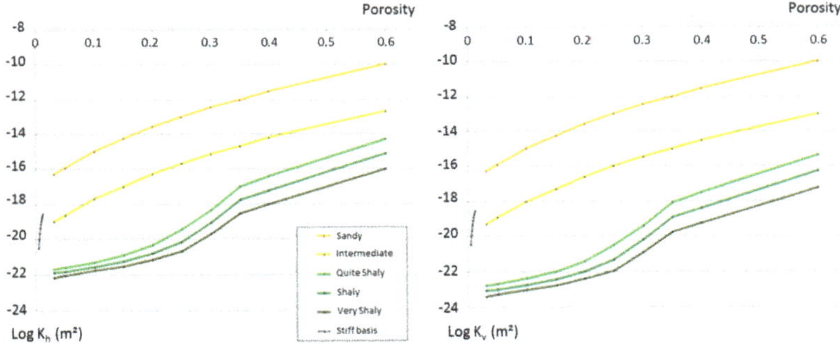

Figure A2. Horizontal and vertical permeability. For a given porosity, the shalier the facies, the less permeable the sediment.

References

1. Bradley, D.C. Passive margins through earth history. *Earth Sci. Rev.* **2008**, *91*, 1–26. [CrossRef]
2. Levell, B.; Argent, J.; Doré, A.G.; Fraser, S. Passive margins: Overview. In *Geological Society, London, Petroleum Geology Conference Series*; Geological Society of London: London, UK, 2011; Volume 7, pp. 823–830. [CrossRef]
3. Wen, Z.; Xu, H.; Wang, Z.; He, Z.; Song, C.; Chen, X.; Wang, Y. Classification and hydrocarbon distribution of passive continental margin basins. *Pet. Explor. Dev.* **2016**, *43*, 740–750. [CrossRef]
4. Mann, P.; Gahagan, L.; Gordon, M.B. Tectonic setting of the world's giant oil and gas fields. In *Giant Oil and Gas Fields of the Decade 1990–1999, AAPG Memoir*; American Association of Petroleum Geologists: Tulsa, OK, USA, 2003; Volume 78, pp. 15–105. [CrossRef]
5. Canals, M.; Lastras, G.; Urgeles, R.; Casamor, J.L.; Mienert, J.; Cattaneo, A.; De Batist, M.; Haflidason, H.; Imbo, Y.; Laberg, J.; et al. Slope failure dynamics and impacts from seafloor and shallow sub-seafloor geophysical data: case studies from the COSTA project. *Mar. Geol.* **2004**, *213*, 9–72. [CrossRef]
6. Vanneste, M.; Sultan, N.; Garziglia, S.; Forsberg, C.F.; L'Heureux, J.-S. Seafloor instabilities and sediment deformation processes: The need for integrated, multi-disciplinary investigations. *Mar. Geol.* **2014**, *352*, 183–214. [CrossRef]
7. Stow, D.A.; Johansson, M. Deep-water massive sands: nature, origin and hydrocarbon implications. *Mar. Pet. Geol.* **2000**, *17*, 145–174. [CrossRef]
8. Shanmugam, G. New perspectives on deep-water sandstones: Implications. *Pet. Explor. Dev.* **2013**, *40*, 316–324. [CrossRef]

9.   Magara, K. *Compaction and Fluid Migration*; Elsevier: Amsterdam, The Netherlands, 1978; Volume 9, ISBN 0-444-41654-4.

10. Du Rouchet, J. Stress Fields, A Key to Oil Migration. *AAPG Bull.* **1981**, *65*, 74–85.

11. Zhang, J. Effective stress, porosity, velocity and abnormal pore pressure prediction accounting for compaction disequilibrium and unloading. *Mar. Pet. Geol.* **2013**, *45*, 2–11. [CrossRef]

12. Li, S.; George, J.; Purdy, C. Pore-Pressure and Wellbore-Stability Prediction to Increase Drilling Efficiency. *J. Pet. Technol.* **2012**, *64*, 98–101. [CrossRef]

13. Faille, I.; Thibaut, M.; Cacas, M.-C.; Havé, P.; Willien, F.; Wolf, S.; Agelas, L.; Pegaz-Fiornet, S. Modeling Fluid Flow in Faulted Basins. *Oil Gas Sci. Technol. Rev. l'IFP* **2014**, *69*, 529–553. [CrossRef]

14. Available online: https://www.code-aster.org (accessed on 18 October 2019).

15. Watts, A. Models for the evolution of passive margins. In *Regional Geology and Tectonics: Phanerozoic Rift Systems and Sedimentary Basins*; Elsevier: Amsterdam, The Netherlands, 2012; Volume 1, pp. 32–57.

16. Catuneanu, O. *Principles of Sequence Stratigraphy*; Elsevier: Amsterdam, The Netherlands, 2006; ISBN 978-0-444-51568-1.

17. O'Grady, D.B.; Syvitski, J.P.M.; Pratson, L.F.; Sarg, J.F. Categorizing the morphologic variability of siliciclastic passive continental margins. *Geology* **2000**, *28*, 207. [CrossRef]

18. Dugan, B.; Sheahan, T.C. Offshore sediment overpressures of passive margins: Mechanisms, measurement, and models. *Rev. Geophys.* **2012**, *50*. [CrossRef]

19. Osborne, M.J.; Swarbrick, R.E. Mechanisms for Generating Overpressure in Sedimentary Basins: A Reevaluation. *AAPG Bull.* **1997**, *81*, 1023–1041.

20. Grauls, D. Overpressures: Causal Mechanisms, Conventional and Hydromechanical Approaches. *Oil Gas Sci. Technol. Rev. l'IFP* **1999**, *54*, 667–678. [CrossRef]

21. Shaker, S. Sequence stratigraphy: Key to geopressure profile assessment. *CSEG Rec.* **2002**, *9*, 88–91. Available online: https://csegrecorder.com/articles/view/sequence-stratigraphy-key-to-geopressure-profile-assessment (accessed on 5 November 2019).

22. Flemings, P.B. Flow focusing in overpressured sandstones: Theory, observations, and applications. *Am. J. Sci.* **2002**, *302*, 827–855. [CrossRef]

23. Xie, X.; Li, S.; Dong, W.; Zhang, Q. Overpressure development and hydrofracturing in the Yinggehai basin, South China Sea. *J. Pet. Geol.* **1999**, *22*, 437–454. [CrossRef]

24. Hustoft, S.; Mienert, J.; Bünz, S.; Nouzé, H. High-resolution 3D-seismic data indicate focussed fluid migration pathways above polygonal fault systems of the mid-Norwegian margin. *Mar. Geol.* **2007**, *245*, 89–106. [CrossRef]

25. Hampton, M.A.; Lee, H.J.; Locat, J. Submarine landslides. *Rev. Geophys.* **1996**, *34*, 33–59. [CrossRef]

26. Wynn, R.B.; Kenyon, N.H.; Weaver, P.P.E.; Evans, J. Continental margin sedimentation, with special reference to the north-east Atlantic margin. *Sedimentology* **2000**, *47*, 239–256.

27. Hühnerbach, V.; Masson, D.; Costa-Project, P.O.T. Landslides in the North Atlantic and its adjacent seas: an analysis of their morphology, setting and behaviour. *Mar. Geol.* **2004**, *213*, 343–362. [CrossRef]

28. Krastel, S.; Lehr, J.; Winkelmann, D.; Schwenk, T.; Preu, B.; Strasser, M.; Wynn, R.B.; Georgiopoulou, A.; Hanebuth, T.J.J. Mass wasting along Atlantic continental margins: A comparison between NW-Africa and the de la Plata River region (northern Argentina and Uruguay). In *Submarine Mass Movements and Their Consequences*; Springer: Cham, Switzerland, 2014; pp. 459–469. [CrossRef]

29. Le Bouteiller, P.; Lafuerza, S.; Charléty, J.; Reis, A.T.; Granjeon, D.; Delprat-Jannaud, F.; Gorini, C. A new conceptual methodology for interpretation of mass transport processes from seismic data. *Mar. Pet. Geol.* **2019**, *103*, 438–455. [CrossRef]

30. Camargo, J.M.; Silva, M.V.; Ferreira Júnior, A.V.; Araújo, T. Marine Geohazards: A Bibliometric-Based Review. *Geosciences* **2019**, *9*, 100. [CrossRef]

31. Rowan, M.G.; Peel, F.J.; Vendeville, B.C. Gravity-driven fold belts on passive margins, In Thrust Tectonics and Hydrocarbon Systems. *AAPG Memoir* **2004**, *82*, 157–182. [CrossRef]

32. Morley, C.; King, R.; Hillis, R.; Tingay, M.; Backé, G.; Morley, C. Deepwater fold and thrust belt classification, tectonics, structure and hydrocarbon prospectivity: A review. *Earth Sci. Rev.* **2011**, *104*, 41–91. [CrossRef]

33. Locat, J.; Leroueil, S.; Locat, A.; Lee, H. Weak layers: Their definition and classification from a geotechnical perspective. In *Submarine Mass Movements and Their Consequences*; Springer: Cham, Switzerland, 2014; pp. 3–12. [CrossRef]

34. Dugan, B. Flemings Overpressure and fluid flow in the new jersey continental slope: implications for slope failure and cold seeps. *Science* **2000**, *289*, 288–291. [CrossRef]
35. Sultan, N.; Cochonat, P.; Canals, M.; Cattaneo, A.; Dennielou, B.; Haflidason, H.; Laberg, J.; Long, D.; Mienert, J.; Trincardi, F.; et al. Triggering mechanisms of slope instability processes and sediment failures on continental margins: a geotechnical approach. *Mar. Geol.* **2004**, *213*, 291–321. [CrossRef]
36. Behrmann, J.H.; Flemings, P.B.; John, C.M. Rapid Sedimentation, Overpressure, and Focused Fluid Flow, Gulf of Mexico Continental Margin. *Sci. Drill.* **2006**, *3*, 12–17. [CrossRef]
37. Urgeles, R.; Locat, J.; Dugan, B. Recursive Failure of the Gulf of Mexico Continental Slope: Timing and Causes. In *Submarine Mass Movements and Their Consequences*; Springer: Dordrecht, The Netherlands, 2007; pp. 209–219. [CrossRef]
38. Wolinsky, M.A.; Pratson, L.F. Overpressure and slope stability in prograding clinoforms: Implications for marine morphodynamics. *J. Geophys. Res. Space Phys.* **2007**, *112*. [CrossRef]
39. Ai, F.; Strasser, M.; Preu, B.; Hanebuth, T.J.J.; Krastel, S.; Kopf, A. New constraints on oceanographic vs. seismic control on submarine landslide initiation: a geotechnical approach off Uruguay and northern Argentina. *Geo-Mar. Lett.* **2014**, *34*, 399–417. [CrossRef]
40. Xie, X.; Müller, R.D.; Ren, J.; Jiang, T.; Zhang, C. Stratigraphic architecture and evolution of the continental slope system in offshore Hainan, northern South China Sea. *Mar. Geol.* **2008**, *247*, 129–144. [CrossRef]
41. Mauduit, T.; Guerin, G.; Brun, J.-P.; Lecanu, H. Raft tectonics: the effects of basal slope angle and sedimentation rate on progressive extension. *J. Struct. Geol.* **1997**, *19*, 1219–1230. [CrossRef]
42. Mourgues, R.; Lacoste, A.; Garibaldi, C. The Coulomb critical taper theory applied to gravitational instabilities. *J. Geophys. Res. Solid Earth* **2014**, *119*, 754–765. [CrossRef]
43. Brun, J.-P.; Fort, X. Salt tectonics at passive margins: Geology versus models. *Mar. Pet. Geol.* **2011**, *28*, 1123–1145. [CrossRef]
44. Yuan, X.P.; Leroy, Y.M.; Maillot, B. Reappraisal of gravity instability conditions for offshore wedges: Consequences for fluid overpressures in the Niger Delta. *Geophys. J. Int.* **2017**, *208*, 1655–1671. [CrossRef]
45. Dahlen, F.A. Noncohesive critical Coulomb wedges: An exact solution. *J. Geophys. Res. Space Phys.* **1984**, *89*, 10125–10133. [CrossRef]
46. Leynaud, D.; Sultan, N.; Mienert, J. The role of sedimentation rate and permeability in the slope stability of the formerly glaciated Norwegian continental margin: The Storegga slide model. *Landslides* **2007**, *4*, 297–309. [CrossRef]
47. Sultan, N.; Gaudin, M.; Berné, S.; Canals, M.; Urgeles, R.; Lafuerza, S. Analysis of slope failures in submarine canyon heads: An example from the Gulf of Lions. *J. Geophys. Res. Space Phys.* **2007**, *112*, 1–29. [CrossRef]
48. Dan, G.; Sultan, N.; Savoye, B.; Deverchere, J.; Yelles, K. Quantifying the role of sandy–silty sediments in generating slope failures during earthquakes: Example from the Algerian margin. *Int. J. Earth Sci.* **2009**, *98*, 769–789. [CrossRef]
49. Urlaub, M.; Zervos, A.; Talling, P.J.; Masson, D.G.; Clayton, C.I. How do ~2° slopes fail in areas of slow sedimentation? A sensitivity study on the influence of accumulation rate and permeability on submarine slope stability. In *Submarine Mass Movements and Their Consequences*; Springer: Dordrecht, The Netherlands, 2012; pp. 277–287. [CrossRef]
50. Ai, F.; Kuhlmann, J.; Huhn, K.; Strasser, M.; Kopf, A. Submarine slope stability assessment of the central Mediterranean continental margin: The Gela Basin. In *Submarine Mass Movements and their Consequences*; Springer: Cham, Switzerland, 2014; pp. 225–236. [CrossRef]
51. Yuan, X.P.; Leroy, Y.M.; Maillot, B. Tectonic and gravity extensional collapses in overpressured cohesive and frictional wedges. *J. Geophys. Res. Solid Earth* **2015**, *120*, 1833–1854. [CrossRef]
52. Berthelon, J.; Yuan, X.; Bouziat, A.; Souloumiac, P.; Pubellier, M.; Cornu, T.; Maillot, B. Mechanical restoration of gravity instabilities in the Brunei margin, N.W. Borneo. *J. Struct. Geol.* **2018**, *117*, 148–162. [CrossRef]
53. Rudkiewicz, J.L.; Penteado, H.D.B.; Vear, A.; Vandenbroucke, M.; Brigaud, F.; Wendebourg, J.; Duppenbecker, S. Integrated Basin Modeling Helps to Decipher Petroleum Systems. In Petroleum Systems of South Atlantic Margins. *AAPG Memoir* **2000**, *73*, 27–40. [CrossRef]
54. Schneider, F.; Wolf, S.; Faille, I.; Pot, D. A 3d Basin Model for Hydrocarbon Potential Evaluation: Application to Congo Offshore. *Oil Gas Sci. Technol. Rev. l'IFP* **2000**, *55*, 3–13. [CrossRef]
55. Xie, X.; Bethke, C.M.; Li, S.; Liu, X.; Zheng, H. Overpressure and petroleum generation and accumulation in the Dongying Depression of the Bohaiwan Basin, China. *Geofluids* **2001**, *1*, 257–271. [CrossRef]

56. Tuncay, K.; Ortoleva, P. Quantitative basin modeling: present state and future developments towards predictability. *Geofluids* **2004**, *4*, 23–39. [CrossRef]

57. Hantschel, T.; Kauerauf, A.I. *Fundamentals of Basin and Petroleum Systems Modeling*; Springer Science & Business Media: Berlin/Heidelberg, Germany, 2009. [CrossRef]

58. Thibaut, M.; Jardin, A.; Faille, I.; Willien, F.; Guichet, X. Advanced Workflows for Fluid Transfer in Faulted Basins. *Oil Gas Sci. Technol. Rev. l'IFP* **2014**, *69*, 573–584. [CrossRef]

59. Fjeldskaar, W.; Andersen, Å.; Johansen, H.; Lander, R.; Blomvik, V.; Skurve, O.; Michelsen, J.K.; Grunnaleite, I.; Mykkeltveit, J.; Mykkeltveit, J. Bridging the gap between basin modelling and structural geology. *Reg. Geol. Metallog* **2017**, *72*, 65–77.

60. Crook, A.J.; Obradors-Prats, J.; Somer, D.; Peric, D.; Lovely, P.; Kacewicz, M. Towards an integrated restoration/forward geomechanical modelling workflow for basin evolution prediction. *Oil Gas Sci. Technol. Rev. l'IFP* **2018**, *73*, 18. [CrossRef]

61. Burgreen-Chan, B.; Meisling, K.E.; Graham, S. Basin and petroleum system modelling of the East Coast basin, New Zealand: A test of overpressure scenarios in a convergent margin. *Basin Res.* **2016**, *28*, 536–567. [CrossRef]

62. Woillez, M.N.; Souque, C.; Rudkiewicz, J.L.; Willien, F.; Cornu, T. Insights in fault flow behavior from onshore Nigeria petroleum system modelling. *Oil Gas Sci. Technol. Rev. d'IFP Energies Nouvelles* **2017**, *72*, 31. [CrossRef]

63. Schneider, F.; Bouteca, M.; Sarda, J.P. Hydraulic Fracturing at Sedimentary Basin Scale. *Oil Gas Sci. Technol. Rev. l'IFP* **1999**, *54*, 797–806. [CrossRef]

64. Hauser, M.R.; Couzens-Schultz, B.A.; Chan, A.W. Estimating the influence of stress state on compaction behavior. *Geophysics* **2014**, *79*, D389–D398. [CrossRef]

65. Prats, J.O.; Rouainia, M.; Aplin, A.C.; Crook, A.J. Stress and pore pressure histories in complex tectonic settings predicted with coupled geomechanical-fluid flow models. *Mar. Pet. Geol.* **2016**, *76*, 464–477. [CrossRef]

66. Tuncay, K.; Park, A.; Ortoleva, P. Sedimentary basin deformation: an incremental stress approach. *Tectonophysics* **2000**, *323*, 77–104. [CrossRef]

67. Obradors-Prats, J.; Rouainia, M.; Aplin, A.C.; Crook, A.J. Assessing the implications of tectonic compaction on pore pressure using a coupled geomechanical approach. *Mar. Pet. Geol.* **2017**, *79*, 31–43. [CrossRef]

68. Campher, C.; Kuhlmann, G.; Adams, S.; Anka, Z.; Di Primio, R.; Horsfield, B. 3D petroleum systems modelling within a passive margin setting, orange basin, blocks 3/4, offshore south Africa—Implications for gas generation, migration and leakage. *S. Afr. J. Geol.* **2011**, *114*, 387–414.

69. Ducros, M.; Carpentier, B.; Wolf, S.; Cacas, M.C. Integration of biodegradation and migration of hydrocarbons in a 2D petroleum systems model: Application to the Potiguar Basin, NE Brazil. *J. Pet. Geol.* **2016**, *39*, 61–78. [CrossRef]

70. Barabasch, J.; Ducros, M.; Hawie, N.; Daher, S.B.; Nader, F.H.; Littke, R. Integrated 3D forward stratigraphic and petroleum system modeling of the Levant Basin, Eastern Mediterranean. *Basin Res.* **2019**, *31*, 228–252. [CrossRef]

71. Guy, N.; Colombo, D.; Frey, J.; Cornu, T.; Cacas-Stentz, M.C. Coupled Modeling of Sedimentary Basin and Geomechanics: A Modified Drucker–Prager Cap Model to Describe Rock Compaction in Tectonic Context. *Rock Mech. Rock Eng.* **2019**, *52*, 3627–3643. [CrossRef]

72. Morency, C.; Huismans, R.S.; Beaumont, C.; Fullsack, P. A numerical model for coupled fluid flow and matrix deformation with applications to disequilibrium compaction and delta stability. *J. Geophys. Res. Space Phys.* **2007**, *112*, 1–25. [CrossRef]

73. Mainguy, M.; Longuemare, P. Coupling Fluid Flow and Rock Mechanics: Formulations of the Partial Coupling Between Reservoir and Geomechanical Simulators. *Oil Gas Sci. Technol. Rev. l'IFP* **2002**, *57*, 355–367. [CrossRef]

74. Guy, N.; Enchéry, G.; Renard, G. Numerical Modeling of Thermal EOR: Comprehensive Coupling of an AMR-Based Model of Thermal Fluid Flow and Geomechanics. *Oil Gas Sci. Technol. Rev. l'IFP* **2012**, *67*, 1019–1027. [CrossRef]

75. Guy, N.; Colombo, D.; Frey, J.; Vincke, O.; Gout, C. Method for Exploitation of Hydrocarbons from a Sedimentary Basin by Means of a Basin Simulation Taking Account of Geomechanical Effects. U.S. Patent No. 10296679, 19 December 2016.

76. Schneider, F.; Potdevin, J.L.; Wolf, S.; Faille, I. Elastoplastic and Viscoplastic Compaction Model for the Simulation of Sedimentary Basins. *Oil Gas Sci. Technol.* **1994**, *49*, 141–148. [CrossRef]

77. Schneider, F.; Potdevin, J.; Wolf, S.; Faille, I. Mechanical and chemical compaction model for sedimentary basin simulators. *Tectonophysics* **1996**, *263*, 307–317. [CrossRef]

78. Zhang, K.; Zhou, H.; Shao, J. An Experimental Investigation and an Elastoplastic Constitutive Model for a Porous Rock. *Rock Mech. Rock Eng.* **2013**, *46*, 1499–1511. [CrossRef]

79. Brüch, A.; Maghous, S.; Ribeiro, F.; Dormieux, L. A thermo-poro-mechanical constitutive and numerical model for deformation in sedimentary basins. *J. Pet. Sci. Eng.* **2018**, *160*, 313–326. [CrossRef]

80. Pinheiro-Moreira, J.L.; Nalpas, T.; Joseph, P.; Guillocheau, F. Stratigraphie sismique de la marge éocène du Nord du bassin de Santos (Brésil): Relations plate-forme/systèmes turbiditiques; distorsion des séquences de dépôt. *Comptes Rendus Acad. Sci. Ser. IIA Earth Planet. Sci.* **2001**, *332*, 491–498. [CrossRef]

81. Dingle, R.V. Continental margin subsidence: A Comparison between the east and west coasts of Africa. In *Reflection Seismology: A Global Perspective*; American Geophysical Union (AGU): Washington, DC, USA, 1982; Volume 6, pp. 59–71.

82. Stow, D.A.V.; Tabrez, A.R. *Hemipelagites: Processes, Facies and Model*; Geological Society: London, UK, 1998; Volume 129, pp. 317–337.

83. Perrier, R.; Quiblier, J. Thickness changes in sedimentary layers during compaction history; methods for quantitative evaluation. *AAPG Bull.* **1974**, *58*, 507–520. [CrossRef]

84. Allen, P.A.; Allen, J.R. *Basin Analysis: Principles and Application to Petroleum Play Assessment*; John Wiley & Sons: Hoboken, NJ, USA, 2013; ISBN 978-0-470-67377-5.

85. Available online: http://www.beicip.com/petroleum-system-assessment (accessed on 18 October 2019).

86. Manger, G.E. *Porosity and Bulk Density of Sedimentary Rocks*; Bulletin (US Geological Survey), 1144-E; U.S. Government Publishing Office: Washington, DC, USA, 1963; pp. 1–55. [CrossRef]

87. Gerçek, H. Poisson's ratio values for rocks. *Int. J. Rock Mech. Min. Sci.* **2007**, *44*, 1–13. [CrossRef]

88. Mondol, N.H.; Castagna, T.P.C.J. Porosity and permeability development in mechanically compacted silt-kaolinite mixtures. In *SEG Technical Program Expanded Abstracts 2009*; Society of Exploration Geophysicists: Tulsa, OK, USA, 2009; pp. 2139–2143.

89. Yang, Y.; Aplin, A.C. A permeability–porosity relationship for mudstones. *Mar. Pet. Geol.* **2010**, *27*, 1692–1697. [CrossRef]

90. Chilingar, G.V. Relationship Between Porosity, Permeability, and Grain-Size Distribution of Sands and Sandstones. In *Developments in Sedimentology*; Elsevier: Amsterdam, The Netherlands, 1964; Volume 1, pp. 71–75.

91. Doyen, P.M. Permeability, conductivity, and pore geometry of sandstone. *J. Geophys. Res. Space Phys.* **1988**, *93*, 7729. [CrossRef]

92. Luijendijk, E.; Gleeson, T. How well can we predict permeability in sedimentary basins? Deriving and evaluating porosity–permeability equations for noncemented sand and clay mixtures. *Geofluids* **2015**, *15*, 67–83. [CrossRef]

93. Lanson, B.; Beaufort, D.; Berger, G.; Bauer, A.; Cassagnabère, A.; Meunier, A. Authigenic kaolin and illitic minerals during burial diagenesis of sandstones: A review. *Clay Miner.* **2002**, *37*, 1–22. [CrossRef]

94. Bolton, A.J.; Maltman, A.J.; Fisher, Q. Anisotropic permeability and bimodal pore-size distributions of fine-grained marine sediments. *Mar. Pet. Geol.* **2000**, *17*, 657–672. [CrossRef]

95. Hu, K.; Issler, D.R. *A Comparison of Core Petrophysical Data with Well Log Parameters, Beaufort-Mackenzie Basin*; Geological Survey of Canada: Ottawa, ON, Canada, 2009. [CrossRef]

96. Selvadurai, A.P.S.; Boulon, M.J.; Nguyen, T.S. The Permeability of an Intact Granite. *Pure Appl. Geophys.* **2005**, *162*, 373–407. [CrossRef]

97. Brace, W.F.; Walsh, J.B.; Frangos, W.T. Permeability of granite under high pressure. *J. Geophys. Res. Space Phys.* **1968**, *73*, 2225–2236. [CrossRef]

98. Moore, D.E.; Lockner, D.A.; Byerlee, J.D. Reduction of Permeability in Granite at Elevated Temperatures. *Science* **1994**, *265*, 1558–1561. [CrossRef]

99. Flemings, P.B.; Dugan, B. Fluid flow and stability of the US continental slope offshore New Jersey from the Pleistocene to the present. *Geofluids* **2002**, *2*, 137–146.

100. Stigall, J.; Dugan, B. Overpressure and earthquake initiated slope failure in the Ursa region, northern Gulf of Mexico. *J. Geophys. Res. Space Phys.* **2010**, *115*, 1–11. [CrossRef]

101. Terzaghi, K.; Peck, R.B.; Mesri, G. *Soil Mechanics in Engineering Practice*; John Wiley & Sons: Hoboken, NJ, USA, 1996; ISBN 978-0-471-08658-1.
102. Nauroy, J.F. *Geomechanics Applied to the Petroleum Industry*; Editions Technip: Paris, France, 2011; ISBN 978-2-710-80932-6.
103. Drucker, D.C.; Prager, W. Soil mechanics and plastic analysis or limit design. *Q. Appl. Math.* **1952**, *10*, 157–165. [CrossRef]
104. Granjeon, D.; Joseph, P.; Harbaugh, J.W.; Watney, W.L.; Rankey, E.C.; Slingerland, R.; Goldstein, R.H.; Franseen, E.K. Concepts and Applications of A 3-D Multiple Lithology, Diffusive Model in Stratigraphic Modeling. In *Numerical Experiments in StratigraphyRecent Advances in Stratigraphic and Sedimentologic Computer Simulations*; Society for Sedimentary Geology: Tulsa, OK, USA, 1999.
105. Busson, J. Caractérisation et Modélisation des Transferts Gravitaires de la Plate-Forme au Basin en Contexte Carbonaté. Ph.D. Thesis, University of Bordeaux, Bordeaux, France, 2018.
106. Cornu, T.; Gout, C.; Cacas-Stenz, M.C.; Woillez, M.N.; Guy, N.; Bouziat, A.; Colombo, D.; Frey, J. NOMBA an integrated project for coupling basin modeling and geomechanical simulations. In Proceedings of the AAPG Hedberg Conference: The Future of Basin and Petroleum System Modelin, Santa Barbara, CA, USA, 3–8 April 2016.

 *geosciences*

Article

Tilting and Flexural Stresses in Basins Due to Glaciations—An Example from the Barents Sea

Ingrid F. Løtveit *, Willy Fjeldskaar and Magnhild Sydnes

Tectonor AS, P.O. Box 8034, N-4068 Stavanger, Norway; wf@tectonor.com (W.F.); ms@tectonor.com (M.S.)
* Correspondence: ifl@tectonor.com

Received: 21 September 2019; Accepted: 7 November 2019; Published: 11 November 2019

Abstract: Many of the Earth's sedimentary basins are affected by glaciations. Repeated glaciations over millions of years may have had a significant effect on the physical conditions in sedimentary basins and on basin structuring. This paper presents some of the major effects that ice sheets might have on sedimentary basins, and includes examples of quantifications of their significance. Among the most important effects are movements of the solid Earth caused by glacial loading and unloading, and the related flexural stresses. The driving factor of these movements is isostasy. Most of the production licenses on the Norwegian Continental Shelf are located inside the margin of the former Last Glacial Maximum (LGM) ice sheet. Isostatic modeling shows that sedimentary basins near the former ice margin can be tilted as much as 3 m/km which might significantly alter pathways of hydrocarbon migration. In an example from the SW Barents Sea we show that flexural stresses related to the isostatic uplift after LGM deglaciation can produce stress changes large enough to result in increased fracture-related permeability in the sedimentary basin, and lead to potential spillage of hydrocarbons out of potential reservoirs. The results demonstrate that future basin modeling should consider including the loading effect of glaciations when dealing with petroleum potential in former glaciated areas.

Keywords: Glaciations; isostasy; flexural stress; faults; hydrocarbon migration

---

1. Introduction

It is commonly accepted that major changes in Earth's climate started in Gelasian (~2.6 million years ago), and initiated the growth of ice sheets in the Northern Hemisphere (e.g., [1]). Over the last 2.6 million years, there have probably been more than 30 glaciations with cycles of ice sheets advancing and retreating on time scales of 40,000 to 100,000 years. A glaciation cycle of 40,000 years dominated in the early period, while the 100,000 year cycle has dominated in the past million years or so [2]. The most extensive glaciations are traditionally referenced in the last million years (Late Quaternary; cf. [3]) and the covers on North America and Europe have been estimated to be around 2–4 km thick near the centers of maximum ice accumulation (cf. e.g., [4]), and tapered towards the ice sheet margins. Repeated episodes of growth and withdrawal of huge ice sheets, glacial erosion, and associated sediment deposition lead to geologically dramatic changes in surface loading. The lithosphere responds to these changes by subsiding under loading, and by uplifting when loading is removed. The driving mechanism for this is isostasy.

Differential vertical movement of the lithosphere related to glacial isostasy lead to repeated tilting of sedimentary formations and potential petroleum reservoirs therein, which may have greatly affected hydrocarbon migration pathways in the former glaciated areas [5]. In addition, the upward and downward bending of the lithosphere due to glaciation leads to flexural stresses [6] likely to affect faults and their permeability, which could add to the changes in hydrocarbon migration pathways.

In parts of the Barents Sea, there are clear indications of hydrocarbon spillage (e.g., [7]). Some of the hydrocarbon spillage is linked to isostatic movements due to cycles of Pliocene-Pleistocene ice sheet loading/unloading and glacial sediment redistribution [8–12]. It is suggested that the recent discovery of the giant Johan Sverdrup oil field in the Norwegian North Sea was oil charged during Quaternary, and that the area more than likely underwent tilting and possible leakage several times over the last one million years [13]. Other research, related to the North Sea, showed that influence of ice sheet loading does not result in a significant tilting [14], which may be the result of smaller gradients of the ice thickness. Anyway, detailed control on the glacial isostasy is important and so far an insufficiently utilized factor for identification of the remaining hydrocarbon resources in sedimentary basins formerly covered by ice sheets.

Faulting related to the last glaciation is reported from northern Scandinavia, where several fault with up to 30 m high fault scarps [15]. Examples include the 80 km long Stuoragurra fault in northern Norway [16] and the 150 km long Pärvie fault in northern Sweden [17]. There is also evidence of such faults in Denmark [18], Germany, Poland, and the United Kingdom ([19–23]). Thus, such faults are found in the center of former ice sheets, as well as at their margins and beyond [24,25].

For the offshore areas, the situation is more complicated. However, the number of seismic events recorded in the Barents Sea in the last 20 years is significant ([26,27]). For the Barents Sea, the impact of the glacial isostasy on triggering earthquakes has not yet been investigated. A better understanding of the origin of the seismicity will help to recognize which faults and related areas exhibit a risk of leakage of hydrocarbons from traps.

Stress changes induced by glaciations are related to two main effects: 1) direct load introduced by the weight of the ice on the surface and 2) flexural load caused by the subsidence and uplift of the lithosphere due to glacial isostasy [10,28]. Horizontal stress due to stress migration, which is depending on loading time and mantle viscosity, is an additional effect that is modeled by Steffen et al. [29]. Grollimund and Zoback [30] investigated, by a 2D approach, lithospheric flexure as an alternative mechanism to explain the local stress perturbations observed in the northern North Sea. They found that flexural stresses due to glaciations/deglaciation could explain the observed lateral stress variations.

Johnston [31] used the Mohr–Coulomb theory for the study of faulting and explained the lack of seismic activity in areas of large continental ice sheets. Turpeinen [32] use finite-element models of rheologically layered lithosphere with a thrust fault to investigate its slip evolution during the glacial/ postglacial period. Their modelling results suggest that the rate of thrusting decreases during ice sheet loading and strongly increases during deglaciation. Steffen et al. [33] introduced a new approach that is based on a Glacial Isostatic Adjustment (GIA) model. They combined the GIA model with a fault surface to investigate the fault slip and fault activation time during a glacial cycle. Their model allows estimation of the fault throw for former glaciated areas. One of their conclusions is that faults start to move close to the end of deglaciation.

Our approach is flexural isostasy modelling, and we are dealing with two effects of the glaciations which we believe future basin modelling has to consider:

1. Possible glacially induced tilting of reservoirs. Here, we calculate the glacially induced tilting of the Norwegian Continental Shelf due to the last ice age, with particular focus on the SW Barents Sea.
2. Assessment of the flexural stresses induced by the lithospheric flexure related to glacial isostasy and tilt, and their effect on faults during glaciations and interglacials, here illustrated by an example from the SW Barents Sea where the stress modelling is limited to the flexural stress effects of deglaciation after LGM. The stress effects of vertical ice load and related stress migration are not calculated here.

2. Glaciations

The last glacial period began about 110,000 years ago and ended 10,000 years ago. A model of the deglaciation history for our area is given in [34]. The glaciations that occurred during this glacial period covered many areas of the Northern Hemisphere. The extent of the Eurasian Ice Sheet during the Last Glacial Maximum (LGM) has been long debated in the literature. Researchers [35] argue for a highly asynchronous ice sheet with regional differences in the age of maximum ice-sheet positions. However, in this paper we use the LGM (~20,000 BP) reconstruction of [36], which suggests the SW Barents Sea was glaciated from 24,000 to 18,000 BP. In the following calculations, we assume the Earth reached isostatic equilibrium during this period.

2.1. Ice Thickness

Glacial isostasy is calculated based on the palaeo ice extent (from dated marginal moraines) and models of the ice thicknesses. Observational data increasingly constrain the extent of the Quaternary ice sheets. The thickness and volume of these ice sheets are much harder to reconstruct and generally need to be inferred from indirect evidence and modeling.

Several models of ice thickness of LGM and the late-glacial deglaciation are published; the loading scenarios are produced using different methods and sets of constraints. Some of them consider thermomechanical coupling (e.g., [37]). A number of models of the LGM ice thickness are presented in [38]. We follow the work of [36]; their LGM extent model is shown in Figure 1.

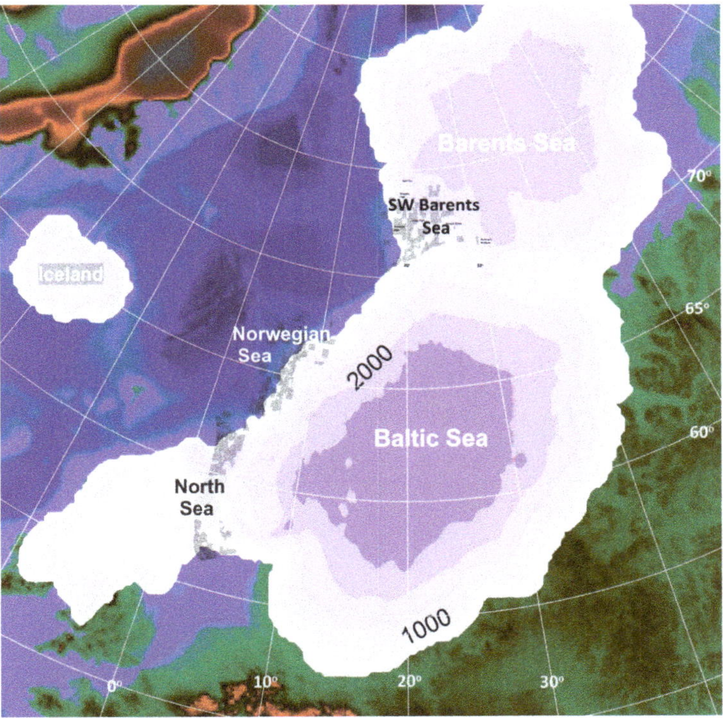

Figure 1. Extent (based on [36]) and maximum thickness model for the Last Glacial Maximum (LGM), from [3]. The production licenses (from www.npd.no) in the Norwegian continental shelf are shown by grey polygons.

2.2. Glacial Isostasy and Tilting of Reservoirs

The Earth's response to deglaciation shows that the lithosphere acts as an elastic shell overlying a viscous mantle. If a load is applied to the elastic lithosphere, part of the applied load will be supported by the elastic lithosphere, and part by buoyant forces of the mantle underneath, acting through the lithosphere.

In the literature, there is disagreement on the elastic thickness of the lithosphere; the elastic thickness of the lithosphere of Fennoscandia in GIA (Glacial Isostatic Adjustment) modeling varies from 30 to 160 km [39]. Our modeling of the post-glacial isostatic response, calibrated with the observations onshore Norway, gave best fit with an elastic thickness of 30 km; see [34]. Calibration with the Barents Sea gave similar results; see [40]. The method used in the isostatic calculations is described in [41] and in the Appendix A. The spatial resolution is 10 km.

There will be significant subsidence of potential sedimentary basins formerly covered by ice sheets. Sedimentary basins that are located in the periphery of the former ice sheet will be notably affected by tilting of the Earth's surface as the subsidence (or uplift after the melting) is gradually decreasing towards the periphery of the former ice sheet. Therefore, the largest tilts of the Earth's surface are found in the more peripheral areas of former ice sheets. This is clearly seen along the entire Norwegian coast and in the Barents Sea (Figure 2). Sedimentary basins located in the peripheral areas in the SW Barents Sea could be tilted by up to 2.7 m/km due to the LGM ice sheet (Figure 2), and even more in the previous greater glaciations [3]. The ice sheets will act as seesaw during the total glaciation period (Quaternary); the Earth will subside under the ice sheets and rise again in interglacial periods.

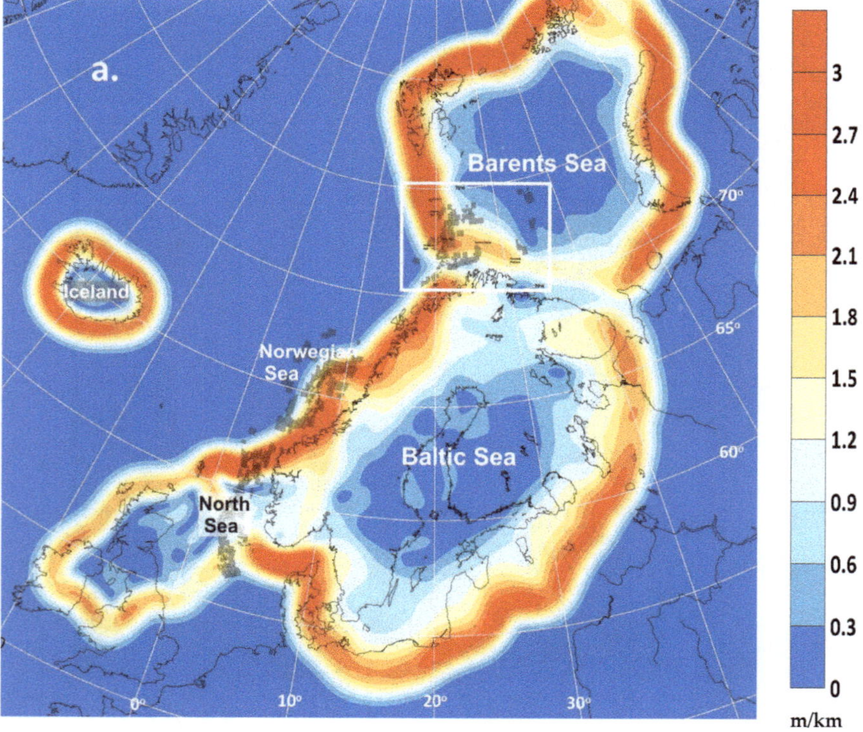

Figure 2. *Cont.*

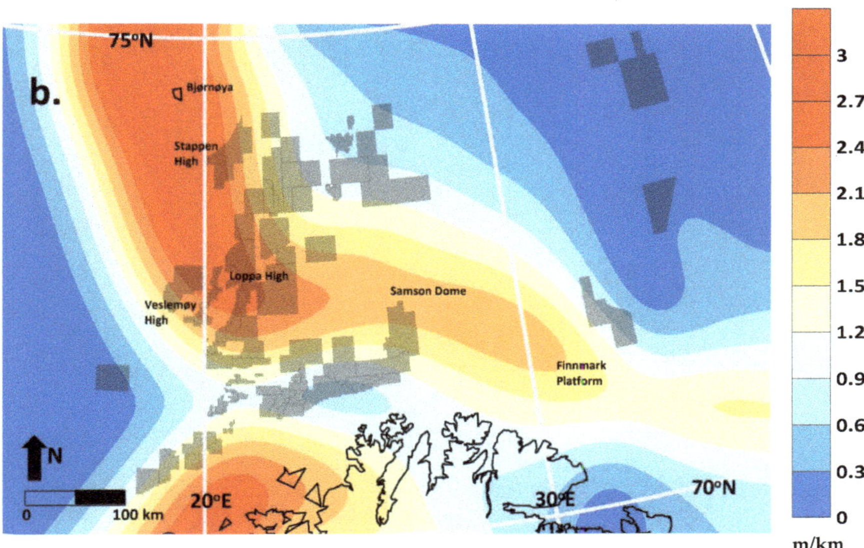

Figure 2. Calculated tilts (m/km) due to glacial isostasy for the LGM ice sheet (based on [3]). The white box in **a**) marks the Norwegian part of the Barents Sea that is open for petroleum exploration (production licenses are marked in grey). The tilts of this area are shown in **b**).

This process is, of course, important for migration modelling. Figure 3 shows how a simple structure behaves during one glacial cycle. It is assumed, for illustration purposes, that the structure is filled with hydrocarbons to the spill point at the onset of glaciation. When the lithosphere and the structure subside due to the glacial load, the reservoir is tilted hydrocarbons start to spill out of the structure. This goes on until the maximum tilting is reached (Figure 3, upper right). The closure volume will be reduced.

When the lithosphere and structure are uplifted due to deglaciation, a new hydrocarbon-water contact (HCWC) will be established (Figure 3, bottom). Paleo-HCWCs can theoretically be found at the spill point level for the positions before and after deglaciation. Residual oil can be found in the water zone down to the original spill point level. In a gas-filled reservoir with an oil leg originally, residual oil can also be found in the gas-filled part, limited by the uppermost HCWC.

The effect of the glacially induced hydrocarbon migration is determined by several factors:

- The number of glacial events. Tilting will result from each isostatic event. A number of glacial events will create chaotic patterns of paleo contacts and residual oil which will be very difficult to interpret.
- The geometry of the receiving area. If a saddle area is opened, the fluids can migrate into higher lying traps or be lost to the surface. A surrounding flat area, on the other hand, can become part of the trap during the subsidence, and fluid can migrate back into the trap under uplift.
- Additional potential factors that determine the migration effect are: i) the geometry of the deformation, ii) the geometry of the hydrocarbon filled structures, iii) the initial fill of the trap, and iv) the spill point location and orientation of the structure.

The isostatic response on the Pleistocene sediment redistribution and ice loading has been evaluated for the Bjørnøyrenna Fault Complex in the Barents Sea [11]. The isostatic impact on hydrocarbon trap capacity changes and hydrocarbon maximum spillage was assessed in that study and the conclusion was that tilting together with gas volume expansion might have been responsible for some part of the hydrocarbon loss during the Cenozoic.

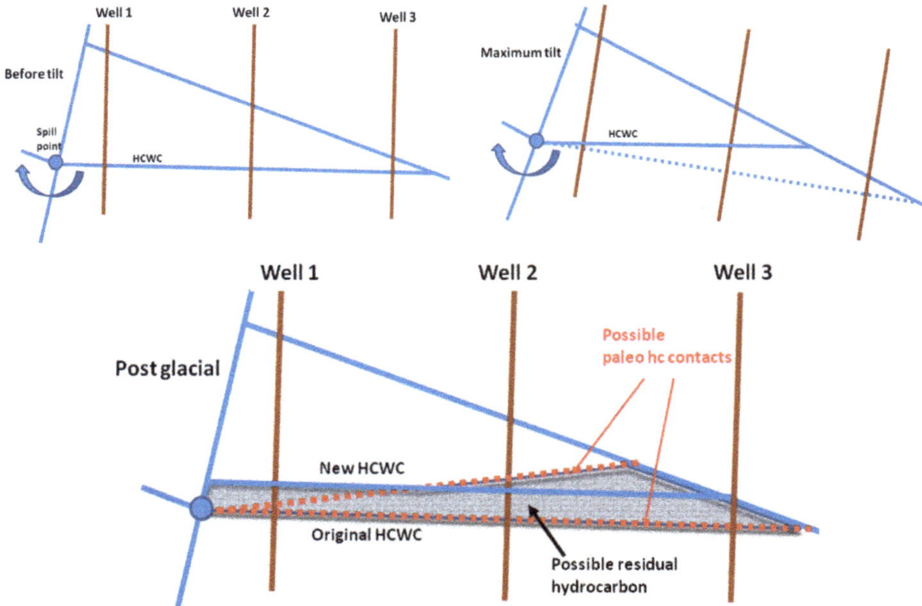

Figure 3. Illustration of tilting during one glacial cycle (redrawn from [5]). **Upper left** shows the situation prior to tilting; **upper right** during tilting. **Lower panel** shows the situation after deglaciation. "Spill point" is the structurally lowest point in a hydrocarbon trap that can retain hydrocarbons. Once a trap has been filled to its spill point, further storage or retention of hydrocarbons will not occur. The illustration assumes the trap filled to the spill point. HCWC is the hydrocarbon-water contact.

3. Flexure-Related Stress Effects

As mentioned above, the flexural stresses that occur during glaciations will be related to the isostatic response of the lithosphere, which is a time-dependent process. The lithosphere is subsiding under the weight of ice, and uplifted back to original state during unloading. During this process, the stresses in the elastic part of the lithosphere will change.

The estimation of stress regime in the lithosphere due to flexural stress during glaciations depends on the timing of crustal stress equilibrium. Two elastic models have been proposed for the estimation of stress changes by flexural stresses during glaciations. Stephansson [42] assumes that the lithosphere is in equilibrium before the ice load is introduced (Figure 4, left). The resulting stress state during ice loading then becomes compressive underneath the ice sheet, and extensional outside the margin of the ice sheet (forebulge area). During ice withdrawal, the horizontal stresses slowly diminish until the uplift ceases and a new equilibrium is reached. Stein et al. [43] present a model where the lithosphere is assumed to have reached equilibrium after some time of ice loading (Figure 4, right). Thus, during melting and uplift the area originally covered by ice experiences extensional stresses, and the area outside the former margin of the ice sheet becomes subject to compressive stresses. These stress changes decrease with depth and are inverted in the lower half of the lithosphere.

If we assume that the duration of ice loading and unloading is long enough for the lithosphere to reach equilibrium during both glaciation and deglaciation, then both models are applicable during one glacial cycle. The average duration of glaciations and interglacials during Quaternary has been 40,000 to 100,000 years. The last ice age ended around 10,000 years ago, and at present the crust is nearly in isostatic equilibrium. The assumption of crustal equilibrium during both glaciations and interglacials is therefore not unrealistic, as the relaxation time for the Fennoscandian uplift is less than 4000 years [34,44].

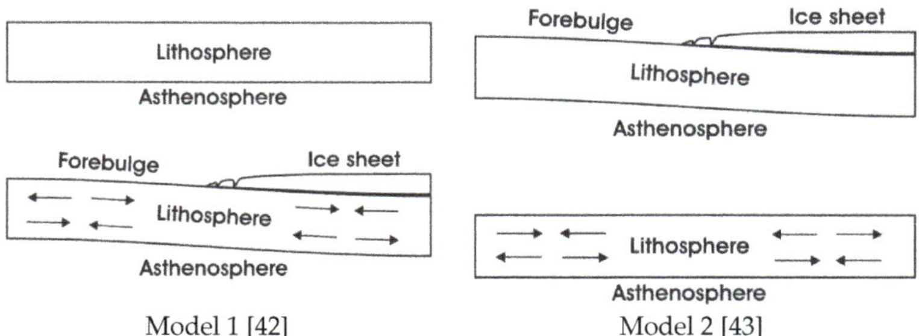

Figure 4. Left: Flexural stress model ([42]) assuming crustal stress equilibrium prior to the ice load. The resulting stresses are large during loading and diminish during unloading. The resulting stress state is compressional beneath the ice sheet and extensional in the hinge zones. **Right:** Flexural stress model ([43]) assuming crustal stress equilibrium after a period of ice loading. From [28].

3.1. Identification of Glacially Flexured Areas

To identify if a certain study area has been affected by glacially-induced flexure, we need information of the glaciation history. The lithospheric flexure is calculated based on the glacial isostasy (Figure 5), which is derived from the ice extent and thickness, as well as the properties and thickness of the lithosphere and mantle.

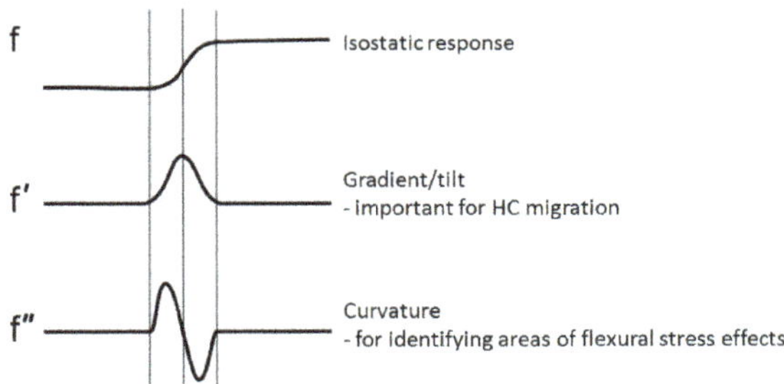

Figure 5. Derivative of the glacial isostasy used to identify areas of high flexural stresses.

Here, we will provide numerical models of flexure-related stress effects in the Hoop Fault Complex, located in the SW Barents Sea (Figure 6). The modelling will be limited to the stress effects related to the isostatic response during the deglaciation after LGM. For identifying the areas of SW Barents Sea that are affected by flexural stresses, we use the glacial isostasy of the LGM ice sheet calculated in Section 2 as input.

Based on the resulting glacial isostasy, we can identify the areas of maximum bending of the crust, and the hinge zones with minimum flexure. The derivation results of a function are illustrated in Figure 5. The first function illustrates the glacial isostasy on a vertical section through the lithosphere near the ice margin. The gradient (or tilt) can be derived from the first derivative of the glacial isostasy, the curvature or flexure of the lithosphere can be found by the second derivative.

The second derivative of the glacial isostasy of the LGM ice for the SW Barents Sea is shown in Figure 6. According to this, it is clear that the SW Barents Sea was greatly affected by lithospheric flexure during the deglaciation after LGM. There is a pronounced hinge zone (white area) extending from west of Bjørnøya over the Stappen High and into Loppa High. From here it extends eastwards over the Samson Dome onto the Finnmark Platform. Areas in the hinge zone are less likely to be affected by the flexural stress after the LGM deglaciation.

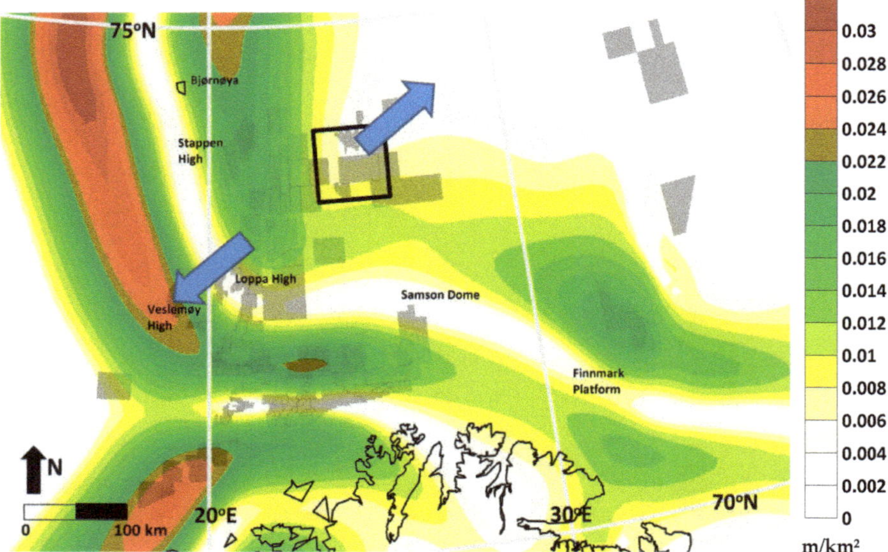

Figure 6. Second derivative, or curvature (m/km²), of the isostatic response of the LGM ice sheet (Figure 1) in SW Barents Sea license area (white square of Figure 2). Yellow, orange, and green areas have high curvature and are likely to be influenced by stress effects during and after the glaciation. (cf. Figure 5). The black square marks the area included in the stress modeling, and the blue arrows show the orientation of maximum flexure-related extension after LGM deglaciation within that area.

3.2. Model Setup

The stress modelling is done in 2D using the state of plane stress, and carried out using the commercial software Comsol Multiphysics 5.2 and the add-on structural mechanics module. Comsol Multiphysics is a FEA (Finite Element Analysis) software, where the problem domain is divided by a mesh of triangular elements in which the calculations are done. The set-up of a stress model in Comsol Multiphysics includes defining model geometry, then adding material properties, loads and boundary conditions.

The model geometry includes faults geometries based on a published continental shelf map of Norway [45] with scale 1:2 mill, interpreted at the base of Top Jurassic. The model includes an area of approximately 300 × 300 km in the SW Barents Sea that was covered by ice during LGM. All fault geometries are manually traced in a CAD tool, and then imported into Comsol Multiphysics.

Every geometry unit of the model is assigned material properties, that is, Young's modulus and Poisson's ratio for the lithologies included. In the models presented here, the host rock is modelled as soft shale with a Young's modulus of 5 GPa and a Poisson's ratio of 0.25, which is within the typical values for shales [46]. Fault zones normally consist of numerous fractures and smaller faults which lower the general stiffness, or Young's modulus, of the host rock [47]. The faults are therefore given a Young's modulus of 0.1 GPa and Poisson's ratio of 0.25.

Next, boundary conditions are assigned to the model. As the model represents the base of a geological unit at depth, the sides of the model cannot deform freely. Two opposite boundaries of the model are therefore fixed. The other two boundaries are assigned a load representing the flexure-related extension originating from the isostatic response during the deglaciation after LGM. The orientation of the flexure-related extension is perpendicular to the curvature trends, with an orientation ~SW–NE. (Figure 6). In a study of the post-glacial lithospheric flexure in the North Sea, analytical and numerical calculation of glacial unloading (assuming isostatic equilibrium before the onset of unloading) of a 1 km thick ice sheet and a lithosphere with elastic thickness of 30 km resulted in bending stresses in the order of 20 MPa [30]. We have used 5 MPa, which is a more conservative magnitude for this effect. The model does not take into consideration possible tectonic background stress at the time of glacial unloading. This modelling may therefore be considered as a study of the effect of stress changes due to the glacially induced flexure.

3.3. Results

In the resulting models (Figure 7), the stress is calculated based on the assumption of a linear elastic material. Rocks normally behave elastic up to 1%–2% strain at low temperature and pressure [46,48,49]. The elastic behavior of rocks is limited by their strength, that is, the maximum applied compressive or tensile stress the rock can withstand before failure occurs. The tensile strength of most rocks is in the range 0.5 to 6 MPa, most commonly 2–3 MPa [46,50]. This means that crustal tensile stress exceeding these values is not very common, as it is likely to result in tensile failure of the rock. The shear strength of rocks is twice their tensile strength, as follows from the Griffith failure criterion [46,51]. Thus, in the models, the tensile stress is truncated at 6 MPa, and the shear stress at 12 MPa. Areas concentrating stress above these values are likely to fail in tension or shear. However, due to uncertainties related to the material properties for the involved sedimentary units and the flexure-related load, the stress magnitudes in the models below should be read in terms of relative values. The stress is likely to be released by failure in the high magnitude areas (marked in yellow and red), and less likely to lead to failure in low magnitude areas (marked in blue).

The first model (Figure 7, top) shows the resulting maximum principal tensile stress (σ_3) when the area is subject to extension during LGM deglaciation. Tensile stress can be described as stretching stress, and can lead to opening up of new extension fractures (when the tensile strength of the rock is exceeded) or opening up of preexisting fractures. The SW–NE extension causes large areas between the faults to concentrate high magnitude tensile stress, whereas other parts of the host rock are located in a "stress shadow". Here, the host rock surrounding SW–NE trending faults accumulate high magnitude tensile stresses, whereas the fault zones themselves show minimal effect. This indicates that due to the LGM deglaciation, the host rock surrounding SW–NE trending faults is likely to experience increased local fracture-related permeability, which may lead potential hydrocarbons towards and along the faults. Focusing on the exploration wells, the result suggests that wells 7324/2-1 and 7324/8-2 are placed within areas of high fracture-related permeability during the LGM deglaciation. Potential hydrocarbon may have migrated along faults out of a potential reservoir during the deglaciation. According to the results, the other wells are located in areas of low tensile stress, and therefore less likely to have been affected by flexural stresses during and after LGM.

The von Mises shear stress results (Figure 7, bottom) show similar tendency. Most of the wells are located in areas of low magnitude shear stress, whereas wells 7324/2-1 and 7324/8-2 are located in areas of higher shear stress magnitudes. Well data for exploration wells drilled in the area show that all wells included here are oil or gas discoveries, except for well 7324/2-1 and 7324/8-2 which are dry with shows, indicating former presence of hydrocarbons.

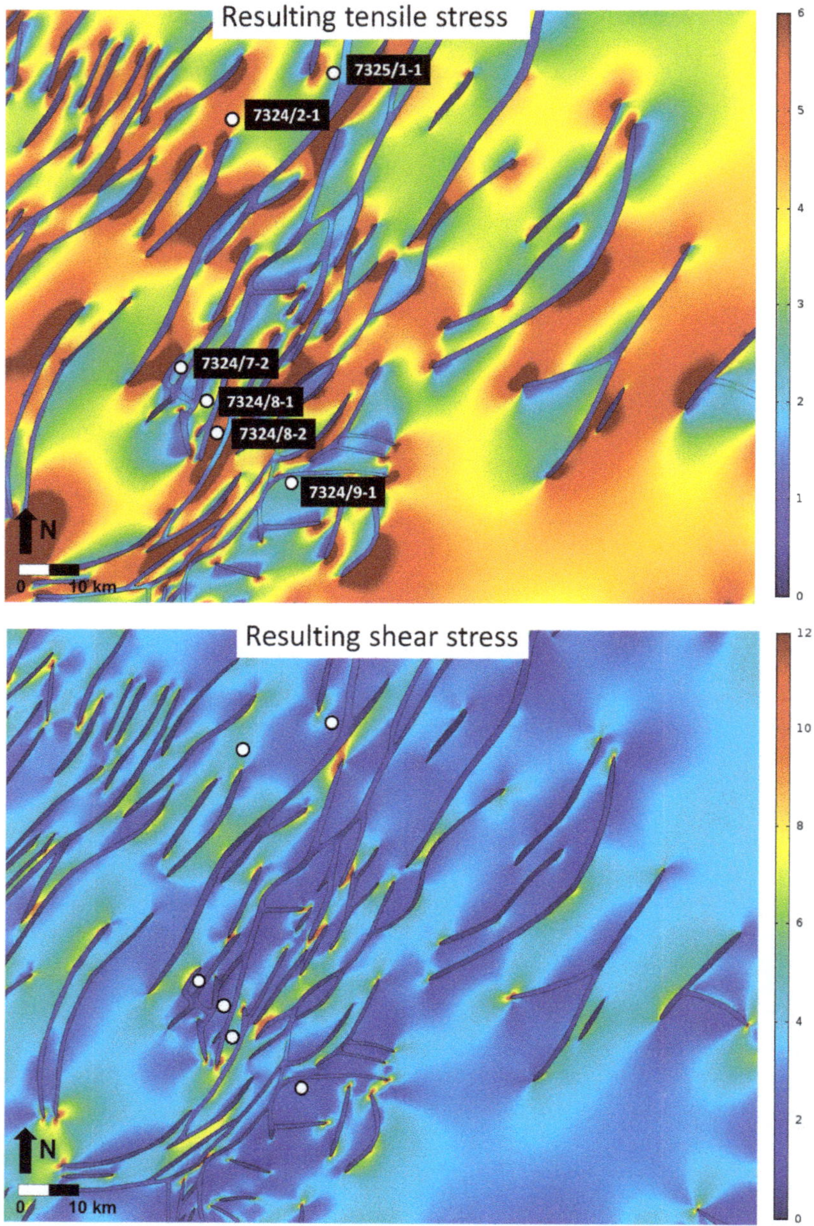

Figure 7. Stress models covering an area of 120 × 90 km in the Hoop Fault Complex, SW Barents Sea. The results include related exploration wells with data made public prior to 2017. **Top**: Resulting distribution of maximum principal tensile stress (σ_3) during SW–NE extension due to deglaciation of LGM. **Bottom**: Resulting distribution of von Mises shear stress during flexure-related SW–NE extension. Scale in megapascals.

4. Discussion

4.1. Glaciation Model

Our glaciation model is not intended to include time-varying ice sheet geometries. However, LGM was the ice sheet configuration that was latest influencing the SW Barents Sea. The SW Barents Sea was glaciated around 24,000 BP and starting to melt away at 18,000 BP [36]. The LGM ice sheet is thus a reasonable ice sheet configuration for modelling the flexural stress effects in SW Barents Sea.

4.2. Earth Model

In the isostatic calculations we have assumed an elastic lithosphere thickness of 30 km (flexural rigidity 5×10^{23} Nm). As mentioned above this Earth model has been shown to match other Fennoscandian post-glacial data, but most of the published GIA Earth models of Fennoscandia have a thick elastic lithosphere, between 80 and 160 km [34,39]. However, a thick lithosphere is not a viable option for data on tilted palaeo shorelines of coastal southwestern Norway and not for the Barents Sea [34,52]. However, a recent study of the Tapes transgression along the Norwegian coast indicates a thicker effective elastic lithosphere thickness of 70 km (flexural rigidity 3×10^{24} Nm) in Northern Norway [53].

The glacial isostatic tilt caused by a 70 km thick elastic lithosphere is shown in Figure 8. Sedimentary basins located in SW Barents Sea could be tilted by up to 2.0 m/km due to the LGM ice sheet, which is about 70% of the tilt with lithospheric thickness of 30 km (cf. Figure 2). The flexural stress due to a 70 km thick lithosphere is thus somewhat lower than for a thinner lithosphere, but the stresses will be located basically in the same area. We believe, however, that a model with thin elastic lithosphere of 30 km better describes the Earth rheology of the Barents Sea than a model with thicker elastic lithosphere (see [52,54]).

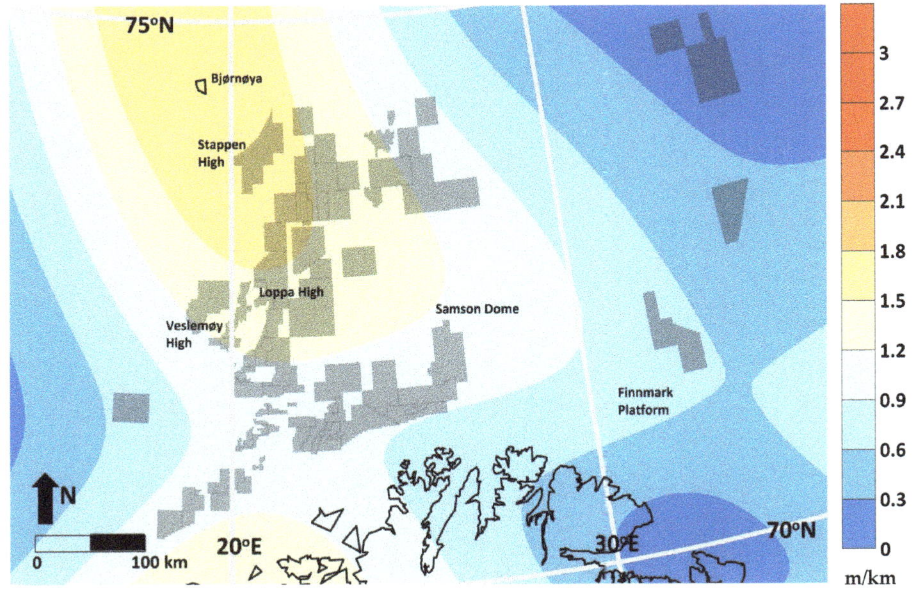

Figure 8. The isostatic tilts for lithosphere thickness of 70 km.

4.3. Stress Models

The way the stress distributes in a material is depending on existing fractures and weaknesses and their orientation to the added load. In the stress model examples we have shown here, a published regional interpreted fault map was used as basis. A more detailed fault interpretation will be needed for a more local study of the stress effects of glacially induced flexure.

4.4. Implications for Basin Modeling

Repeated ice loading during the Quaternary could have significant impact on sedimentary basins and add a degree of difficulty to petroleum exploration. The isostatic effects of ice sheets, (in addition to possible glacial erosion/deposition) could give significant tilting of sedimentary basins in the peripheral areas of former glaciated area. The tilting of the reservoirs at the Norwegian Continental Shelf could be up to almost 3 m/km and could alter hydrocarbon migration pathways. The ice sheets will act as seesaw during glacial/interglacial period, while the isostatic response of erosion and sedimentation will grow iteratively during the glaciations, depending on the stability of the depocenters.

Differential vertical movement of the lithosphere related to glacial isostasy leads to repeated tilting of sedimentary formations and potential petroleum reservoirs therein, which may have greatly affected hydrocarbon migration pathways in the former glaciated area [3]. In addition, the upward and downward bending of the lithosphere lead to flexural stresses likely to affect faults and their permeability, which could add to changes in hydrocarbon migration pathways. The flexural stresses are adding to background stresses in the area.

The map of the flexural effect based on glacial isostasy shows that the SW Barents Sea is affected by flexural stresses due to the last ice age. Stress models suggest that this effect may have caused high magnitude stresses to accumulate and therefore lead to an increased fracture-related permeability and escape of hydrocarbons in certain areas.

Most of the production licenses on the Norwegian Continental Shelf are located inside the former Last Glacial Maximum ice sheet (Figure 1). The results of our study demonstrate the importance of including the effect of glaciations in basin modeling studies of areas of petroleum potential. Adding to this are the temperature effects of glaciations and the effects of glacial erosion on temperature and stress discussed in more detail in [3].

Author Contributions: Conceptualization, I.F.L., W.F.; methodology, I.F.L., W.F., M.S.; software, I.F.L., W.F.; investigation, I.F.L., W.F., M.S.; writing—original draft preparation, I.F.L., W.F.; writing—review and editing, I.F.L., W.F.; M.S.; project administration, I.F.L.

Funding: Part of this study was supported through project no. 200657 "Neogene Uplift of the Barents Sea" by Norwegian Research Council and the oil companies ConocoPhillips, Det Norske, E.ON E&P, GDF Suez E&P Norge AS, NORECO, OMV (Norge), Statoil, Total E&P and Wintershall.

Acknowledgments: The authors would like to thank four anonymous reviewers for constructive and fruitful feedback, which improved the paper.

Conflicts of Interest: The authors declare no conflict of interest. The funders had no role in the design of the study; in the collection, analyses, or interpretation of data; in the writing of the manuscript, or in the decision to publish the results.

Appendix A

Glacial Isostatic Model

The ultimate isostatic compensation achieved by any harmonic component is given by [44,55]:

$$h(k) \;=\; \frac{F(k)\alpha^{-1}}{\rho g} \tag{A1}$$

where F(k) = transformed ice load, ρ = density of the upper mantle, and g = gravity.

The 'lithosphere filter' is $\alpha = \frac{2\mu k}{\rho g}\left(S^2 - k^2 H^2\right) + (CS + kH)]/(S + kHC)$, where k = wavenumber, H = elastic thickness of the lithosphere, μ = Lame's parameter, S = sinh kH and C = cosh kH. The relation between the flexural rigidity (D) and the elastic thickness of the lithosphere (H) is given by the equation $D = \frac{EH^3}{12(1-v^2)}$, where E = Young's modulus and v = Poisson's ratio.

References

1. Gibbard, P.L.; Head, M.J.; Walker, M.J.C. Formal ratification of the quaternary system/period and the pleistocene series/Epoch with a base at 2.58 Ma. *J. Quat. Sci.* **2009**, *25*, 96–102. [CrossRef]
2. Mangerud, J.; Gyllencreutz, R.; Lohne, Ø.; Svendsen, J.I. Glacial history of Norway. In *Developments in Quaternary Science*; Ehlers, J., Gibbard, P.L., Hughes, P.D., Eds.; Elsevier: Amsterdam, The Netherlands, 2011; Volume 15, pp. 279–298, ISBN 978-0-444-53447-7.
3. Fjeldskaar, W.; Amantov, A. Effects of glaciations on sedimentary basins. *J. Geodyn.* **2018**, *118*, 66–81. [CrossRef]
4. Argus, D.F.; Drummond, R.; Peltier, W.R. Space geodesy constrains ice age terminal deglaciation: The global ICE-6G_C (VM5a) model. *J. Geophys. Res. Solid Earth* **2015**, *120*, 450–487. [CrossRef]
5. Kjemperud, A.; Fjeldskaar, W. Pleistocene glacial isostasy-implications for petroleum geology. In *Structural and Tectonic Modelling and its Application to Petroleum Geology*; Larsen, R.M., Brekke, H., Larsen, B.T., Talleraas, E., Eds.; Elsevier: Amsterdam, The Netherlands, 1992; pp. 187–195.
6. Grunnaleite, I.; Fjeldskaar, W.; Wilson, J.; Faleide, J.; Zweigel, J. Effect of local variations of vertical and horizontal stresses on the Cenozoic structuring of the mid-Norwegian shelf. *Tectonophysics* **2009**, *470*, 267–283. [CrossRef]
7. Vadakkepuliyambatta, S.; Bünz, S.; Mienert, J.; Chand, S. Distribution of subsurface fluid-flow systems in the SW Barents Sea. *Mar. Pet. Geol.* **2013**, *43*, 208–221. [CrossRef]
8. Cavanagh, A.J.; Di Primio, R.; Scheck-Wenderoth, M.; Horsfield, B. Severity and timing of Cenozoic exhumation in the southwestern Barents Sea. *J. Geol. Soc.* **2006**, *163*, 761–774. [CrossRef]
9. Doré, A.G.; Jensen, L.N. The impact of late Cenozoic uplift and erosion on hydrocarbon exploration: offshore Norway and some other uplifted basins. *Glob. Planet. Change* **1996**, *12*, 415–436. [CrossRef]
10. Duran, E.R.; Di Primio, R.; Anka, Z.; Stoddart, D.; Horsfield, B. 3D-basin modelling of the Hammerfest Basin (southwestern Barents Sea): A quantitative assessment of petroleum generation, migration and leakage. *Mar. Pet. Geol.* **2013**, *45*, 281–303. [CrossRef]
11. Zieba, K.J.; Grøver, A. Isostatic response to glacial erosion, deposition and ice loading. Impact on hydrocarbon traps of the southwestern Barents Sea. *Mar. Pet. Geol.* **2016**, *78*, 168–183. [CrossRef]
12. Lerche, I.; Yu, Z.; Tørudbakken, B.; Thomsen, R. Ice loading effects in sedimentary basins with reference to the Barents sea. *Mar. Pet. Geol.* **1997**, *14*, 277–338. [CrossRef]
13. Stoddart, D.; Jørstad, A.; Rønnevik, H.-C.; Fjeldskaar, W.; Løtveit, I.F. Recent glacial events in the Norwegian North Sea-implications towards a better understanding of charging/leakage of oil fields and its impact oil exploration. In Proceedings of the IMOG 2015—27th International Meeting on Organic Geochemistry, Prague, Czech Republic, 13–18 September 2015. [CrossRef]
14. Medvedev, S.; Hartz, E.H.; Schmid, D.W.; Zakariassen, E.; Varhaug, P. *Influence of Glaciations on North Sea Petroleum Systems*; Geological Society: London, UK, Special Publications; 2019; Volume 494, p. SP494-2018-2183. [CrossRef]
15. Mikko, H.; Smith, C.A.; Lund, B.; Ask, M.V.; Munier, R. LiDAR-derived inventory of post-glacial fault scarps in Sweden. *GFF* **2015**, *137*, 334–338. [CrossRef]
16. Dehls, J.F.; Olesen, O.; Olsen, L.; Blikra, L.H. Neotectonic faulting in northern Norway; the Stuoragurra and Nordmannvikdalen postglacial faults. *Quat. Sci. Rev.* **2000**, *19*, 1447–1460. [CrossRef]
17. Lagerbäck, R. Neotectonic structures in northern Sweden. *Geol. Foren. Stockh. Förh.* **1979**, *100*, 271–278. [CrossRef]
18. Brandes, C.; Winsemann, J.; Roskosch, J.; Meinsen, J.; Tanner, D.C.; Frechen, M.; Steffen, H.; Wu, P. Activity along the Osning Thrust in Central Europe during the Lateglacial: Ice-sheet and lithosphere interactions. *Quat. Sci. Rev.* **2012**, *38*, 49–62. [CrossRef]
19. Sandersen, P.B.E.; Jørgensen, F. Neotectonic deformation of a Late Weichselian outwash plain by deglaciation-induced fault reactivation of a deep-seated graben structure. *Boreas* **2015**, *44*, 413–431. [CrossRef]
20. Grube, A. Palaeoseismic structures in Quaternary sediments, related to an assumed fault zone north of the Permian Peissen-Gnutz salt structure (NW Germany)—Neotectonic activity and earthquakes from the Saalian to the Holocene. *Geomorphology* **2019**, *328*, 15–27. [CrossRef]
21. Hoffmann, G.; Reicherter, K. Soft-sediment deformation of late Pleistocene sediments along the southwestern coast of the Baltic Sea. *Intern. J. Earth Sci.* **2012**, *101*, 351–363. [CrossRef]

22. Van Loon, A. (Tom); Pisarska-Jamroży, M. Sedimentological evidence of Pleistocene earthquakes in NW Poland induced by glacio-isostatic rebound. *Sediment. Geol.* **2014**, *300*, 1–10. [CrossRef]
23. Stewart, I.S.; Firth, C.R.; Rust, D.J.; Collins, P.E.; Firth, J.A. Postglacial fault movement and palaeoseismicity in western Scotland: A reappraisal of the Kinloch Hourn fault, Kintail. *J. Seism.* **2001**, *5*, 307–328. [CrossRef]
24. Fejerskov, M.; Lindholm, C. Crustal stress in and around Norway: an evaluation of stress-generating mechanisms. *Geol. Soc. London Spéc. Publ.* **2000**, *167*, 451–467. [CrossRef]
25. Druzhinina, O.; Bitinas, A.; Molodkov, A.; Kolesnik, T. Palaeoseismic deformations in the Eastern Baltic region (Kaliningrad District of Russia). *Estonian J. Earth Sci.* **2017**, *66*, 119–129. [CrossRef]
26. Antonovskaya, G.; Konechnaya, Y.; Kremenetskaya, E.O.; Asming, V.; Kvaerna, T.; Schweitzer, J.; Ringdal, F. Enhanced earthquake monitoring in the European Arctic. *Polar Sci.* **2015**, *9*, 158–167. [CrossRef]
27. Gibbons, S.J.; Antonovskaya, G.; Asming, V.; Konechnaya, Y.V.; Kremenetskaya, E.; Kværna, T.; Schweitzer, J.; Vaganova, N.V. The 11 October 2010 Novaya Zemlya Earthquake: Implications for Velocity Models and Regional Event Location. *Bull. Seismol. Soc. Am.* **2016**, *106*, 1470–1481. [CrossRef]
28. Neuzil, C.E. Hydromechanical effects of continental glaciation on groundwater systems. *Geofluids* **2012**, *12*, 22–37. [CrossRef]
29. Steffen, R.; Steffen, H.; Wu, P.; Eaton, D.W. Reply to comment by Hampel et al. on "Stress and fault parameters affecting fault slip magnitude and activation time during a glacial cycle.". *Tectonics* **2015**, *34*, 2359–2366.
30. Grollimund, B.; Zoback, M.D. Post glacial lithospheric flexure and induced stresses and pore pressure changes in the northern North Sea. *Tectonophysics* **2000**, *327*, 61–81. [CrossRef]
31. Johnston, A.C. Suppression of earthquakes by large continental ice sheets. *Nature* **1987**, *330*, 467–469. [CrossRef]
32. Turpeinen, H.; Hampel, A.; Karow, T.; Maniatis, G. Effect of ice sheet growth and melting on the slip evolution of thrust faults. *Earth Planet. Sci. Lett.* **2008**, *269*, 230–241. [CrossRef]
33. Steffen, R.; Steffen, H.; Wu, P.; Eaton, D.W. Stress and fault parameters affecting fault slip magnitude and activation time during a glacial cycle. *Tectonics* **2014**, *33*, 1461–1476. [CrossRef]
34. Fjeldskaar, W.; Amantov, A. Tilted Norwegian post-glacial shorelines require a low viscosity asthenosphere and a weak lithosphere. *Reg. Geol. Metallogeny* **2017**, *70*, 48–59.
35. Larsen, E.; Fredin, O.; Lyså, A.; Amantov, A.; Fjeldskaar, W.; Ottesen, D. Causes of time-transgressive glacial maxima positions of the last Scandinavian Ice Sheet. *Nor. J. Geol.* **2016**, *96*.
36. Hughes, A.L.C.; Gyllencreutz, R.; Lohne, Ø.S.; Mangerud, J.; Svendsen, J.I. The last Eurasian ice sheets—A chronological database and time-slice reconstruction, DATED-1. *Boreas* **2016**, *45*, 1–45. [CrossRef]
37. Zweck, C.; Huybrechts, P. Modeling of the northern hemisphere ice sheets during the last glacial cycle and glaciological sensitivity. *J. Geophys. Res. Space Phys.* **2005**, *110*, 1984–2012. [CrossRef]
38. Auriac, A.; Whitehouse, P.L.; Bentley, M.J. Glacial isostatic adjustment associated with the Barents Sea ice sheet: A modeling inter-comparison. *Quat. Sci. Rev.* **2016**, *147*, 122–135. [CrossRef]
39. Steffen, H.; Wu, P. Glacial isostatic adjustment in Fennoscandia—A review of data and modeling. *J. Geodyn.* **2011**, *52*, 169–204. [CrossRef]
40. Amantov, A.; Fjeldskaar, W. Meso-Cenozoic exhumation and relevant isostatic process: The Barents and Kara shelves. *J. Geodyn.* **2018**, *118*, 118–139. [CrossRef]
41. Fjeldskaar, W. The flexural rigidity of Fennoscandia inferred from the post-glacial uplift. *Tectonics* **1997**, *16*, 596–608. [CrossRef]
42. Stephansson, O. *Stress Measurements And Modelling of Crustal Rock Mechanics in Fennoscandia*; In Earthquakes at North-Atlantic Passive Margins: Neotectonics and Postglacial, Rebound, Gregersen, S., Basham, P.W., Eds.; Springer: Dordrecht, The Netherlands, 1989; pp. 213–229.
43. Stein, S.; Cloetingh, S.; Sleep, N.H.; Wortel, R.; Gregersen, S.; Basham, P.W. Passive Margin Earthquakes, Stresses and Rheology. In *Earthquakes at North-Atlantic Passive Margins: Neotectonics and Postglacial Rebound*; Springer: Dordrecht, The Netherlands, 1989; pp. 231–259.
44. Cathles, L.M. *The Viscosity of the Earth's Mantle*; Princeton University Press: Princeton, NJ, USA, 1975; p. 386.
45. Brekke, H.; Kalheim, J.E.; Riis, F.; Egeland, B.; Blystad, P.; Johnsen, S.; Ragnhildstveit, J. *Two-Way Time-Map of the Unconformity at the Base of Top Jurassic (North of 69° N) and the Unconformity at the Base of Cretaceous (South of 69°N), Offshore Norway, Including the Main Geological Trends Onshore. NPD Continental Shelf Map No. 1*; The Norwegian Petroleum Directorate/The Geological Survey of Norway: Stavanger/Trondheim, Norway, 1992.

46. Gudmundsson, A. *Rock Fractures in Geological Processes*, 1st ed.; Cambridge University Press: New York, NY, USA, 2011.

47. Gudmundsson, A.; Simmenes, T.H.; Larsen, B.; Philipp, S.L. Effects of internal structure and local stresses on fracture propagation, deflection, and arrest in fault zones. *J. Struct. Geol.* **2009**, *32*, 1643–1655. [CrossRef]

48. Paterson, M.S.; Paterson, D.M.S. *Experimental Rock Deformation—The Brittle Field*; Springer: Berlin, Germany, 1978; Volume 13.

49. Farmer, I. *Engineering Behaviour of Rocks*; Chapman and Hal: Londor, UK, 1983.

50. Amadei, B.; Stephansson, O. *Rock Stress and its Measurement*; Chapman and Hall: London, UK, 1997; p. 490.

51. Jaeger, J.C.; Cook, N.G.W.; Zimmerman, R.W. *Fundamentals of Rock Mechanics*, 2nd ed.; Chapman and Hall: London, UK, 2007.

52. Fjeldskaar, W.; Bondevik, S.; Amantov, A. Glaciers on Svalbard survived the Holocene thermal optimum. *Quat. Sci. Rev.* **2018**, *199*, 18–29. [CrossRef]

53. Fjeldskaar, W.; Bondevik, S. The Early-Mid-Holocene transgression (Tapes) at the Norwegian coast–comparing observations with numerical modelling. *Quat. Sci. Rev.* (submitted).

54. Medvedev, S.; Hartz, E.H.; Faleide, J.I. Erosion-driven vertical motions of the circum Arctic: Comparative analysis of modern topography. *J. Geodyn.* **2018**, *119*, 62–81. [CrossRef]

55. Fjeldskaar, W.; Cathles, L. Rheology of Mantle and Lithosphere Inferred from Post-Glacial Uplift in Fennoscandia. In *Glacial Isostasy, Sea-Level and Mantle Rheology*; Springer: Berlin, Germany, 1991; pp. 1–19.

Article

Multiscale Modeling of Glacial Loading by a 3D Thermo-Hydro-Mechanical Approach Including Erosion and Isostasy

Daniele Cerroni [1], Mattia Penati [1], Giovanni Porta [2], Edie Miglio [1], Paolo Zunino [1] and Paolo Ruffo [3,*]

[1] MOX, Department of Mathematics, Politecnico di Milano, 20133 Milano, Italy;
    daniele.cerroni@polimi.it (D.C.); mattia.penati@polimi.it (M.P.); edie.miglio@polimi.it (E.M.);
    paolo.zunino@polimi.it (P.Z.)
[2] Department of Civil and Environmental Engineering, Politecnico di Milano, 20133 Milano, Italy;
    giovanni.porta@polimi.it
[3] Eni-Upstream and Technical Services, San Donato Milanese, 20097 Milano, Italy
* Correspondence: paolo.ruffo@eni.com; Tel.: +39-02-520-61777

Received: 12 September 2019; Accepted:23 October 2019 ; Published: 30 October 2019

Abstract: We present a computational framework that allows investigating the Thermo-Hydro-Mechanical response of a representative part of a sedimentary basin during a glaciation cycle. We tackle the complexity of the problem, arising by the mutual interaction among several phenomena, by means of a multi-physics, multi-scale model with respect to both space and time. Our contribution addresses both the generation of the computational grid and the algorithm for the numerical solution of the problem. In particular we present a multi-scale approach accounting for the global deformation field of the lithosphere coupled with the Thermo-Hydro-Mechanical feedback of the ice load on a representative part of the domain at a finer scale. In the fine scale model we also include the erosion possibly caused by the ice melting. This methodology allows investigating the evolution of the sedimentary basin as a response to glaciation cycle at a fine scale, taking also into account the large spatial scale movement of the lithosphere due to isostasy. The numerical experiments are based on the analysis of simple scenario, and show the emergence of effects due to the multi-physics nature of the problem that are barely captured by simpler approaches.

Keywords: multiscale/multiphysics basin modeling; thermo-hydro-mechanical model; isostatic adjustment; computer simulations; finite element method

---

1. Introduction

Reconstructing the stress and deformation history of a sedimentary basin is a challenging and important problem in the geosciences and a variety of applications [1]. The mechanical response of a sedimentary basin is the consequence of complex multi-physics processes involving mechanical, geochemical, geophysical, geological and thermal aspects [2]. The strongly coupled nature of the deformation problem may be understood in terms of the feedbacks underlying crustal dynamics. The pore fluid pressure affects stress, stress changes can lead to fracturing, and fracturing can affect pore fluid pressure [3,4].

Basin scale compaction processes involve mechanical and chemically induced transformations that take place during the accumulation of sediments [2]. In this context a number of approaches have considered the geochemical and mechanical compaction problem from a one-dimensional perspective, i.e., by considering mass, momentum and energy balances along the vertical direction, applied to fluid and solid phases [3–6]. These simplified one-dimensional approaches may be effective in interpreting qualitatively well data (e.g., [7]), however, they cannot capture inherently three dimensional processes that may arise due to the coupling of mechanical deformations and fluid mechanics in geological bodies that play an important role in the presence of glaciations [8].

Hydro-mechanical effects of continental ice sheets are widely recognized to cause movements and stresses of overridden terrains by ice load. The effect of the ice load on top of the sedimentary basin can be represented by the combination of two effects. The first is a large scale effect where we consider the action of the ice load on the entire lithosphere. The second is a fine scale analysis where we take into account the thermo-hydro-mechano-chemical (THMC) effects of the ice load into a small portion of the crust, such as a sedimentary basin [8].

In the global large scale framework, the interaction between the lithosphere and the glaciation cycle is modeled by means of a viscoelastic model. This choice is based on significant previous efforts devoted to define a proper mathematical model for the description of glacial isostatic adjustment. Initially this problem has been considered by Rayleigh [9] which studied the problem of a pre-stressed elastic compressible layer as an approximation of a "flat" planet. After Rayleigh's work other authors enriched his theory, including many other details, like the effect of viscosity or the stratified structure of the Earth. First of all, Love [10] gave a more detailed theory and defined the basic concepts which are included in more recent works. Peltier and his coauthors in a series of articles [11–13] gave a detailed description of a more realistic viscoelastic model of stratified Earth. The Peltier's model is essentially an extension of Love's model, where a viscoelastic rheology is used instead of an elastic one. All the mathematical details of this theory are contained in the works of Biot and, more recently, Ogden [14,15].

In this work we apply these models to describe the global deformation field of the lithosphere and to extract from it the information of the movement of a selected part of the sedimentary basin. Since the spatial scale of such region is very small compared to the global scale, we describe it as a rigid motion. More precisely, the rigid motion of the fine scale basin model is extrapolated from the lithosphere displacement fields and used at run time to move the computational grid in the simulations. In what follows we describe in more details this the workflow of this multiscale approach.

A number of numerical simulation tools have been presented to model THMC processes [16–18]. While considering a similar mathematical approach, our work introduces the following new features with respect to previous studies:

(i)   it combines the global deformation of the lithosphere with the local simulation of pressure and temperature fields;
(ii)   it is built on available geophysical and geological information on the whole sedimentary system and relying on information available at selected wells;
(iii) it allows considering the effects of erosion induced by glaciation, which is often neglected in previous studies.

Our work focuses on the integration of THM simulation of a single glaciation cycle with larger scale information available on basin scale compaction and lithosphere dynamics, thus the proposed THM simulation of glaciations can be cast within a multi-scale geological simulation framework. A visual sketch of the multiscale model outline is provided in Figure 1.

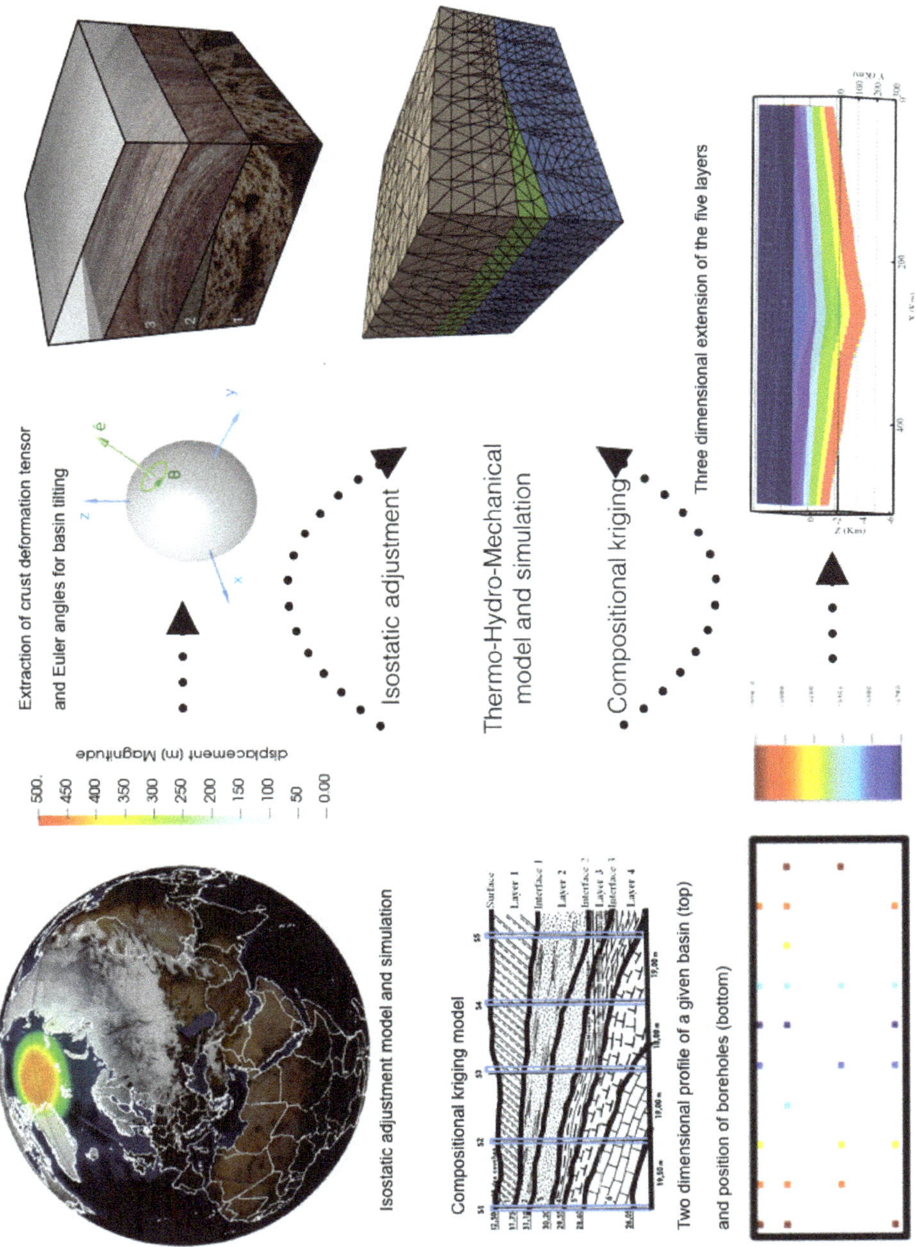

Figure 1. A visual sketch of the multiscale model outline.

2. Isostatic Glacial Rebound Model

From the mechanical point of view the interior of the Earth can be considered as composed of four main layers: the inner and outer core, the mantle and the lithosphere [19]. During the growth of a continental ice sheet, the lithosphere under the ice load is deformed into the mantle and the removal of the ice load during deglaciation initiates a rebound process. The uplift is well known in formerly glaciated areas, e.g., North America and Scandinavia, and in currently deglaciating areas, e.g., Alaska, Antarctica, and Greenland. Compared to water, the mantle viscosity is 10^{22}–10^{25} times higher, therefore the uplifting will be slowed down and continue long time after the ice has gone. The entire process of subsidence during the glacial growth, followed by uplift during and after deglaciation, is referred to as glacial isostatic adjustment.

The glacial isostatic adjustment process is dependent on the viscosity structure of the mantle, as well as the elastic thickness of the lithosphere. Observations of this process can therefore be used to gain insight into these properties of the Earth and this is important for an understanding of the dynamics of the Earth's interior.

A well established assumption for the computation of the solid Earth response to surface ice loads over glacial timescales is that the Earth can be considered as a viscoelastic body ([20]). In particular the lithosphere can be assumed to be elastic and the solid mantle beneath behaves as a viscous fluid.

A complete review of the state of the art concerning the modeling and simulation of the glacial rebound can be found in [21,22] whereas the importance of this phenomenon in the context of basin simulation is discussed in [23–25].

In the next section the physical model adopted for the simulation of the post-glacial rebound is presented; the details of the numerical method can be found in Appendix A.

A Viscoelastic Model for the Earth

In accordance with the previously cited works the Earth has been modeled has a linear viscoelastic spherical shell $\Omega \subset \mathbb{R}^3$ since the dynamics due to the glacial isostatic adjustment does not involve the core of the planet. Following the approach presented in [26] we assume that the viscoelastic stress tensor is given $\sigma = \sigma^e - q$, where σ^e is the elastic stress and tensor q is an internal variable used to model the effects of viscosity. Denoting with $\mathbf{u}$ the displacement field, the elastic stress tensor σ^e is given by the sum of the deviatoric and the volumetric parts

$$\sigma^e = 2\mu\, e(\mathbf{u}) - pI\,,$$

where I denotes the identity matrix. The deviatoric part is the product between the shear modulus μ and the deviatoric strain $e(\mathbf{u})$ defined as

$$e(\mathbf{u}) := \frac{1}{2}(\nabla\mathbf{u} + \nabla\mathbf{u}^t) - \frac{1}{3}(\nabla \cdot \mathbf{u})I.$$

The volumetric part depends on the pressure p which is defined by the equation

$$\nabla \cdot \mathbf{u} + \frac{3}{2\mu}\frac{1-2v}{1+v}p = 0, \tag{1}$$

where v is the Poisson ratio.

The internal variable q is defined by the evolution equation

$$\begin{cases} \dot{q} + \dfrac{1}{\tau}q = \dfrac{1}{\tau}2\mu\, e(\mathbf{u}), \\[2mm] q(0) = 0, \end{cases} \tag{2}$$

where $\dot{q}$ denotes the derivative of the quantity q with respect to time and τ is called relaxation time and it is related to the viscosity η through the relation $\tau = \frac{\eta}{2\mu}$.

The Equation (2) can be rewritten in the integral form as:

$$q(t) = \int_0^t \frac{1}{\tau}e^{-\frac{t-s}{\tau}}2\mu\, e(\mathbf{u}(s))\, ds.$$

By means of the previous expressions the viscoelastic stress tensor σ is evaluated through a convolution integral defined as

$$\sigma(t) = -pI + \int_0^t e^{-\frac{t-s}{\tau}} 2\mu\, e(\dot{\mathbf{u}}(s))\, ds\,.$$

This equation coupled with the Equation (1) and with the equation of conservation of linear momentum give us a system of partial differential equations that describe the motion of a viscoelastic body:

$$\begin{cases} \nabla \cdot \sigma + \mathbf{f} = 0 & \text{in } \Omega, \\[2mm] \nabla \cdot \mathbf{u} + \dfrac{3}{2\mu}\dfrac{1-2v}{1+v}p = 0 & \text{in } \Omega, \\[2mm] \sigma = -pI + \displaystyle\int_0^t e^{-\frac{t-s}{\tau}} 2\mu\, e(\dot{\mathbf{u}}(s))\, ds & \text{in } \Omega. \end{cases} \tag{3}$$

The unknowns of this system of equations are the displacement field $\mathbf{u}$, the pressure field p and the stress tensor field σ. The volumetric force field $\mathbf{f}$ is the gravitational force field, how this term has been modeled will be discussed later.

Since our domain is a spherical shell its boundary is the union of two connected components: the inner and the outer surfaces of the shell. The outer surface Γ_{out} is the surface of the Earth and on this portion of the boundary we assume to know the history of the load due to the presence of the ice or other type of loads, like sediments. According to these data the following boundary condition is assumed: $\sigma\mathbf{n} = \mathbf{s}_{\text{load}}$ on Γ_{out}, where $\mathbf{n}$ denotes the outer normal defined on the boundary. The inner surface Γ_{in} of the shell is chosen in such a way it corresponds to the core-mantle boundary (about 2900 km of depth). On this portion of the boundary the displacement $\mathbf{u}$ is assumed to be equal to zero, since the deformation due to the glacial isostatic adjustment involves only the shallow part of the mantle, until few hundreds of kilometer, namely $\mathbf{u} = 0$ on Γ_{in}.

The force field $\mathbf{f}$ is given by the product between the density field ρ and the acceleration gravity $\mathbf{g}$. Since we are dealing with a model of the whole Earth we cannot assume a constant value for the gravity and its value must be computed in accordance with the density field using the Gauss's law for the gravity

$$\begin{cases} \mathbf{g} = -\nabla\phi, \\ \nabla \cdot \mathbf{g} + 4\pi G\rho = 0, \end{cases} \tag{4}$$

where G is the universal gravitational constant and the boundary conditions are $\mathbf{g} = -g_{\text{surface}}\mathbf{n}$ on Γ_{out} and $\phi = 0$ on Γ_{in}. Even though the displacement $\mathbf{u}$ is small if compared to the characteristic length of problem (1 km vs. 6371 km) the gravity acceleration acting on the point change in accordance with the displacement field, so the force field $\mathbf{f}$ is given by $\rho\mathbf{g}(\mathbf{x} + \mathbf{u}) \approx \rho\mathbf{g}(\mathbf{x}) + \rho(\nabla_{\mathbf{x}}\mathbf{g})\mathbf{u}$. From a physical point of view the first term $\rho\mathbf{g}(\mathbf{x})$ is a static component, it does not change in time, on the other side the second term $\rho(\nabla_{\mathbf{x}}\mathbf{g})\mathbf{u}$ is the buoyancy term that determine the uplift and subsidence processes. This physical interpretation can be justified from a mathematical point of view. Exploiting the linearity of the problem the stress tensor σ is decomposed as a sum of two components: a static stress term σ^0 (which is not important for our purposes) and a dynamic term σ^d which is the solution of Problem Equation (3) with $\mathbf{f} = \rho(\nabla_{\mathbf{x}}\mathbf{g})\mathbf{u}$ and the boundary conditions.

3. The Thermo-Hydro-Mechanical Model Including Erosion

In this section we focus the attention on the mathematical framework describing the mechanical and thermal evolution of the basin. First, we show how the basin model is built and how we basin tilting is extracted from the large scale isostasy model. Second, we use mathematical models to describe the thermal and mechanical evolution of the basin under the effect of a glaciation cycle.

The combination of these models consist in the multiscale modeling of glacial loading illustrated in Figure 1.

In the next sections the THMC model is introduced; the details of the numerical method can be found in Appendix B.

3.1. The Geological Model of a Sedimentary Basin

We assume that information about sedimentary units thicknesses, porosity and/or mineral composition are available at selected locations across a sedimentary system, typically from well logs, i.e., along the vertical direction (as sketched in Figure 1). In this framework we employ the stochastic inverse modelling procedure implemented in [7] to interpret vertical distributions of system properties with a one-dimensional model, which was developed in [5] starting from classical approaches to vertical compaction modeling (e.g., [2]). At each location such one-dimensional model provides an approximation of layers interface locations, whose characterization under uncertainty is investigated in detail in [27]. These interface locations can then be approximated in the whole domain starting from these data. This task is here performed using an interpolation based on ordinary kriging [28]. Our procedure assumes a smooth spatial variation of the sedimentary unit thicknesses. While this hypothesis can be restrictive in practical cases (e.g., in the presence of fault zones), our procedure is still able to handle geological settings of interest such as the occurrence of pinch out layers.

3.2. Extraction of the Basin Tilting from the Isostasy Model

The isostatic movement of the basin is taken into account as a rigid motion. In particular, we locate the computational cell that embed the fine scale geometrical model from the isostatic adjustment simulation, as shown in Figure 3. From the displacement field of the selected cell we evaluate the deformation gradient and the vertical rigid motion. We then isolate the rotational part of the deformation gradient that is uniquely defined by the polar decomposition [29], $F = R \cdot U$, where F is the deformation gradient, U is the symmetric stretch tensor and R, such that $\det(R) = 1$, is the rotation matrix that we want to determine. In particular, from the deformation gradient we evaluate the Right Cauchy-Green Deformation Tensor namely matrix $G := F^T \cdot F$. Using the definition of F, we obtain that $G = (RU)^T \cdot (RU) = (U^T \cdot R^T) \cdot (R \cdot U)$, and using the orthogonality of the rotation matrix, namely $R^T = R^{-1}$, we obtain $G = U^T \cdot U$. We recall that U is a diagonalizable matrix so can be represented by $U = Q^{-1} \cdot \Lambda \cdot Q$, where Q is the square matrix in which the ith column is the eigenvector q_i of U and Λ is the diagonal matrix composed by the corresponding eigenvalues, namely $\Lambda_{ii} = \zeta_i$. We recall that U is a symmetric matrix so $U^T \cdot U = U \cdot U$, so using the spectral decomposition we obtain that $G = (Q^{-1} \cdot \Lambda^2 \cdot Q)$ where Λ^2 is a diagonal matrix defined by $\Lambda_{ii}^2 = \zeta_i^2$. From the previous considerations, it follows that by computing the eigenvalues and the eigenvectors of G, which is a known matrix, we calculate the matrix Q, Λ^2, and, as a consequence, Λ. Using $U = Q^{-1} \cdot \Lambda \cdot Q$ we compute the matrix U so that we finally retrieve the rotation matrix R using $R = F \cdot U^{-1}$.

The collection of the rotation matrices together with the axial displacement evaluated at every time step of the isostatic adjustment simulation, defines the rigid motion that we consider in fine scale model. In particular we use a Lagrangian and an Euelarian approach for the mechanical and the thermal problems, respectively. More precisely, in the mechanical problem the imposed rigid motion does not causes any additional stress so that the calculation of stress and the strain can be performed in the reference (static) configuration, while the rotation matrix is used to change the orientation of the gravity vector. In this framework, we let pressure, flow and displacement evolve under the action of the weight of the basin and of the ice. Concerning the thermal evolution of the basin we account of the variation of the heat flux with the isostatic movement of the sedimentary basin by means of the Eulerian approach, that is we actually move the domain in the computational model. In this way, both the vertical displacement and the rotation matrix are taken into account in the evaluation of the thermal source, as it will be discussed later.

3.3. A Poromechanical Approach to Coupled Hydro-Mechanical Effects

For the mechanical evolution of the basin we rely on the theory of poroelasticity introduced by Biot in [30] under the quasi-static assumption for modeling a linearly elastic fully saturated porous medium. Such approach is widely used in literature, see [1,16,31,32] for a non exhaustive list of examples. Given a domain $\Omega \in \mathbb{R}^d$, we consider for simplicity an isotropic material (named the *skeleton*) filled with an isothermal single-phase fluid. A sketch of a typical domain together with the frame of reference is shown in Figure A1. In this framework the momentum equation reads

$$-\nabla \cdot \sigma(\mathbf{u}) + \alpha \nabla p = \mathbf{f} \qquad \text{in } \Omega(t), \tag{5}$$

$$\partial_t \left(\frac{p}{M} + \alpha \nabla \cdot \mathbf{u} \right) - \nabla \cdot K \nabla p = 0 \qquad \text{in } \Omega(t), \tag{6}$$

where (with little abuse of notation with respect to the isostatic adjustment model) here $\mathbf{u}$ denotes the solid matrix displacement vector and p is the variation of pore pressure from the hydrostatic load. We notice that ∂_t denotes the standard partial derivative with respect to time in the Eulerian framework. The parameters α, M and K are the Biot coefficient, the Biot modulus and the hydraulic conductivity, respectively. We recall that the hydraulic conductivity is defined as the ratio between the permeability k_s and the dynamic viscosity of the fluid μ_f, namely $K = k_s / \mu_f$. Finally $\mathbf{f}$ is the gravity load of the porous material evaluated as $\mathbf{f} = (\rho_s - \rho_f)\mathbf{g}$, where ρ_s, ρ_f and $\mathbf{g}$ are the fluid density, the solid density and the gravity vector, respectively. We remark that in the mechanical model the isostasy movement coming from the isostatic adjustment simulation is taken into account by means of a Lagrangian approach. As a consequence, the gravity vector $\mathbf{g}$ varies during the simulation according to a prescribed profile. Such profile is defined by the angles evaluated from the polar decomposition of the deformation gradient of the computational cell, of the large scale simulation, that contains the portion of the sedimentary basin we consider. To complete the definition of the problem we recall the linear elasticity behavior for the skeleton. This implies that the stress tensor σ, appearing in (5), is defined by $\sigma(\mathbf{u}) := 2\mu\varepsilon(\mathbf{u}) + \lambda \nabla \cdot \mathbf{u}$, where μ and λ are the Lamé coefficients and $\varepsilon(\mathbf{u})$ is the symmetric gradient of the skeleton displacement. For further details on poromechanical modeling, the interested reader is referred to e.g., [32,33].

For a well-posed problem we must complement the previous governing equations with appropriate boundary and initial conditions. Concerning the initial condition, the following constraints $\mathbf{u} = 0$ and $p = 0$ are considered at the initial time $t = t_0$. Let us label with Γ the top surface of the basin, while $\partial\Omega_b$ and $\partial\Omega_l$ are the bottom and the lateral boundary of the domain Ω, as show in Figure A1. According to this notation, we consider the following boundary conditions, $p = 0, \sigma(\mathbf{u}) \cdot \mathbf{n} = \sigma_{ice}(t)$ on $\Gamma(t)$, $\mathbf{u} = 0, \nabla p \cdot \mathbf{n} = 0$ on $\partial\Omega_b(t)$, $\mathbf{u} \cdot \mathbf{n} = 0, \nabla p \cdot \mathbf{n} = 0$ on $\partial\Omega_l(t)$, where $\mathbf{n}$ is the unit outward normal to the boundary and $\sigma_{ice}(t)$ is the load resulting from the ice sheet on top of the basin. We further assume that the load relative to the solid component of the ice sheet transfers only to the solid matrix of the basin, while the fluid phase at the top surface is subject to the hydrostatic load.

3.4. Thermal Effects

The heat transfer is modeled through the following advection diffusion equation

$$\rho_b c_b \frac{\partial T}{\partial t} + \rho_f c_f \mathbf{v}_D \cdot \nabla T - \nabla \cdot (b \nabla T) = q_r, \tag{7}$$

where T is the temperature field, ρ_f and c_f are the fluid density and specific heat, respectively; ρ_b and c_b are the bulk density and the bulk specific heat defined as $\rho_b c_b = \phi \rho_f c_f + (1 - \phi)\rho_s c_s$ with ϕ, ρ_s and c_s being the porosity, the solid density and specific heat capacity; $\mathbf{v}_D$ is the Darcy velocity defined as $\mathbf{v}_D = -K\nabla p$, b is the bulk thermal conductivity which is an average of the conductivity of the solid and the fluid phase and finally q_r is the heat source. Concerning the initial condition, we consider an homogeneous temperature field, namely $T = 0 \, \forall \mathbf{x} \in \Omega(t = t_0)$. According to the nomenclature shown in Figure A1, we consider the following boundary conditions, $T = T_{ice}(t)$ on $\Gamma(t)$, $\nabla T \cdot \mathbf{n} = 0$ on $\partial\Omega(t) \setminus \Gamma(t)$, where $\mathbf{n}$ is the unit outward normal to the boundary and $T_{ice}(t)$ is the basin top temperature. Following a widely used approach in literature, see for example [16,25], the presence of the ice on top of the basin is taken into account by means of a variation of the top temperature of the basin ($T_{ice}(t)$) during the glaciation cycle.

4. Results and Discussion

4.1. Results of the Glacial Rebound Model

The glacial rebound simulation addresses the deformation of the lithosphere of the whole Earth, based on a viscoelastic model. The forcing term of such simulation is the load of the ice sheet under a glaciation cycle.

The approach presented in this work is different from the one presented in the previously cited articles [11–13]. In these articles the authors describe a model for the deglaciation, the initial condition of their model is a glaciated Earth, the authors take into account the presence of this load at the initial condition defining a prestressed configuration. We rather simulate a full cycle of glaciation-deglaciation process to avoid the definition of a prestressed configuration, in our model the initial condition considered is a fully relaxed configuration, without enforced load.

In this case we consider a benchmark problem, where a circular sheet of radius $R = 1111$ km (equivalent to an angular sector of $10°$) centered at the North pole is formed and melt over a time window of 26×10^3 years, according to the variable thickness profile shown in Figure 2. More precisely, we assume that the glaciation phase starts 26×10^3 years ago and ends at present. We split this time window in three uniform intervals of 8.6×10^3 years each. In the first one we assume formation of the ice sheet up to a maximal thickness of 4 km. In the central phase, we assume that the ice is static, while in the last term we model ice melting with a linearly decreasing profile of the ice thickness. The results of such simulation are reported in Figure 3. In Figure 4 we show a zoom of the computational cells on the crust layer of the glacial rebound simulations. The zoom is taken in correspondence of the computational cell used to calculate the deformation gradient F. More precisely, the numerical calculation of the deformation gradient involves the displacement field in the x, y and z direction at the nodal points of of the selected cell. The time history of the displacement at these time points is shown in Figure 5. Even if the presented test is just a synthetic example, the order of magnitude of the obtained vertical displacement is reasonable and in good agreement with the results reported in [24,25].

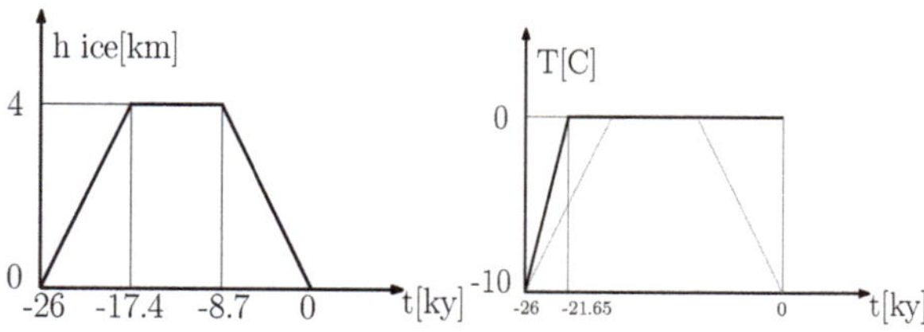

Figure 2. The evolution of the height of the ice sheet is shown on the left and the prescribed temperature field on top on the basin is reported on the right.

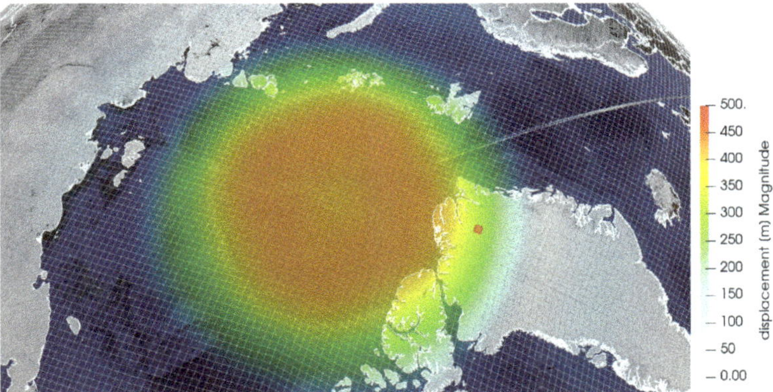

Figure 3. The simulation of isostatic adjustment with visualization of the selection of the location where the deformation tensor is extracted.

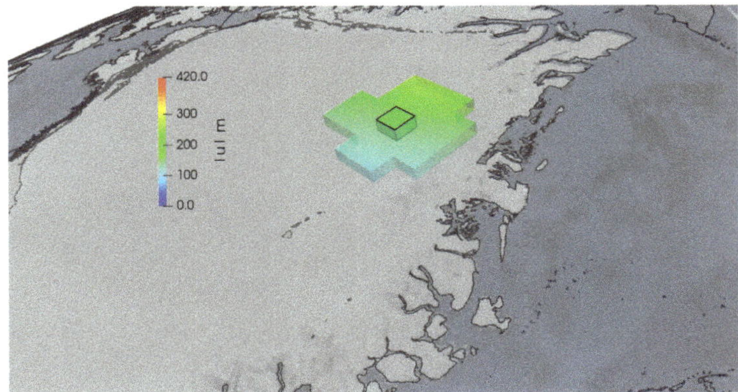

Figure 4. A zoom of the computational cells on the crust layer of the glacial rebound simulations. The magnitude of the vertical displacement is shown (meters). The black edges highlight the computational cell used to calculate the deformation gradient F.

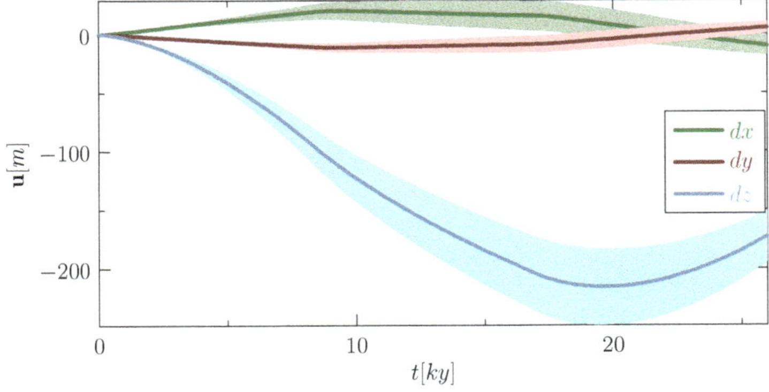

Figure 5. We show the average displacement field over time in the x, y and z direction at the nodal points of of the selected cell. Precisely, the light regions mark the minimum and the maximum displacement sampled at the nodes of the selected computational cell.

4.2. Results of the Kriging Algorithm for the Geological Basin Model Setup

An example of a simple reconstructed basin geometry is represented in Figure 6. These results are obtained on a synthetic example which considers a domain of 50 by 20 km in planar dimension. The one-dimensional simulations consider sediment deposition for a total period of time of 40 Ma. The sedimentation velocity is assumed to be fixed in time and is assigned at each x, y location between a maximum value 90 m/Ma (at the domain center) and and a minimum of 40 m/Ma (at the domain boundaries, i.e., at $x = 0$ and $x = 50$ km). Interfaces between different layers are approximated in such a domain by applying ordinary kriging starting from interface depths calculated at 20 locations randomly placed in the computational domain. Four interfaces are considered in total, including the top and bottom surfaces. We assume that two interface collapse on each other, i.e., the second and third interface correspond for $x > 35$ km. This means that one of the layers is not found at locations $x > 35$ km, and that a layer pinch-out occurs.

Note that in the simulation approach presented in Section 3 the effects of chemical compaction processes, such as quartz precipitation and smectite to illite transformation [6,34,35], are not explicitly included. However, the effects of chemical processes are considered in the one-dimensional model employed to approximate the interfaces [5,36] at selected locations, and therefore are implicitly embedded in the system geometry. The geometrical reconstruction of the sedimentary system obtained in this way can then be further enriched by mapping mineral compositions, porosity, permeability or other properties which may be available at selected location, e.g., through compositional kriging. These data are not considered in the following for simplicity.

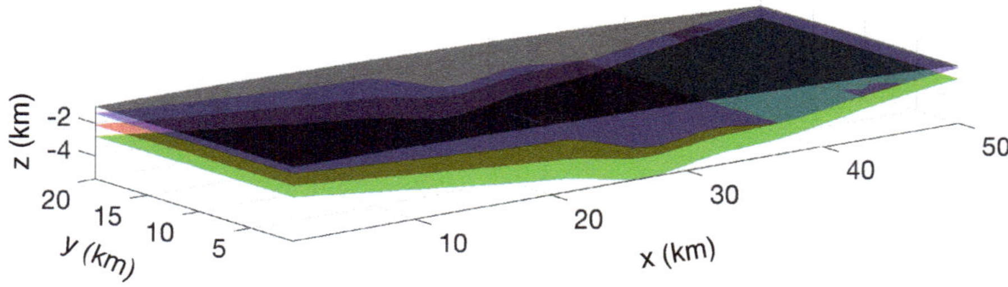

Figure 6. Kriging-based basin-scale reconstruction of layers interfaces.

4.3. Results of the Thermo-Hydro-Mechanical Effects of Glaciation

In this section we consider the application of the solver described in Section 3. As introduced in Section 3.3, the mechanical effect of the glaciation is taken into account by means of a variable load. This load is related to the height of the ice accumulated on top of the basin and follows the curve shown in the left part of Figure 2. Concerning the thermal problem, adiabatic conditions are considered at the bottom and the lateral surfaces of the physical domain while a time-dependent temperature profile is imposed in the top boundary of the basin. This profile models the thermal effect of the ice and it is shown in the right panel of Figure 2.

The configuration of the basin simulated in this section and the physical parameters used to initialize the THM model are reported in Figure 7. More precisely, to perform the three-dimensional THM simulation, we consider a portion of the basin of 4×4 km size with an average depth of of 2.8 km located at the x, y coordinates $(32.5, 36.5) \times (8, 12)$ km of the basin shown in Figure 6. The domain is split into three layers (numbered as 1,2 and 3 from bottom to top, as shown in the left part of Figure 7) and one of them is not continuous across the whole planar domain size resulting in a geological model with a pinch-out, consistent with the data in Figure 6. The material properties of all the different components of the basin are summarized in the right panel of Figure 7. We assume that the intermediate layer has a permeability that is significantly higher than the other ones and we consider the evolution of the basin during a time window of 26 ky with a time step of 0.14 ky chosen to follow with enough detail glaciation evolution.

Young Modulus	E_1	10^{11}	Pa
	E_2	$2\,10^{10}$	Pa
	E_3	10^{10}	Pa
Rock density	ρ_s	$2.2\,10^3$	kg/m^3
Rock permeability	K_1	10^{-18}	m^2
	K_2	10^{-12}	m^2
	K_3	10^{-16}	m^2
Thermal diffusivity	b_1	10^{-6}	m^2/s
	b_2	$2\,10^{-6}$	m^2/s
	b_3	10^{-6}	m^2/s
Water density	ρ_l	10^3	kg/m^3
Water viscosity	μ_l	10^{-3}	Pa s
Dimension	l	$4\,10^3$	m
Radiogenic source	q	10^{-7}	W/m^3K

Figure 7. On the left we show a sketch of the physical domain, ot top of which is superposed a layer visualizing the ice sheet. The labels 1, 2, 3 mark the different materials. On the right the list of materials properties is reported.

In the example of this section, the surface erosion on top of the basin is active during the last 8.7 ky of the simulation. It is modeled as the prescribed evolution of the upper part of the sedimentary basin. More precisely, we assume that during the last part of the simulation a certain amount of material is removed from the upper part of the basin. As a consequence the top surface of the physical domain evolves from S_1 to S_2, as shown in Figure 8. From the modeling standpoint, the motion of the top surface is described by means of a the level set function, that is defined in terms of spatial coordinates x, y and time t,

$$ls(x, z, t) = -(-0.285x + 1.6z + 12353) + 400(1 + a) \,, \tag{8}$$

$$a = 0 \qquad \text{if } t < -8.7Ky \,, \tag{9}$$

$$a = \frac{t + 8.7Ky}{8.7ky} \qquad \text{if } t \geq -8.7Ky \,. \tag{10}$$

Finally, in the simulations presented in this section, we consider that the basin is exposed to a spatially dependent heat source q_r, determined by the heat flux from the mantle and by the internal radiogenic thermal source. The combination of these factors is accounted as a volumetric thermal source of this form $q_r(\mathbf{x}) = q_r(z) = q_0 f(z)$. Following [37], the baseline source q_0 is determined by means of a heat balance equation that distributes over the basin volume the heat flux coming form the bottom surface (i.e., the one closer to the mantle) and the radiogenic source. Precisely, we set $q_0 = \phi S/V + q_{r,0}$ where ϕ is the surface heat flux from the bottom surface, S, V is the basin volume and $q_{r,0}$ is the baseline radiogenic source. In this way we obtain $q_0 = 0.1 \ \mu W/m^3$ .To account for the exponentially decreasing radioactivity with depth, this source is evaluated as a function of the actual position of the domain according to the following empirical formula $q = q_0 e^{-z/D}$ where the parameter $D = 5$ km.

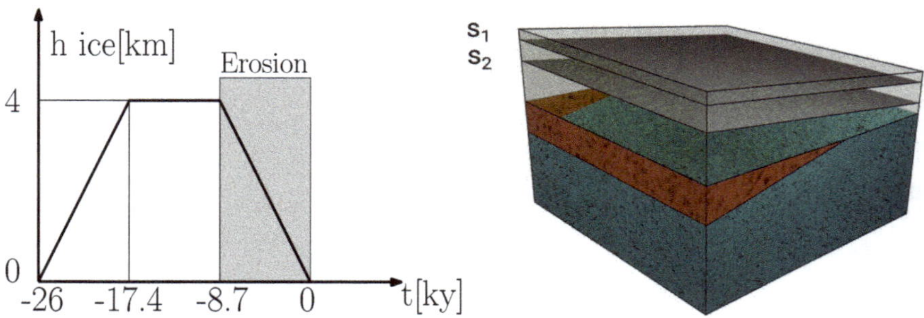

Figure 8. The evolution of the top surface of the basin due to erosion form configuration $S1$ to $S2$ is shown during the last part of the simulation, from -8.7 ky to the present.

The kriging-based horizon reconstruction addressed before approximates the interface locations on a user-defined spatially uniform grid (with uniform size equal to 0.5 km in our example), so that each interface is represented by a point cloud. Such point clouds are used to generate the internal surfaces of the basin as shown in the left panel of Figure 9. Then the boundaries of the horizons are used for the creation of the lateral surfaces of the geological model as shown in the middle panel. These operations result in the definition of a water tight geological model and are performed using the platform GOCAD. The geological model is then used as the input of the mesh generator RINGMesh [38] that produces the 3D a labeled computational grid shown in Figure 9, right panel.

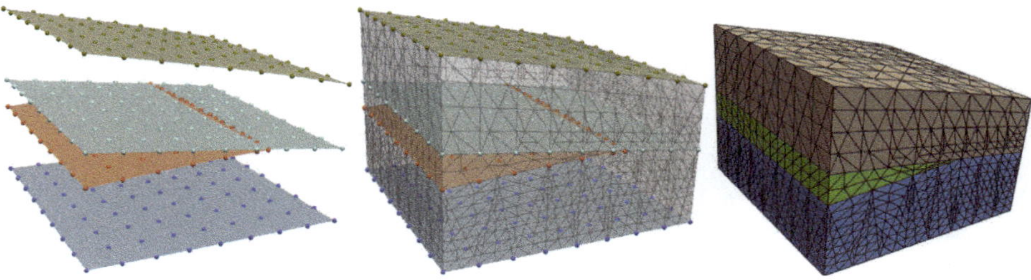

Figure 9. A sketch of the pipeline to build the geological model and the mesh. On the left we show the point cloud and corresponding reconstruction of the horizons. In the middle we show the lateral surfaces together with the internal horizon. On the right we report the final Computational grid of the geological model.

For the analysis and the interpretation of the results we subdivide the simulation in three different phases: phase A (formation of the ice sheet) from -26 to -17.4 ky where the ice is growing, phase B (isostatic adjustment) from -17.4 to -8.7 ky where the ice load is steady and most of the isostatic motion takes place and the phase C (erosion) from -8.7 to 0 ky where the ice on top of the basin vanishes and the erosion takes place. A schematic of this temporal subdivision is reported in Figure 10. The current configuration of the domain $\Omega(t)$, rotated according to the rigid motion coming from the isostatic adjustment simulation at different time steps, is shown in Figure 11. The evolution of displacement and pressure along the phases A, B and C of the simulation is shown in Figure 12. In the first phase (A), the action of the ice load generates the mechanical compaction of the basin. According to the Biot model the compaction leads to an increase in the pore pressure in the different layers of the basin. This effect is more evident in the top layers, 2 and 3, while layer 1 is almost unperturbed, because it is the most impermeable and it is subject to zero displacement conditions on the bottom surface. Moreover, we notice that the pressure field in the pinch out layer (layer 2) is almost uniform and equal to the value at the interface with the upper layer (layer 3). This effect is due to the fact that layer 2 is the most permeable. According to the Darcy law, this is the region where the smallest pressure gradients are observed.

In the second phase (B) we can observe the effect of the isostatic adjustment. The interaction between the isostatic motion end the poromechanical problem is illustrated in Figures 12 and 13, where we show at $t_B = -13$ ky the pressure field in an internal slice of the domain, together with the displacement direction inside layer 3. We observe that the imposed rotation generates a tangential component of the load applied on the top of layer 3, namely $\sigma_n^{ice} \cdot \mathbf{t} \neq 0$. This effect, combined with the boundary conditions enforcing zero normal displacement on the lateral surfaces of the basin, generates a mechanical compression effect along the tangential direction of the top and bottom surface planes of layer 3. Because of the poromechanic coupling, this tangent stress induces the pressure peak observed at the left corner of the basin. Furthermore, in the right part of Figure 13, an internal slice along the xz plane is shown. From this view, the transition from layer 1 to layer 2 (at the pinch out) is visible. In this visualization, the effect of the permeability jump can be appreciated. More precisely, the pressure increases only along the interface between layers 1 and 3, while the pressure gradients are significantly smeared out along the contact line with layer 2 as a consequence of the high permeability of it. Finally, we notice that these effects are difficult to appreciate in the bottom layer, where the high young modulus limits the displacement field and the low permeability almost blocks the diffusion of the pressure peak occurring in the top layer.

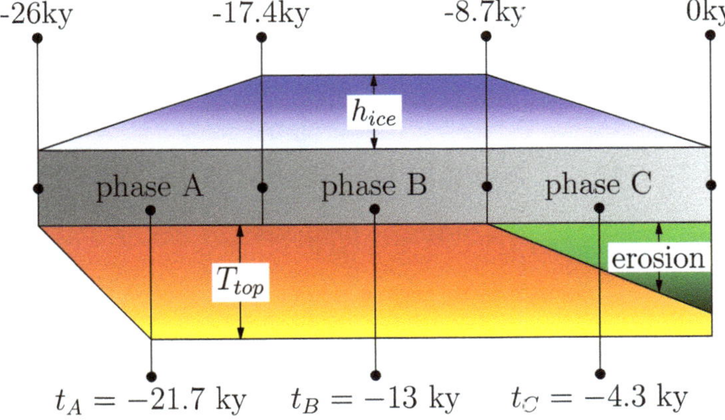

Figure 10. Timeline of the simulation. On the top we show the height of the ice, on the bottom temperature profile (red) and the thickness of the eroded material (green).

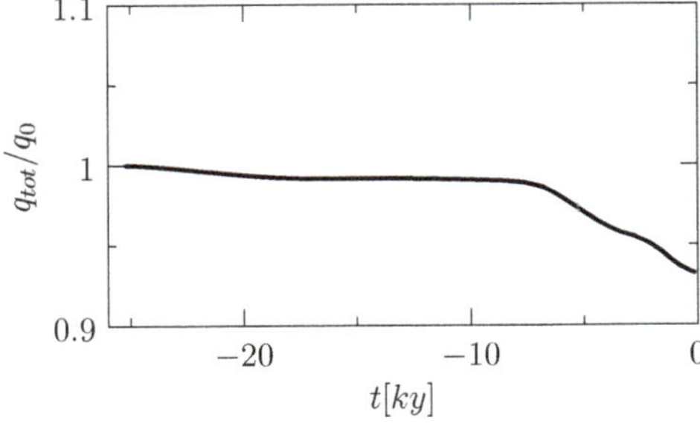

Figure 11. Time variation of the relative total thermal source (q_r) during the evolution of the domain because of isostasy and erosion.

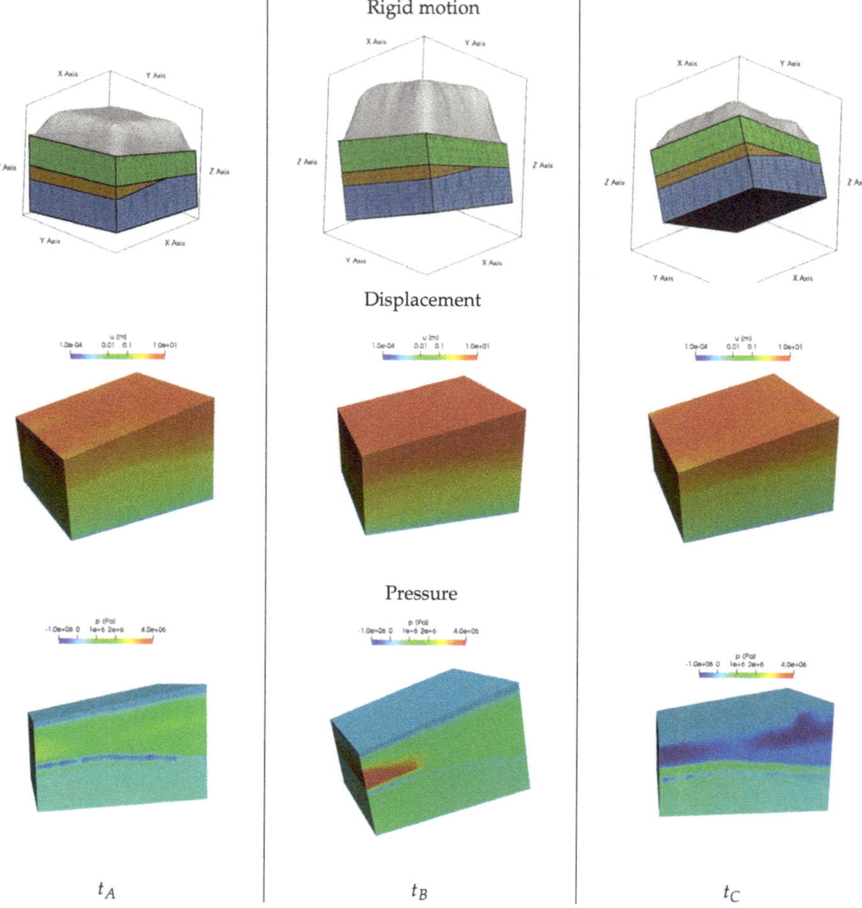

Figure 12. Evolution of the isostatic adjustment, the displacement and the pressure fields along the different phases of the simulation. In particular, we show on the top we show an the imposed rigid motion. The displacement (middle row) and the pressure fields (the variation from the hydrostatic pressure profile is shown, bottom row) are reported at times $t_A = -21.7ky$ $t_B = -13Ky$ and $t_C = -4.3Ky$, from left to right.

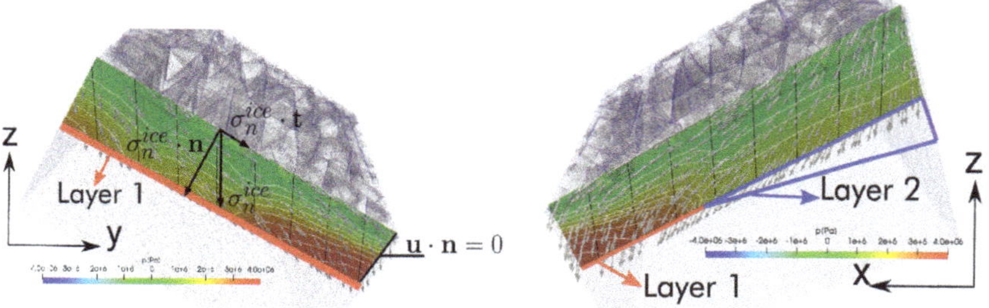

Figure 13. Front (on the left) and lateral (on the right) view of the basin. The color marks the pressure field, the arrows shows the direction of the solid displacement (in the interior and on the boundary of the domain, that is $\mathbf{u} \cdot \mathbf{n} = 0$) and the black lines show the direction of gravity. The configuration of the basin layers is also shown, to facilitate the interpretation of the results.

To appreciate the impact of taking into account the isostatic response, we compare the pressure field in the top layer with the one obtained through a different numerical experiment in which we neglect the isostatic adjustment and the erosion. In the second scenario the basin model is completely static. In such case, because the load on top of the basin is constant, the general temporal trend of the pressure is to progressively reach a steady state in all the layers of the basin according to the characteristic temporal scale determined by the different permeability values.

The comparative study of the two scenarios described above is reported in Figure 14. In this Figure we compare the pressure field along two transversal lines r_1 and r_2, obtained switching on (first scenario, solid line) and off (second scenario, dashed line) the isostatic motion of the basin. As previously observed in the central panel of Figure 12, we notice that in the first scenario the pressure reaches a peak in the left corner of the layer 3 (as also illustrated in Figure 14). Such peak is not present in the second scenario, where we neglect the isostatic movement of the basin, as we can see from Figure 14 (dotted line). From that comparison we notice that limited spatial variations of the pressure are generate without taking into account the rigid motion of the basin, Conversely, the introduction of the isostatic rebound significantly increases the pressure variability, generating local peaks that depend on the morphology of the basin. Then, we conclude that the motion due to the isostatic adjustment simulation is the primary reason of the overpressure generation.

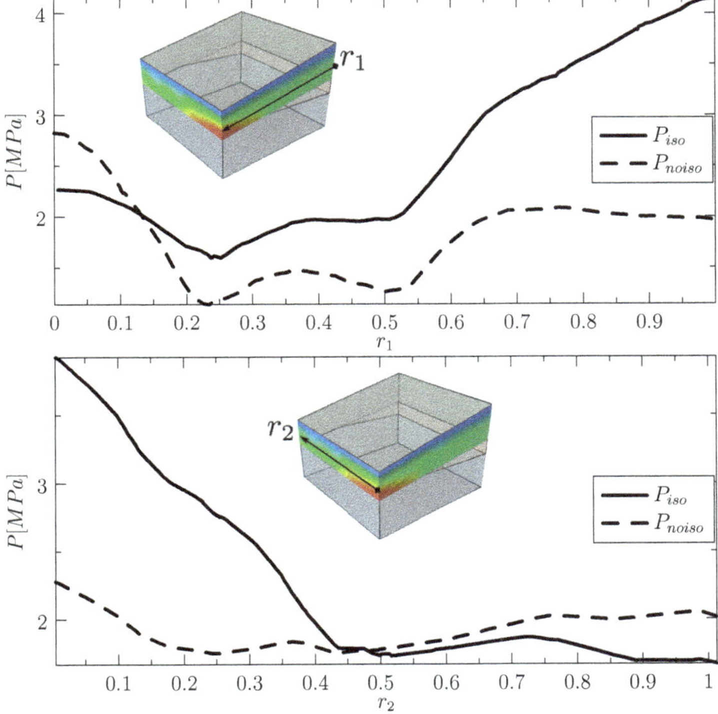

Figure 14. Pressure field on the lateral (on top) and front (on the bottom) lines r_1 and r_2, respectively. Solid (P_{iso}) and dashed line (P_{no-iso}) mark the results obtained considering and neglecting the effect of erosion and of the isostasy movements, respectively.

We now focus on the final phase (namely phase C) when the ice sheet is disappearing from the top of the basin and the erosion takes place. In this phase the mechanical load due to the ice weight goes to zero. The reduction of mechanical compaction is also augmented by the erosion of the top part of the basin. The combination of these effects leads to a general decrease of the pore pressure in all layers, as shown in the right panel of Figure 12, right column. More precisely, the lowest values of pressure (in terms of variations from the hydrostatic load) are located in the softest layer.

Concerning the temperature field, the model is initiated with a temperature field at thermal equilibrium and it evolves according to the (time dependent) heat equation under the action of a space dependent heat source. To analyze and validate the simulation of the thermal field we address two indicators: the overall thermal gradient from the top to the bottom of the basin and the temporal variations of the temperature from the steady state. Concerning the vertical temperature gradient, we observe from Figure 15 that a difference of 60 degC along the vertical axis is established as a consequence of thermal balance between the thermal source and the imposed temperature at the top of the basin. This results is in good agreement with the thermal gradients expected in literature, see for example [37,39] for thermal properties and, in particular [37] for expected temperature profiles. For a more detailed analysis of the thermal variations from the initial equilibrium state, we report, in Figure 16, difference of the temperature field at time $t_A = -21.7ky$ $t_B = -13Ky$ from the one at $t_0 = -26Ky$. We notice that at t_A and t_B, the main temperature variations are driven by the increase of surface temperature from -10 degC to 0 degC at the top of the basin, due to the protective effect of the ice cap, that is progressively forming in this phase. Comparing more in detail the profiles at t_A and t_B, we observe that at time t_B the temperature in the central part of the basin has slightly increased with respect to the one at t_A, by the effect of the thermal source . The profile at t_C differs significantly form the others, because of the erosion, active in the time interval from t_B to t_C. During erosion, the reference temperature of 0 degC enforced at the contact interface with the ice, progressively shifts downwards, swiping a region of the basin that was previously hotter than 0 degC. For this reason, the temperature in the basin after erosion significantly decreases with respect to the initial time. Finally, in Figure 11 we report the variation of the total thermal source term, relative to the initial state, that is the spatial integral during the evolution of the domain $\Omega(t)$ of the heat source $\int_{\Omega(t)} q(\mathbf{x})d\omega / \int_{\Omega(t_0)} q(\mathbf{x})d\omega$. These data suggest that variation of the basin volume due to erosion decreases the thermal source more significantly than the progressive increase of basin depth due to isostasy.

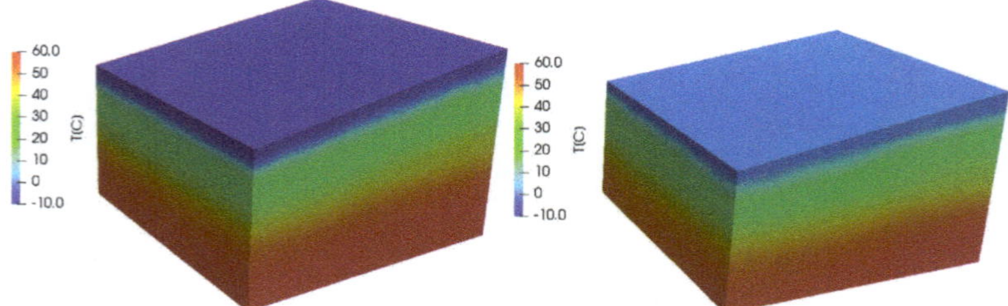

Figure 15. Comparison of the initial ($t = -26$ ky on the left) and final ($t = 0$ ky on the right) temperature fields. It is observed, at the depth of 4 km, temperature gradient of approximately 60°C as qualitatively expected in [37].

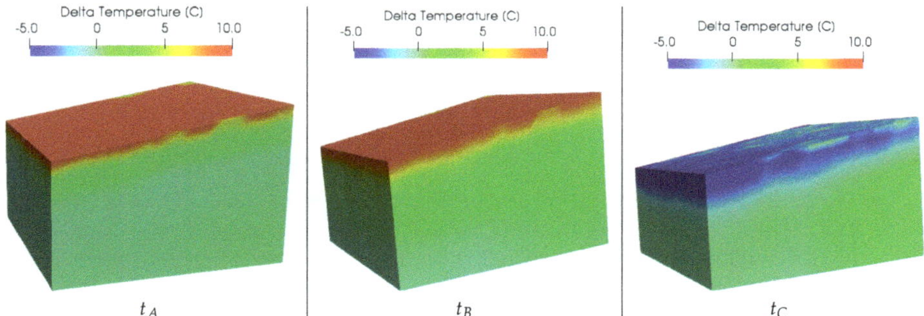

Figure 16. The difference of the temperature field at time $t_A = -21.7\,\text{ky}$, $t_B = -13\,\text{ky}$ and $t_C = -4.3\,\text{ky}$ from that at $t_0 = -26\,\text{ky}$ is shown from left to right.

The analysis of the thermal field confirms that, in the short time window of one cycle of glaciation, the surface temperature, possibly modulated by erosion, is the main diving factor that determines temperature variations from the equilibrium state. This finding is in accordance with the detailed analysis of the effects of glaciations on sedimentary basins provided in [25]. It should also be observed that using a time dependent heat equation correctly models the natural inertia of the basin to change its equilibrium state. As a result, it is expected that the repetition of many glaciation cycles is required to significantly cool down the entire basin.

5. Conclusions

We have developed a fully coupled multiphysics and multiscale description of the evolution of a sedimentary basin under the effect of a glaciation cycle. To the best of our knowledge, thermo-hydro-mechanical effects are combined with isostatic adjustment and erosion within a fully time dependent three-dimensional simulation for the first time. Although the geological model that has been considered does not represent yet the complexity of a real sedimentary basin, we have described a pipeline of steps that could handle the most complex cases as well, thanks to a multiscale approach that decouples phenomena occurring at very different space and time scales. For example, our methodology establishes a quantitative framework to transfer information from the definition of a large scale geological architecture to local fluid displacement and deformation dynamics. Preliminary numerical results suggest that the combination of all these phenomena reveals the emergence of effects that were not expected or predictable using simpler approaches. In the considered synthetic test case we have quantified the effects of large scale isostatic displacement on local pressure and displacement field, in the presence of layer a pinch-out. Moreover, our results demonstrate the effect of erosion on the temperature dynamics of the sedimentary system. We believe that our results show the predictive potential of this holistic description of sedimentary basins subject to glaciations.

Author Contributions: The initial idea of the work was given by P.R., work conceptualization was then jointly performed by all authors. The methodology of the THM model was developed by D.C., G.P. and P.Z. The methodology of the isostatic adjustment model was developed by E.M. and M.P. The software used for numerical simulations was developed by D.C. and M.P. The writing of the manuscript was performed by D.C., M.P., G.P., E.M. and P.Z. and the manuscript review was performed by P.R., that validated also all the results.

Funding: This research was funded by Eni s.p.a. under the project *STREAM THMC* grant number 4310304289, contract number 2500032655, of the cooperation agreement 4400007601 between Eni s.p.a. and Politecnico di Milano.

Conflicts of Interest: The authors declare no conflict of interest.

Abbreviations

The following abbreviations are used in this manuscript:

THM   Thermo-Hydro-Mechanical

Appendix A. The Numerical Solver for the Isostatic Glacial Rebound Model

The numerical solver for the glacial isostatic adjustment is implemented using the deal.II library [40,41]. This library provides the tools for an efficient implementation of a parallel solver based on the Discontinuous Galerkin method. Let $\mathcal{T}_h$ be a subdivision of the geometric domain Ω consisting of non-overlapping hexahedra with characteristic mesh size h, we introduce the finite dimensional spaces

$$\mathbf{V}_h^k = \{\mathbf{v} \in L^2(\Omega; \mathbb{R}^3) : \ \mathbf{v}|_K \in \mathbb{P}^k(K; \mathbb{R}^3) \ \forall K \in \mathcal{T}_h\},$$
$$Q_h^k = \{q \in L^2(\Omega) : \ q|_K \in \mathbb{P}^k(K) \ \forall K \in \mathcal{T}_h\}.$$

Let $\mathcal{E}_h = \bigcup_{K \in \mathcal{T}_h} \partial K$ denote the set of the faces of $\mathcal{T}_h$ then $\mathcal{E}_h^0 = \mathcal{E}_h \backslash \partial \Omega$ is the set of internal faces. Let $e \in \mathcal{E}_h^0$ be a face shared by two elements K^+ and K^-, define the unit normal vector $\mathbf{n}^+$ and $\mathbf{n}^-$ on the e pointing exterior to K^+ and K^-, respectively. With $\phi^{\pm} = \phi|_{K^{\pm}}$ we set the average operators as

$$\{\mathbf{v}\} = \frac{\mathbf{v}^+ + \mathbf{v}^-}{2} \qquad \{q\} = \frac{q^+ + q^-}{2},$$

together with the jump operators:

$$[\![\mathbf{v}]\!] = \mathbf{v}^+ \otimes \mathbf{n}^+ + \mathbf{v}^- \otimes \mathbf{n}^- \qquad [\![q]\!] = q^+ \mathbf{n}^+ + q^- \mathbf{n}^-.$$

Equation (4) is discretized using the standard SIPG method [42]: find $\phi_h \in Q_h^k$ such that

$$\int_{\Omega} \nabla_h \phi_h \cdot \nabla_h \psi_h + 4\pi G \rho \psi_h \, d\mathbf{x} +$$
$$+ \int_{\mathcal{E}_h^0} \gamma [\![\phi_h]\!] \cdot [\![\psi_h]\!] - \{\nabla_h \phi_h\} \cdot [\![\psi_h]\!] - [\![\phi_h]\!]\{\nabla_h \psi_h\} \, ds + \text{B.C.} = 0 \qquad \forall \psi_h \in Q_h^k.$$

where B.C. is the term related to the boundary conditions and γ is a proper penalization factor. The algebraic system obtained from this weak formulation is solved using the standard conjugate gradient method preconditioned with a geometric multigrid method, available in deal.II library [43].

The Equations (3) are discretized using a standard second order accurate, one-step and unconditionally stable scheme [26] and the integral is approximated using the mid-point formula. Using this approach the equations describing the motion of the viscoelastic Earth model can be rewritten in semi-discrete form using only the variables $\mathbf{u}$, p and h:

$$\begin{cases} \nabla \cdot \left(2\mu e^{-\frac{\Delta t_n}{2\tau}} e(\mathbf{u}_n - \mathbf{u}_{n-1}) - p_n I + e^{-\frac{\Delta t_n}{\tau}} h_{n-1} \right) + \rho (\nabla_\mathbf{x} \mathbf{g}) \mathbf{u}_n = 0 & \text{in } \Omega, \\[2ex] \nabla \cdot \mathbf{u}_n + \dfrac{3}{2\mu} \dfrac{1 - 2\nu}{1 + \nu} p_n = 0 & \text{in } \Omega, \\[2ex] h_n = 2\mu e^{-\frac{\Delta t_n}{2\tau}} e(\mathbf{u}_n - \mathbf{u}_{n-1}) + e^{-\frac{\Delta t_n}{\tau}} h_{n-1} & \text{in } \Omega. \end{cases}$$

The first two equations of this system have the same structure of the linear elastic problem and they are discretized using the Discontinuous Galerkin scheme. In order to keep the notation simpler the shear modulus will be redefined as $\hat{\mu}_n = \mu e^{-\frac{\Delta t_n}{2\tau}}$ and all the known terms in the first equation are written as a unique term $\mathbf{F}_n$:

$$\mathbf{F}_n = \nabla \cdot \left(2\hat{\mu}_n e(\mathbf{u}_{n-1}) - e^{-\frac{\Delta t_n}{\tau}} h_{n-1} \right).$$

These equations can be rewritten in a full discrete form using the standard SIPG-method for the linear elasticity [44]: find $(\mathbf{u}_{h,n}, p_{h,n}) \in \mathbf{V}_h^{k+1} \times Q_h^k$ such that

$$\begin{cases} a(\mathbf{u}_{h,n}, \mathbf{v}_h) + b(\mathbf{v}_h, p_{h,n}) = \int_\Omega \mathbf{F}_n \cdot \mathbf{v}_h \, d\mathbf{x} + \text{B.C.} \quad \forall \mathbf{v}_h \in \mathbf{V}_h^{k+1}, \\[2mm] b(\mathbf{u}_{h,n}, q_h) - c(p_{h,n}, q_h) = \text{B.C.} \qquad\qquad\qquad \forall q_h \in Q_h^k. \end{cases}$$

where the bilinear forms are defined by

$$a(\mathbf{u}_h, \mathbf{v}_h) = \int_\Omega 2\hat{\mu}_n e(\mathbf{u}_h) : e(\mathbf{v}_h) + \rho(\nabla_\mathbf{x}\mathbf{g})\mathbf{u}_h \cdot \mathbf{v}_h \, d\mathbf{x} +$$
$$+ \int_{\mathcal{E}_h^0} \gamma [\![\mathbf{u}_h]\!] : [\![\mathbf{v}_h]\!] - \{2\hat{\mu}_n e(\mathbf{u}_h)\} : [\![\mathbf{v}_h]\!] - [\![\mathbf{u}_h]\!] : \{2\hat{\mu}_n e(\mathbf{v}_h)\} \, ds,$$
$$b(\mathbf{u}_h, q_h) = -\int_\Omega \nabla \cdot \mathbf{u}_h q_h \, d\mathbf{x} + \int_{\mathcal{E}_h^0} [\![\mathbf{u}_h]\!] : \{q_h I\} \, ds,$$
$$c(p_h, q_h) = \int_\Omega \frac{3}{2\mu} \frac{1 - 2\nu}{1 + \nu} p_h q_h \, d\mathbf{x},$$

The terms B.C. are related to the boundary conditions and γ is a proper penalization factor. This problem is equivalent to the algebraic block structured linear system

$$\begin{bmatrix} 2\hat{\mu}_n A & B^T \\[2mm] B & -\dfrac{3}{2\mu}\dfrac{1-2\nu}{1+\nu} C \end{bmatrix} \begin{bmatrix} U_n \\ P_n \end{bmatrix} = \begin{bmatrix} F_n \\ G_n \end{bmatrix}$$

This system is solved using the GMRES method and it is preconditioned with the block-triangular preconditioner

$$P = \begin{bmatrix} 2\hat{\mu}_n \hat{A} & -B^T \\[2mm] 0 & \dfrac{3}{2\mu} M \end{bmatrix}$$

where $\hat{A}$ is the matrix associated to the Discontinuous Galerkin approximation of the vector-valued Laplace operator in $\mathbf{V}_h^{k+1}$ and M is the mass matrix in Q_h^k. The implementation of P^{-1} requires the computation of the inverse matrix for $\hat{A}$ and M. The mass matrix M can be inverted easily, because a proper choice of the base functions for the space Q_h^k and the quadrature rule to evaluate the integral leads to a diagonal mass matrix. The inverse of matrix $\hat{A}$ is replaced with a geometric multigrid preconditioner [45].

Even if the numerical solver is general with respect to the polynomial order k, the results presented in this work are obtained using quadratic polynomials for the gravity potential field (in order to evaluate the gradient of gravity acceleration) and the stable pair of spaces $\mathbf{V}_h^2 \times Q_h^1$ is used for the displacement and the pressure unknowns in the viscoelastic problem.

Appendix B. The Numerical Solver for the THM Model

The surface erosion on top of the basin is taken into account as a prescribed evolution of the upper part of the sedimentary basin. To model erosion, we use the cut finite element method, briefly CutFEM, in which the boundary of the physical domain is represented on a background grid using a level set function see for example [46]. This approach requires that the computational domain must embed all possible eroded configurations. As a consequence, we immerse the physical domain that describes the basin $\Omega(t)$, into a larger computational domain $\Omega(t) \cup \Omega_{out}$ as shown in Figure A1. The background or computational grid is also used to approximate the solution of the governing problem. We ideally divide the computational grid into three regions, $\Omega(t)$, Ω_{out} and Γ, where the level set is lower, grater and equal to zero, respectively; $\Omega(t)$ is the physical domain where the poromechanical problem is solved; Ω_{out} is a dummy zone that does not affect the solution, while Γ is the top surface of the basin where the top boundary conditions are enforced.

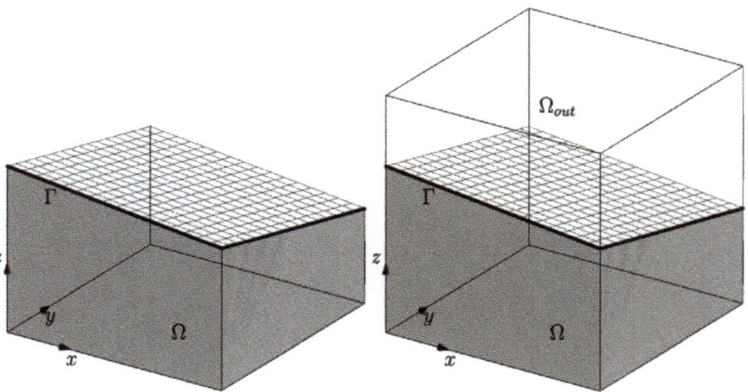

Figure A1. On the left we show the physical domain. On the right the physical domain (gray) is embedded in a larger computational domain ($\Omega \cup \Omega_{out}$). Γ marks the top surface of the basin.

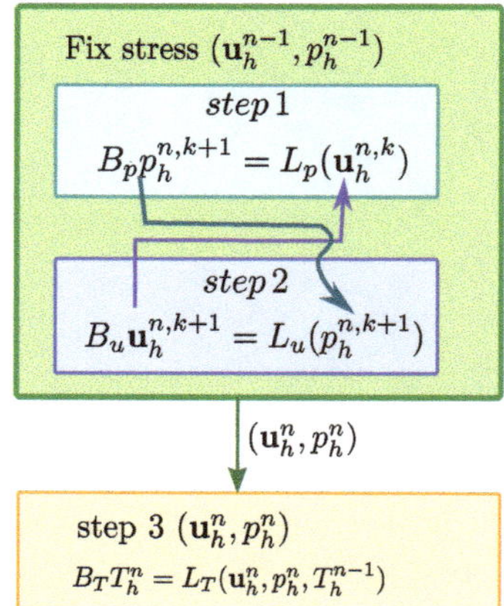

Figure A2. A schematic of the complete algorithm for the solution of the thermal poro mechanical problem. Matrices B_u, B_p, B_T correspond to the discretization of Equations (A1) a, b, c respectively.

Since Equations (5)–(7) are solved numerically, we briefly introduce the corresponding finite element discretization, which is based on the weak formulation of the problem.

Let $\mathcal{T}_h := \{K\}$ denote a triangulation of $\Omega_{\mathcal{T}} = \Omega(t) \cup \Omega_{out}$ that does not necessarily conform to the surface Γ and let us introduce the discrete spaces

$$\mathbf{V}_h := \{\mathbf{v}_h \in H^1(\Omega_{\mathcal{T}}, \mathbb{R}^3) : \mathbf{v}_h|_T \in \mathbb{P}^1(K, \mathbb{R}^3), \forall K \in \mathcal{T}_h\},$$

$$Q_h := \{q_h \in H^1(\Omega_{\mathcal{T}}) : q_h|_T \in \mathbb{P}^1(K), \forall K \in \mathcal{T}_h\},$$

$$W_h := \{w_h \in H^1(\Omega_{\mathcal{T}}) : w_h|_T \in \mathbb{P}^1(K), \forall K \in \mathcal{T}_h\},$$

where $\mathbb{P}^1$ denotes the space of scalar piecewise linear polynomials on $\mathcal{T}_h$. We introduce the following bilinear forms: $a(\mathbf{u}, \mathbf{v}) := 2 \int_{\Omega(t)} \mu\varepsilon(\mathbf{u}) : \varepsilon(\mathbf{v}) d\Omega + \int_{\Omega(t)} \lambda(\nabla \cdot \mathbf{u})(\nabla \cdot \mathbf{v}) d\Omega$ and $c(p, q, K) := \int_{\Omega_{in}} K\nabla p \cdot \nabla q d\Omega$. Let us also introduce the operators $D_p(p, q, \mathbf{u}) = 1/M(p, q) + \alpha(\nabla \cdot \mathbf{u}, q)$, $D_T(T, w, \mathbf{v}) = \rho_b c_b(T, w) + \rho_f c_f(\mathbf{v} \cdot \nabla T, w)$ and $\Theta\langle p, q\rangle_\Gamma^K = -\int_\Gamma K\nabla p \cdot \mathbf{n}q d\Gamma + \gamma h^{-1} \int_\Gamma pq d\Gamma - \int_\Gamma K\nabla q \cdot \mathbf{n}p d\Gamma$, where $(\cdot, \cdot)$ is the standard inner product in the space $L^2(\Omega(t))$, h is the characteristic size of the quasi-uniform computational mesh and $\gamma > 0$ denotes a penalization parameter. For the imposition of the pressure and temperature boundary conditions on the internal unfitted interface Γ, we rely to the Nitsche's method following the approaches proposed in [47–50]. This technique allows to weakly enforce interface conditions at the discrete level by adding to the variational formulation of the problem appropriate penalization terms (γ). Finally, considering a backward-Euler time discretization scheme, the fully discretized problem at time t^n, $n = 1, 2, ..., N$ can be written as follows: find $(\mathbf{u}_h, p_h, T_h) \in \mathbf{V}_h \times Q_h \times W_h$ such that:

$$\begin{cases} a(\mathbf{u}_h^n, \mathbf{v}_h) - \alpha(p_h^n, \nabla \cdot \mathbf{v}_h) = (\mathbf{f}, \mathbf{v}_h) & \forall \mathbf{v}_h \in \mathbf{V}_h, \\ D_p(p_h^n, q_h, \mathbf{u}_h^n) + \tau\Lambda(p_h^n, q_h, K) = D_p(p_h^{n-1}, q_h, \mathbf{u}_h^n) & \forall q_h \in Q_h, \\ D_T(T_h^n, w_h, \mathbf{v}_{D,h}^n) + \tau\Lambda(T_h^n, w_h, b) = D_T(T_h^{n-1}, w_h, 0) + (q_r^n, w_h) & \forall w_h \in W_h. \end{cases} \tag{A1}$$

where $\Lambda(p, q, k) = c(p, q, K) + \Theta\langle p, q\rangle_\Gamma^K$ and τ is the computational time step.

The solution of Equations (A1) is not trivial. We decouple the solution of poromechanical problem (i.e., the first two Equations of (A1)) from the solution of the thermal problem (last Equation of (A1)). However the solution of the poromechanical problem is still challenging due to the tight coupling between deformation and flow. To solve such problem we adopt the fixed stress iterative scheme as proposed in [31,51]. This algorithm is a sequential procedure where the flow is solved first followed by the solution of the mechanical problem. In particular, in every time steps, the algorithm is iterated until the solution converges within an acceptable tolerance. In the first step the fixed stress algorithm, given $(\mathbf{u}_h^{n,k}, p_h^{n,k}) \in \mathbf{V}_h^r \times Q_h^s$ we evaluate $p_h^{n,k+1} \in Q_h^s$ where $\cdot^k$ is the index of the fixed stress iteration. In the second step, given the new pressure $p_h^{n,k+1} \in Q_h^s$ we find the new $\mathbf{u}_h^{n,k+1} \in \mathbf{V}_h^r$. The iterative steps are performed until the following convergence criterion is fulfilled

$$\frac{p_h^{n,k+1} - p_h^{n,k}}{p_h^{n,0}} < \eta_p, \qquad \frac{\mathbf{u}_h^{n,k+1} - \mathbf{u}_h^{n,k}}{\mathbf{u}_h^{n,0}} < \eta_u, \tag{A2}$$

where η_p and η_u are the desired tolerances (here chose equal to 10^{-7}). A schematic of the complete algorithm for the solution of the thermal poro mechanical problem in a compact form is shown in Figure A2. For further details about the performance of this method we remand to [52].

References

1. Tuncay, K.; Park, A.; Ortoleva, P. Sedimentary basin deformation: An incremental stress approach. *Tectonophysics* **2000**, *323*, 77–104. [CrossRef]
2. Wangen, M. *Physical Principles of Sedimentary Basin Analysis*; Cambridge University Press: Cambridge, UK, 2010.
3. Bethke, C. A numerical-model of compaction-driven groundwater-flow and heat-transfer and its application to the paleohydrology of intracratonic sedimentary basins. *J. Geophys. Res.* **1985**, *90*, 6817–6828. [CrossRef]
4. Tuncay, K.; Ortoleva, P. Quantitative basin modeling: Present state and future developments towards predictability. *Geofluids* **2004**, *4*, 23–39. [CrossRef]
5. Formaggia, L.; Guadagnini, A.; Imperiali, I.; Lever, V.; Porta, G.; Riva, M.; Scotti, A.; Tamellini, L. Global sensitivity analysis through polynomial chaos expansion of a basin-scale geochemical compaction model. *Comput. Geosci.* **2013**, *17*, 25–42. [CrossRef]
6. Giovanardi, B.; Scotti, A.; Formaggia, L.; Ruffo, P. A general framework for the simulation of geochemical compaction. *Comput. Geosci.* **2015**, *19*, 1027–1046. [CrossRef]

7.   Colombo, I.; Porta, G.; Ruffo, P.; Guadagnini, A. Uncertainty quantification of overpressure buildup through inverse modeling of compaction processes in sedimentary basins. *Hydrogeol. J.* **2017**, *25*, 385–403. [CrossRef]

8.   Neuzil, C.E. Hydromechanical effects of continental glaciation on groundwater systems. *Geofluids* **2012**, *12*, 22–37. [CrossRef]

9.   Rayleigh, L. On the Dilatational Stability of the Earth. In *Proceedings of the Royal Society of London. Series A, Containing Papers of a Mathematical and Physical Character*; Royal Society: London, UK, 1906, Volume 77, pp. 486–499.

10. Love, A.E.H. *Some Problems of Geodynamics*; Cambridge University Press: Cambridge, UK, 1967.

11. Peltier, W.R.; Andrews, J.T. Glacial-Isostatic Adjustment—I. The Forward Problem. *Geophys. J. Int.* **1976**, *46*, 605–646. [CrossRef]

12. Peltier, W.R.; Wu, P.; Yuen, D.A., The Viscosities of the Earth's Mantle. In *Anelasticity in the Earth*; American Geophysical Union: Washington, DC, USA, 1981; pp. 59–77. [CrossRef]

13. Wu, P.; Peltier, W.R. Viscous gravitational relaxation. *Geophys. J. Int.* **1982**, *70*, 435–485. [CrossRef]

14. Biot, M.A. *Mechanics of Incremental Deformations. Theory of Elasticity and Viscoelasticity of Initially Stressed Solids and Fluids, Including Thermodynamic Foundations and Applications to Finite Strain*; John Wiley & Sons, Inc.: New York, NY, USA; London, UK; Sydney, Australia, 1965; p. xvii+504.

15. Ogden, R.W. Incremental statics and dynamics of pre-stressed elastic materials. In *Waves in Nonlinear Pre-Stressed Materials*; Springer: Vienna, Austria, 2007; Volume 495, pp. 1–26; doi:10.1007/978-3-211-73572-5_1. [CrossRef]

16. Nasir, O.; Fall, M.; Nguyen, S.; Evgin, E. Modeling of the thermo-hydro-mechanical-chemical response of sedimentary rocks to past glaciations. *Int. J. Rock Mech. Min. Sci.* **2013**, *64*, 160–174. [CrossRef]

17. Ruhaak, W.; Bense, V.; Sass, I. 3D hydro-mechanically coupled groundwater flow modelling of Pleistocene glaciation effects. *Comput. Geosci.* **2014**, *67*, 89–99. [CrossRef]

18. Sterckx, A.; Lemieux, J.M.; Vaikmäe, R. Representing glaciations and subglacial processes in hydrogeological models: A numerical investigation. *Geofluids* **2017**, *2017*. [CrossRef]

19. Turcotte, D.; Schubert, G. *Geodynamics*; Cambridge University Press: Cambridge, UK, 2002.

20. Peltier, W. Impulse response of a Maxwell Earth. *Rev. Geophys.* **1974**, *12*, 649–669. [CrossRef]

21. Whitehouse, P. Glacial isostatic adjustment and sea-level change. State of the art report. Technical report; Svensk Kärnbränslehantering AB Swedish Nuclear Fuel and Waste Management Co.: Solna, Sweden, 2009.

22. Whitehouse, P. Glacial isostatic adjustment modelling: Historical perspectives, recent advances and future directions. *Earth Surf. Dynam.* **2018**, *6*, 401–429 . [CrossRef]

23. Kjemperud, A.; Fjeldskaar, W. Pleistocene glacial isostasy—implications for petroleum geology. In *Structural and Tectonic Modelling and its Application to Petroleum Geology*; Elsevier: Amsterdam, The Netherlands, 1992; pp. 187–195.

24. Zieba, K.; Grøver, A. Isostatic response to glacial erosion, deposition and ice loading. Impact on hydrocarbon traps of the southwestern Barents Sea. *Mar. Pet. Geol.* **2016**, *78*, 168–183. [CrossRef]

25. Fjeldskaar, W.; Amantov, A. Effects of glaciations on sedimentary basins. *J. Geodyn.* **2018**, *118*, 66–81. [CrossRef]

26. Simo, J.; Hughes, T. Computational Inelasticity. In *Interdisciplinary Applied Mathematics*; Berlin/Heidelberg, Germany, 1998.

27. Colombo, I.; Nobile, F.; Porta, G.; Scotti, A.; Tamellini, L. Uncertainty Quantification of geochemical and mechanical compaction in layered sedimentary basins. *Comput. Methods Appl. Mech. Eng.* **2018**, *328*, 122–146. [CrossRef]

28. Remy, N. S-GeMS: The Stanford Geostatistical Modeling Software: A Tool for New Algorithms Development. *Geostat. Banff 2004* **2005**, 865–871._89. [CrossRef]

29. Gurtin, M. *An Introduction to Continuum Mechanics*; Elsevier Science: Amsterdam, The Netherlands, 1982.

30. Biot, M. General theory of three-dimensional consolidation. *J. Appl. Phys.* **1941**, *12*, 155–164. [CrossRef]

31. Both, J.; Borregales, M.; Nordbotten, J.; Kumar, K.; Radu, F. Robust fixed stress splitting for Biot's equations in heterogeneous media. *Appl. Math. Lett.* **2017**, *68*, 101–108. [CrossRef]

32. Coussy, O. *Poromechanics*; John Wiley & Sons: Hoboken, NJ, USA, 2004.

33. Cheng, A.D. *Poroelasticity*; Springer: Berlin/Heidelberg, Germany, 2016; Volume 27.

34. Taylor, T.; Giles, M.; Hathon, L.; Diggs, T.; Braunsdorf, N.; Birbiglia, G.; Kittridge, M.; Macaulay, C.; Espejo, I. Sandstone diagenesis and reservoir quality prediction: Models, myths, and reality. *AAPG Bull* **2010**, *94*, 1093–1132. [CrossRef]

35. Fowler, A.; Yang, X.S. Dissolution/precipitation mechanisms for diagenesis in sedimentary basins. *J. Geophys. Res. Solid Earth* **2003**, *108*, EPM 13-1–EPM 13-14. [CrossRef]

36. Ruffo, P.; Porta, G.; Colombo, I.; Scotti, A.; Guadagnini, A. Global sensitivity analysis of geochemical compaction in a sedimentary Basin. In Proceedings of the 1st EAGE Basin and Petroleum Systems Modeling Workshop: Advances in Basin and Petroleum Systems Modeling in Risk and Resource Assessment, Dubai, United Arab Emirates, 19–22 October 2014.

37. Čermák, V.; Laštovičková, M. Temperature profiles in the Earth of importance to deep electrical conductivity models. In *Electrical Properties of the Earth's Mantle*; Springer: Berlin/Heidelberg, Germany, 1987; pp. 255–284.

38. Pellerin, J.; Botella, A.; Bonneau, F.; Mazuyer, A.; Chauvin, B.; Lévy, B.; Caumon, G. RINGMesh: A programming library for developing mesh-based geomodeling applications. *Comput. Geosci.* **2017**, *104*, 93–100. [CrossRef]

39. Gibert, B.; Seipold, U.; Tommasi, A.; Mainprice, D. Thermal diffusivity of upper mantle rocks: Influence of temperature, pressure, and the deformation fabric. *J. Geophys. Res. Solid Earth* **2003**, *108*. [CrossRef]

40. Alzetta, G.; Arndt, D.; Bangerth, W.; Boddu, V.; Brands, B.; Davydov, D.; Gassmoeller, R.; Heister, T.; Heltai, L.; Kormann, K.; et al. The deal.II Library, Version 9.0. *J. Numer. Math.* **2018**, *26*, 173–183. [CrossRef]

41. Bangerth, W.; Hartmann, R.; Kanschat, G. deal.II – a General Purpose Object Oriented Finite Element Library. *ACM Trans. Math. Softw.* **2007**, *33*, 1–27. [CrossRef]

42. Arnold, D.N.; Brezzi, F.; Cockburn, B.; Marini, L.D. Unified analysis of discontinuous Galerkin methods for elliptic problems. *SIAM J. Numer. Anal.* **2001**, *39*, 1749–1779. [CrossRef]

43. Kanschat, G. Multilevel methods for discontinuous Galerkin FEM on locally refined meshes. *Comput. Struct.* **2004**, *82*, 2437–2445. [CrossRef]

44. Rivière, B. *Discontinuous Galerkin Methods for Solving Elliptic And Parabolic Equations*; Frontiers in Applied Mathematics; Society for Industrial and Applied Mathematics (SIAM): Philadelphia, PA, USA, 2008; Volume 35, p. xxii+190. [CrossRef]

45. Carriero, S. Tidal Forces Influence on Earth's Crust Deformation. A Massive Parallel Solver for the Solid Earth Tide Phenomenon. Master's Thesis, Politecnico di Milano, Milan, Italy, 2018.

46. Burman, E.; Hansbo, P. Fictitious domain finite element methods using cut elements: II. A stabilized Nitsche method. *Appl. Numer. Math.* **2012**, *62*, 328–341. [CrossRef]

47. Bukač, M.; Yotov, I.; Zakerzadeh, R.; Zunino, P. Partitioning strategies for the interaction of a fluid with a poroelastic material based on a Nitsche's coupling approach. *Comput. Methods Appl. Mech. Eng.* **2015**, *292*, 138–170. [CrossRef]

48. Lehrenfeld, C.; Reusken, A. Optimal preconditioners for Nitsche-XFEM discretizations of interface problems. *Numer. Math.* **2017**, *135*, 313–332. [CrossRef]

49. Hansbo, A.; Hansbo, P. An unfitted finite element method, based on Nitsche's method, for elliptic interface problems. *Comput. Methods Appl. Mech. Eng.* **2002**, *191*, 5537–5552. [CrossRef]

50. Hansbo, P.; Larson, M.; Zahedi, S. A cut finite element method for a Stokes interface problem. *Appl. Numer. Math.* **2014**, *85*, 90–114. [CrossRef]

51. Mikelić, A.; Wheeler, M. Convergence of iterative coupling for coupled flow and geomechanics. *Comput. Geosci.* **2013**, *17*, 455–461. [CrossRef]

52. Cerroni, D.; Radu, F.A.; Zunino, P. Numerical solvers for a poromechanic problem with a moving boundary. *Math. Eng.* **2019**, *1*, 824. [CrossRef]

Article

The Impact of Salt Tectonics on the Thermal Evolution and the Petroleum System of Confined Rift Basins: Insights from Basin Modeling of the Nordkapp Basin, Norwegian Barents Sea

Andrés Cedeño *, Luis Alberto Rojo, Néstor Cardozo, Luis Centeno and Alejandro Escalona

Department of Energy Resources, University of Stavanger, 4036 Stavanger, Norway
* Correspondence: andres.f.cedenomotta@uis.no

Received: 27 May 2019; Accepted: 15 July 2019; Published: 17 July 2019

Abstract: Although the thermal effect of large salt tongues and allochthonous salt sheets in passive margins is described in the literature, little is known about the thermal effect of salt structures in confined rift basins where sub-vertical, closely spaced salt diapirs may affect the thermal evolution and petroleum system of the basin. In this study, we combine 2D structural restorations with thermal modeling to investigate the dynamic history of salt movement and its thermal effect in the Nordkapp Basin, a confined salt-bearing basin in the Norwegian Barents Sea. Two sections, one across the central sub-basin and another across the eastern sub-basin, are modeled. The central sub-basin shows deeply rooted, narrow and closely spaced diapirs, while the eastern sub-basin contains a shallower rooted, wide, isolated diapir. Variations through time in stratigraphy (source rocks), structures (salt diapirs and minibasins), and thermal boundary conditions (basal heat flow and sediment-water interface temperatures) are considered in the model. Present-day bottom hole temperatures and vitrinite data provide validation of the model. The modeling results in the eastern sub-basin show a strong but laterally limited thermal anomaly associated with the massive diapir, where temperatures in the diapir are 70 °C cooler than in the adjacent minibasins. In the central sub-basin, the thermal anomalies of closely-spaced diapirs mutually interfere and induce a combined anomaly that reduces the temperature in the minibasins by up to 50 °C with respect to the platform areas. Consequently, source rock maturation in the areas thermally affected by the diapirs is retarded, and the hydrocarbon generation window is expanded. Although subject to uncertainties in the model input parameters, these results demonstrate new exploration concepts (e.g., deep hydrocarbon kitchens) that are important for evaluating the prospectivity of the Nordkapp Basin and similar basins around the world.

Keywords: salt; thermal modeling; basin modeling; source rock maturation; petroleum system

---

1. Introduction

The occurrence of evaporitic intervals in sedimentary basins and their subsequent mobilization play an important role in the evolution of the petroleum system [1,2]. Salt mobilization and diapirism control the spatial and temporal distribution of suprasalt reservoirs and source rocks [3–5], and they influence the style and timing of stratigraphic and structural traps. Salt's low permeability also inhibits the vertical migration of hydrocarbons, deflecting migration pathways [6]. Likewise, local salt depletion by salt withdrawal may lead to the formation of welds, which can provide migration pathways between subsalt source rocks and suprasalt reservoirs [7].

Salt has a thermal conductivity that is 2 to 3 times higher than sedimentary formations [6,8,9]. Accordingly, salt structures can modify the spatial and temporal thermal regime of the basin through focusing and defocusing of heat [8,10]. Salt domes create a dipole-shaped thermal anomaly with a negative thermal anomaly towards their base and a positive thermal anomaly in the suprasalt strata [9]. When salt bodies reach the surface, the dipole-shaped anomaly becomes a negative monopole, creating a conduit of low thermal resistance for heat conduction out of the basin [10]. Due to the difference in thermal conductivity between the salt and the surrounding sedimentary formations, thermal anomalies are also induced in the vicinity of salt bodies. The size and shape of these anomalies is controlled by the size of the salt bodies [8,10]. Maturation of kerogen within source rocks and reservoir diagenesis are temperature-controlled processes; therefore, any salt-related temperature deviation from the regional trend may have a significant impact on these processes.

The Nordkapp basin is a NE-SW trending rift basin of Late Paleozoic age located in the Norwegian Barents Sea (Figure 1A,B). Thick Pennsylvanian-Lower Permian layered evaporite sequences (LES) and their subsequent Mesozoic and Cenozoic mobilization generated numerous and closely spaced salt diapirs along the basin axis, and salt pillows generated along the basin margins [11–14]. Hydrocarbon exploration in the basin dates back to the 1980s. However, exploration has exclusively focused on the western sub-basin, while the central and eastern sub-basins remain underexplored (Rojo and Escalona, 2018). Only one non-commercial discovery in the western sub-basin, the Pandora discovery (well 7228/7-1A, Figure 1B), has been made in Triassic sediments which flank a salt diapir [15].

The Norwegian Barents Sea is known to host various petroleum systems sourced by Upper Paleozoic and Mesozoic organic rich intervals [16,17]. In the Nordkapp Basin, however, Paleozoic and Mesozoic strata are deeply buried due to Triassic halokinesis [18]. Therefore, it is tempting to assume that these source rocks became overmature for hydrocarbon generation in the Mesozoic. Interestingly, 2D structural restorations by [18,19] show that diapirs reached the seafloor since the Triassic, which may have cooled the basin and delayed maturation of the source rocks, as documented in offshore Mexico and Brazil [6,9]. Hence, this can open the possibility for a deeper prospectivity of the Nordkapp Basin and other salt-bearing basins in the Barents Sea.

Although there is potential for commercial discoveries in the Nordkapp Basin, there is an imperative need for understanding the impact of halokinesis on the thermal history of the basin and source rock maturation. In particular, the effect of closely spaced diapirs on the thermal evolution of this confined basin must be addressed. Therefore, in this study we explore the dynamic history of salt movement and its thermal effect by integrating 2D structural restorations with thermal modeling in order to: (1) evaluate how halokinesis impacted the thermal distribution of the basin through time, and (2) explore the implications of the modeled thermal history on the petroleum system and prospectivity of the basin. We use the structural restorations and selected model parameter values as a reasonable scenario to accomplish these objectives. Testing the sensitivity of different restoration or thermal parameters is beyond the scope of this work.

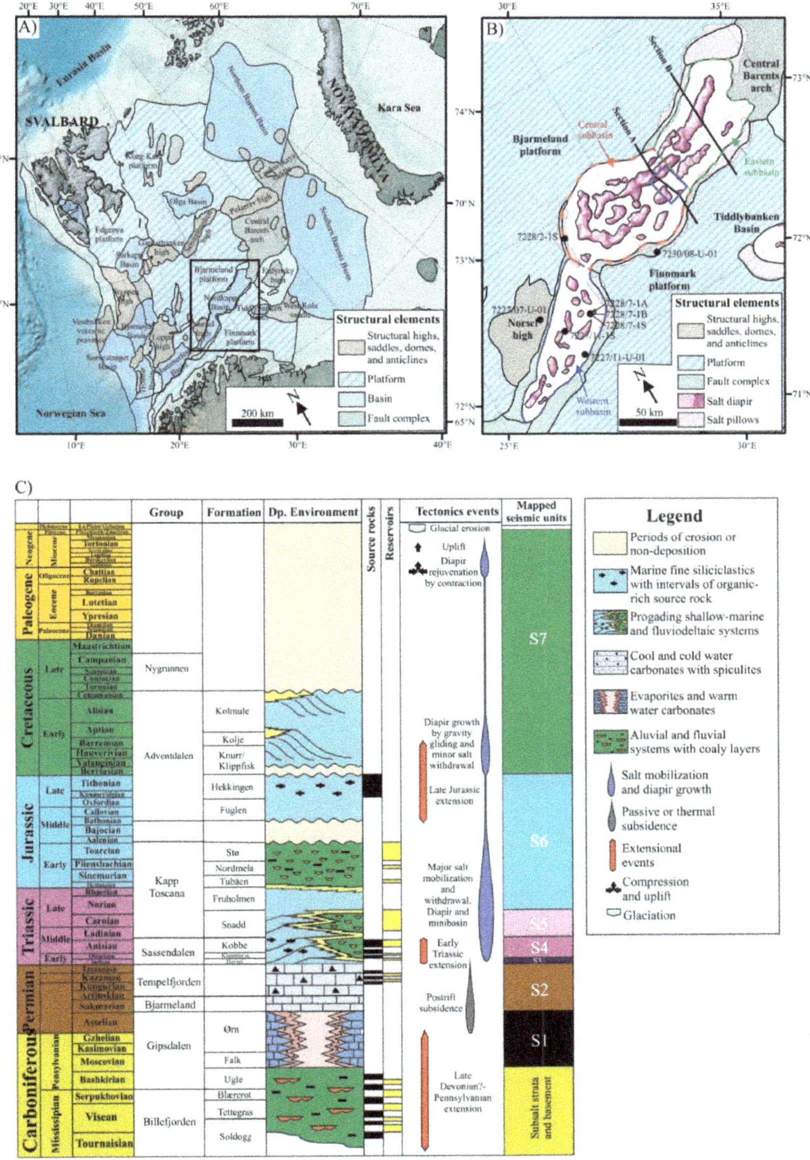

Figure 1. (**A**) Main structural elements of the Barents Sea. The Nordkapp Basin is indicated by the black rectangle. (**B**) Main structural elements of the Nordkapp Basin (modified from Rojo et al. [18]. The basin is divided in three sub-basins: western, central and eastern sub-basins. Black dots are exploration and shallow stratigraphic wells, whereas black lines show the location of the studied sections through the central (section **A**) and eastern (section **B**) sub-basins. (**C**). A chronostratigraphic chart illustrating the main stratigraphic units, depositional environments, tectonic events, and mapped seismic units of the Barents Sea and Nordkapp Basin (based on [17,18]).

2. Geologic Evolution of the Nordkapp Basin

2.1. Late Paleozoic

The Nordkapp Basin formed as the result of two extensional events of different orientation [19–23] (Figure 1B): (1) pre-Mississippian NE-SW extension, which reactivated previous NW-SE Caledonian structures and formed the central sub-basin; and (2) Pennsylvanian NW-SE extension which reactivated NE-SW Caledonian structures, forming the western and eastern sub-basins. Based on outcrops in Svalbard [24,25] and wells in the Finnmark platform [26], the syn-rift section of the Nordkapp basin is expected to contain siliciclastics interbedded with coal of the Mississippian Billefjorden Group (Figure 1C). Potential Mississippian reservoirs include sandstones deposited within meandering and braided fluvial systems, and interlayered coals are potential gas-prone source rocks.

Basin extension continued from the Pennsylvanian to the Early Permian, and it was followed by a period of thermal subsidence until the end of the Paleozoic [17,27]. Basin restriction favored the precipitation of syn-rift to early post-rift evaporites along the basin axis, whereas deposition of warm water carbonate buildups and gypsum occurred at the basin boundaries (Gipsdalen Group) [28–30] (Figure 1C). These deposits were overlain by cool-water carbonates of the Bjarmeland Group and cold water carbonates and spiculites of the Tempelfjorden Group [17,27] (Figure 1C). Based on the Alta and Ghota discoveries in the Loppa High, Upper Permian carbonaceous mudstones and limestones of the Tempelfjorden Group are expected to contain intervals of oil-prone source rocks, and possible reservoirs associated with karstified carbonates in the Nordkapp Basin [17].

2.2. Mesozoic

The onset of salt mobilization and diapirism occurred during the earliest Triassic, and it was triggered by basement-involved extension and differential loading induced by prograding siliciclastic sediments sourced from the Uralides [14,18,19,31]. Minibasin growth and diapir uplift ceased by the end of the Middle Triassic due to welding of the underlying salt [14,18]. Diapir growth after the Middle Triassic is attributed to evacuation of remaining salt adjacent to diapirs and basinward suprasalt gliding and contraction [14]. The Triassic minibasins of the Nordkapp Basin record the NW-SE transgressive-regressive fluviodeltaic systems of the Sassendalen and Kapp Toscana groups, which progressively prograded towards the NW of the Barents Shelf (Figure 1C) [32–35]. Based on the Goliat discovery in the Hammerfest Basin, oil and gas-prone source rocks may be present in the Sansendalen Group, whereas shallow marine and fluviodeltaic reservoirs can be present in the Sassendalen and Kapp Toscana groups [5,17].

The Late Triassic-Early Jurassic was marked by a regional decrease in accommodation in the Barents Shelf, including the Nordkapp Basin [36]. This resulted in a complex drainage system characterized by erosion and reworking of previous Triassic deposits, which favored the deposition of shallow marine and fluviodeltaic Lower Jurassic reservoirs of good quality (e.g., Stø Formation in the Kapp Toscana Group) [17]. A new episode of extension associated with the initial opening of the North Atlantic to the west of the Barents Shelf [21], partially affected the Nordkapp Basin during the Late Jurassic-Early Cretaceous [12] (Figure 1C). Normal fault systems at basin boundaries caused basinward gravity gliding of suprasalt strata and subsequent thin-skinned contraction and growth of pre-existing diapirs [14]. Lower Jurassic deposits were flooded and overlain by the Upper Jurassic marine black shales of the Hekkingen Formation, which is considered to be a potential source rock in the Barents Sea [16,17]. These marine shales were in turn covered by Lower Cretaceous, SSW prograding sediments sourced from the northern Barents Shelf [37,38].

2.3. Late Cretaceous-Cenozoic

During the Late Cretaceous-Cenozoic, the Nordkapp Basin underwent several contractional events, which are attributed to plate tectonic reorganizations related to the opening of the North Atlantic (Figure 1C; [21]). Consequently, pre-existing diapirs in the Nordkapp Basin were rejuvenated by contractional diapirism [13,14]. Finally, successive events of uplift and erosion, including the Pleistocene glacial erosion, removed about 1.5 km of Cretaceous and Cenozoic strata in the basin [39,40].

3. Methodology

In order to model the thermal evolution of the Nordkapp Basin, we considered two processes: (1) the evolving basin geometry, and (2) the evolution of thermal boundary conditions, i.e., basal heat flow and sediment-water interface temperatures. The evolving basin geometry was reconstructed from structural restoration of two sections, one across the central sub-basin (section A) and another across the eastern sub-basin (section B) (Figure 2). Section A has deeply rooted, narrow and closely spaced diapirs (Figure 2A), whereas section B contains a shallower rooted, wider, isolated diapir (Figure 2B). Before the restoration process, the sections were depth-converted using the velocity model from Rojo et al. [18], which uses interval velocities and k factors (change in interval velocities vs depth) from wells 7228/9-1, 7228/7-1, 7125/1-1, 7124/3-1, and 7229/11-1 (Figure 3).

Details of the restoration are described in Rojo et al. [18]. In short, kinematic restorations were performed using the *Move* software (Midland Valley). These restorations account for uplift and erosion in the Late Cenozoic, restore the Cenozoic and pre-diapir units using flexural slip, restore the syn-diapir units using vertical shear, and assume flexural isostasy with a lithospheric elastic thickness of 20 km [41].

Thermal modeling of selected restoration stages (paleo-geometries) was performed using the *PetroMod* software (Schlumberger). Each stage was imported into the software, and polygons representing each seismic unit (Figure 2B,D) were digitized. At each stage, the lower boundary condition of the thermal model is the heat flow at the basement-sediment interface. The evolution of this basal heat flow through time was reconstructed from the present thermal gradient of the basin from nearby wells, the history of rifting and evolution of the stretching (β) factor [42], and inverse modeling using a modified McKenzie model [43]. The upper boundary condition of the thermal model is the sediment-water interface temperature (SWIT). This surface temperature was reconstructed from the paleo-latitude of the basin and water depths over time. The following section describes the input paleo-geometries, thermal boundary conditions, and model parameters in detail.

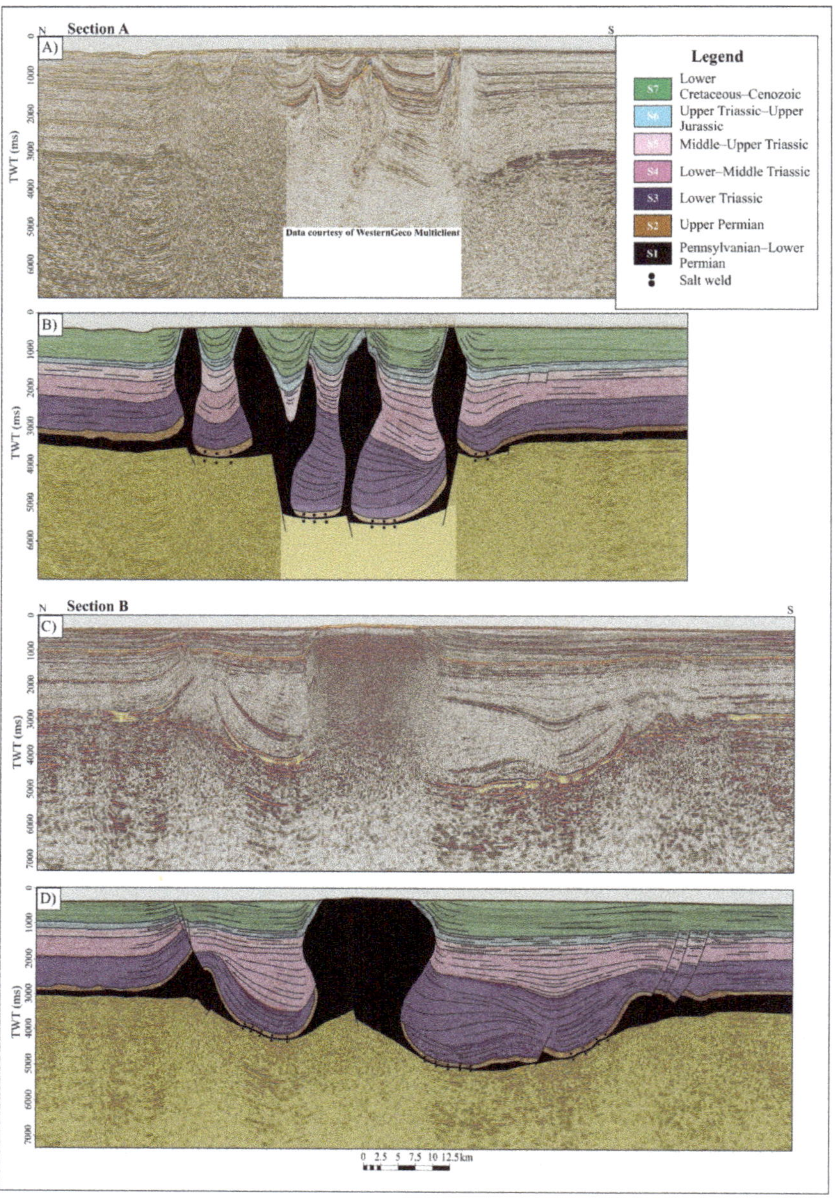

Figure 2. (**A**) Uninterpreted seismic section A across the central sub-basin. The central seismic panel is from a higher-resolution, full azimuth seismic data provided by WesternGeco Multiclient. (**B**) Interpreted section A. The base salt was estimated based on parallel seismic sections that extend to 7000 ms. (**C**) Uninterpreted seismic section B across the eastern sub-basin. (**D**) Interpreted section B. Legend shows the interpreted seismic units (Figure 1C). See Figure 1B for location of these sections. Figures modified from Rojo et al. [18].

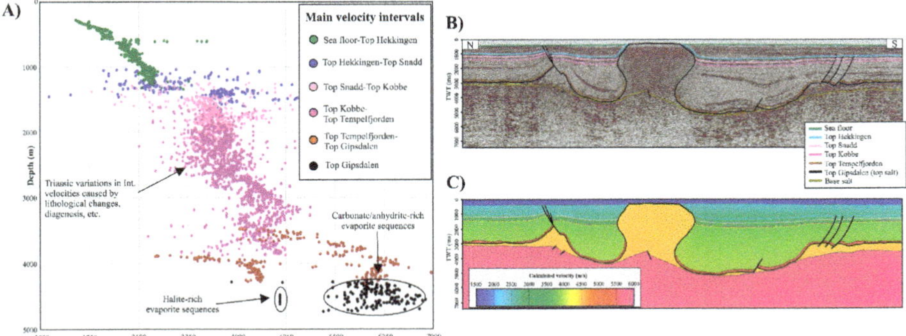

Figure 3. (**A**) Plot showing calculated interval velocities vs depth from wells 7228/9-1, 7228/7-1A, 7125/1-1, 7124/3-1, 7229/11-1. (**B**) Seismic section B showing the location of the main velocity transitions. (**C**) Seismic section B displaying the interval velocities used to depth convert the seismic profile (modified from Rojo et al. [18]).

4. Paleo-Geometries, Boundary Conditions, and Model Parameters

4.1. Paleogeometries

Figure 4 displays the restoration of the two cross sections in the central (section A, Figure 4A) and eastern (section B, Figure 4B) sub-basins. The restoration stages are the paleo-geometries input to the thermal model. It should be noted that in Figure 4 and later figures (Figures 7 and 9), for the purpose of display and comparison between sections, section A is not vertically exaggerated while section B has a vertical exaggeration of 2. Below we present a short description of the geologic evolution portrayed by the restorations.

4.1.1. Section A, Central Sub-basin

Syn-rift to early post-rift layered evaporate sequences (LES) precipitated in a symmetric graben and were overlain by pre-kinematic Upper Permian carbonates (Figure 4A, VIII). During the earliest Triassic, thick-skinned extension accompanied by differential loading of Triassic sediments sourced from the Urals, created a structural style consisting of NW shifting patterns in salt withdrawal and ENE-WSW-trending passive diapirs (Figure 4A, V–VII). Passive diapirism and welding occurred diachronously across the section, and first occurred in the northern part of the graben due to the preferential loading of salt in this region (Figure 4A, V). In the southern part, minor loading caused the formation of a salt pillow. During the Early-Middle Triassic, differential loading focused on the southern part of the graben, which caused a shift in salt withdrawal and subsequent salt expulsion towards the south (Figure 4A, IV). This favored the formation of a half turtle structure and a passive diapir above the southern boundary fault. By the end of this period, the minibasins grounded the base salt forming salt welds (Figure 4A, IV). During the Late Triassic to Late Jurassic, thick-skinned extension induced diapir collapse of the northern diapir since the underlying salt was almost totally evacuated (Figure 4A, III). This formed a minibasin above the diapir crest, filled by Upper Triassic and Jurassic strata. Minor differential loading of the remaining salt in the south favored the continued growth of the southern and central salt diapirs (Figure 4A, III). During the Cretaceous and Cenozoic, higher sedimentation rates than diapir growth rates led to burial of the salt diapirs (Figure 4A, II). This episode was then followed by Late Cenozoic contraction, diapirism and uplift, which eroded approximately 1.5 km of Cretaceous and Cenozoic strata (Figure 4A, I).

4.1.2. Section B, Eastern Sub-basin

LES in this section precipitated in a more sag-type basin, which was later covered by Upper Permian carbonates (Figure 4B, VIII). Earliest Triassic sediment loading accompanied by thick-skinned extension generated expulsion rollovers towards the north and a central passive diapir (Figure 4B, V–VII). Differential loading occurred preferentially at the basin axis forming the first salt weld by the earliest Triassic, whereas a significant amount of salt remained on the northern and southern basin boundaries (Figure 4B, V–VI). During the Middle Triassic, differential loading focused mostly on the north, forming the northern minibasin and favoring the growth of the northern salt pillow and the central passive diapir (Figure 4B, IV). To the south, differential loading caused a shift in salt withdrawal, which resulted in the welding of the southern minibasin and the formation of a half-turtle structure. Continuous basin subsidence by thick-skinned extension and salt withdrawal produced the flexure and extension of suprasalt strata at basin boundaries, resulting in the generation of suprasalt fault complexes (Figure 4B, IV). By the end of the Middle Triassic, the northern and southern minibasins grounded the base salt (Figure 4B, IV). Even though most of the underlying salt was evacuated, the central salt diapir continued to grow during the rest of the Mesozoic by gliding of suprasalt strata towards the basin axis and thin-skinned contraction (Figure 4B, II–IV). Higher sedimentation rates than diapir growth rates caused the burial of salt structures during the Cretaceous and Cenozoic (Figure 4B, II). During the Cenozoic, salt structures were rejuvenated by contraction. This episode was followed by Late Cenozoic uplift and erosion, which removed approximately 1.5 km of Cretaceous and Cenozoic strata (Figure 4B, I).

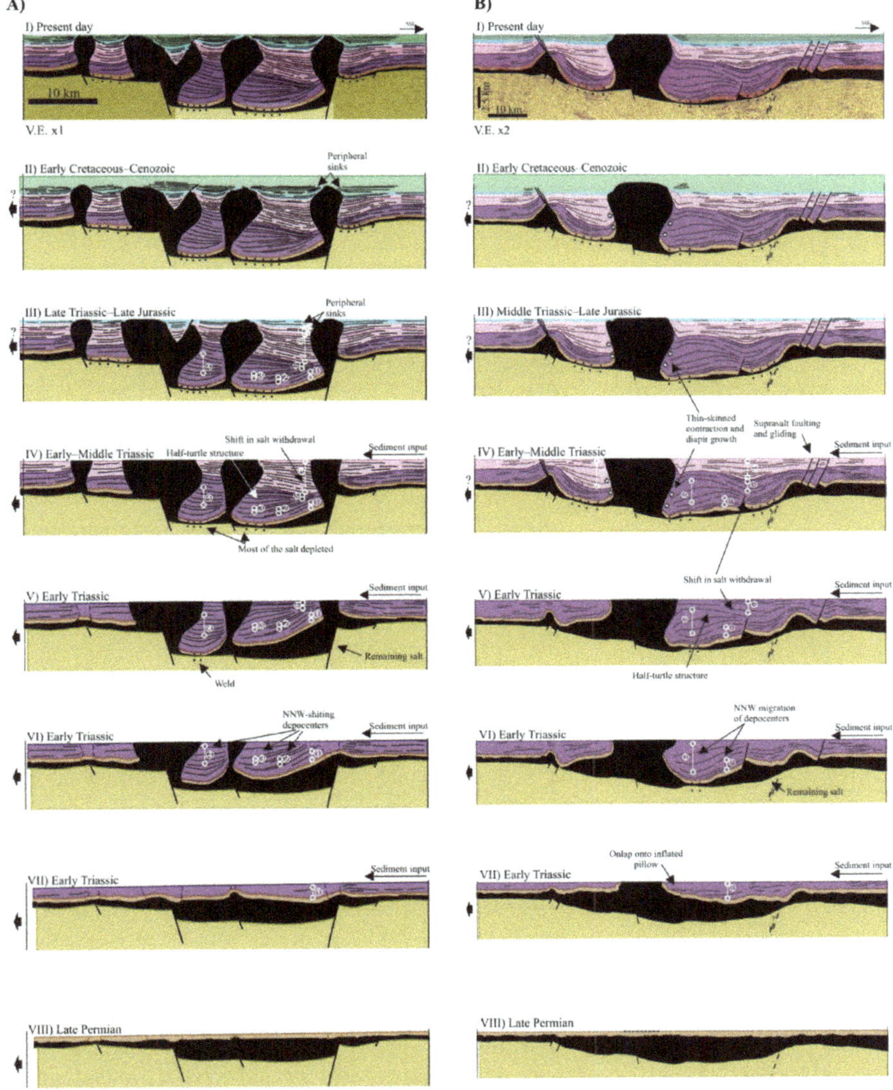

Figure 4. (**A**) Structural restoration of section A (Figure 2B) in the central sub-basin. (**B**) Structural restoration of section B (Figure 2D) in the eastern sub-basin. The different restoration stages are the paleo-geometries input to the thermal model. Colored rock units correspond to the interpreted seismic units in Figure 2 (and Figure 1C). See Figure 1B for location of the sections. Restorations modified from Rojo et al. [18].

4.2. Thermal Boundary Conditions

Thermal boundary conditions determine the primary energetic inputs to reproduce the temperature history of the basin and, consequently, for the maturation of source rocks. As mentioned above, the sediment-water interface temperature (SWIT) represents the upper boundary condition whereas the basal heat flow represents the lower boundary condition. These parameters were constrained to ensure that the modeled thermal field best-fits temperature data in the basin.

For the SWIT, we used the automatic SWIT tool in PetroMod, which extracts a standard temperature at sea level over geological time based on the basin's present-day geographic location and paleo-latitude (Figure 5A). A transformation that corrects the surface temperature against the paleo-water depth was applied and the SWIT estimated. The latitudinal position assigned to the Nordkapp basin was 72°N, and paleo-water depths (Figure 4B) were taken from [18].

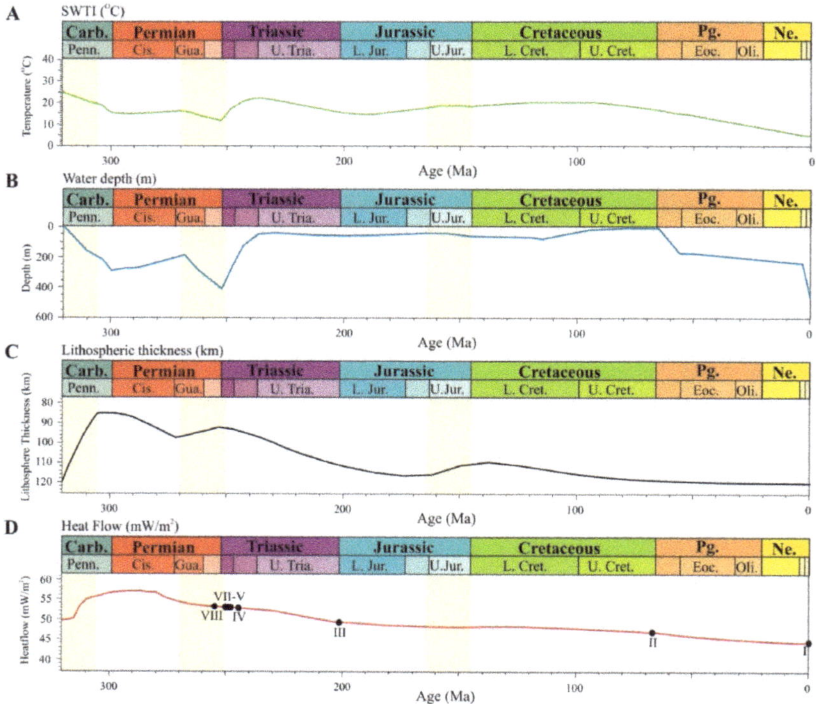

Figure 5. Modeled (**A**) sediment-water interface temperature (SWIT), (**B**) water depth, (**C**) lithospheric thickness, and (**D**) basal heat flow through time in the Nordkapp Basin. Yellow rectangles show the three rifting periods. Dots in D are the selected restoration stages in Figure 4.

The basal heat flow was calculated by reconstructing the lithospheric thickness from the Late Carboniferous to the present (Figure 5C). In order to do this, we adopted mean values for the stretching factor (β) of each rift phase (Figure 5, yellow rectangles): Late Carboniferous (320–305 Ma, β = 1.6), Late Permian (270–250 Ma, β = 1.25), and Late Jurassic–Early Cretaceous (165–145 Ma, β = 1.15), as proposed by [42]. The Post-Caledonian lithosphere and crustal thicknesses were defined at the rift initiation. An initial lithospheric thickness of 120 km and an initial crustal thickness of 35 km (17.5 km upper crust, 17.5 km lower crust) were adopted from Clark et al. (2014). The present-day crustal thickness is 18 km (16 km upper crust, 2 km lower crust) as documented by [41,42,44]. The temperature at the lithosphere-asthenosphere boundary was set to 1300 °C and the mantle heat flow to 30 mW/m^2. Figure 5C,D depict the modeled lithospheric thickness and basal heat flow through time, respectively.

A good fit between measured and calculated temperature values of five exploration wells (7228/7-1B, 7228/7-1S, 7228/2-1S, 7228/7-1A, 7228/2-1S) and three shallow wells in the Nordkapp Basin (7227/11-U-02, 7227/07-U-01, 7230/08-U-01) was achieved using estimated basal heat flow values of ~45 mW/m^2 (Figure 6A). Extrapolating the thermal model calculated in the western sub-basin to the entire Nordkapp Basin is unrealistic due to possible variations in the initial thickness of the Post-Caledonian crust and lithosphere [42], and the magnitude of rifting across the basin. Nevertheless, we chose to do so it in order keep the model as simple as possible and to avoid biasing the model with poorly-constrained inputs. In addition, the effects of episodic glacial loading-unloading and erosion during the Late Cenozoic documented by several authors [45–49] are not considered in the model.

Unlike temperature, vitrinite reflectance data display higher maturities than the calculated trend (Figure 6B). Vitrinite measurements from wells 7228/7-1A and 7228/7-1S in the central part of the basin plot closer to the calculated trend than corresponding values from well 7228/2-1S in the northern rim of the basin (Figure 6B, see Figure 1B for well locations). This most likely reflects different amounts of erosion at different locations in the basin. Well 7228/2-1 S shows a characteristic pattern of increasing thermal maturity with stratigraphic age from Cenozoic to Middle Triassic (Figure 6B). However, at ~3600 m depth, a sudden increase in vitrinite values is observed near the top of the Lower Triassic Havert Formation. Here, the well values define a steeper vitrinite trend, implying higher temperature gradient. Igneous activity could explain this higher thermal gradient, but there is no evidence of such activity. The modeled paleo-heat flow (Figure 5D) depicts higher flow rates (~54 mW/m^2) during and immediately after the Late Permian rifting, which seem to be a plausible explanation to the higher thermal gradient. Thus, the modeled heat flow (Figure 5D) is a reasonable scenario for the Nordkapp Basin, and together with the SWIT (Figure 5A), it defines the boundary conditions for the thermal model. Transient heat-flow conditions were assumed at all times.

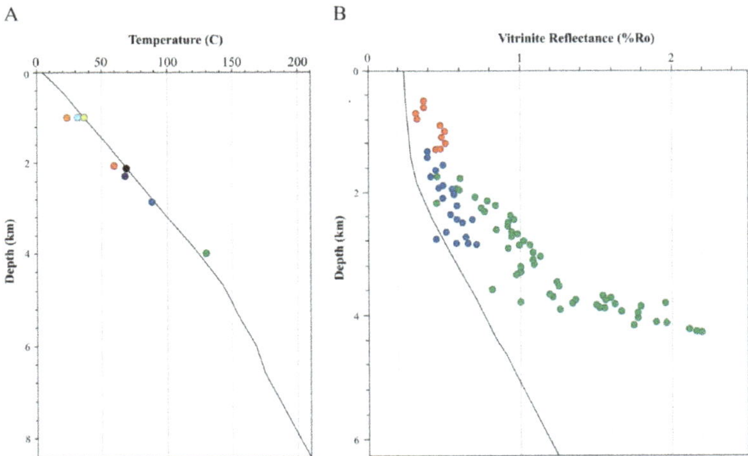

Figure 6. (**A**) Bottom hole temperature measurements of wells 7228/7-1B (black), 7227/11-U-02 (orange), 7227/07-U-01 (light blue), 7230/08-U-01 (light green), 7228/7-1S (red), 7228/2-1S (purple), 7228/7-1A (dark blue), and 7228/2-1S (dark green). Line shows the modeled, present temperature versus depth trend. (**B**) Vitrinite reflectance data from wells 7228/7-1S (red), 7228/7-1A (blue), and 7228/2-1S (green). Line shows the modeled vitrinite reflectance. Figure 1B shows the location of the wells.

4.3. Model Parameters

The restoration stages in Figure 4 were imported and digitized in Petromod for thermal modeling. Sediment types and ages were defined based on the seismic units in Figure 1C. Rock properties were assigned to each unit (Table 1). Lithologies were user-defined considering the proportion of different sediments, i.e., sandstone, siltstone, shale, limestone, coal, and salt. Each defined lithology was assigned an initial porosity value and a porosity versus depth trend (factor c in Table 1) that decreases exponentially with the greatest porosity loss happening at shallow depths [50,51]. The thermal conductivity of the various units was set to vary linearly as a function of porosity. The model also accounts for variations in the thermal conductivity with increasing temperature following the model described by [52], i.e., the conductivity of salt drops from 6.5 (W/mk) at 20 °C to 4.14 (W/mk) at 220 °C.

Finally, the thermal model was computed numerically using finite elements on a regular grid with $300(x) \times 150(y)$ cells.

Five source rock intervals were modeled: The Upper Jurassic Hekkingen Formation (S6), the Upper Triassic Snadd Formation (S5), the Lower to Middle Triassic Kobbe Formation (S4), the Permian Tempelfjorden Group (S2), and a Carboniferous pre-salt coaly source rock (Figure 1C, green rows in Table 1). Since S4 is remarkably thick (Figure 2), we considered the source rock interval, presumably correlatable with the Botneheia Formation in the western Barents Sea [33,35,53], to be only in the uppermost 300 m of this unit. The large stratigraphic interval encompassed by the modeled source rocks reflects the uncertainty in source rock distribution and thickness in the basin.

Table 1. Rock units and their parameter values for thermal modeling. For lithology definition, the percentage of sandstone (ss), siltstone (slt), shale (sh), limestone (ls), coal (co) and salt (sl) is defined. Source rock intervals are colored green. Values of thermal conductivity and heat capacity are given for each unit at 20 °C.

Unit	Lithology	Density (kg/m^3)	Surface Porosity (%)	Factor c (1/km)	Thermal Conductivity (W/mK)	Radiogenic Heat (microW/m^3)	Specific Heat Capacity (Kcal/K/Kg)
S7	ss20, sh80	2730	53.5	0.57	2.01	1.31	0.21
S6	ss10, shs20,sh70	2721	59.5	0.65	2.11	1.37	0.21
S5	ss70. sh30	2714	49.7	0.47	3.03	0.88	0.2
S4	ss30, sh70	2566	61.3	0.67	2.13	1.31	0.21
S3	ss30, sh70	2706	61.3	0.67	2.13	1.31	0.21
S2	sh10, ls90	2709	52.9	0.55	2.51	0.73	0.2
S1	sl100				6.5	0.01	0.21
Pre-salt	ss25, sh50, co25	2540	63.65	0.64	1.34	0.97	0.23
Basement	Gneiss				2.7	2	0.19

5. Results

5.1. Thermal Evolution

Figure 7 depicts the modeled evolution of temperature through time in the central sub-basin (Figure 7A) and eastern sub-basin (Figure 7B). During the initial Late Permian stage prior to salt mobilization, in the central part of the basin the thermal gradient is reduced within the salt (up to 3 km thick) as depicted by widely spaced isotherms (Figure 7A,B, VIII). Temperatures are elevated above the salt, implying enhanced heat flow. Towards the shoulders of the basin, the regional thermal gradient is reestablished.

In the Early to Middle Triassic stages, when salt was mobilized and reached the surface (Figure 7, IV to VII), the temperature distribution was altered as salt diapirs provided vertical conduits for conducting heat out of the basin, inducing a negative thermal anomaly in the interior of the diapirs. This thermal anomaly is highest at the center of the diapirs and gradually decreases outwards. Isotherms within the diapirs are widely spaced and, as a consequence, isotherms are deeper below salt than in the adjacent minibasins. Around the salt diapirs, the temperature is also affected by the reduced geothermal gradient inside the diapir. The wide, isolated diapir in the eastern sub-basin shows this effect more clearly (Figure 7B, IV). In this diapir, the temperature contrast between the salt interior and surrounding minibasins at ~3 km is as much as 30 °C. In the central sub-basin (section A), the negative thermal anomaly of a single diapir cannot be seen as the diapirs are narrower and closer together (Figure 7A, IV). In this section, the thermal effect of each diapir mutually interferes, resulting in a combined effect that lowers the isotherms below the regional trend. In the center of the basin, temperatures are as much as 60 °C lower than those in the platform areas far from the salt effect.

After deposition of the Upper Jurassic and Cretaceous-Cenozoic sediments, the salt diapirs are no longer connected to the surface (Figure 7, stages II and III). A positive thermal anomaly developed above the closely spaced diapirs in section A and above the single diapir in section B owing to the focusing of heat by the underlying salt. In section A, the temperature rises ~10 °C above the largest diapir compared to the surrounding sediments (Figure 7A, II) while in section B the temperature rises ~15 °C above the massive diapir (Figure 7B, II). The negative thermal anomaly within the salt diapirs is still present, but at a lower intensity than in previous stages.

Several pseudo-wells along the sections demonstrate the thermal effect of the salt at present day. In the central sub-basin (section A), we extracted temperatures in pseudo-wells through a small diapir at 9 km from the northern edge of the cross-section, and through a wider diapir at 17 km. Temperatures were also extracted in pseudo-wells at 27 km and at 56 km in a minibasin and a platform area, respectively (Figure 7A, I). Figure 8A depicts the distribution of temperature in these wells. The thermal gradient in the small diapir (9 km well) is greater than in the large diapir (17 km well). The temperature difference between these wells is ~15 °C at a depth of 5 km. In both diapirs, the thermal gradient beneath the salt increases and depicts a similar trend to the one observed at the well in the platform area (56 km well). The temperature beneath the small diapir (9 km well) is still ~25 °C higher than below the large diapir (17 km well) (Figure 8A). These differences are most likely related to the size of the diapirs, with the wider diapir conducting heat more efficiently. Significant thermal differences exist between the two wells outside salt structures (27 and 56 km wells). The maximum temperature difference between these two wells is ~35 °C at a depth of 5 km (Figure 8A). Although both wells are outside the salt diapirs, the well in the minibasin (27 km) is between closely spaced diapirs whose mutually interfering effect induces a broad negative thermal anomaly in the central part of the basin. Therefore, temperatures in the minibasin resemble those in the salt diapirs rather than those in the platform areas.

In the eastern sub-basin (section B), we extracted temperatures in pseudo-wells at 29 km from the northern edge of the cross-section in a minibasin, at 45 km through the massive diapir, and at 100 km in the platform area (Figure 7B, I). Figure 8B displays the distribution of temperature in these wells. The temperature gradient in the salt diapir (45 km well) is considerably lower than in the minibasin and platform area (29 and 100 km wells). The greatest temperature difference is ~110 °C at a depth of 7 km between the wells in the diapir (45 km) and the platform area (100 km). The well in the minibasin (29 km) shows intermediate temperatures, although still it is ~70 °C warmer than the diapir at a depth of 7 km (Figure 8B). These significant thermal differences are most likely related to the large size (width) of the salt diapir.

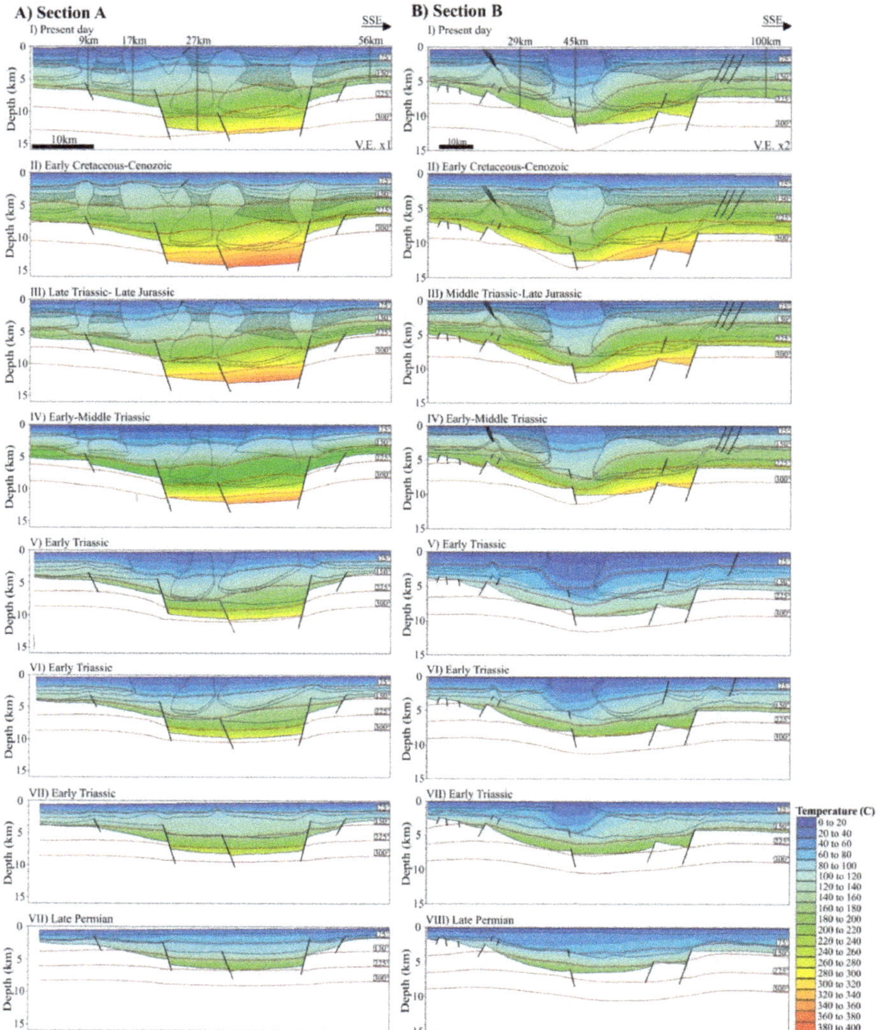

Figure 7. Evolution of temperature through time in (**A**) section A in the central sub-basin, and (**B**) section B in the eastern sub-basin. Black lines are unit contacts and red lines are isotherms. For guidance, stippled unit is the Lower to Middle Triassic Kobbe Formation (S4). Figure 1B shows the location of the sections.

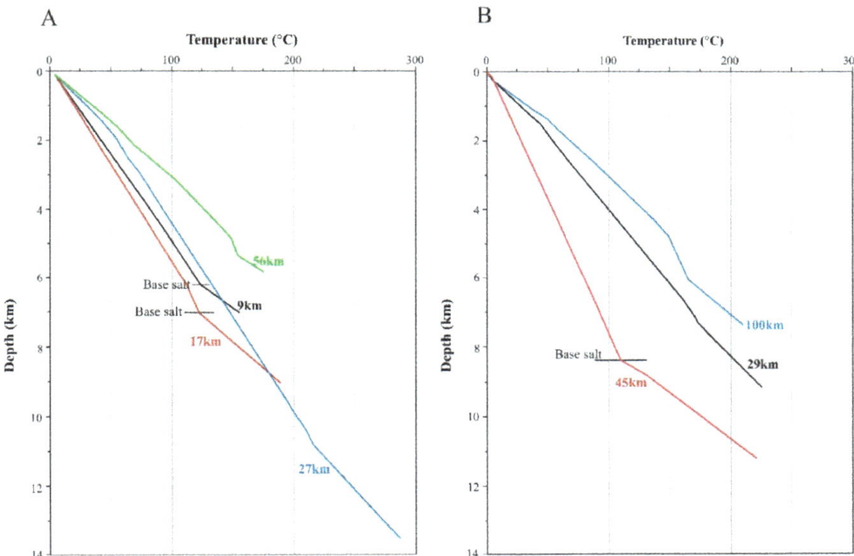

Figure 8. Modeled, present temperature distribution in pseudo-wells along (**A**) section A in the central sub-basin, and (**B**) section B in the eastern sub-basin. Thick lines in the minibasins and platforms wells show the temperature distribution in potential reservoirs. Figure 6 (I) shows the location of the pseudo-wells.

5.2. Source Rock Maturation

Figure 9 depicts the evolution in thermal maturity that each of the modeled source rock units experienced through time. The vitrinite reflectance model developed by Sweeney and Burnham [54] was implemented for maturation modeling of the source rocks. It simulates the onset of the oil window at 0.55%Ro, and the upper limit of thermogenic gas (dry) generation at 4%Ro.

In the Late Permian, pre-salt source rocks in the central part of the basin were buried at depths of more than 4 km, and maturities of 0.7–1.0%Ro were attained (Figure 9, VIII). In the basin's margins, the same rocks were shallower at ~2 km and accordingly, they were immature (<0.5%Ro). Upper Permian (S2) source rocks were shallowest and immature.

From the Early Triassic until the Late Jurassic, salt mobilization was confined to the central part of the basin (Figure 9, III–VII). Salt evacuation resulted in minibasin subsidence and infill, causing progressive maturation along the axes of the rapidly subsiding minibasins. Pre-salt and Upper Permian (S2) source rock units were deeply buried in the central part of the basin where their maturities were highest. Maturity in these units systematically decreases updip away from the salt bodies. In the central sub-basin, source rock maturity exceeded 2.0%Ro and reached maturities beyond any hydrocarbon generation (>4%Ro, Figure 9A, III). In the eastern sub-basin, the thermal maturities generally were lower and ranged from 0.55 to 4.0%Ro, although locally these source rocks exceeded values of 4.0%Ro (Figure 9B, III). The base of the source rocks modeled at the top of the Lower to Middle Triassic (S4) unit locally attained maturity values higher than 0.55%Ro in the central sub-basin (Figure 9A, III).

From the Early Cretaceous through the Cenozoic, widespread sedimentation increased the burial of source rocks to their maximum maturity. In the central sub-basin, Paleozoic (pre-salt and S2) source rocks mostly became overmature (>4%Ro) in the central part of the basin, whereas over the platforms they lied within the gas window (1.3–4%Ro, Figure 9A, II). In the eastern sub-basin, the Paleozoic source rocks lied within the gas window, except in the southern mini-basin where they were overmature (Figure 9B, II). The maturity of the Triassic (S4 to S5) and Jurassic (S6) source rocks generally exceeded 0.55%Ro, and locally reached a maximum of 1.0–1.3%Ro in the central sub-basin (Figure 9A, II). In the eastern sub-basin, Upper Triassic to Jurassic source rocks (S5 and S6) were immature near the diapir, while Upper Permian source rocks (S2) were still in the late oil to wet gas window adjacent to this diapir (Figure 9B, II). A similar although less pronounced effect is observed along the flanks of the northern, widest diapir of the central sub-basin (Figure 9A, II).

In the Late Cenozoic, compression caused widespread uplift and exhumation, and subsequently the Mesozoic-Cenozoic section underwent erosion, freezing maturation. At present-day, exceptionally low maturity in the Upper Permian to Jurassic source rocks exists along the flanks of the salt diapirs (Figure 9, I). This reduction in thermal maturation in the vicinity of salt diapirs is ubiquitous, indicating that source rock maturation was not only controlled by burial, but also by the thermal anomalies induced by the salt structures. In the central part of the basin, pre-salt and Upper Permian (S2) source rocks are deeper and mostly overmature in the central sub-basin, while in the eastern sub-basin they are in the late oil to dry gas window (Figure 9, I).

In order to better visualize the effect of salt on source rock maturation, we generate vitrinite reflectance versus depth trends in two pseudo-wells through minibasin locations, at 27 km in section A and at 29 km in section B (Figure 9, I). 1D modeling was performed both with the presence of the salt diapirs (continuous lines, Figure 10) and with the salt diapirs substituted by sediments (dashed lines, Figure 10). It should be noted that maturation freezes at ~23 Ma when regional uplift of the basin is simulated.

At the 27 km location in the central sub-basin, pre-salt and Upper Permian (S2) source rocks experienced a rapid maturation that drove them into the dry (pre-salt) and wet gas (S2) window at ~240 Ma (Figure 10A). Thermal maturation continued, and both intervals became overmature (>4%Ro) at around 150 Ma (pre-salt) and 50 Ma (S2). Without the negative thermal effect of salt structures, these source rocks would have entered the oil window as early as 315 Ma (pre-salt) and 255 Ma (S2), and they would have become overmature at ~255 Ma. The Lower-Middle Triassic (S4) source rocks entered the oil window at ~220 Ma, and gradually maturated to present values of ~1%Ro (Figure 10A). In the absence of salt structures, these rocks would have reached the oil window at ~250 Ma, and they would presently be in the wet gas window (~1.7%Ro). The Middle-Upper Triassic (S5) source rocks entered the oil window at ~85 Ma and reached a maximum vitrinite reflectance of ~0.8%Ro. On the other hand, The Upper Triassic-Upper Jurassic (S6) source rocks barely reached the oil window at ~30 Ma (Figure 10A). These two organic rich intervals would have been oil mature at ~135 Ma and 75 Ma, and they would presently be in the late (1.15%Ro) and main (0.75%Ro) oil window, respectively, if no salt structures existed in the basin.

In the eastern sub-basin at the 29 km location, the pre-salt and Upper Permian (S2) source rocks experienced a rapid transition from immature to the wet gas window in the Early Triassic (~250 Ma, Figure 10B). Maturation continued without interruption and these units reached the dry gas window (pre-salt = 3.7 and S2 = 3.35%Ro) before the Oligocene uplift (~23 Ma). In the absence of salt structures, these potential source rocks would have been overmature at ~170 Ma and 85 Ma, respectively. Mesozoic source rocks (S4, S5, and S6) are overall marginally to mid-mature (Figure 10B). The Lower-Middle Triassic (S4) source rocks entered the oil window at ~105 Ma, and reached a maximum maturity of ~0.7%Ro. The Middle-Upper Triassic (S5) interval entered the oil window at ~60 Ma and attained a maturity of ~0.65%Ro. The youngest Triassic-Upper Jurassic (S6) source rocks are marginally mature with vitrinite values of ~0.53%Ro. Without the massive salt diapir, the maturity of these three organic rich intervals would be 0.9, 0.75, and 0.65%Ro, respectively.

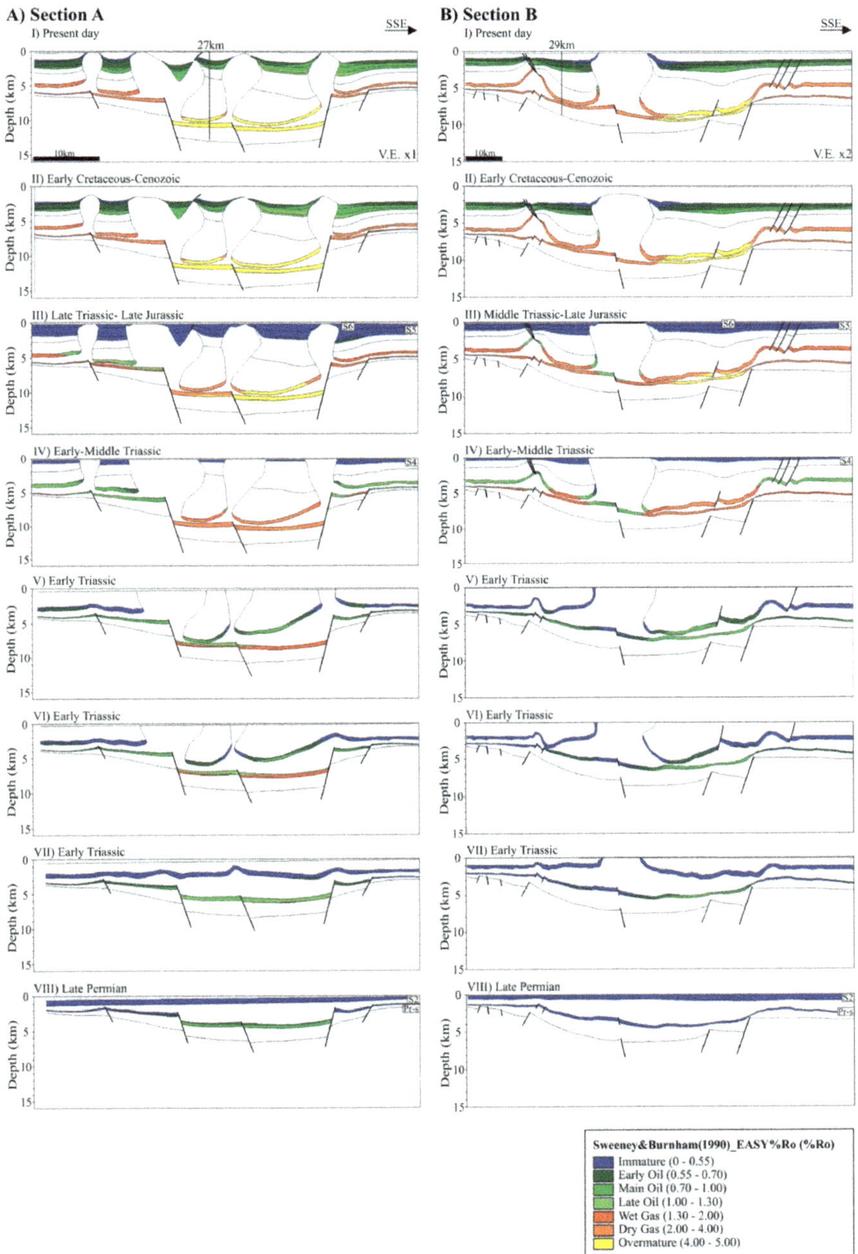

Figure 9. Source rock maturation through time in (**A**) section A in the central sub-basin, and (**B**) section B in the eastern sub-basin. Figure 1B shows the location of the sections.

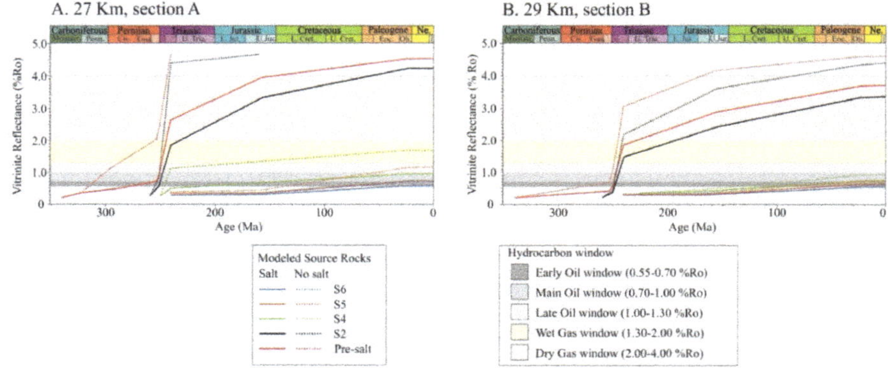

Figure 10. Source rock maturation through time in a pseudo-well perforating a minibasin in (**A**) section A in the central sub-basin, and (**B**) section B in the eastern sub-basin. Two scenarios are compared: One with salt structures as observed today (continuous lines), and another without salt structures (dashed lines). Figure 9 (I) shows the location of the pseudo-wells.

6. Discussion

6.1. Uncertainties

In confined salt-bearing basins, such as the Nordkapp Basin, seismic imaging and interpretation of salt bodies are certainly challenging due to the steeply dipping diapir flanks and complex ray paths of the seismic waves travelling through the salt [55,56]. Consequently, poor seismic imaging of salt structures can lead to incorrect interpretation of their shapes, which undoubtedly has negative consequences for determining the progressive evolution and thermal effect of salt structures through time. Uncertainty also arises from depth conversion of the seismic profiles. Figure 3A displays a wide range of interval velocities due to lateral variations in lithology and different degrees of compaction and diagenesis. In addition, the velocities of deep sediments within the Nordkapp minibasins are unknown because there are no exploration wells through the entire minibasins stratigraphy. Despite these uncertainties, the velocity model used in this study provides similar results (depths) to previous magnetic and gravity studies by Gernigon et al. [23] in the Eastern Barents Sea.

2D structural reconstructions involve several uncertainties associated with the type of unfolding method (flexural slip vs simple shear), decompaction curves, water depth, and elastic thickness. Testing different restoration parameters will indeed result in different paleogeometries. However, the objective of this paper is not to test the sensitivity of model parameters but rather use reasonable parameter values. For example, we use simple shear to remove the deformation caused by passive diapirism because the length loss adjacent to salt diapirs is negligible compared to the length of the section. On the other hand, we use flexural slip in sequences SU2 and SU3 because it preserves length in these parallel-folded units [57]. Based on studies by Klausen and Helland-Hansen [58], we use the Sclater and Christie [51] decompaction curve because it fits well the porosity versus depth trends observed on borehole data in the Barents Sea. Finally, an elastic thickness of 20 km was chosen based on Gac et al. [41].

In terms of thermal modeling, most of the crucial stages (IV to VIII) for the formation of the Nordkapp Basin encompass a relatively short (~20 Ma) and old (Permian-Early Triassic) time interval (Figure 5D). Therefore, restoring sensible boundary conditions (basal heat flow and SWTI) for this period is crucial. Inevitably, assumptions are intrinsic to the model due to the general lack of calibration data, with the exception of a few bottom well temperatures and vitrinite data (Figure 6). Extrapolating boundary conditions calculated in the western sub-basin to the central and eastern sub-basins can be unrealistic, since the initial thickness of the Post-Caledonian crust and lithosphere, and the amount of rifting defined by β most likely varied along the basin [42]. Additional uncertainty arises from the need to simplify lithologies, particularly for the deepest units, given the importance of their thermal conductivities and heat capacities in the simulations.

On the other hand, the model, as designed, would greatly benefit from assessing the distribution and thickness of the source rocks in each of the minibasins. For the petroleum system, the evaluation of the actual presence of good reservoir levels in the basin is a must. Despite these limitations, the modeling results represent the Nordkapp Basin geology and tectonic evolution, and they can be used to develop further exploration concepts in this basin and other basins alike.

6.2. The Importance of Thermal Modeling in Confined Salt-bearing Basins

The thermal effect of salt structures has been documented by previous studies in passive margins such as the Gulf of Mexico [6,9]. In this tectonic setting, salt deposition occurs in unconfined large areas where accommodation is controlled by thermal subsidence [59]. This also produces tilting of the margin, which in turn triggers downslope salt gliding and structures such as salt stocks, salt tongues, and allochthonous salt sheets [2]. These structures have received special attention due to their sealing capacity and their impact on maturation of underlying source rocks [6,8–10].

In the case of salt-bearing rift basins, syn-rift salt deposition is really limited by the rift geometry [59]. Salt mobilization by either extension and/or differential loading results in a structural style consisting of sub-vertical and closely spaced salt structures, which commonly coincide with the presence of subsalt faults [60–62]. Factors such as diapir shape and spacing play an important role in the thermal evolution of these basins [9].

Combined structural restorations and thermal modeling from the Nordkapp Basin indicate that the shape of salt diapirs and their inherent thermal anomalies vary through time and display a characteristic negative thermal effect (i.e., downward shift of isotherms), which is directly proportional to the width of the salt diapir (Figure 7). This is clearly observed in the eastern sub-basin (section B), where the presence of a wider and isolated salt diapir induces a strong, but laterally limited, negative anomaly. Temperatures along the diapir flanks could be up to 70 °C cooler and exceptionally low (~150 °C) at depths of ~9 km beneath the diapir. This integrated approach also highlights that closely-spaced diapirs in the central sub-basin mutually interfere and produce a combined negative thermal anomaly, which lowers the temperature in the minibasins by up to 50 °C with respect to the adjacent platform areas. Thus, although large salt tongues and allochthonus salt sheets are absent in confined basins, sub-vertical and closely spaced salt structures still generate a combined thermal anomaly that extends over large areas of the basin.

6.3. Implications for the Petroleum System

Salt diapirs in confined basins impact the petroleum system by retarding maturation of organic rich sediments, expanding the hydrocarbon generation window, and hindering diagenetic processes in reservoir levels. Figure 11 summarizes the implications of our results for the petroleum system of the Nordkapp Basin.

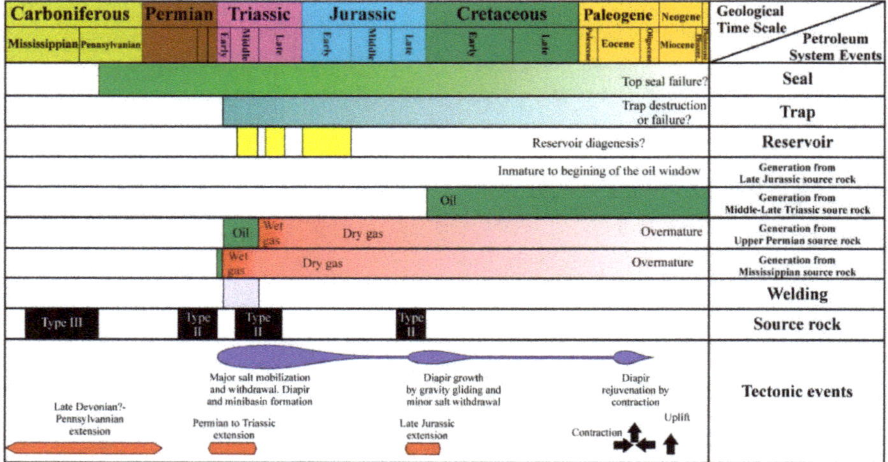

Figure 11. Petroleum systems chart for the Nordkap Basin. Insights from the modeling results are incorporated in the chart.

6.3.1. Source rocks

Present day vitrinite reflectance values in the Nordkapp Basin indicate that when maturation ceased upon uplift in the Oligocene, Mississippian (gas-prone) and Permian (oil-prone) source rock units were mostly overmatured (Figure 9A, I, and Figure 11). In areas where pre-salt source rocks lied at shallower depths (e.g., eastern sub-basin and basin shoulders), Paleozoic source rocks were still within the gas window (Figure 9B, I, and Figure 11). The modeled Permian source rocks show exceptional low maturity at the flanks of the massive salt diapir in the eastern sub-basin. The Middle-Upper Triassic oil and gas prone source rocks were most probably able to generate hydrocarbons as they entered well into the oil window (Figures 10 and 11). Hydrocarbon generation from the Upper Jurassic oil-prone source rocks, however, is limited since these rocks were marginally to early mature. It is noteworthy that in the vicinity of salt structures, i.e., minibasins, thermal maturation rates are diminished (Figure 10). In the absence of such structures, the timing of maturation and generation of the different hydrocarbon phases is much earlier than what is implied in Figure 11, substantially affecting the petroleum system evolution.

6.3.2. Reservoirs

The negative thermal anomaly caused by salt diapirs may have prevented temperature-driven diagenetic processes (e.g., quartz cementation) in potential reservoirs of Triassic and Jurassic age (Figure 7, II). Nevertheless, diapir widening must also be considered when assessing the impact of diagenesis in flanking reservoirs. This process can naturally enhance the stress at diapir flanks, causing quartz pressure dissolution, and subsequent decrease of reservoir quality [2]. This needs to be given attention in the Nordkapp Basin since structural restorations suggest significant diapir widening from the Middle Triassic to the Late Jurassic (Figure 4, III–IV).

6.3.3. Traps

Based on the structural restorations, near-diapir structural and stratigraphic traps were present since the end of the Early Triassic (Figures 4 and 11). Megaflaps [5,63] and halokinetic sequences [5,64,65], which formed in response to the active and passive stages of diapirism from the Early Triassic to Cretaceous (Figure 4, IV–VII, and Figure 11), are present at different stratigraphic levels. Potential traps also include Early–Middle Triassic half turtle structures (Figure 4, IV) and suprasalt fault complexes at the basin boundaries (Figure 4B, I).

6.3.4. Seal and migration

Impermeable salt can act as a seal rock for vertical and lateral migration of hydrocarbons in the Nordkapp Basin. The structural restorations illustrate that salt welds were present since the end of the Early Triassic (Figure 4, V). This welding may have allowed gas migration from Mississippian gas-prone source rocks into suprasalt Mesozoic reservoirs, which in turn may have favored reservoir porosity preservation at high depths, as documented by McBride et al. [6] for the Gulf of Mexico. Additionally, closely spaced diapirs in the central sub-basin can generate laterally sealed minibasins, which if capped by fine-grained rocks could create favorable scenarios for hydrocarbon entrapment.

Continuous diapir growth and successive reactivation of suprasalt fault complexes during the Mesozoic and Cenozoic could have also modified and/or destroyed structural traps and breached seals, causing migration of hydrocarbons to shallower traps or escape from the system (Rojo et al. [18]; their Figure 20). Late Cenozoic regional uplift and erosion [39] may have led to hydrocarbon phase separation, top seal failure, and remigration. The modification and destruction of traps, together with deep hydrocarbon kitchens, could have resulted in a complex petroleum system, where migration of petroleum, flushing of older traps, and mixing of hydrocarbons of different maturity and ages are dominant features. These observations are consistent with the current understanding of the petroleum system and geochemical data by Ohm et al. [16] and Lerch et al. [66] for the Norwegian Barents Sea.

7. Conclusions

In this study, we have integrated 2D structural restorations with thermal modeling to investigate how halokinesis impacted the thermal evolution of the basin through time and to explore the implications of the modeled thermal history on the petroleum system of the Nordkapp Basin, a confined, salt-bearing rift basin.

Combined structural restorations and thermal modeling show that the shape of salt diapirs and their negative thermal effect change through time. In the case of an isolated salt diapir, it induces a strong, but laterally limited, negative anomaly, which is directly proportional to its width. Temperatures along the diapir flanks are 70 °C cooler and are exceptionally low (~150 °C) at depths of ~9 km beneath the salt. On the other hand, the thermal anomalies of closely-spaced diapirs mutually interfere and generate a combined negative thermal anomaly that reduces the temperature in the minibasins by up to 50 °C with respect to the adjacent platform areas.

Although large salt tongues and allochthonus salt sheets are generally absent in confined rift basins, sub-vertical and closely spaced salt structures still generate a combined anomaly that extends over large areas of the basin. As a result, thermal maturation of the source rocks in the minibasins is retarded, and the hydrocarbon generation window is expanded. Thus, laterally-sealed minibasins offer favorable scenarios for deeper than normal hydrocarbons kitchens and entrapment in various near-diapir structural and stratigraphic traps.

Author Contributions: Conceptualization, A.C., L.A.R., N.C. and L.C.; methodology, A.C., L.A.R., L.C. and N.C.; software, A.C., L.A.R., L.C. and N.C.; validation, A.C., L.A.R., L.C. and N.C.; formal analysis, A.C., L.A.R., N.C. and L.C.; investigation, A.C., L.A.R., N.C. and L.C.; resources, A.C., L.A.R., L.C. and N.C.; writing—original draft preparation, A.C., L.A.R., N.C. and L.C.; writing—review and editing, A.C., L.A.R., N.C., L.C. and A.E.; visualization, A.C., L.A.R., N.C. and L.C.; supervision, A.C., L.A.R. and N.C.; project administration, A.C., L.A.R. and N.C.

Funding: This research received no direct funding. However, Andrés Cedeño was supported by the JuLoCrA project.

Acknowledgments: This work evolved from a Master thesis by Luis Centeno at the University of Stavanger. We are grateful to WesternGeco Multiclient for providing full azimuth 3D seismic data from the Nordkapp Basin, and the Norwegian Petroleum Directorate for kindly providing the NPD-BA-11 2D seismic survey in the eastern Norwegian Barents Sea. Thanks to Schlumberger and Midland Valley for providing academic licenses of their softwares PetroMod and Move, respectively. The authors are grateful to two anonymous reviewers and to the guest editor Willy Fjeldskaar for their valuable contributions. Andres Cedeno thanks the JuLoCrA project for economic support.

Conflicts of Interest: The authors declare no conflict of interest.

References

1. Hudec, M.R.; Jackson, M.P. Terra infirma: Understanding salt tectonics. *Earth-Sci. Rev.* **2007**, *82*, 1–28. [CrossRef]
2. Jackson, M.P.; Hudec, M.R. *Salt Tectonics: Principles and Practice*; Cambridge University Press: Cambridge, UK, 2017; 498p.
3. Matthews, W.J.; Hampson, G.J.; Trudgill, B.D.; Underhill, J.R. Controls on fluviolacustrine reservoir distribution and architecture in passive salt-diapir provinces: Insights from outcrop analogs. *AAPG Bull.* **2007**, *91*, 1367–1403. [CrossRef]
4. Banham, S.G.; Mountney, N.P. Controls on fluvial sedimentary architecture and sediment-fill state in salt-walled mini-basins: Triassic Moenkopi Formation, Salt Anticline Region, SE Utah, USA. *Basin Res.* **2013**, *25*, 709–737. [CrossRef]
5. Rojo, L.A.; Escalona, A. Controls on minibasin infill in the Nordkapp Basin: Evidence of complex Triassic synsedimentary deposition influenced by salt tectonics. *AAPG Bull.* **2018**, *102*, 1239–1272. [CrossRef]
6. McBride, B.C.; Weimer, P.; Rowan, M.G. The effect of allochthonous salt on the petroleum systems of northern Green Canyon and Ewing Bank (offshore Louisiana), northern Gulf of Mexico. *AAPG Bull.* **1998**, *82*, 1083–1112.
7. Rowan, M.G.; Lawton, T.F.; Giles, K.A. Anatomy of an exposed vertical salt weld and flanking strata, La Popa Basin, Mexico. In *Salt Tectonics, Sediments and Prospectivity*; Alsop, G.I., Archer, S.G., Hartley, A.J., Grant, N.T., Hodgkinson, R., Eds.; Geological Society: London, UK, 2012; Volume 363, pp. 33–57.
8. Lerche, I.; O'brien, J.J. *Dynamical Geology of Salt and Related Structures*; Academic Press: San Diego, CA, USA, 1987; p. 832.
9. Mello, U.T.; Karner, G.D.; Anderson, R.N. Role of salt in restraining the maturation of subsalt source rocks. *Mar. Pet. Geol.* **1995**, *12*, 697–716. [CrossRef]
10. Yu, A.; Lerche, I. Thermal impact of salt: Simulation of thermal anomaly in the Gulf of Mexico. *Pure Appl. Geophys.* **1992**, *2*, 180–192. [CrossRef]
11. Gabrielsen, R.; Kløvjan, O.; Rasmussen, A.; Stølan, T. Interaction between halokinesis and faulting: Structuring of the margins of the Nordkapp Basin, Barents Sea region. In *Structural and Tectonic Modeling and Its Implication to Petroleum Geology*; Larsen, B., Brekke, H., Larsen, B., Talleraas, E., Eds.; Norwegian Petroleum Society (NPF): Oslo, Norway, 1992; Volume 1, pp. 121–131.
12. Jensen, L.N.; Sørensen, K. Tectonic framework and halokinesis of the Nordkapp Basin, Barents Sea. In *Structural and Tectonic Modeling and Its Application to Petroleum Geology*; Larsen, R.M., Brekke, H., Larsen, B.T., Talleraas, E., Eds.; Norwegian Petroleum Society (NPF), Special Publications: Oslo, Norway, 1992; Volume 1, pp. 109–120.
13. Koyi, H.; Talbot, C.J.; Tørudbakken, B.O. Salt tectonics in the Northeastern Nordkapp basin, Southwestern Barents Sea. In *Salt Tectonics: A Global Perspective*; Jackson, M.P.A., Roberts, D.G., Snelson, S., Eds.; American Association of Petroleum Geologists: Tulsa, OK, USA, 1995; AAPG Memoir 65; pp. 437–447.
14. Nilsen, K.T.; Vendeville, B.C.; Johansen, J.-T. Influence of regional tectonics on halokinesis in the Nordkapp Basin, Barents Sea. In *Salt Tectonics: A Global Perspective*; Jackson, M.P.A., Roberts, D.G., Snelson, S., Eds.; American Association of Petroleum Geologists: Tulsa, OK, USA, 1995; AAPG Memoir 65; pp. 413–436.
15. Stadtler, C.; Fichler, C.; Hokstad, K.; Myrlund, E.A.; Wienecke, S.; Fotland, B. Improved salt imaging in a basin context by high resolution potential field data: Nordkapp Basin, Barents Sea. *Geophys. Prospect.* **2014**, *62*, 615–630. [CrossRef]
16. Ohm, S.E.; Karlsen, D.A.; Austin, T. Geochemically driven exploration models in uplifted areas: Examples from the Norwegian Barents Sea. *AAPG Bull.* **2008**, *92*, 1191–1223. [CrossRef]
17. Henriksen, E.; Bjørnseth, H.; Hals, T.; Heide, T.; Kiryukhina, T.; Kløvjan, O.; Larssen, G.; Ryseth, A.; Rønning, K.; Sollid, K. Uplift and erosion of the greater Barents Sea: Impact on prospectivity and petroleum

systems. In *Arctic Petroleum Geology*; Spencer, A.M., Embry, A.F., Gautier, D.L., Stoupakova, A.V., Sørensen, K., Eds.; Geological Society, London, Memoirs: Lodon, UK, 2011; Volume 35, pp. 271–281.

18. Rojo, L.A.; Cardozo, N.; Escalona, A.; Koyi, H. Structural style and evolution of the Nordkapp Basin, Norwegian Barents Sea. *AAPG Bull.* **2019**, in press. [CrossRef]

19. Rowan, M.G.; Lindsø, S. Chapter 12—Salt Tectonics of the Norwegian Barents Sea and Northeast Greenland Shelf A2. In *Permo-Triassic Salt Provinces of Europe, North Africa and the Atlantic Margins*; Soto, J.I., Flinch, J.F., Tari, G., Eds.; Elsevier: Amsterdam, The Netherlands, 2017; pp. 265–286.

20. Dengo, C.; Røssland, K. Extensional tectonic history of the western Barents Sea. In *Structural and Tectonic Modeling and Its Application to Petroleum Geology*; Larsen, R.M., Brekke, H., Larsen, B.T., Talleraas, E., Eds.; Norwegian Petroleum Society (NPF): Oslo, Norway, 1992; Volume 1, pp. 91–108.

21. Faleide, J.I.; Tsikalas, F.; Breivik, A.J.; Mjelde, R.; Ritzmann, O.; Engen, O.; Wilson, J.; Eldholm, O. Structure and evolution of the continental margin off Norway and the Barents Sea. *Episodes* **2008**, *31*, 82–91.

22. Gernigon, L.; Brönner, M.; Roberts, D.; Olesen, O.; Nasuti, A.; Yamasaki, T. Crustal and basin evolution of the southwestern Barents Sea: From Caledonian orogeny to continental breakup. *Tectonics* **2014**, *33*, 347–373. [CrossRef]

23. Gernigon, L.; Brönner, M.; Dumais, M.-A.; Gradmann, S.; Grønlie, A.; Nasuti, A.; Roberts, D. Basement inheritance and salt structures in the SE Barents Sea: Insights from new potential field data. *J. Geodyn.* **2018**, *119*, 82–106. [CrossRef]

24. Worsley, D. The post-Caledonian development of Svalbard and the western Barents Sea. *Polar Res.* **2008**, *27*, 298–317. [CrossRef]

25. Smyrak-Sikora, A.; Johannessen, E.P.; Olaussen, S.; Sandal, G.; Braathen, A. Sedimentary architecture during Carboniferous rift initiation—The arid Billefjorden Trough, Svalbard. *J. Geol. Soc.* **2018**, *176*, 225–252. [CrossRef]

26. Bugge, T.; Mangerud, G.; Elvebakk, G.; Mork, A.; Nilsson, I.; Fanavoll, S.; Vigran, J. Upper Paleozoic succession on the Finnmark platform, Barents Sea. *Nor. J. Geol.* **1995**, *75*, 3–30.

27. Gudlaugsson, S.; Faleide, J.; Johansen, S.; Breivik, A. Late Palaeozoic structural development of the south-western Barents Sea. *Mar. Pet. Geol.* **1998**, *15*, 73–102. [CrossRef]

28. Stemmerik, L.; Elvebakk, G.; Worsley, D. Upper Palaeozoic carbonate reservoirs on the Norwegian arctic shelf; delineation of reservoir models with application to the Loppa High. *Pet. Geosci.* **1999**, *5*, 173–187. [CrossRef]

29. Stemmerik, L. Late Palaeozoic evolution of the North Atlantic margin of Pangea. *Palaeogeogr. Palaeoclimatol. Palaeoecol.* **2000**, *161*, 95–126. [CrossRef]

30. Stemmerik, L.; Worsley, D. 30 years on-Arctic Upper Palaeozoic stratigraphy, depositional evolution and hydrocarbon prospectivity. *Nor. J. Geol.* **2005**, *85*, 151–168.

31. Koyi, H.; Talbot, C.J.; Torudbakken, B. Analogue models of salt diapirs and seismic interpretation in the Nordkapp Basin, Norway. *Pet. Geosci.* **1995**, *1*, 185–192. [CrossRef]

32. Glørstad-Clark, E.; Faleide, J.I.; Lundschien, B.A.; Nystuen, J.P. Triassic seismic sequence stratigraphy and paleogeography of the western Barents Sea area. *Mar. Pet. Geol.* **2010**, *27*, 1448–1475. [CrossRef]

33. Klausen, T.G.; Ryseth, A.E.; Helland-Hansen, W.; Gawthorpe, R.; Laursen, I. Regional development and sequence stratigraphy of the Middle to Late Triassic Snadd Formation, Norwegian Barents Sea. *Mar. Pet. Geol.* **2015**, *62*, 102–122. [CrossRef]

34. Eide, C.H.; Klausen, T.G.; Katkov, D.; Suslova, A.A.; Helland-Hansen, W. Linking an Early Triassic delta to antecedent topography: Source-to-sink study of the southwestern Barents Sea margin. *GSA Bull.* **2017**, *130*, 263–283. [CrossRef]

35. Klausen, T.; Aas, T.J.; Haug, E.C.; Behzad, A.; Snorre, O.; Domenico, C. Clinoform development and topset evolution in a mud-rich delta—The Middle Triassic Kobbe Formation, Norwegian Barents Sea. *Sedimentology* **2018**, *65*, 1132–1169. [CrossRef]

36. Anell, I.; Braathen, A.; Olaussen, S. The Triassic—Early Jurassic of the northern Barents Shelf: A regional understanding of the Longyearbyen CO_2 reservoir. *Nor. J. Geol.* **2014**, *94*, 83–98.

37. Grundvåg, S.A.; Marin, D.; Kairanov, B.; Śliwińska, K.K.; Nøhr-Hansen, H.; Jelby, M.E.; Escalona, A.; Olaussen, S. The Lower Cretaceous succession of the northwestern Barents Shelf: Onshore and offshore correlations. *Mar. Pet. Geol.* **2017**, *86*, 834–857. [CrossRef]

38. Marin, D.; Escalona, A.; Nøhr-Hansen, H.; Kasia, K.Ś.; Mordasova, A. Sequence stratigraphy and lateral variability of Lower Cretaceous clinoforms in the SW Barents Sea. *AAPG Bull.* **2017**, *101*, 1487–1517. [CrossRef]

39. Henriksen, E.; Ryseth, A.; Larssen, G.; Heide, T.; Rønning, K.; Sollid, K.; Stoupakova, A. Tectonostratigraphy of the greater Barents Sea: Implications for petroleum systems. In *Arctic Petroleum Geology*; Spencer, A.M., Embry, A.F., Gautier, D.L., Stoupakova, A.V., Sørensen, K., Eds.; Geological Society, London, Memoirs: London, UK, 2011; Volume 35, pp. 163–195.

40. Baig, I.; Faleide, J.I.; Jahren, J.; Mondol, N.H. Cenozoic exhumation on the southwestern Barents Shelf: Estimates and uncertainties constrained from compaction and thermal maturity analyses. *Mar. Pet. Geol.* **2016**, *73*, 105–130. [CrossRef]

41. Gac, S.; Klitzke, P.; Minakov, A.; Faleide, J.I.; Scheck-Wenderoth, M. Lithospheric strength and elastic thickness of the Barents Sea and Kara Sea region. *Tectonophysics* **2016**, *691*, 120–132. [CrossRef]

42. Clark, S.A.; Glørstad-Clark, E.; Faleide, J.I.; Schmid, D.; Hartz, E.H.; Fjeldskaar, W. Southwest Barents Sea rift basin evolution: Comparing results from backstripping and time-forward modeling. *Basin Res.* **2014**, *26*, 550–566. [CrossRef]

43. He, Z.; Crews, S.G.; Corrigan, J. Rifting and Heat Flow: Why the Mckenzie model is only part of the story: AAPG Hedberg Conference. In Proceedings of the Basin Modeling Perspectives: Innovative Developments and Novel Applications, The Hague, The Netherlands, 6–9 May 2007.

44. Klitzke, P.; Faleide, J.I.; Scheck-Wenderoth, M.; Sippel, J. A lithosphere-scale Structural model of the Barents Sea and Kara Sea region. *Solid Earth* **2015**, *6*, 153–172. [CrossRef]

45. Gac, S.; Hansford, P.A.; Faleide, J.I. Basin modeling of the SW Barents Sea. *Mar. Pet. Geol.* **2018**, *95*, 167–187. [CrossRef]

46. Riis, F. Dating and measuring of erosion, uplift and subsidence in Norway and the norwegian shelf in glacial periods. *Norsk Geologisk Tidsskrift* **1992**, *72*, 325–331.

47. Fjeldskaar, W.; Lindholm, C.; Dehls, J.F.; Fjeldskaar, I. Postglacial uplift, neo- tectonics and seismicity in Fennoscandia. *Quat. Sci. Rev.* **2000**, *19*, 413–1422. [CrossRef]

48. Cavanagh, A.J.; di Primio, R.; Scheck-Wenderoth, M.; Horsfield, B. Severity and timing of Cenozoic exhumation in the Southwestern Barents Sea. *J. Geol. Soc.* **2006**, *163*, 761–774. [CrossRef]

49. Fjeldskaar, W.; Amantov, A. Effects of glaciations on sedimentary basins. *J. Geodyn.* **2018**, *118*, 66–81. [CrossRef]

50. Athy, L.F. Density, porosity and compaction of sedimentary rocks. *AAPG Bull.* **1930**, *14*, 1–24.

51. Sclater, J.G.; Christie, P.A.F. Continental stretching: An explanation of the post-mid Cretaceous subsidence of the Central North Sea Basin. *J. Geophys. Res.* **1980**, *85*, 3711–3739. [CrossRef]

52. Sekiguchi, K. A method for determining terrestrial heat flow in oil basin areas. In *Terrestrial Heat Flow Studies and Structure of the Lithosphere*; Rybach, C.L., Chapman, D.S., Eds.; Tectonophysics, Science Direct: Liblice, Czechoslovakia, 1984; Volume 103, pp. 67–79.

53. Mørk, A.; Dallmann, W.K.; Dypvik, H.; Johannessen, E.P.; Larssen, G.B.; Nagy, J.; Nøttvedt, A.; Olaussen, S.; Pchelina, T.M.; Worsley, D. Mesozoic lithostratigraphy. In *Lithostratigraphic Lexicon of Svalbard. Review and Recommendations for Nomenclature Use. Upper Palaeozoic to Quaternary Bedrock*; Dallmann, W.K., Ed.; Norsk Polarinstitutt: Tromsø, Norway, 1999; pp. 127–214.

54. Sweeney, J.J.; Burnham, A.K. Evaluation of a simple model of vitrinite reflectance based on chemical kinetics (1). *AAPG Bull.* **1990**, *74*, 1559–1570.

55. Jones, I.F.; Davison, I. Seismic imaging in and around salt bodies. *Interpretation* **2014**, *2*, 1–20. [CrossRef]

56. Rojo, L.A.; Escalona, A.; Schulte, L. The use of seismic attributes to enhance imaging of salt structures in the Barents Sea. *First Break* **2016**, *34*, 41–49. [CrossRef]

57. Rowan, M.G.; Ratliff, R.A. Cross-section restoration of salt-related deformation: Best practices and potential pitfalls. *J. Struct. Geol.* **2012**, *41*, 24–37. [CrossRef]

58. Klausen, T.; Helland-Hansen, W. Methods for Restoring and Describing Ancient Clinoform Surfaces. *J. Sediment. Res.* **2018**, *88*, 241–259. [CrossRef]

59. Warren, J.K. Evaporites through time: Tectonic, climatic and eustatic controls in marine and nonmarine deposits. *Earth-Sci. Rev.* **2010**, *98*, 217–268. [CrossRef]

60. Koyi, H.; Petersen, K. Influence of basement faults on the development of salt structures in the Danish Basin. *Mar. Pet. Geol.* **1993**, *10*, 82–94. [CrossRef]

61. Stewart, S.; Ruffell, A.; Harvey, M. Relationship between basement-linked and gravity-driven fault systems in the UKCS salt basins. *Mar. Pet. Geol.* **1997**, *14*, 581–604. [CrossRef]

62. Withjack, M.O.; Callaway, S. Active normal faulting beneath a salt layer: An experimental study of deformation patterns in the cover sequence. *AAPG Bull.* **2000**, *84*, 627–651.

63. Rowan, M.G.; Giles, K.A.; Hearon, T.E., IV; Fiduk, J.C. Megaflaps adjacent to salt diapirs. *AAPG Bull.* **2016**, *100*, 1723–1747. [CrossRef]

64. Giles, K.A.; Rowan, M.G. Concepts in halokinetic-sequence deformation and stratigraphy. In *Salt Tectonics, Sediments and Prospectivity*; Alsop, G.I., Archer, S.G., Hartley, A.J., Grant, N.T., Hodgkinson, R., Eds.; Geological Society: London, UK, 2012; Volume 363, pp. 7–31.

65. Hearon, T.E., IV; Rowan, M.G.; Giles, K.A.; Hart, W.H. Halokinetic deformation adjacent to the deepwater Auger diapir, Garden Banks 470, northern Gulf of Mexico: Testing the applicability of an outcrop-based model using subsurface data. *Interpretation* **2014**, *2*, SM57–SM76. [CrossRef]

66. Lerch, B.; Karlsen, D.A.; Matapour, Z.; Seland, R.; Backer-Owe, K. Organic geochemistry of Barents Sea petroleum: Thermal maturity and alteration and mixing processes in oils and condensates. *J. Pet. Geol.* **2016**, *39*, 125–148. [CrossRef]

 *geosciences*

Article

On the Significance of Salt Modelling—Example from Modelling of Salt Tectonics, Temperature and Maturity Around Salt Structures in Southern North Sea

Ivar Grunnaleite [1,*] and Arve Mosbron [2]

[1] Tectonor AS, P.O. Box 8034, NO-4068 Stavanger, Norway
[2] LOTOS Exploration & Production Norge AS, P.O.Box 132, NO-4068 Stavanger, Norway
* Correspondence: ig@tectonor.com

Received: 2 July 2019; Accepted: 19 August 2019; Published: 22 August 2019

Abstract: Salt structures are attractive targets for hydrocarbon exploration. Salt can flow as a viscous fluid, act as hydrocarbon seal, and salt-related deformation may create reservoir traps. The high conductivity of salt can be crucial for hydrocarbon maturation in a basin. Here, we present results from the study of salt structures on the Eastern flank of Central Graben, on the Norwegian sector of the North Sea. By using our in-house basin modeling software (BMT^{TM}), we modelled the salt structure evolution and the effects of salt on temperature and maturation. Our results show up to 85 °C cooling due to the salt heat pipe effect. An integrated impact of cooling is the depression of vitrinite Ro by up to 1.0% at the base of a large salt balloon. Our work shows that it is of critical importance to correctly identify salt volumes and to have a good geological model, and to understand the timing and geometrical evolution of salt structures. This study is, to our knowledge, the most specific analysis of the impact of salt on basin temperature and maturation published so far, and is an example of how basin modeling in the future should be an integrated part of exploration.

Keywords: salt structures; modeling principles; geohistory evolution; temperature effects; conductivity effects on maturation

1. Introduction

All over the world, salt structures are found to be potential targets for hydrocarbon exploration [1]. The low density and the viscous nature of salt enable it to deform by buoyancy flow, distorting, and perhaps penetrating sedimentary sequences above it. Hydrocarbons may be trapped under large domes above salt, along salt flanks, under salt over-hangs, or along salt associated faults. The structural traps we see on the present seismic sections are unlikely to be the structures that were in place when hydrocarbons migrated into the area. Thus, reconstructing the evolution of salt structures may be crucial for constraining the filling and spilling of hydrocarbon accumulations. Due to the high thermal conductivity of salt, salt structures act like heat pipes and can dramatically change subsurface temperatures, potentially affecting the timing of hydrocarbon maturation.

For a general overview of principles of salt properties and the concept of salt tectonics, the new textbook of from Reference [2] is recommended.

Simulating the evolution of a salt structure is not straightforward because many different processes and deformation mechanisms are involved. Salt deforms as a viscous fluid, whereas the surrounding sediments typically deform by brittle and/or viscous and elastic mechanisms. Salt movement may be triggered by local faulting or regional extension, compression or strike-slip, differential loading, erosion, dissolution, etc. [3,4], and the mobilization of the salt invokes several other processes that can reinforce the salt flowage. Salt flowage and accumulation of salt in central pillows, domes, or diapirs assume withdrawal and thinning of the salt from the surrounding area (e.g., Reference [5]). This again causes increased subsidence and collapse of the overburden, which provides accommodation space for sedimentation, in turn enhancing the salt drive. Likewise, doming and uplift of overburden above the evolving salt may lead to erosion, increasing the gravity instability by removing mass above the thickest part of the salt and loading erosive material on the flanks. The upward movement of salt will apply an upward force on the sedimentary overburden. This effect can create overpressure in certain areas and affect the sealing and fluid-flow properties of a reservoir.

Salt has very unique properties compared to the surrounding sediments.

- It behaves as a plastic material under stress. If the applied stress is equal to or larger than a characteristic yield stress, it will flow or creep. Rock creep needs high temperatures, but salt can creep even at room temperature if it contains water [2].
- Salt has high seismic velocity (~4500 m/sec. [6]) leading to unreal "velocity pull-up structures" beneath the salt in seismic time-sections.
- The thermal conductivity of salt (K) is up to three times greater than the surrounding sediments [7].
- Salt mobility is dependent on temperature and the amount of water in salt [2]. Dry, cold salt is immobile [8], whereas wet salt at 60 °C and 280 bar confining pressure only can support 30% of the confining stress of dry salt (see Reference [9], referred in Reference [10]).
- Salt does not compact like sediments, but have a more or less constant density of ~2.15–2.20 g/cm^3 as a function of depth [11–13] (see Figure 1).

Clastic sediments have lower density at deposition, but compacts rapidly by burial due to the loss of porosity. At 1000–1500 m depth, sediment density equals the density of salt. Thus, with a sediment cover of more than 1000–1500 m salt movement is enhanced by buoyancy. In Reference [12], it is stated that a burial below at least 1600 m and more typically 3000 m of siliciclastic sediments is required before the average density of the entire overburden exceeds that of salt, and a diapir is able to reach the surface driven by buoyancy alone.

Salt movement is generally initiated by instability and as long as the sediment load is equally distributed over a flat salt surface, no salt movement will occur.

The density relationship with the depth for salt and surrounding sediments are shown in Figure 1. Lateral instability of the salt could be caused by tectonic events, differential loading, erosion, dissolution, tilting, or folding of the overburden by compression or drag by moving overburden, etc. These factors could create a relief on the surface of the salt sufficient to start movement of salt towards the locally highest point of the salt surface. When the flow of salt has been initiated, it will continue and be self-supported by buoyancy.

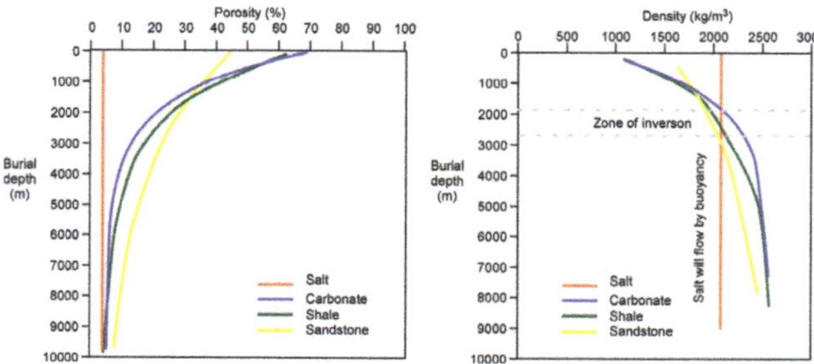

Figure 1. Salt has constant porosityand density, whereas sediments compact and lose porosity as they are buried. When the surrounding sediments have a higher density than salt, buoyancy will drive the upward movement of salt. The curves show the properties of sediments used in the modelling in this paper.

Salt movement will increase the initial relief, and withdrawal of salt from an area will create accommodation space for more sediments, which again will enhance differential loading-stress. Salt movement will slow, and eventually halt, as the diapir grows tall and reaches levels close to the surface. The lithostatic stress from above is lowered and accordingly plasticity of salt is reduced. At the same time, the effect of buoyancy is lost, and the drag effect and friction from surrounding sediments is increased. The structural evolution of salt has been modeled with both physical and numerical models, but very little is published on the temperature and maturation effects of salt structures except for Reference [14] that modelled these effects on very simple geometries.

This paper reports modeling of salt evolution in Southern North Sea and quantification of the temperature effects around salt diapirs. The modeling was done to quantify the effects on maturation and petroleum potential around the modeled salt structures, and in general to highlight the importance of addressing these effects when exploring for hydrocarbons around salt structures.

2. Study Area

The modeled salt structures are situated on the Hidra High (Model 1), and Sørvestlandet High (Model 2) on the platform just East of Central Graben in the Southwestern part of the Norwegian North Sea shelf (Figure 2).

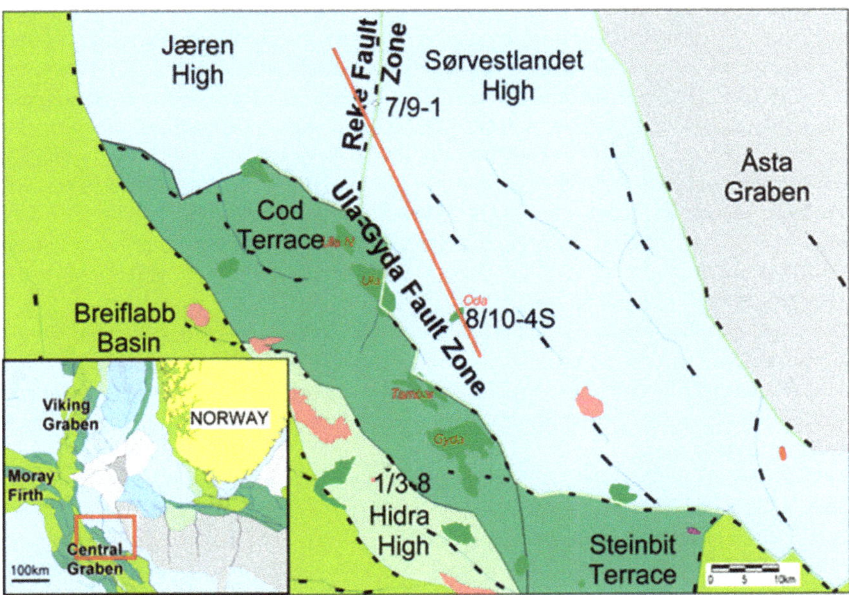

Figure 2. Overview map of the study area situated on the eastern flank of the Central Graben seen as the green basement trend between the blue platform areas. Maps are modified from Norwegian Petroleum Directorate (NPD) maps. The inset map shows the study area situated centrally in the North Sea between mainland UK and Norway. Model 1 is positioned close to Well 1/3-8 on the Hidra High. Model 2 is shown by the red line. Hydrocarbon discoveries are shown in pink and light green.

Several oil fields (Mime, Ula, Tambar and Gyda) on the Cod Terrace and Sørvestlandet High and other discoveries in the areas are situated over salt structures with reservoirs in Upper Jurassic Ula sandstone [15]. On the Hidra High, several hydrocarbon discoveries are made including gas and oil in Late Paleocene Forties Formation (Oselvar, Ipswich), gas in Late and Early Cretaceous (1/3-1) and oil and gas condensate in Jurassic reservoirs (King Lear, Romeo). The Sørvestlandet area is dominated by pod-interpod salt structures [16–19]. Pods are irregular rounded basin filled with Triassic sediments and distributed in a matrix of Zechstein salt forming walls, diapirs, ridges, and highs in a polygonal hive-pattern between the pods (Figures 3 and 4). The pods are capped by Jurassic to Cretaceous sediments. A system of collapse grabens in the Cretaceous unit follows the salt ridges surrounding the pod basins as can be seen in Figures 5 and 6. Their positions indicate a development related to vertical movement of salt structures and coeval subsidence in the pod basins.

The hydrocarbon potential in this area is related to Upper Jurassic reservoir rocks above structures defined by salt tectonics. A thick, more or less uniform salt unit was deposited over the entire area in Permian time, and has gradually developed into the complex salt structures we see today. Their development is related to episodes of salt movement from Early Triassic to Neogene time (Figure 4). The Model 2 profile crosses the Oda structure (well 8/10-4S) where oil is discovered in the Ula sandstone above a large salt structure. Other structures in the area have proven to be dry, thus it is of great importance to increase the understanding of the evolution and stratigraphy of similar undrilled structures in this area.

3. Geology

3.1. Stratigraphy

A standardized stratigraphic column for the Central Trough area is shown in Figure 3. The geological development in the greater Central Through area starts with clastic deposits of Paleozoic rocks, Devonian and Permian Rotliegend strata, as observed in the Embla field (block 2/7). The Paleozoic strata are generally interpreted to be continental and arid deposits. At the end of the Paleozoic (Late Permian time), the entire Southern North Sea basin developed into a large evaporitic basin; the northern Zechstein basin, with relatively thick and widespread salt and evaporite deposits.

The Mesozoic commenced with Triassic deposition, believed to be dominantly continental, and additional accommodation space is generated from Permo-Triassic rifting. The Triassic sediment thickness is generally significant. On the UK "J-block" (on Josefine Ridge), just West of the study area, productive Triassic intervals are exploited, and productive Triassic sandstones are present in the Ula field (7/12) and at the Gaupe field (block 6/3). The significant thickness of the Triassic actually commenced halokinesis of the Zechstein group salt by the Middle to Late Triassic time.

The Jurassic saw the introduction of marine sedimentary environments in the Central Trough area, but after the deposition of the terrestrial Bryne Formation. The Upper Jurassic is clearly marine and the Ula Formation yield several producing fields. The Mandal Formation, plus partly the Farsund and Haugesund Formations, are the dominant source rocks for the petroleum in the area. Strong rifting occurred in the Late Jurassic.

The Cretaceous is totally marine with marly and shaly deposits of the Cromer Knoll Group (Lower Cretaceous) and chalky deposits of the Shetland (or Chalk) Group (Upper Cretaceous). The Cretaceous sediments were deposited in a passive setting, without rifting. The Chalk strata exhibit significant production, notably from the Ekofisk Field in block 2/4. In the Tertiary, the passive and relatively deep marine sedimentation continued, shale is the dominant lithology. However, sandy systems, especially the Forties Formation system, reached the study area and comprise several fields; Blane, Cod and Oselvar have all produced from the Paleocene Forties Formation.

3.2. Triassic Pod Basins

It is known from literature that salt movement in this area started already in Early Triassic and the large salt-structures were clearly defined before deposition of the interpreted Cretaceous and younger horizons, see e.g., References [3,19,20].

Our interpretation and analysis of the Triassic sequences also revealed that the major salt structures were established already in the Early Triassic time (Figure 4). Their position is closely related to faults in the pre-salt sequence. Late Permian to Early Triassic tension, possibly related to movement along the Tornquist Zone [20,21], leads to extensional faulting. The faulting created a surface relief on the salt, which focused the deposition into depressions that gradually developed into deep basins. As a surface relief was created, and the sediment load was enough to apply a stress above the yield stress, the lower part of the salt deformed plastically and started to move. Once the areas of deposition and areas of salt growth were fixed, the structures were exaggerated by repeating this salt flow pattern for each deposited sedimentary sequence in a self-supporting system. During Triassic time, the deposition concentrated in circular sag-basins or pods between more or less evenly spaced salt diapirs and ridges. This is seen especially in the North of the Ula Fault (Figures 5 and 6) on the Cod Terrace and Sørvestlandet High. Between the sag basins, relatively lower "salt-saddles" acted as salt-feeders into the main salt structures that develop where salt walls and saddles join up in the corners of a polygon pattern. The modeled profile in this study crosses several pod-basins and salt saddles north of the Ula –Gyda Fault Zone (Figures 2 and 5), close to the top of the salt structures 7/9-1 and Oda.

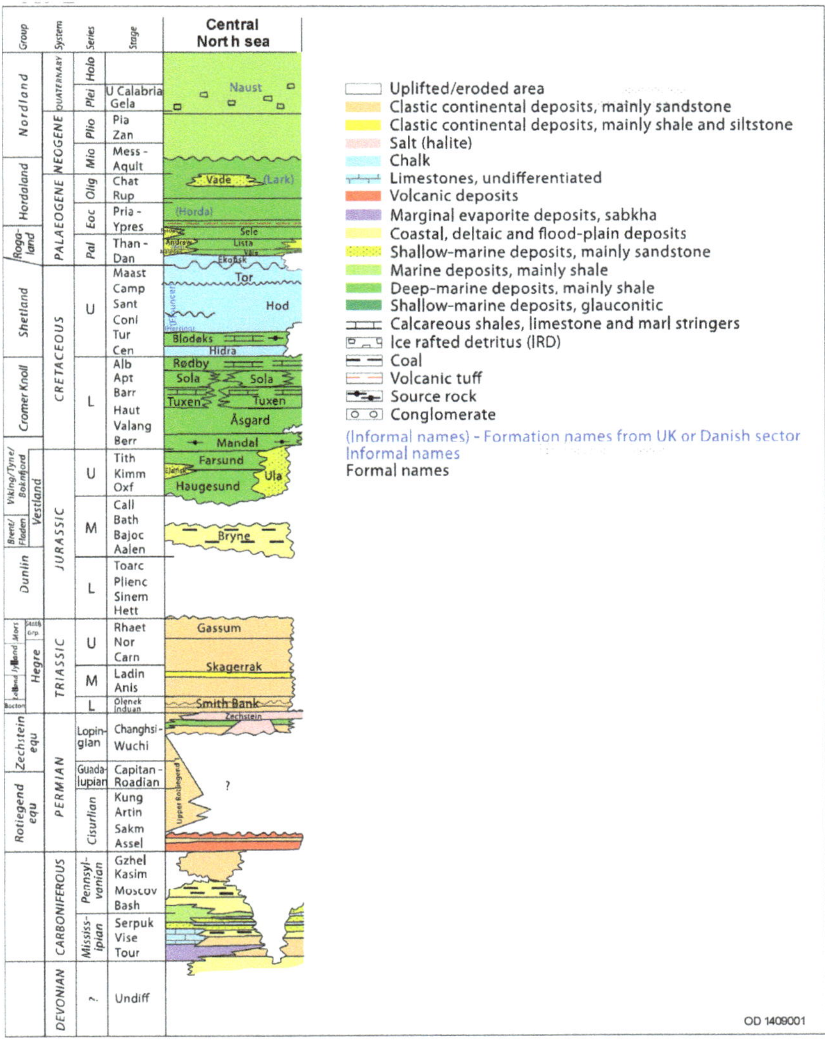

Figure 3. Stratigraphic column from the Central Trough area, modified from Norwegian Petroleum Directorate (www.npd.no).

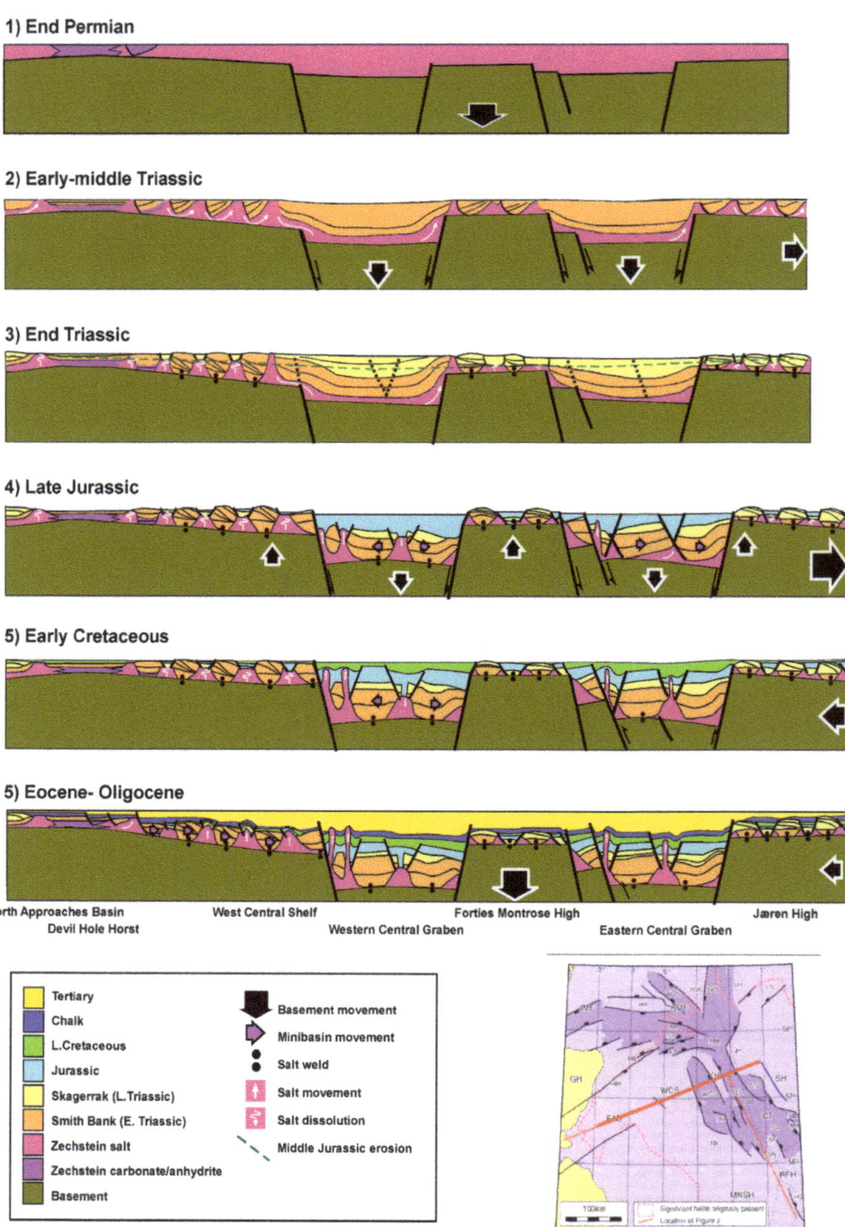

Figure 4. General section across the Central Graben showing the sequential evolution of salt structures. The study area is situated on the Jæren High to the right on this figure. The inset map shows the position of the regional transect in red. The Figure is modified from Reference [3]. The modeled structures of this study are based on typical structures from Eastern Central Graben (Model 1) and Jæren High (Model 2) whose position is shown as a blue line.

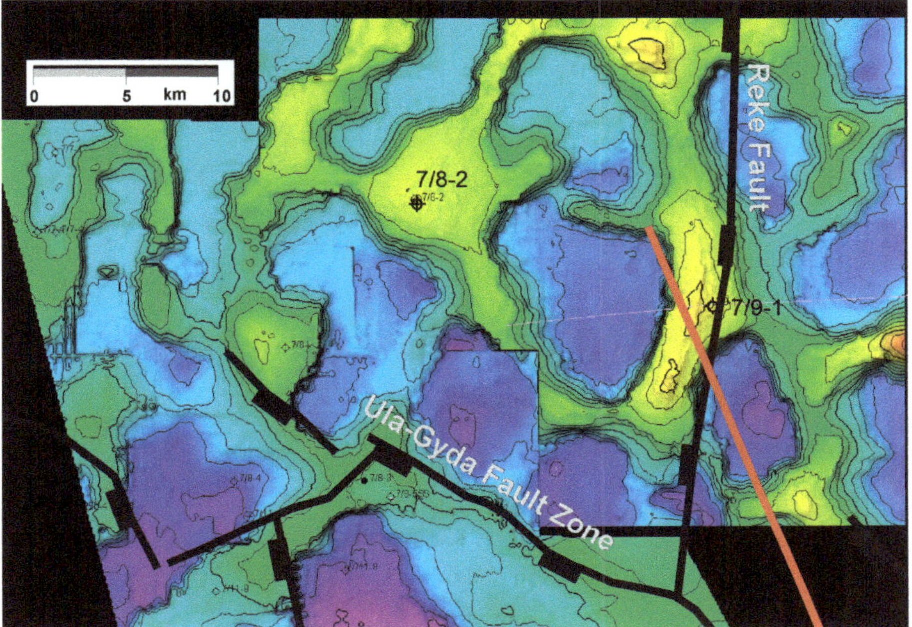

Figure 5. Top Zechstein salt map showing typical Triassic pod basins (in blue colours) surrounded by a polygonal pattern of salt walls and highs (in green-yellow colour). The red line shows the position of the Northern part of Model 2.

3.3. Cretaceous Collapse Grabens

On the Base Cretaceous map (Figure 6), a characteristic polygonal pattern of grabens between the salt structures defines the flexure between polygonal sag basins or "pods" around the salt structures. The collapse grabens all appear in the Cretaceous above the "salt-saddles" or supra salt [19], and this may be related to a Cretaceous phase of regional tilting and inversion (Figure 7), as suggested by Reference [3] and others. Figure 4 shows a change from extension to compression in Early Cretaceous.

The WNW-ESE striking grabens seem to be the best developed, indicating a tensional stress direction of SSW-NNE. If this stress was compressive, the relative high saddle-ridges acting as salt feeders between the sag basins would be compressed and experience longitudinal extensional faulting in the upper part due to bending. Concurrently, subsidence would be increased in the sag-basins ("pods") and sedimentation would resume (Figure 7). This inversion phase would thus lead to increased salt movement and basin subsidence. Furthermore, this may have resulted in salt withdrawal from the pod-defining ridges when salt moved into the major salt diapirs and later caused graben collapse in these areas. Longitudinal narrow horsts are observed on top of some of the major salt structures (Figure 8b). These may have developed as normal faults simultaneously with the graben development, but later developed into horsts due to upward salt movement. As salt moves into the highest structure, salt is withdrawing from the lower connected saddle ridges. As a consequence, the extensional faults on these saddles developed into grabens due to net salt withdrawal here.

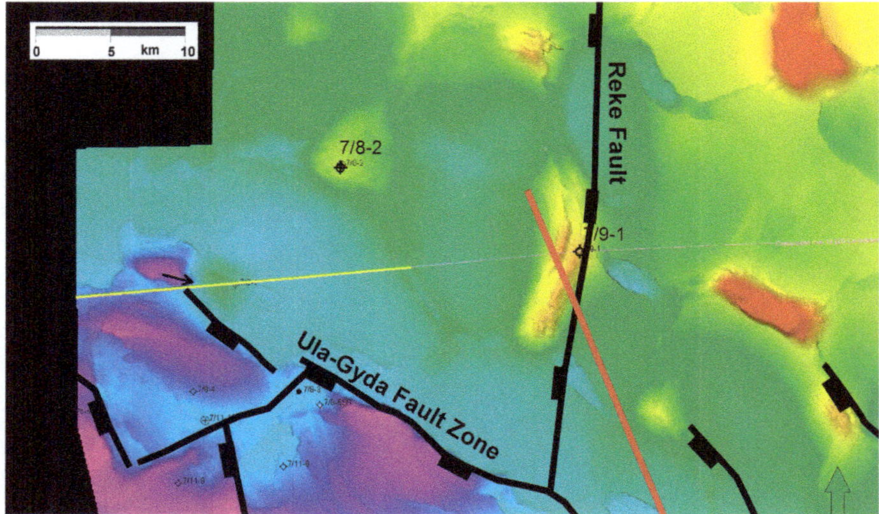

Figure 6. Base Cretaceous (BCU) time map provided by Lotos showing Cretaceous collapse grabens North of the Ula Fault. The Triassic sag-basins or pods are the polygonal flat areas between positive salt-structures and supra-salt collapse grabens. The arrows indicate the direction of salt movement away from the collapsing grabens accumulating into the growing salt diapirs. Red colors define the shallowest areas, and purple defines the deepest areas. Red line shows the Northern part of Model 2.

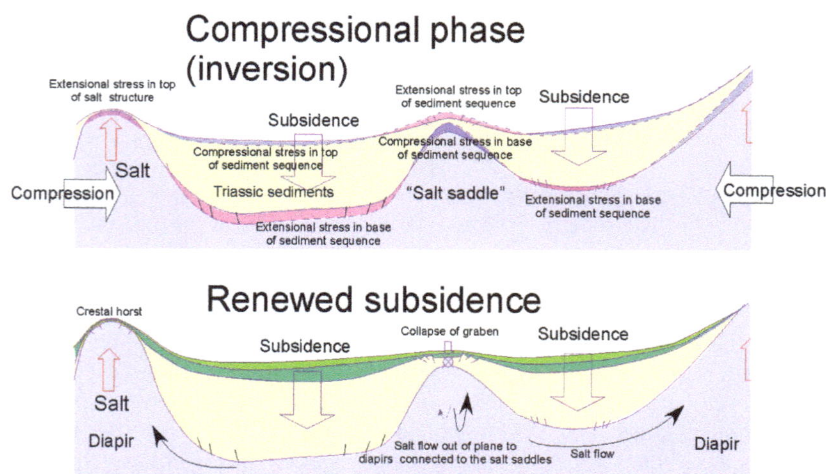

Figure 7. Principle sketch showing how a compressional stress regime would affect basins, salt saddle and diapir area. Extensional stresses above the saddle area act as a hinge between subsiding basins. Extensional faults will follow the length axis of the saddle areas and could collapse when compression ceases and renewed subsidence occurs. This could also explain the existence of crestal horsts seen on the major salt structures (Figure 8).

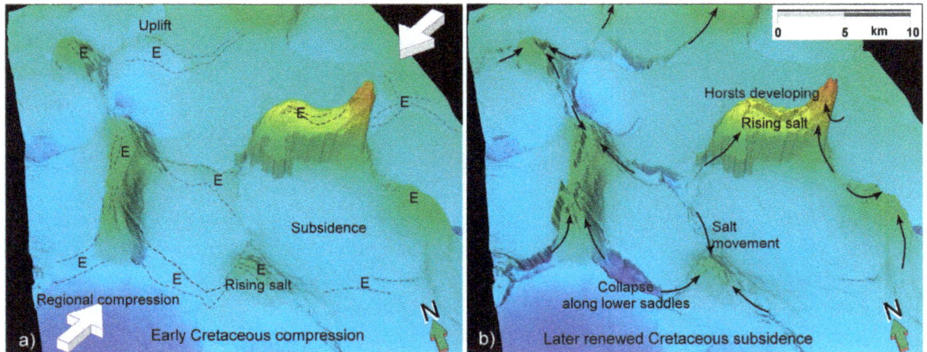

Figure 8. (**a**) Effect of Early Cretaceous compression from SW on an unfaulted surface. Blue arrows indicate areas with increased subsidence, red arrows show areas with increased uplift. In the saddle areas between basins and over the highest salt structures extensional faults will occur, as indicated by the stippled lines and marked by E. The background figure is based on the BCU surface and is manipulated in a photo editor to mimic a smoother pre-faulting surface. (**b**) Effect of renewed subsidence. Collapse grabens (most prominent along the WNW-ESE trend) have developed along the salt saddles. Horsts are developed on the highest salt structures. Background figure is an unedited BCU surface interpreted by LOTOS.

4. Salt Reconstruction Modelling

4.1. Methods

The 2D modeling reported in this section was performed by the use of our in-house basin modeling software BMT (see References [22,23], for more details). The following functionalities in BMT have been used here:

i. Reconstruction of the geohistory. The geohistory reconstruction is carried out by a decompaction technique in which the layers are removed one-by-one and corrections are made for the present day compacted thicknesses. The decompaction technique (generally combined with fault-restoration) gives a number of 2D 'time-slices' of the basin development.

ii. Calculation of temperature and maturity history. The calculation is based on the reconstructed geohistory (in i.) using a finite-difference grid. Input for the modeling is paleo heat flow from the mantle, surface temperature history, thermal conductivities, and heat capacities of the sediments. The calculation of vitrinite reflectance is using EASY%RO model.

From an initial sediment composition at the surface, the porosity in a sedimentary basin will be reduced and the density increased as a function of increasing stress and temperature. Empirically, the final porosities in many lithologies show an exponential decline with depth. The calculation of porosities in sedimentary basins is thus often done by using exponential lithology-specific porosity–depth relations. These are typically given the form:

$$\phi = \phi_0 \exp(-cz)$$

where ϕ is porosity (fraction), ϕ_0 is the surface porosity (fraction), c is a constant, and z is the depth in km (see e.g., Reference [24]). The porosity parameters used in this modeling study are shown in Table 1.

The salt movement is simulated during backstripping by manual editing of the salt geometry at every appropriate time step. This is accomplished by two methods (Figure 9) that can be combined: (1) Changing the lithology of a polygon at a given time ("litho-switching") and (2) "inflating" or "deflating" salt polygons ("mass editing").

Litho-switching allows adding or removing portions of a salt body to mimic salt growth or withdrawal. It is used when a salt body completely pierces overlying sediments or when the salt body grows horizontally, for example. For more details, see Reference [23].

Inflation and deflation technique allows the size and shape of a single salt polygon or a set of salt polygons to be changed (Figure 9c). A polygon is inflated or deflated by increasing or decreasing the length of the selected grid line, which is connected along the upper and lower boundaries of a polygon. For more details, see Reference [23]. When a selected grid line is inflated or deflated, it will remain fixed with respect to one of these boundaries while moving the other boundary upwards or downwards. The polygons may be vertically inflated to many times their original height or deflated to a thickness of less than one meter.

The mass added or removed by litho-switching is not accounted for in the modeling. For example, if the lithology of a polygon is switched from salt to shale during reconstruction, the added shale mass is not automatically deducted from elsewhere in the shale unit. In essence, the mass of a polygon that has been litho-switched spontaneously appears or disappears. However, BMT quantifies the area of selected polygons allowing the user to keep track of (and account for) the area added or removed by litho-switching. Anyway, as salt movement by nature is three-dimensional with mass moving in and out in all directions, there is no reason to keep volume-edited area constant in these 2D profiles.

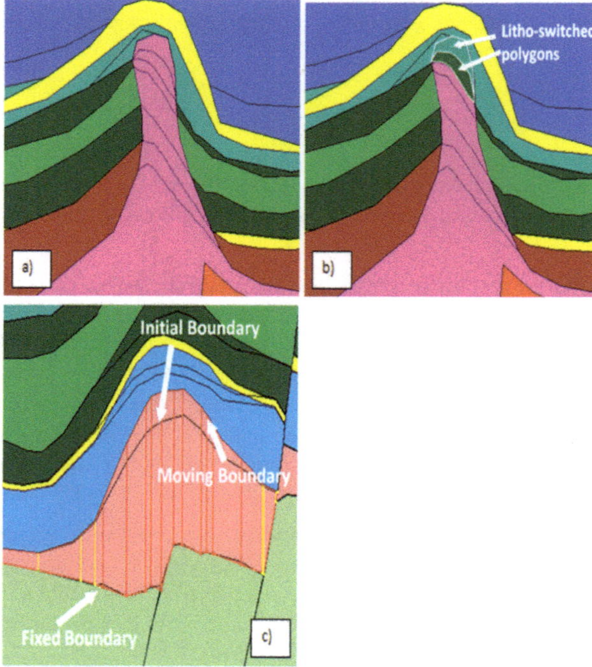

Figure 9. Illustrations of methods for modeling salt movement in the BMT modeling software. (**a,b**) litho-switching. The salt lithology for a polygon is changed at a given time. (**c**) Mass-editing by polygon inflation. The selected vertical in red color, are extended above the upper (moving) boundary while remaining fixed along the lower boundary. Note how overlying polygons are pushed upwards.

4.2. Geohistory Reconstruction

We are modeling the geohistory, temperature, and maturity development on two profiles from the Eastern margin of the Central Graben, in the Norwegian part of the North Sea. The first profile

(Model 1) is only 11 km long and includes a single salt structure. The second profile (Model 2) is 56 km long and includes five salt structures (see Figure 2 for locations).

4.2.1. Model 1

Model 1 is one single balloon-shaped salt structure from the Hidra High (Figure 10). The salt structure geometry of Model 1 was based on an interpreted seismic section close to a well 1/8-3 on the Hidra High on the North-Eastern flank of the Central Graben (Figure 2).

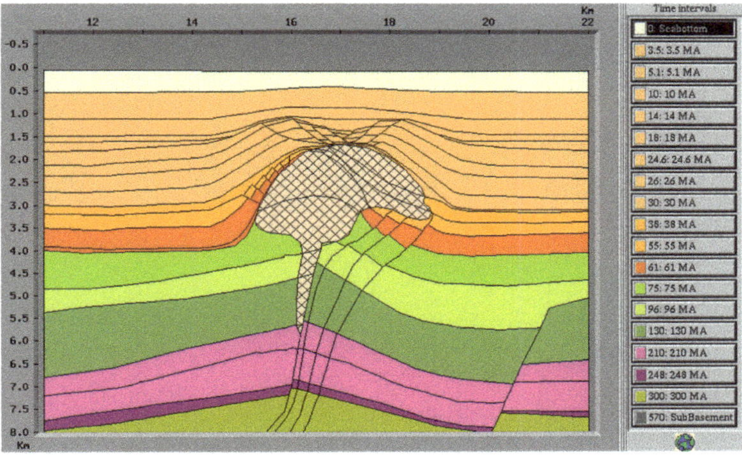

Figure 10. Interpreted present-day geometry of Model 1. Units are coloured by age. Ex situ salt volume is marked by cross-hatching. In situ salt coloured dark purple as defined by age interval (248–300 Ma).

Figure 11 shows the reconstructed geohistory combining litho-switching and mass-editing to enable both upward and later horizontal movement of the salt volume into an overhang. The lithology input parameters are shown in Table 1. The salt movement is initiated along a fault zone and once a relief on the salt surface is established in Triassic time upward movement of salt continuous through Cretaceous time when a vertical salt diapir of more than 5 km height was established. Increasingly differential loading lead to a skewed development of the top salt bulb, bending it over and developing it into a hemisphere asymmetrically placed relative to its now dramatically thinned and broken stem during Paleogene to Cenozoic evolution.

Table 1. Lithology parameters used in the basin modeling.

Lithology	Surface Porosity F_o	Exponential Porosity Constant c	Thermal Conductivity (W/mK)	Specific Heat Capacity (J/kg K)
Claystone and siltstone	0.63	0.51	1.7 (6% porosity) 1.0 (60% porosity)	940
Shale	0.63	0.51	2.8 (6% porosity) 1.2 (60% porosity)	1190
Chalk	0.70	0.71	2.04 (13% porosity) 1.23 (38% porosity)	1090
Sandstone Jurassic	0.37	0.58	2.04 (13% porosity) 1.23 (36% porosity)	1090
Sandstone Triassic	0.49	0.27	3.4 (6% porosity) 2.5 (40% porosity)	1080
Salt	0.04	0.05	5.1 (3% porosity) 5.0 (4% porosity)	1060
Basement	0.04	0.05	3.1 (3% porosity) 3.1 (4% porosity)	1100

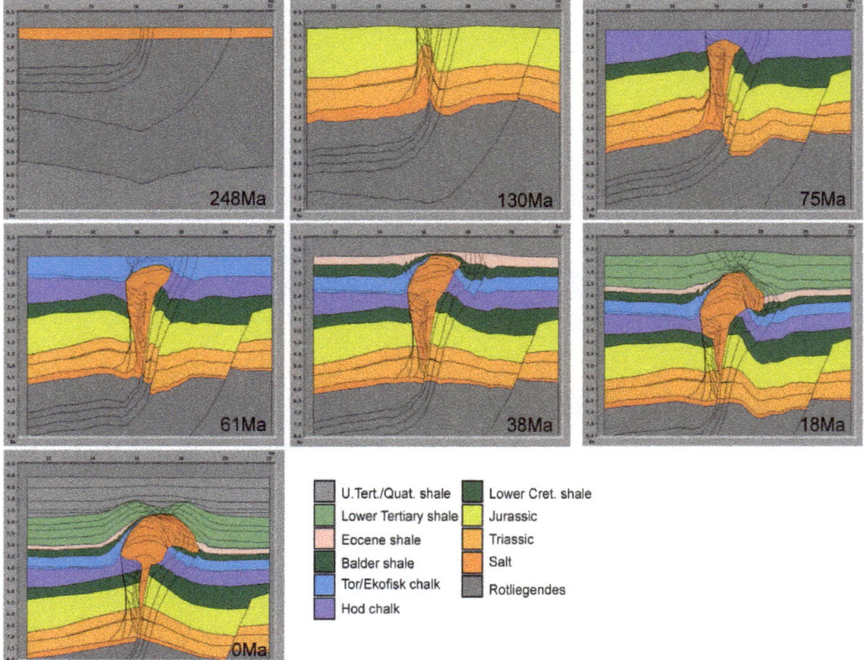

Figure 11. Reconstructed geohistory of Model 1, showing the stepwise evolution from a uniform salt layer to the present day salt balloon. The figure is colored based on the lithologies used in the model. The main salt volume protrudes through the Ekofisk and Hod chalks (in purple and blue) and is capped by Cenozoic shales.

4.2.2. Model 2

Model 2 is based on a NNW-SSE striking seismic section across Jæren High and Sørvestlandet High on the Eastern flank of Central Graben (Figure 2). The profile extends SSE-wards from the salt structure near well 7/9-1 in the Southeastern part of Jæren High, and crosses the N-S striking Reke Fault Zone, continues along the Southeastern part of Sørvestlandet High, until it ends after crossing the Oda salt structure (well 8/10-4S) in the Oda Field.

The model was based on interpreted and depth-converted seismic sections and maps shown in Figure 12. The seismic section was interpreted by LOTOS down to Base Cretaceous (BCU) and top and base of the salt. The interpretations of the salt structures are based on clear folding or penetration of Jurassic to Cenozoic sequences at the top, often with associated high amplitude reflection above the top of the salt structures (caused by the presence of anhydrite). Down-flank, the interpretation is much more uncertain as the seismic data are very transparent both in the salt and in the surrounding Triassic units. The base of major salt structures is defined by the width of the velocity pull-ups observed below the salt (Figure 12a).

The lack of regional seismic correlation markers within the continental Triassic units and the low impedance contrast between Zechstein salt and Triassic pod units make interpretation and correlation between pods difficult [17]. For the Triassic sequences, no ages were known. For the modeling, 14 Triassic sequences were interpreted and given ages evenly distributed between 248 Ma and 205 Ma.

The profile has five separate salt structures becoming increasingly more prominent throughout Triassic times, as sedimentation focus on small semi-circular pod-basins pushing the salt into the ridges between them. The distribution of these pods can be seen on the Top Zechstein map (Figure 5) and less pronounced on the Base Cretaceous (BCU) map of Figure 12b, best developed North of the

Ula-Gyda Fault Zone. The Triassic pod basins probably became welded to the pre-salt unit as salt was completely removed below these pod basins during Jurassic to Cretaceous times. The smallest structures seem to cease moving during Late Triassic to Jurassic times, whereas the largest structures (i.e., 7/9-1 structure and Oda structure (8/10-4)) seem to have been actively growing well into Cenozoic time. Note the grabens in the Cretaceous unit above the central salt structures (Figure 12a), also seen as green lens-shaped depressions on the Base Cretaceous map (Figure 12b).

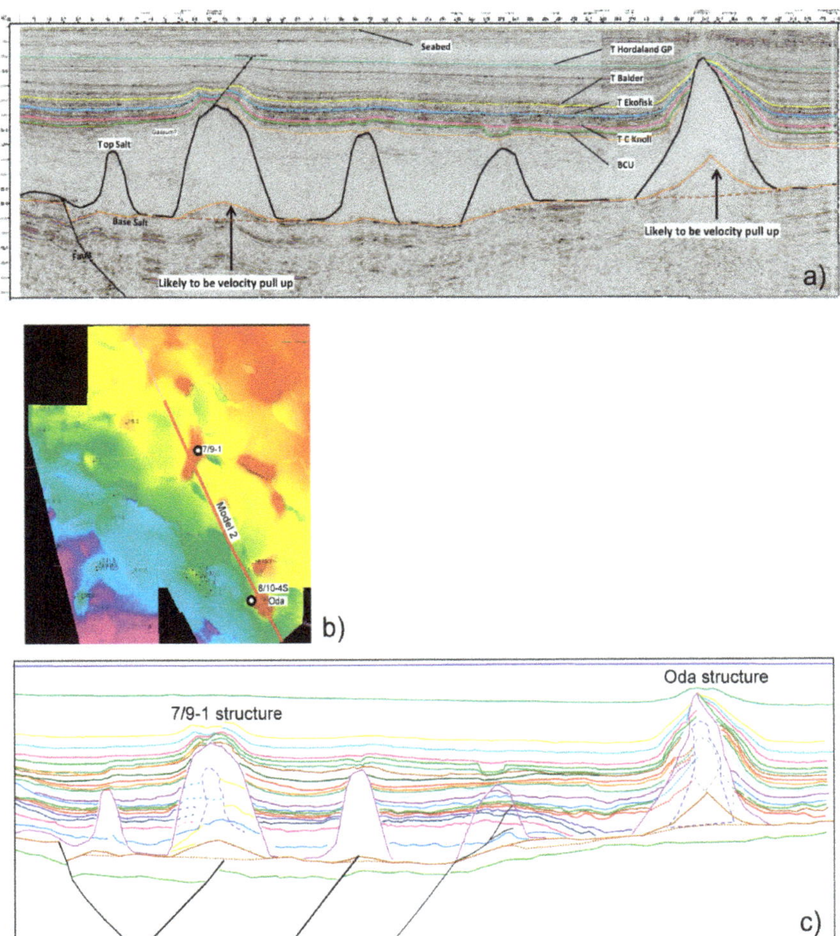

Figure 12. Interpretation of present-day structures on Model 2. (**a**) Interpreted depth-converted seismic section. (**b**) Map of Base Cretaceous horizon with the position of the modeled profile. (**c**) Detailed interpretation of Triassic sequences and possible interpretation within the salt structures. Continuous stratification can be interpreted through the second diapir from the right, and one can speculate if this really is a salt structure. Between this structure and the large Oda structure to the South, the lower Triassic units seem to downlap onto a paleo-slope. A stippled "core" of the largest structures indicates areas where no apparent stratification could be interpreted. In the temperature modeling, this "core" was used as a minimum-salt model.

Strongly dipping and onlapping Triassic sequences are not found on this NW-SE profile, but are commonly seen, especially in profile crossing the dip of the major WSW facing faults. These dipping units are related to asymmetric salt withdrawal in pods [17] and initial subsidence is related to underlying faults. It is shown by numerical modeling that subsidence in a pod basin can affect other pod basins close to it, it can also prevent subsidence, induce tilting, and cause asymmetric subsidence and even lateral translation of entire pod basins [25].

As interpretation of salt structures was uncertain, three cases was modelled, as shown in Figure 13, and used in the further geohistory- and temperature modelling.

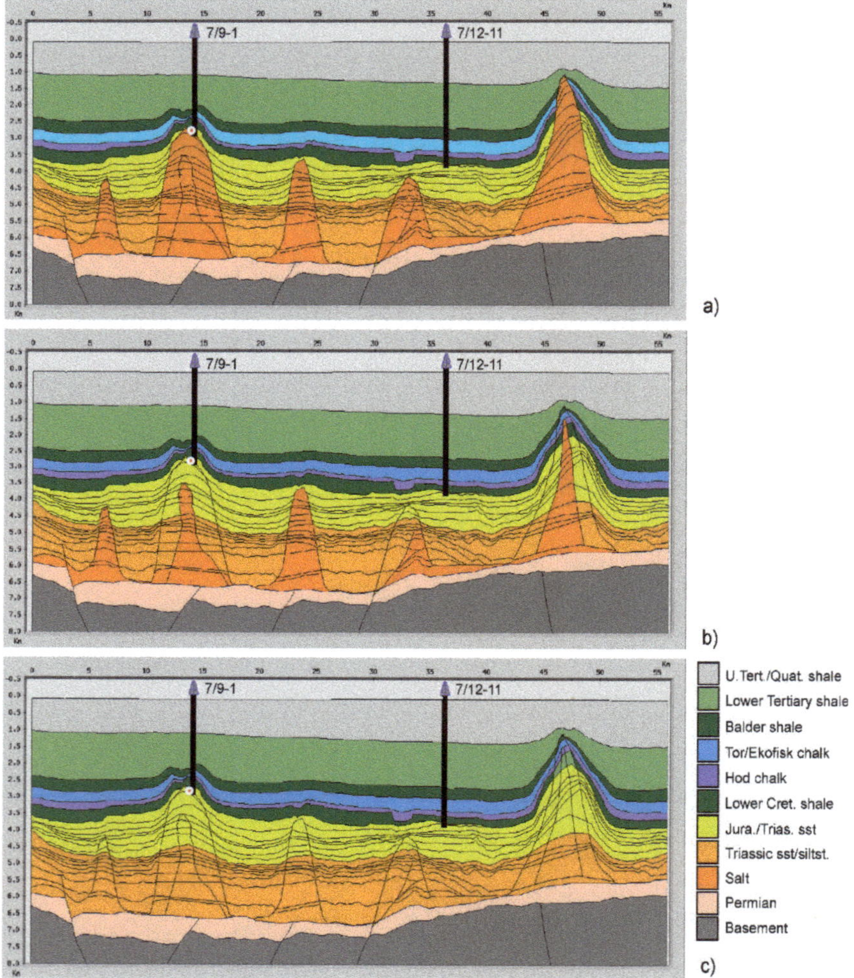

Figure 13. Present day geometry input of Model 2, colored by lithology. From top: (**a**) Maximum-salt model; (**b**) minimum-salt model; and (**c**) no-salt model. The locations of wells 7/9-1 and 7/12-11 are projected onto the profile.

Interpretation and sequence analysis of the Triassic sequences suggest that subsidence and development of Triassic basins (pods) started Northwest of the Oda structure relatively early in Triassic time, probably related to fault movement in the pre-salt units and basement. In Late Triassic, sediment deposition started south of the Oda structure. All the five salt-diapirs on this profile developed in parallel until latest Triassic when the minor diapirs halted, while the Oda and 7/9-1 structure continued to grow. These two structures continued to be actively growing at least into Paleogene times (Eocene-Oligocene). In the model, the structure at ~33 km (second structure from right) reached a maximum height at Base Cretaceous time and lost volume and height in Cretaceous time as the graben above developed and salt-mass was lost to the sides. Some of the time-steps reconstructed in the geohistory are shown in Figure 14.

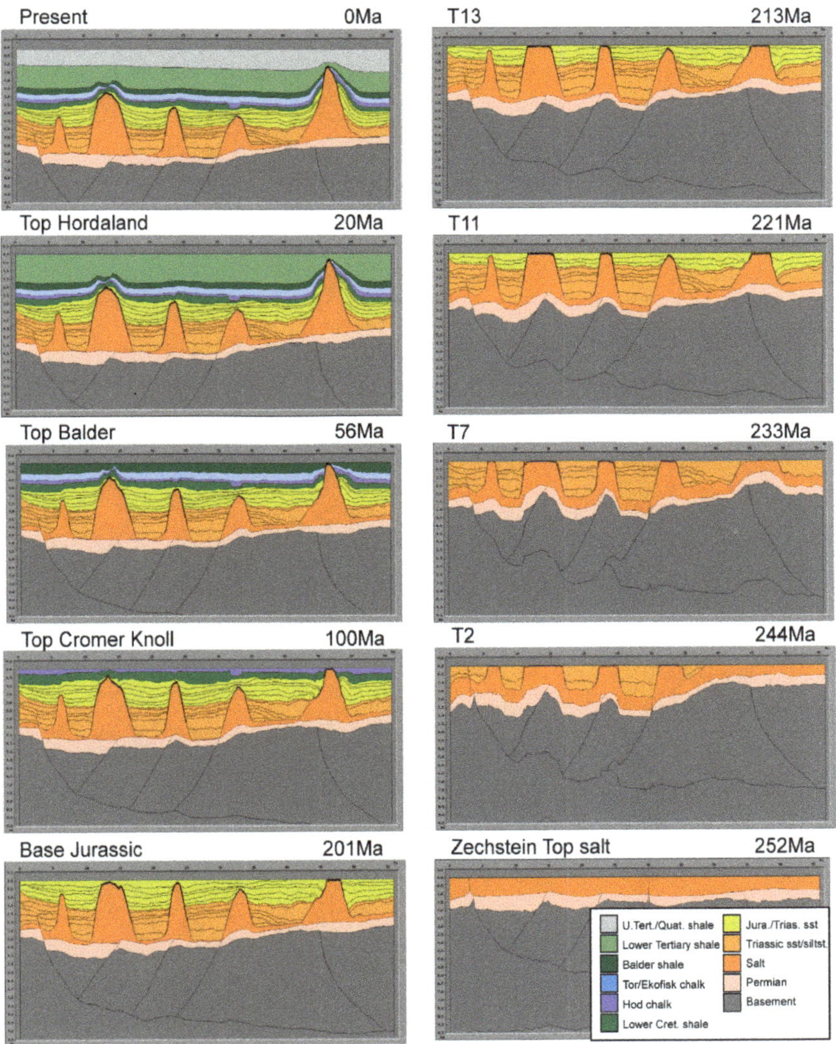

Figure 14. Geohistory reconstruction of Model 2. The figure shows 10 of the total 24 time-steps in the model, including 14 time steps during Triassic time.

5. Modelling of Temperature History and Vitrinite Reflectance

BMT utilizes finite difference calculations by conduction with a rectangular finite difference grid of varying sizes (cf. [23]). For every reconstructed time step in the geohistory, BMT builds a new high-resolution thermal modeling grid. Around small features, the grid size is especially fine to ensure realistic calculations. The difference grid in this study consists of a minimum of 400×400 cells of varying sizes. The spatial variation in rock properties and possible differences from one time step to the next are adjusted for so that appropriate finite difference calculations are maintained. The finite difference calculation by conduction is controlled by the temperatures from the previous time step, thermal conductivity (vertical and horizontal), and specific heat capacity of the basin's lithology/lithologies.

Salt has much higher thermal conductivity than the neighboring sediments (cf. Table 1). This means that a salt diapir that pierces through units of sedimentary rocks will lead heat up through the surrounding, relatively less conductive, sediments that acts as insulation. This could lead to higher temperatures in the units above salt structures and temperature depletion around the base of the salt structure.

5.1. Temperature Effects of Salt

5.1.1. Synthetic Case

To examine the temperature effect of salt, we made a simple model with a vertical salt plug over a thin salt layer (Figure 15). No thinning of the salt layer and no salt withdrawal with increased basin subsidence is used in these models. The top of the model is kept at surface at all time steps. All sediments were defined as shale.

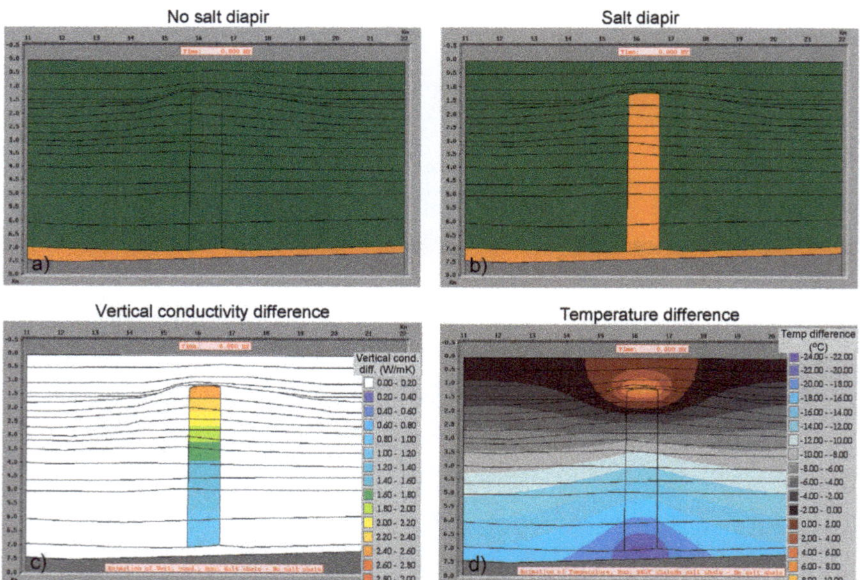

Figure 15. Temperature effect of salt in the synthetic simple model. B) Model without salt diapir, (**a**,**b**) with salt diapir (orange). (**c**) Difference in thermal conductivity between the two models, (**d**) temperature difference between the two models. Grey to blue colors: Colder sediments in salt diapir model. Brown to red colors: Higher temperature in diapir model.

The thermal conductivity of salt is strongly temperature dependent, described by a temperature correction factor α. This factor is used in the following manor to correct thermal conductivity values (equation modified from Reference [26]):

$$\lambda = \lambda ref \frac{[1]}{[1 + \alpha T]}$$

where:

λ is the corrected thermal conductivity in W/m°C

λ_{ref} is the thermal conductivity calculated at the reference temperature in W/m°C.

α is temperature correction factor in 1/°C, assumed here to be 5.0×10^{-3} °C

T is the temperature in °C

Figure 15 shows the input profile with and without salt and the resulting temperature effects. The resulting effect on temperature is clearly seen in the temperature difference plot in Figure 15d where the model with salt diapir has significantly higher temperatures in the sediments above the top of the salt structure (brown to red colors), and colder sediments (grey to blue colors) at depth relative to the case without the salt diapir. These simple models demonstrate that the conductivity difference between salt and sediments play an important role in temperature distribution around salt structures.

5.1.2. Temperature History of Model 1

Model 1 is modeled with and without salt structures. The first model follows the evolution of salt structures of Figure 11. The second model has continuous sedimentary units and a uniformly thin salt unit at the base throughout the geohistory. Figure 16 shows the reconstructed geometries at two time steps (20 Ma and 60 Ma) and the temperature difference between these two models.

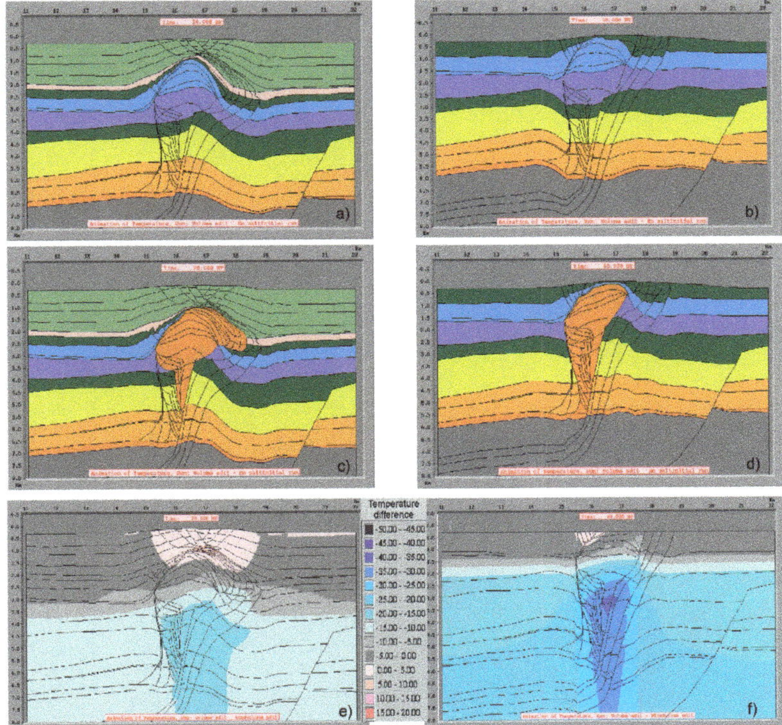

Figure 16. (**a–d**) Reconstructed geometries at 20 Ma (**a,c**) and 60 Ma (**b,d**) without and with salt structures, Figure (**e,f**) shows the temperature differences between the two models; temperature cooling in blue colous and temperature increase as pink to red colors.

At 60 Ma the salt structure is still continuous and a strong vertical cooling effect is seen along the salt. At 20 Ma the salt balloon is broken free from its stem and the vertical cooling is less. The relatively hotter, isolated, salt balloon volume leaks heat upward into the shallow sediment section above.

The temperature differences in these two models could be up to +5–10 °C above the salt at 20 Ma and −35–40 °C in the salt stem. The paleo heat flow was constant across the profile and calculated by two tectonic events in the area: A Permo–Triassic event and an Earliest Cretaceous event.

In the model, the entire width of the profile is affected by the temperature change. Deeper than 2 km, the temperatures are generally lowered with the existence of salt, increasingly towards the stem of the salt structure. Increased temperature due to the salt structure is only seen in the upper part of the salt balloon and directly above it. The largest difference in the sediments is found around the stem of the salt balloon during Latest Cretaceous time (75 Ma) (see Figure 17), whereas maximum heating of the area above salt occurred in Miocene time. The maximum modeled temperature reduction was between 84.7 °C at 75 Ma (Point 3 in Figure 17). A maximum positive temperature difference of only 5.8 °C was modeled directly above the highest point of the salt structure at base Miocene time (point 1 in Figure 17). An intermediate point in the chalk unit (point 2) shows a maximum temperature reduction in the salt model of 24 °C at 75 Ma.

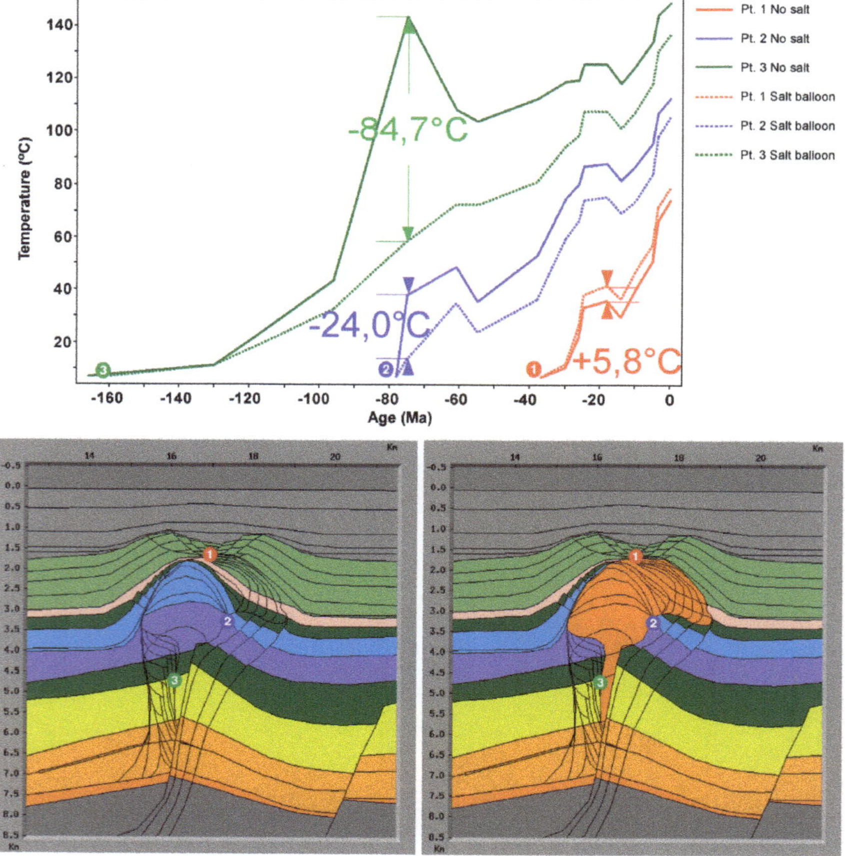

Figure 17. Temperature history difference in models with and without the growth of a salt balloon structure. The curves show temperature history of the points 1, 2, and 3 given on the two model profiles. The temperature of point 1 above the salt increases by 5.8 °C in the salt balloon model, whereas the lowest point below salt shows a decrease of more than 84 °C in salt model.

5.1.3. Model 2

Due to the uncertainty in interpretation of the salt bodies on Model 2, we have made three modeling options for determination of the temperature history and maturation (shown in Figure 13). The temperature over the profile was calculated based on thermal conductivities of the sediments (as a function of porosity; given in Table 1) and the modeled palaeo heat flow.

In Reference [26], the authors have found significant anisotropy in the thermal conductivities of shales in the North Sea. This is taken into account in our study. The vertical and horizontal thermal conductivities are shown in Figure 18a.

The palaeo heat flow was calculated by two tectonic events in the area; a Permo-Triassic event and a Mid-Jurassic to Earliest Cretaceous event; see more details in Reference [22]. The effect of the salt body on temperature history for a selected point (location; Figure 13c) above the salt is shown in Figure 18b. The calculated temperature is, as expected, the highest for the maximum-salt model.

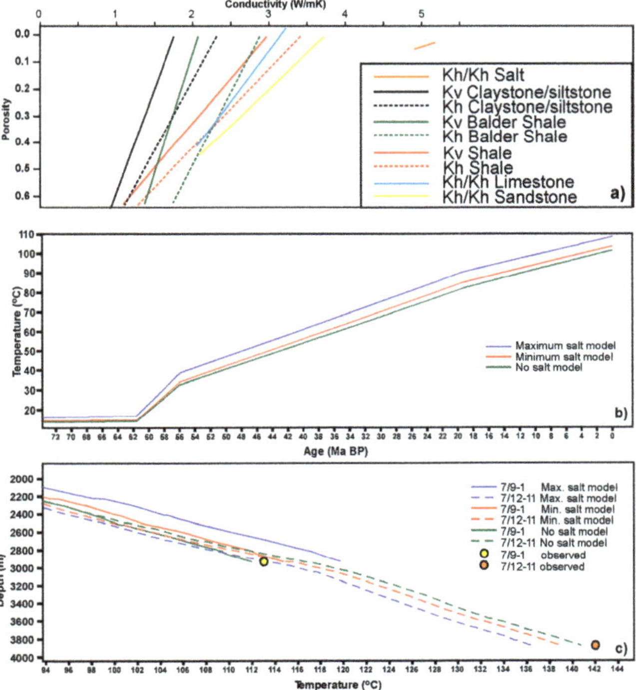

Figure 18. Thermal conductivity and temperature history for the three salt models shown in Figure 13. (**a**) Thermal conductivities used in the modeling. Kv = vertical conductivity, Kh = horizontal conductivity. Note that shale and claystones have different Kv and Kh. Other lithologies have uniform conductivities in both directions. (**b**) Calculated temperature history for the point above the salt close to base of well 7/9-1 (position shown in Figure 13). (**c**) Present day calculated and observed temperature in wells 7/9-1 and 7/12-11.

The present day heat flow is found by calibration to present day temperature. The calculated versus observed present day temperatures for the wells 7/9-1 and 7/12-11 are shown in Figure 18c. The calculated present day temperature difference is approximately 10 °C between the max-salt model and no-salt model. We have used uniform heat flow over the profile and the same matrix thermal conductivities within the various sedimentary units over the profile. Based on these assumptions, we see that the best match with the observed temperature in well 7/9-1 is given by the minimum-salt option and the no-salt option. The largest discrepancy is given by the max-salt option. A better fit for the max-salt option could be achieved with a reduced heat flow; however, a reduced heat flow would give larger discrepancies for well 7/12-11 for all three options; the three options are already at the lower possible range.

Figure 19 compares the present day temperatures on the models with no salt and maximum salt, as defined in Figure 13. In the model with salt, isotherm curves rise above the upper part of the salt columns, but dip down in the deepest part of the salt and below the salt columns. The increase in temperature can be visible at least 2400 m above the top of the large salt column at position 13 km on the modeled profile. At the base of the intra-salt basins, the temperatures are lowered by about 20 °C. At the top of the salt columns, the temperature increases up to 8.5 °C, most above the highest ones.

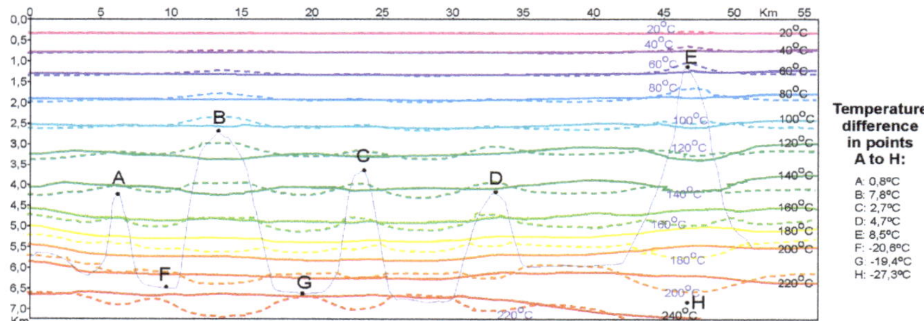

Figure 19. Calculated present-day temperature for models with salt and without salt. The temperature isotherm curves for the model with salt (stippled lines and blue temperature numbers), and the model without salt (continuous lines, black temperature numbers). In the salt model points A–D show higher temperatures than in the model without salt. Points F–H show lower temperatures in the salt model.

5.2. Vitrinite Reflectance

Vitrinite is an organic material from cell walls of woody plant tissue [27]. Vitrinite reflectance (VR) is a measure of the percentage of incident light reflected from the surface of vitrinite particles in a sedimentary rock. It is referred to as %Ro. Results are often presented as a mean Ro value based on all vitrinite particles measured in an individual sample. Vitrinite reflectance is a standard method for measuring the thermal maturity of sedimentary rocks and kinetic models of vitrinite reflectance are commonly used to constrain paleo thermal histories in basin and petroleum system modeling. BMT uses the EASY%RO model to calculate vitrinite reflectance [28]. Generally, the onset of oil generation is correlated with a reflectance of %Ro = 0.5–0.6% and the termination of oil generation with reflectance of 0.85%–1.1%. The onset of gas generation ('gas window') is typically associated with %Ro values of 1.0%–1.3% and terminates around 3.0%. The vitrinite reflectance was here modeled for cases with and without development of piercing salt structures for both models, and the results are shown in Figures 20 and 21.

In model 1, the increased conductivity of the salt and the resultant higher sediment temperatures above salt and lower temperatures below and around the salt clearly affects the modeled vitrinite reflectance (Figure 20). The same can be seen, but less pronounced, in Model 2 with the five columnar diapirs (Figure 21). The hydrocarbon potential may be lower in areas around penetrating salt structures, as cooling of the sediments around the stem of the salt structure lowers the maturation potential locally. This may be enhanced by erosion over the salt diapirs.

The temperature effects in the area above salt are not sufficient to have significant hydrocarbon potential in Model 1 (Figure 20). The model with salt has only <0.05% higher VR at shallow depths and increasingly lower VR towards the stem below salt with a maximum difference of 0.6%–1.0% lower VR at 7 km depth in the lower Triassic sequence.

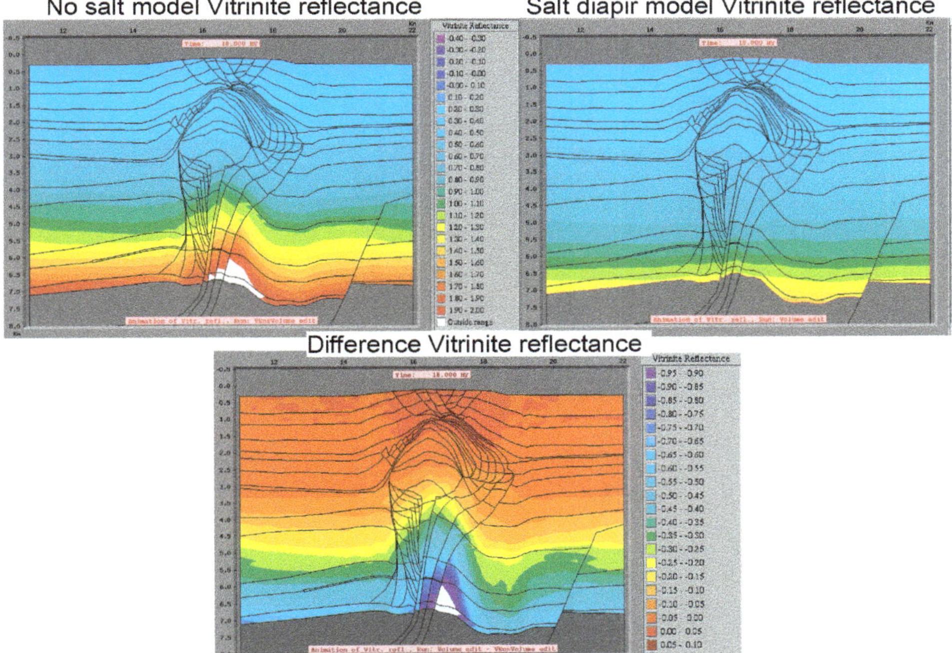

Figure 20. Top: Modeled vitrinite reflectance for Model 1 at 18 Ma, without salt balloon (top left) and model with salt balloon (top right). Lower figure shows the difference in modeled vitrinite reflectance between the two models.

The calculated present day vitrinite reflectance difference between the models with and without salt structures for Model 1 and Model 2 are shown in Figure 21. The modeled "oil windows" and "gas windows" are indicated by light green and red colors for the no-salt models and with hatching on the salt-structure models. Models with salt diapirs have lower maturation in the basins between salt structures than models without salt diapirs. In general, the VR difference increases by depth from around 2 km depth, with lower Ro values (less mature sediments) in the salt-diapir models. Maturation is higher above salt diapirs, but only at shallow depths and has little impact on hydrocarbon potential.

In Model 1, a maximum depth difference of −1065 m at the base of the "gas window" is shown towards the salt stem. At Ro = 1.0 (base of "oil window" the maximum depth difference is −935 m (Figure 21a). The salt modeled option may still be gas generating in the Jurassic and upper Triassic below the salt structure, while the model without the salt balloon has a present day "gas-window" that only reaches down into Lower Cretaceous strata.

Along the profile in Model 2, the VR difference decreases upward, but the salt model has consistently lower %Ro values in the Triassic pod basins relative to the no-salt model. At a position 30 km along the profile %Ro values of 3.0 are found 600 m shallower in the no-salt model. As expected from the temperature modeling, the oil and gas windows are deeper between salt pillars than in the model without salt. Maximum depth of oil window is as much as 264 m deeper centrally in the Triassic pod at position 20 km in the salt model (marked A in Figure 21b). Above the salt structure at 33 km, the salt model reaches base of oil window (i.e., Ro = 1.1) at 203 m shallower depth than in the no-salt model (marked B). The resulting vitrinite reflectance models indicate that the Triassic sequences of Model 2 may still be gas-generating presently.

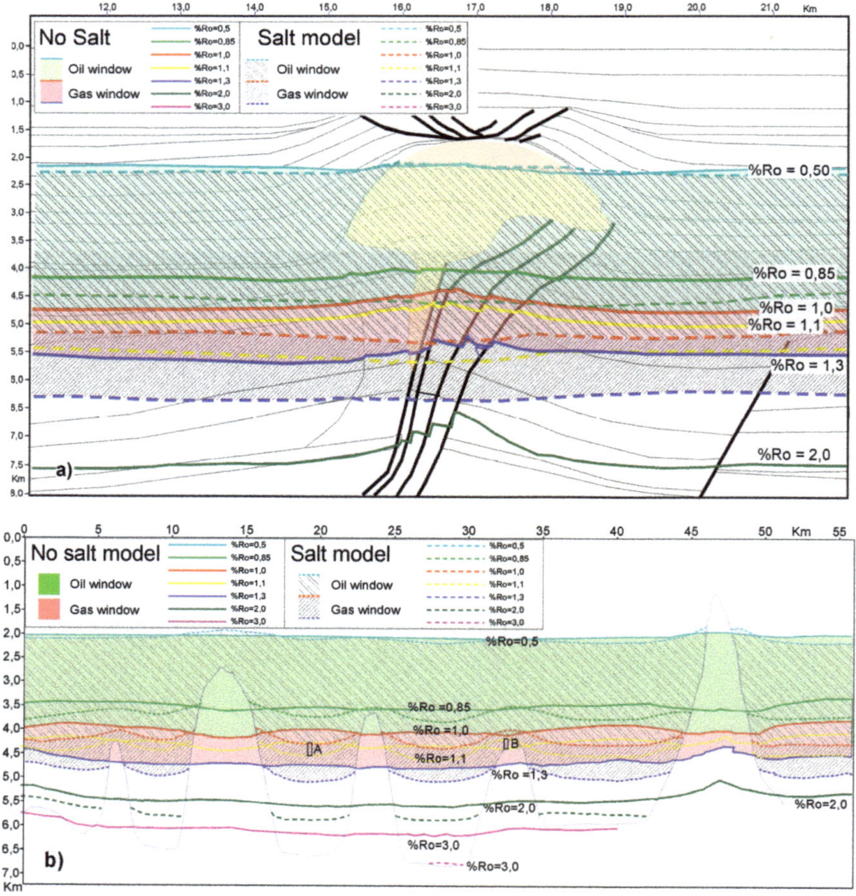

Figure 21. Differences in modeled Vitrinite Reflectance at present day geometry between models with salt (stippled lines) and model without salt diapirs (solid lines). "Oil-window" and "gas-windows" are indicated between %Ro values of 0.5–1.0 and 1.0–1.3, respectively. (**a**) model 1, (**b**) model 2.

6. Discussion

This study has demonstrated the importance of high resolution basin modeling for understanding the dynamic evolution of basin geometry, structures, temperature and maturity- history in areas with salt structures. Development of salt structures and surrounding pod basins is a self–reinforcing process. Increased subsidence starts salt movement and initiates the chain of movements leading to buoyant movement and piercing of salt structures. In this process, salt thins and withdraws from the deepest basins and move towards the salt structures, leaving room for sediment accommodation and subsidence. This structural development may affect the pattern of heat distribution

Since salt has very unique rheological properties it is vital to have the possibility to reconstruct a reasonable geometry and temperature evolution. In modeling of salt structures interpretation is probably the largest uncertainty, and it is important to have a good definition of lithologies in the model, as the conductivities of different lithologies vary a lot (Figure 18a). As shales due to their layered nature of flattened clay minerals, have substantially higher horizontal conductivity than vertical conductivity [29], they transfer heat horizontally and act as insulators vertically. Salt, sandstones, and carbonates have equal horizontal and vertical-conductivity.

As shown in this study and previous studies [2,14], the presence of salt can have a large impact on the local temperature distribution within an area, and the higher conductivity of salt plays an important role both for evolution of reservoir and traps and for defining areas of optimal temperature and maturation. Our results are in good correspondence to the modeling results of Reference [14], which modeled a maximum temperature lowering of 85 °C at the base of a salt diapir reaching the surface, and similar vitrinite reflectance lowering around a salt diapir relative to a similar geometry shale diapir.

For petroleum exploration in a salt environment, the geometric evolution in time and space is important. Results of geohistory reconstruction and basin modeling are important for understanding the evolution of both reservoirs and traps in space and time. A good model for the evolution of basin geometry and salt volume is essential to make a good temperature and maturation model. Modeling of geological processes always involves strong simplification and general assumptions based on experience and knowledge about the area and processes that are modeled. The quality of the model is directly related to the quality of the input data. The input parameters used must be based on the best possible knowledge, but will always represent an amount of uncertainty. In our models, the quality of seismic interpretation and definition and distribution of lithological units are very important. The seismic interpretation is probably the largest uncertainty in this modeling, but other input parameters may contribute to the uncertainties. Repeated modeling with sensitivity analysis of uncertain parameters, with calibration to measured data is important to reduce these uncertainties.

To model a complex evolution of salt-structures, it is necessary to use basin modeling software that allows very high and variable resolution, both in time and space. The BMT modeling software nicely fits to these requirements, which are important to be able to model the evolution of complex geometries of salt structures and their temperature and maturity effects.

Additional important effects of active salt development are better understanding and observation of fracturing above piercing salt structure, faulting and deformation related to salt withdrawal and basin collapse, erosion above salt structure and redistribution of potential reservoir sediments and others.

Late regional phases of erosion could strongly affect the geometry and temperature history in a salt basin and correct modeling of these episodes are important. A salt structure reaching the surface is considered to be a very effective conductor draining a the basin of heat [14].

The lowered temperature below salt can delay and extend the time of hydrocarbon generation and preserve source rocks even in a relatively deep basin. This factor may thus be important in late and rapidly evolving salt areas, such as on continental margins where subsidence and maturation occurred late in parallel to the evolution of the complex salt structures.

Salt movement up through carbonates may also lead to substantial dissolution of carbonate [30] that may precipitate as carbonate–cement in fractured reservoir units, ruining reservoir quality. These processes are, however, beyond the scope for this article.

7. Conclusions

Detailed study of seismic data from our study area revealed a complex development of salt structures since early Triassic time, involving both the extensional and compressional phases.

Our detailed modeling has recreated the salt evolution and shown that the presence of salt structures have a substantial effect on temperature and maturation, both close to the salt itself, and in a larger distance around and below salt structures.

The results show that basin areas at least 3–5 km to the sides of the salt structures are affected by lowered temperatures and maturation. The conductivity effect of salt increases toward the surface and thus salt structures near or at the surface are more efficient at draining heat out of the basin.

Temperatures increase in the sediments above the salt structures, but this has very little impact on maturation, because the heated sediments are too shallow and the temperature increase too small to increase maturation potential. The cooling of the sediments around salt structures may play a larger role as it affect larger areas and in deeper units. Cooling of the sediments due to presence of salt may be a very important factor in hydrocarbon exploration in regions with salt structures.

The geometry of the present day salt and its history of development are very important. A vertical columnar salt diapir will have a symmetrical temperature anomaly around the salt. If an irregular salt geometry develops with a shallow overhanging salt volume, an increased temperature anomaly would occur below the salt overhang, possibly preserving maturation in the position of stratigraphic traps. A larger area above the salt is also affected by heating than for a narrow columnar structure. Our modeling software allows modeling of complex development of such overhanging salt structures.

A simple salt anticline or a vertical salt column still connected to the initial salt layer has the largest salt volumes at depth where the conductivity-contrast to the surrounding sediments is the least. A detached allochtonous salt volume [2] or a canopy or mushroom-shaped salt structures have most of its salt volume at shallower depths where they can more effectively drain the heat from sediments below.

In our modeling, the effects on maturation occur at depths below 2 km, and at depths of more than 3.5–4 km the maturation levels are lowered as much as 500 m in pod-basins between vertical diapirs and as much as 1–1.5 km below and overhanging mushroom-shaped salt structure. This shows that the effect is substantial enough that it should be considered when exploring for hydrocarbons in salt regions.

Author Contributions: Conceptualization, I.G. and A.M.; methodology, I.G.; software, I.G.; validation, I.G. and A.M; investigation, I.G.; writing—original draft preparation, I.G. and A.M.; writing—review and editing, I.G. and A.M.; project administration, I.G.; funding acquisition, I.G.

Funding: This research received no external funding.

Acknowledgments: First we have to thank LOTOS Exploration and Production Norge AS (LEPN), the operator of PL 498/B, for letting us publish results based on their data, for their valuable input and help with expertise and data. Also thanks to the partners in PL 498/B; Skagen 44, Edison, North and Lime. Thanks to Willy Fjeldskaar, Ingrid F. Løtveit and two anonymous reviewers for constructive comments and suggestions.

Conflicts of Interest: The authors declare no conflict of interest.

References

1. Alsop, G.I.; Archer, S.G.; Hartley, A.J.; Grant, N.T.; Hodgkinson, R. *Salt Tectonics, Sediments and Propectivity*; Special Publication, Geological Society: London, UK, 2012; 632p.

2. Jackson, M.P.A.; Hudec, M.R. *Salt Tectonics: Principles and Practice*; Cambridge University Press: Cambridge, UK, 2017; 510p.

3. Stewart, S.A.; Clark, J. Impact of salt on the structure of the Central North Sea hydrocarbon fairways. In *Petroleum Geology of Northwest Europe: Proceedings of the 5th Conference*; Fleet, A.J., Boldy, S.A.R., Eds.; Geological Society: London, UK, 1999; pp. 179–200.

4. Alsop, G.I.; Weinberger, R.; Marco, S.; Levi, T. Faults and fracture patterns around a strike-slip induced salt wall. *J. Struct. Geol.* **2018**, *106*, 103–124. [CrossRef]

5. Trusheim, F. Mechanism of salt migration in Northern Germany. *AAPG Bull.* **1960**, *44*, 1519–1540.

6. Zong, J.; Coskun, S.; Stewart, R.R.; Dyaur, N.; Myers, M.T. Salt densities and velocities with application to Gulf of Mexico salt domes. In *Salt Challenges in Hydrocarbon Exploration SEG Annual Meeting Post-Convention Workshop*; Society of Exploration Geophysicists: New Orleans, LA, USA, 2015; Available online: http://www.agl.uh.edu/pdf/jingjing-salt%20densities.pdf (accessed on 24 March 2019).

7. Blackwell, D.D.; Steele, J.L. Thermal conductivity of sedimentary rocks: Measurements and significance. In *Thermal History of Sedimentary Basins*; Naeser, N.D., McColloh, T.H., Eds.; Springer: New York, NY, USA, 1989; pp. 13–37.

8. Gussow, W.C. Salt diapirism: importance of temperature and energy source of emplacement. In *Diapirism and Diapirs*; Baraunstein, J., O'Brien, G.D., Eds.; AAPG Mem: Tulsa, OK, USA, 1968; Volume 8, pp. 16–52.

9. Urai, J.L. Deformation of Wet Salt Rock. Ph.D. Thesis, University of Utrecht, Holland, The Netherlands, 1983. (Referred in Price and Cosgrove, 1990) Unpublished.

10. Price, N.J.; Cosgrove, J.W. *Analysis of Geological Structures*; Cambridge University Press: Cambridge, UK, 1990; 502p.

11. Jackson, M.P.A.; Talbot, C.J. External shapes, strain rates and dynamics of salt structures. *Geol. Soc. Am. Bull.* **1986**, *97*, 305–323. [CrossRef]

12. Hudec, M.R.; Jackson, M.P.A. Terra infirma: Understanding salt tectonics. *Earth-Sci. Rev.* **2007**, *82*, 1–28. [CrossRef]

13. Brun, J.-P.; Fort, X. Salt tectonics at passive margins: Geology versus models. *Mar. Pet. Geol.* **2011**, *28*, 1123–1145. [CrossRef]

14. Mello, U.T.; Karner, G.D.; Anderson, R.N. Role of salt in restraining the maturation of subsalt source rocks. *Mar. Pet. Geol.* **1995**, *12*, 697–716. [CrossRef]

15. Baniak, G.M.; Gingras, M.K.; Burns, B.A.; Pemberton, S.G. An example of a highly bioturbated, storm-influenced shoreface deposit: Uper Jurassic Ula Formation, Norwegian North Sea. *Sedimentology* **2014**, *61*, 1261–1285. [CrossRef]

16. Hodgson, N.A.; Farnsworth, J.; Fraser, A.J. Salt-related tectonics, sedimentation and hydrocarbon plays in the Central Graben, North Sea, UKCS. *Geol. Soc. Lond. Spec. Publ.* **1992**, *67*, 31–63. [CrossRef]

17. Karlo, J.F.; van Buchem, F.S.P.; Moen, J.; Milroy, K. Triassic-age salt tectonics of the Central North Sea. *Interpretation* **2014**, *2*, SM19–SM28. [CrossRef]

18. Mannie, A.S.; Jackson, C.A.L.; Hampson, G.J. Shallow-marine reservoir development in extensional diapir-collapse minibasins: An integrated subsurface case study from the Upper Jurassic of the Cod terrace, Norwegian North Sea. *AAPG Bull.* **2014**, *98*, 2019–2055. [CrossRef]

19. Mannie, A.S.; Jackson, C.A.L.; Hampson, G.J.; Fraser, A.J. Tectonic controls on the spatial distribution and stratigraphic architecture of net-transgressive shallow- marine synrift succession in a salt-influenced rift basin: Middle to Upper Jurassic, Norwegian Central North Sea. *J. Geol. Soc.* **2016**, *173*, 901–915. [CrossRef]

20. Høiland, O.; Kristensen, J.; Monsen, T. Mesozoic evolution of the Jæren High area, Norwegian Central North Sea. In *Petroleum Geology of Northwest Europe: Proceedings of the 4th Conference*; Parker, J.R., Ed.; Geological Society: London, UK, 1993; pp. 1189–1195.

21. Pegrum, R.M. Structural development of the south-western margin of the Russian-Fennoscandian Platform. In *Petroleum Geology of the Northern European Margin*; Spencer, A.M., Ed.; Graham & Trotman: London, UK, 1984; pp. 173–189.

22. Fjeldskaar, W.; Ter Voorde, M.; Johansen, H.; Christiansson, H.P.; Faleide, J.I.; Cloetingh, S.A.P.L. Numerical simulation of rifting in the northern Viking Graben: the mutual effect of modelling parameters. *Tectonophysics* **2004**, *382*, 189–212. [CrossRef]

23. Fjeldskaar, W.; Andersen, Å.; Johansen, H.; Lander, R.; Blomvik, V.; Skurve, O.; Michelsen, J.K.; Grunnaleite, I.; Mykkeltveit, J. Bridging the gap between basin modelling and structural geology. *Reg. Geol. Metallog.* **2017**, *72*, 65–77.

24. Sclater, J.G.; Christie, P. Continental stretching: an explanation of the post-mid-Cretaceous subsidence of the Central North Sea basin. *J. Geophys. Res.* **1980**, *85*, 3711–3739. [CrossRef]

25. Fernandez, N.; Jackson, C.A.L.; Hudec, M.; Dooley, T.P. The Competition for Salt and Kinematic Interactions between Minibasins during Density-Driven Subsidence Observations from Numerical Models. 2019. Available online: https://www.researchgate.net/publication/331904078_ (accessed on 4 August 2019).

26. Burrus, J.; Audebert, F. Thermal and compaction processes in a young rifted basin containing evaporties: Gulf of Lions, France. *AAPG Bull.* **1990**, *74*, 1420–1440.

27. Crelling, J.C.; Dutcher, R.R. *Principles and Applications of Coal Petrology: SEPM Short*; Course. Society of Economic Paleontologists Mineralogists: Tulsa, OK, USA, 1980; Volume 8, 127p.

28. Sweeney, J.J.; Burnham, A.K. Evolution of a simple model of vitrinite reflectance based on chemical kinetics. *AAPG Bull.* **1990**, *74*, 1559–1570.

29. Fjeldskaar, W.; Prestholm, E.; Guargena, C.; Gravdal, N. Isostatic and tectonic development of the Egersund Basin. In *Basin Modelling: Advances and Applications*; Doré, T., Augustson, J.H., Hermanrud, C., Stewart, D.J., Sylta, Ø., Eds.; Elsevier: Amsterdam, The Netherlands, 1993; pp. 549–562.

30. Davison, I.; Alsop, G.I.; Evans, N.G.; Safarics, M. Overburden deformation patterns and mechanisms of salt diapir penetration in the Central Graben, North Sea. *Mar. Pet. Geol.* **2000**, *17*, 601–618. [CrossRef]

Article

Extensive Sills in the Continental Basement from Deep Seismic Reflection Profiling

Larry D. Brown [1],* and Doyeon Kim [2]

[1]   Department of Earth and Atmospheric Sciences, Cornell University, Ithaca, NY 14853, USA
[2]   Department of Geology, University of Maryland, College Park, MD 20742, USA; dk696@cornell.edu
*   Correspondence: ldb7@cornell.edu

Received: 28 July 2020; Accepted: 21 October 2020; Published: 10 November 2020

Abstract: Crustal seismic reflection profiling has revealed the presence of extensive, coherent reflections with anomalously high amplitudes in the crystalline crust at a number of locations around the world. In areas of active tectonic activity, these seismic "bright spots" have often been interpreted as fluid magma at depth. The focus in this report is high-amplitude reflections that have been identified or inferred to mark interfaces between solid mafic intrusions and felsic to intermediate country rock. These "frozen sills" most commonly appear as thin, subhorizontal sheets at middle to upper crustal depths, several of which can be traced for tens to hundreds of kilometers. Their frequency among seismic profiles suggest that they may be more common than widely realized. These intrusions constrain crustal rheology at the time of their emplacement, represent a significant mode of transfer of mantle material and heat into the crust, and some may constitute fingerprints of distant mantle plumes. These sills may have played important roles in overlying basin evolution and ore deposition.

Keywords: crustal sills; hydrocarbon and mineral resources; sedimentary basins

---

1. Introduction

The geometry, composition, and mechanics of crustal intrusions, including sills, have been subject of numerous field and lab investigations; the associated literature is extensive [1–6]. Although sills are often observed in surface geological outcrops and thus available for direct sampling [7,8], their subhorizontal geometry makes them less likely to outcrop than near-vertical dykes, especially if buried beneath later sedimentary basins. The detection and delineation of magma, especially if still molten, at depth has been an explicit goal of many geophysical surveys using a variety of techniques, each with its own strengths and limitations. Gravity, for example, can be used to identify mass excess or mass deficiency that can be attributed respectively to mafic or granitic materials (molten or frozen) at depth [9–12]. Magnetotelluric methods have been widely used to detect magma at depth due their sensitivity to the high conductivities associated with magma and magmatic fluids [13,14]. However, gravity is notoriously non-unique [15], as are the various electrical methodologies [16,17]. Both offer relatively limited resolution at depth, especially if the target is a thin planar structure. Seismic methods using both artificial and earthquake sources have also been widely used to define crustal structure in general and magmatic additions both hot and cold. Thybo and Artemieva [18]) review many of the controlled source refraction/wide angle results that have been used to infer massive magmatic underplating in the crust. Seismic tomography is also now a commonly used tool to search for magma in all its forms at depth [19–21]. However, even the most recent tomography is limited in spatial resolution to tens of kilometers [22,23]. A somewhat greater resolution can be achieved using receiver functions computed from teleseismic recordings. Receiver functions have been interpreted to indicate an extensive sill beneath the Altiplano-Puna volcanic zone of the central Andes [24,25].

In many applications, the seismic reflection technique [26], also known as seismic reflection profiling or multichannel seismic profiling, offers the highest resolution of any geophysical technique. Seismic reflection surveying, the primary geophysical method used for oil and gas exploration, has become a highly sophisticated tool for imaging the subsurface in both 2D and 3D at depths ranging from the near surface to the upper mantle. Oil industry reflection surveys have already had a major impact on our understanding of the sill distributions in sedimentary basins [27–29].

Here, we focus on published examples of identified or inferred sills in the continental basement hidden beneath the sedimentary cover, with special attention to frozen sills that may be fingerprints of ancient thermal processes and the large-distance lateral transport of magma in the crystalline crust.

2. Observations

2.1. Seismic Bright Spots and Magma in the Crust

The starting point for the recognition of sills on seismic reflection recordings is their expected strong seismic contrast with surrounding rock, as expressed by the reflection coefficient. The reflection coefficient (RC) for vertically incident seismic waves upon a horizontal interface with density (ρ_1) and seismic velocity (V_1) overlying a layer with density (ρ_2) and seismic velocity (V_2) is given by the formula [26]:

$$RC = (\rho_2 V_2 - \rho_1 V_1)/(\rho_2 V_2 - \rho_1 V_1). \tag{1}$$

While this relation applies to both compressional (P) waves and shear (S) waves, the reflection data reported here was collected using P waves only. The representative density and P wave velocity values for rocks most representative to this review are shown in Table 1.

Table 1. Relevant physical properties of representative materials. The velocity measurements correspond to 200 MPa and the magma measurements correspond to 2000 °C.

Material	Vp (km/s) @ 200 Mpa	Density, kg/m^3	RC % against UCC	Source
granite-granodiorite	6.246	2.76	0.3	[30,31]
diabase	6.712	2.87	9.3	[30,31]
andesitic magma	2.5	2.45	−46.3	[32] Mt. Hood andesite
basaltic magma	6.243	2.76	−30.6	[32] Columbia River Basalt
Phyllite	6.243	2.76	0,7	[30,31]
Average upper continental crust (UCC)	6.2	2.76	0.0	[31–33]

If a phyllite body in an average upper continental crust (UCC) is taken as representative of most reflection coefficients encountered in deep reflection surveys, the reflection coefficients for "hot" sills—silicic or mafic—emplaced in the UCC are an order of magnitude larger and imply that the corresponding reflection amplitudes would be anomalously strong. With respect to "cold" magmas, intermediate composition sills are likely to give rise to modest amplitudes at best, while mafic sills would still be expected to give rise to notably strong reflections compared to the surrounding heterogeneities.

However, the observed reflection amplitudes are affected by a number of factors other than the reflection coefficient, including geometrical focusing, layer tuning, transmission loss, and anelastic attenuation [26,29], so that these values should considered as rough guides only. Nevertheless, a reflection with an amplitude that appears to be anomalously strong compared to its neighboring reflections could be a candidate for either a still fluid magma (granitic or mafic) or a frozen mafic sill.

The shear wave reflection coefficients from "hot" magma at depth would be expected to be even larger since shear wave velocities approach 0 in a fluid. Studies in the Rio Grande Rift of New Mexico by Alan Sanford and his colleagues [34] and at volcanoes in northeastern Japan by Hasegawa and colleagues [35] are pioneering examples of detecting and mapping magma at depth using anomalous reflected shear waves from microearthquake sources.

The anomalous shear wave reflectors at midcrustal depths near Socorro, New Mexico, referred to as the Socorro Magma Body (SMB; [36]), attracted the attention of the nascent COCORP (Consortium for Continental Reflection Profiling) project in 1976. COCORP multichannel vibroseis source surveys then imaged an anomalously strong P wave reflector (called the Socorro Bright Spot) that corresponds directly with the anomalous S reflector (Figures 1 and 2; [37]). The unusually strong amplitude P waves from the COCORP controlled source survey (Figure 2) and the anomalous S waves reflections on microearthquake recordings [34,38] are both consistent with a solid–fluid interface [35,39]. A hot magma interpretation is also consistent with the tectonic setting—i.e., a Cenozoic rift characterized by high heat flow—and is supported by MT measurements of high conductivities at mid-crustal depths [40]. Of particular significance are the geodetic and INSAR observations of contemporary surface uplift that suggest active magma inflation at the depth of this reflector [41–43]. The receiver function analysis of teleseismic data [44] confirms that the SMB is a relatively thin layer of magma corresponding to the seismic reflection bright spot.

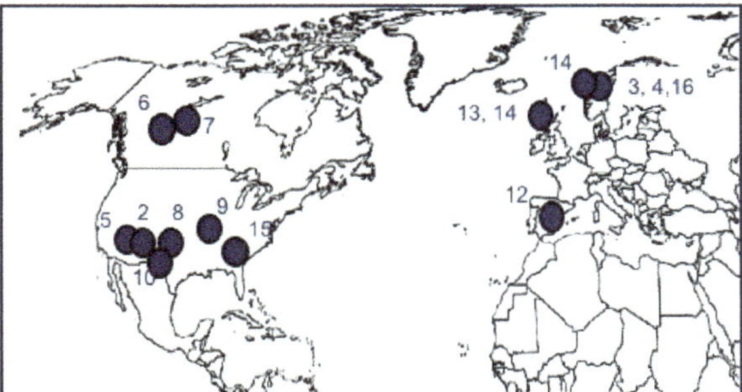

Figure 1. Map showing the locations of seismic data cited in this paper. The numbers refer to the figures in this paper.

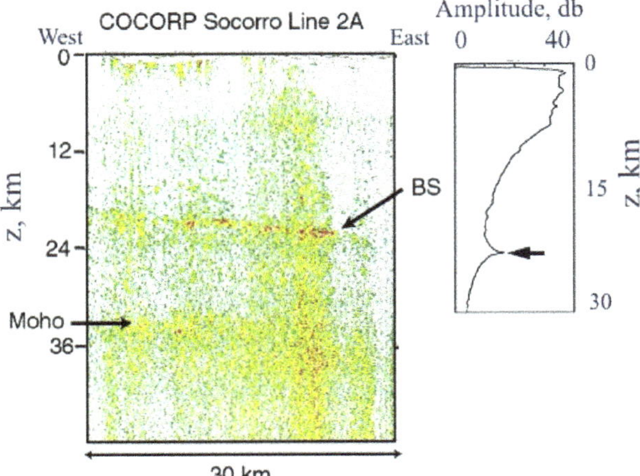

Figure 2. (**Left**) "True Amplitude" deep seismic reflection section showing a "bright spot" interpreted to mark magma in the mid-crust in the central Rio Grande Rift. (**Right**) reflection amplitude vs. depth for the sum of raw seismic traces quantifying the anomalous nature of the reflection amplitude of the Socorro Bright Spot (COCORP Line 1; [36,45]).

The Socorro Bright Spot has served as a model for a magma interpretation of other anomalously strong deep seismic reflections (bright spots) reported from other regions of the world [45]. Examples include Death Valley in southern California [46], the Basin and Range of northwestern Nevada [47], the southern Tibetan Plateau [48], the central Andes [49], and the Taupo Volcanic Zone in central New Zealand [50]. Marine reflection profiling and 3D reflection surveys have also proved effective in mapping of likely magma chambers in the crystalline oceanic crust beneath mid-ocean ridges [51–53]. In most of these examples, the magma interpretation is bolstered by complementary geophysical observations, including MT and wide-angle refraction/reflection surveys and/or tomographic imaging with natural sources. A proper review of such "hot" magma reflections would entail a quantitative comparison of the individual survey results, as well as expansion to include the diverse range of other geophysical observations that have been reported to indicate similar features. Here, we choose instead to focus on the lesser known examples of frozen sills for which reflection surveys may be the only methodology capable of their detection.

2.2. "Frozen" Sills Detected by Reflection Surveys

The geometry and physical contrasts associated with solid intrusions at depth suggest that they should be less easily detectable than fluid magma by many geophysical techniques. However, the reflection coefficient associated with solid mafic sills emplaced in upper continental crust (Table 1) and their subhorizontal geometry both favor detection and mapping by multichannel reflection profiling. The examples reviewed below confirm that expectation. In most cases, the interpretation of strong basement reflections as mafic sills is largely circumstantial and lean heavily on analogy. However, in one particular case the interpretation is unassailable.

2.2.1. Ground Truth: Siljan, Sweden

Of special importance to inferring sills from reflection data is the multichannel seismic reflection profiling at the Siljan Ring in east central Sweden (Figures 1 and 3; [54–56]). The Siljan seismic profiles (Figure 3) reveal a distinctive sequence of strong reflections that have been identified as dolerite sills by drill holes (Figure 4). Analogy with the reflection character of the Siljan reflections (high amplitudes, subhorizontal orientation, linear extent) has been a primary argument for the interpretation of similar appearing reflections elsewhere in the world, including those discussed in this paper.

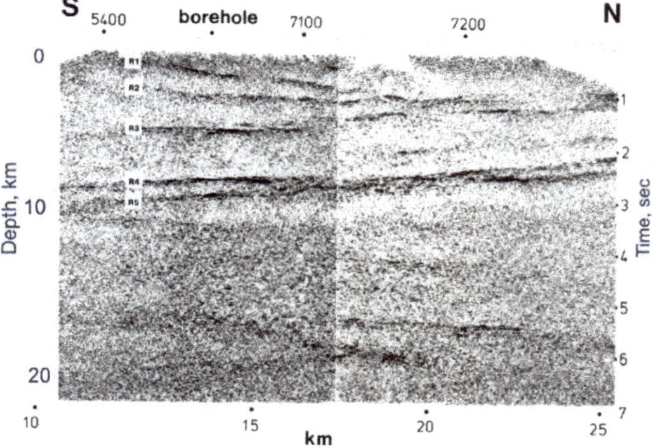

Figure 3. Strong subhorizontal reflections mapped in the upper 10 km of Proterozoic crust by seismic reflection profiling at the Siljan Ring Sweden [55]. Depth assuming the average velocity of 6 km/s. R1–R5 label individual reflectors in this stack.

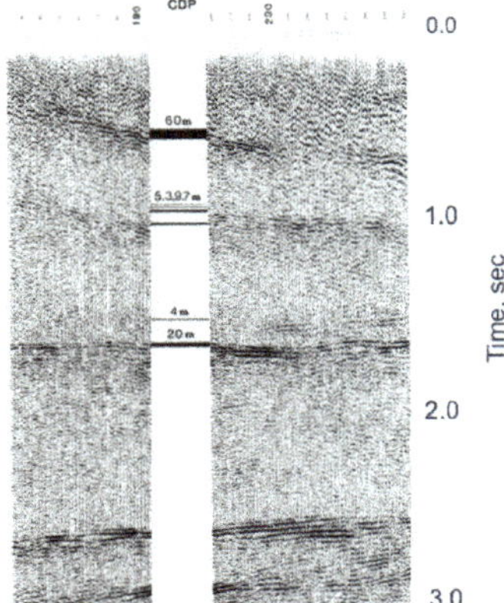

Figure 4. Detail of the Siljan Ring reflections in Figure 3 correlated with the borehole observations of diabase intrusions (solid black lines, with indicated thicknesses [55]). The sills date from 850 to 1700 my. This clear identification of strong reflectivity with mafic sills [55,56] has served as a reference for the interpretation for similar appearing reflection sequences in other areas.

2.2.2. Identification via Outcrop

The Siljan reflections are relatively unique in terms of being tested by drilling. However, outcrop correlations have also been used to infer mafic sills as being responsible for similar reflection sequences. Surveys by the Consortium for Continental Reflection Profiling (COCORP) in central Arizona revealed a relatively thick suite of strong reflections in the upper crust (A in Figure 5) very similar to the Siljan seismic images (Figures 1 and 5 [57]). The seismic modeling of mafic sills exposed in basement outcrops approximately 20 km southeast of Line 3 was found to closely reproduce the character of the reflections seen on the seismic sections, bolstering their interpretation as arising from the subsurface extension of the outcropping cold diabase intrusions. This correlation is significant because the tectonic setting in Arizona would also be consistent with the presence of fluid magma in the subsurface. The deeper reflection C might be a candidate for hot magma, a suggestion based primarily on its discordance with sequence A and similarity in depth to the Socorro Magma body.

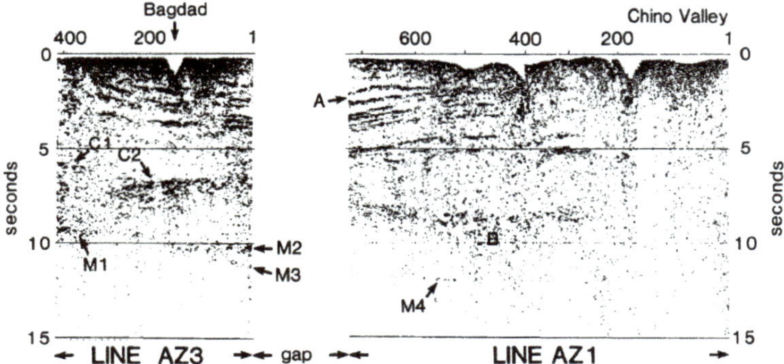

Figure 5. Prominent layered reflections (A) traced by the COCORP seismic reflection profiles in central Arizona [57]. Seismic modeling of mafic intrusions exposed nearby, together with the similarity in reflection character to the Siljan Ring reflectors, supports the interpretation of this sequence as also due to mafic intrusion [58]. The deeper reflections (C, B) may mark still-fluid magma. M indicates possible Moho reflections.

2.2.3. Extensive Sills in the Canadian Craton: Relicts of a Proterozoic Plume?

Among the most distinctive sill-like reflections, at least in terms of their observed extent, are a series of reflectors traced by the seismic surveys collected by the LITHOPROBE deep seismic program in Alberta and Saskatchewan, northwest Canada (Figures 1 and 6). The Winagami Reflection Sequence, first reported by Ross and Eaton [59], is a set of distinctive reflections that bears clear similarities to the Siljan reflections. Cross-cutting relationships between the Winagami reflections and weaker dipping reflections associated with dated tectonic events support an intrusive origin and suggest that the Winagami sequence was emplaced by a thermal event between 1.760 and 1.890 GA [59]. Originally estimated to extend beneath 120,000 km^2 of Paleoproterozoic basement, subsequent profiling reported by Mandler and Clowes [60] traced comparable reflections (the Head-Smashed In sequence) over an additional 6000 km^2 in similar tectonic terrane further south

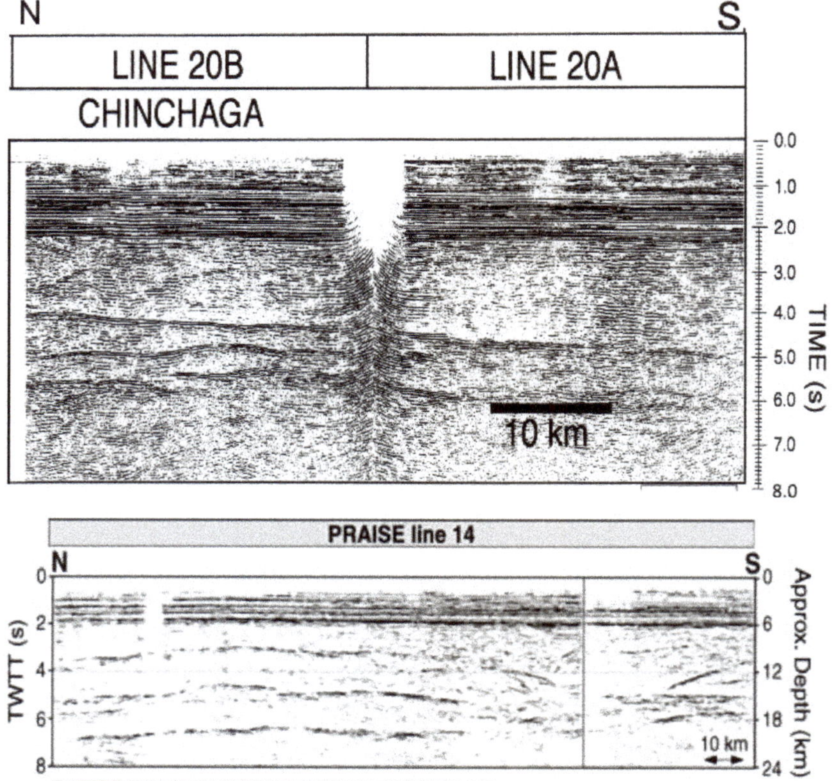

Figure 6. (**Top**) the Winagami reflection sequence revealed by LITHOPROBE reflection profiling in Alberta, NW Canada [59]. (**Bottom**) presumably correlative northward extension of the Winagami reflectors mapped by LITHOPROBE 3D seismic profiling [61].

Even more distinctive than the Winagami Sequence is the Wollaston Lake reflector (Figures 1 and 7 [62]), traced by the LITHOPROBE reflection profiles in the Trans-Hudson hinterland of Saskatchewan, approximately 500 km northeast of the Winagami surveys. The Wollaston Lake Reflector presents as a distinct narrow band of reflections that can be traced for over 160 km. Mandler and Clowes [62] associate the Wollaston Lake reflector with the 1.265 Ga McKenzie thermal event, which is associated with well-known outcroppings of diabase dykes [63]. This is substantially younger than the inferred age of the Winagami sequence and thus implies two distinct thermal events, both with extensive plutonic injections. While the long-distance lateral transport of magma has been documented in outcropping sills and dikes [7,63,64] the continuity of individual sills like the Wollaston Lake over such large distances is made most evident by these seismic images.

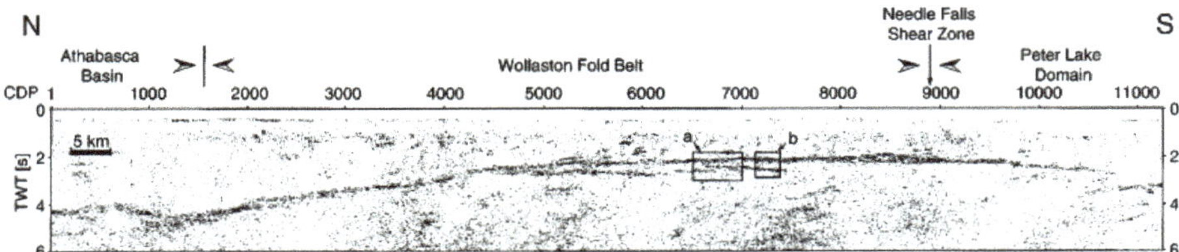

Figure 7. The Wollaston Lake reflector, imaged by LITHOPROBE seismic reflection profiling in Saskatchewan, NW Canada [62]. This feature can be traced as a distinct narrow band of reflections for 160 km within the Early Proterozoic Trans-Hudson hinterland. Based on its reflection character and tectonic context, it has been interpreted as a diabase sill associated with the 1.27 Ga McKenzie igneous event.

2.2.4. Basement Layering across the Central US: Fingerprints from the Keweenawan Plume?

The very first COCORP profiles were carried out in Hardeman County, northern Texas (Figure 1 [65]. The most notable discovery of those surveys was a distinctive sequence of subhorizontal reflections in the uppermost crystalline basement. These reflections were traced by subsequent COCORP surveys well into southern Oklahoma, where they are abruptly truncated by the Wichita Uplift (Figure 8). The extensive, layered nature of these reflections was initially interpreted to suggest a Proterozoic sedimentary or metasedimentary origin [66]. A supracrustal interpretation was reinforced by the observation of similarly layered reflections on COCORP seismic reflection profiling in Indiana, Illinois, and Ohio (Figures 1 and 9 [67]) and oil industry data in eastern New Mexico (Figures 1 and 10 [68,69]). The vast extent implied by correlating these layers from eastern New Mexico to central Ohio (Figure 11) is consistent with a depositional origin, although the apparent spatial correlation of these layers with the 1.5 MYA Granite-Rhyolite province [70] suggest the possibility that volcanic material (rhyolite?) rather than sedimentary rocks is involved [67]. However, the similarity of the Texas reflections to the Siljan images (compare Figure 8 with Figure 5) has continued to raise the question as to whether this midcontinent basement layering is actually a series of diabase sills.

Kim and Brown [69] revisited this issue in their interpretation of basement layering imaged by the reprocessing of 3D oil exploration seismic data in eastern New Mexico. Strong intrabasement reflections were reported from previous work by Adams and Miller [68] on 2D oil industry seismic data located nearby (Figure 10). Both of these papers referenced observations from an oil industry drillhole in southwest Texas that encountered layered ultramafic rocks at depths that seem to correspond to the basement layering in New Mexico (Figure 10 [72]). The ultramafic rocks recovered from the borehole were found to be of Keweenawan age (1.1635 Ga; [71]). Ernst and Buchan [73] describes links between layered ultramafic bodies and large-scale sill/dike intrusions. Kim and Brown ([69]) suggest that the southwest Texas borehole "calibration", the similarity in appearance of the upper basement layering on COCORP seismic data from New Mexico to Ohio, together with their similarity to the Siljan results support their interpretation as mafic sills associated with the Keweenawan plume. If correct, this interpretation indicates lateral injection of Keweenawan magma in the upper crust of the central US on a continental scale (Figure 11). Kim and Brown [69] point out this extent is comparable to the spatial extent of the McKenzie dikes in NW Canada [62] which are presumably from the same plume source as the Wollaston Lake reflector (Figure 11).

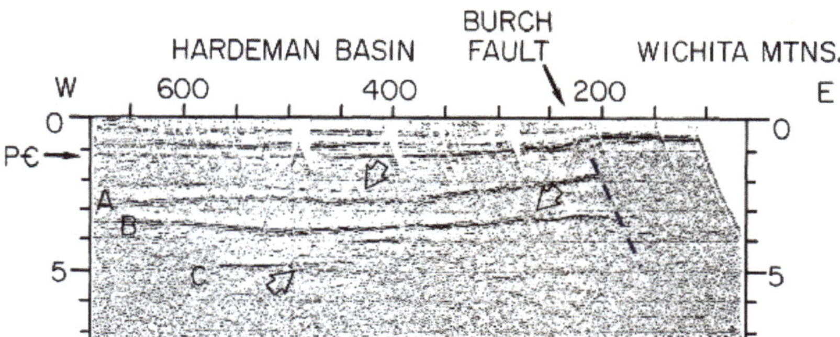

Figure 8. Layered reflectors traced by COCORP seismic profiles in northern Texas and southern Oklahoma [66]. P-C indicates the top of Precambrian basement from local boreholes. Arrows mark the reflective sequence (A, B, and C), here interpreted as mafic sills. Dashed line indicates the apparent truncation of basement reflections along the Burch fault of the Wichita Mountains.

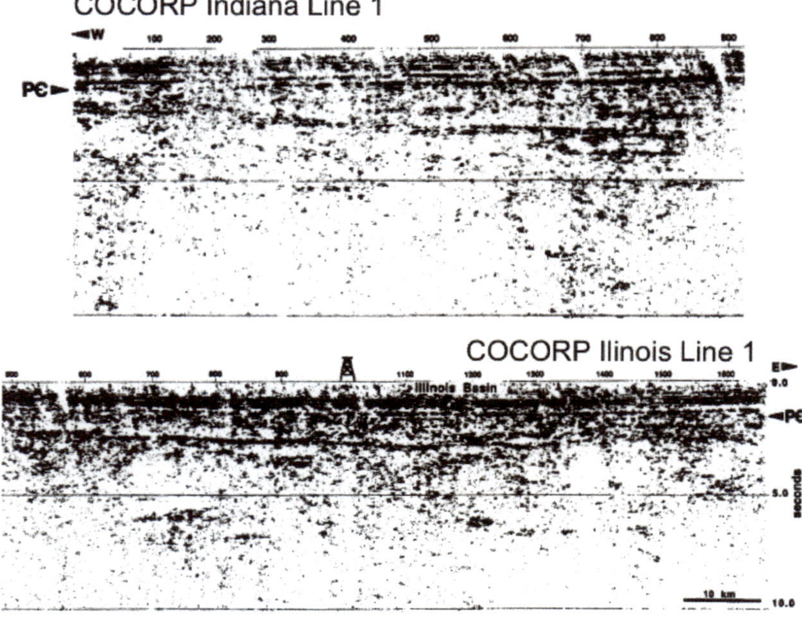

Figure 9. Layered basement reflections from the COCORP reflection profiles in western Ohio [67].

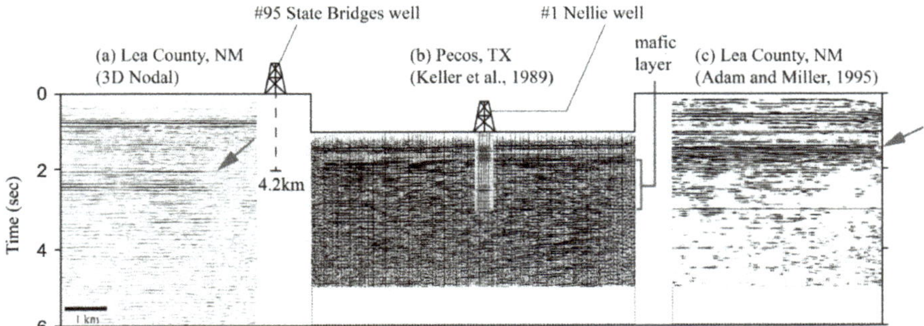

Figure 10. (**Left** and **Right**) layered basement reflections from the reprocessing of oil exploration seismic data from east central New Mexico [68,69]. (**Center**) oil industry drilling has identified correlative layered reflectors in southwest Texas as being from layered ultramafic rocks [71]. After Kim and Brown [69].

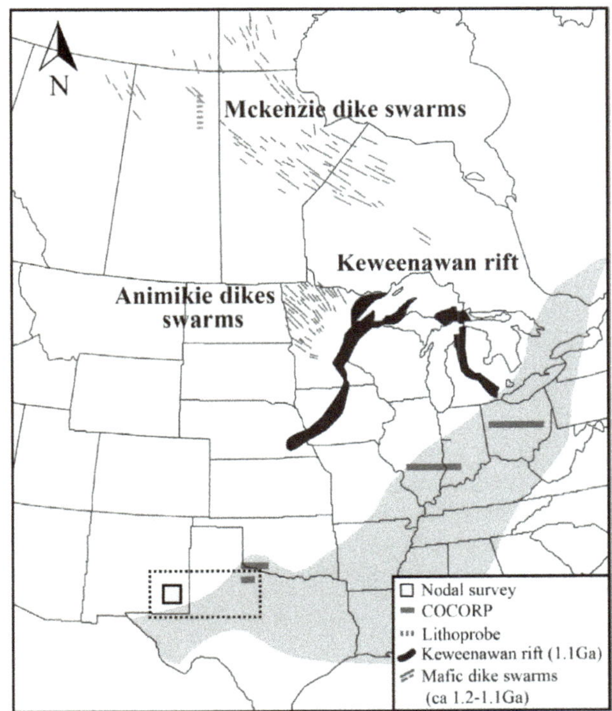

Figure 11. Comparison of the extent of sills, represented by LITHOPROBE seismic lines, and exposed dikes in northwest Canada associated with the Proterozoic McKenzie event with the extent of layered reflectors that appear to be correlative beneath much of the US midcontinent sedimentary rock cover, as represented by COCORP surveys. The lightly shaded area indicates the inferred extent of the Proterozoic Granite-Rhyolite provide [70]. After Kim and Brown [69].

2.2.5. Deep Sills and Ore Deposits—The Iberian Massif

The extensive nature of the aforementioned examples of known and possible sills suggest thermal events of a substantial nature. Both heat and fluid transfer attendant on their emplacement may have been a significant factor in ore development. A possible link between deep sill emplacement and ore deposits is perhaps most strongly suggested by a deep seismic profile in southwestern Spain (Figures 1 and 12). The IBERSEIS seismic profile across the Iberian massif [74,75] revealed a prominent subhorizontal band of strong reflectivity at midcrustal depths (Figure 12). The interpretation of this 140 km-long feature, known as the Iberian Reflective Body (IRB), as a mafic sill is argued from its relatively high reflection amplitudes (20% higher than reflections immediately above and below), its geological setting (exposing late Carboniferous mafic intrusions in an early collision zone), and its correspondence with a relatively high conductivity indicated by MT and high density inferred from gravity data [76]. The IRB also appears to have served as a rheological decollement for structures both above and below [74,76].

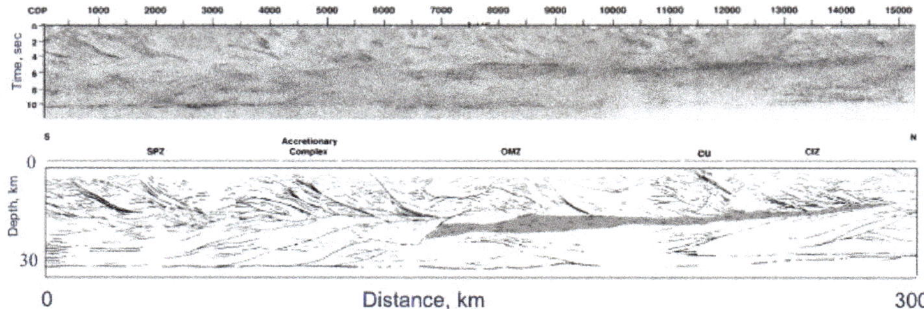

Figure 12. IBERSEIS deep seismic reflection profile tracing a prominent midcrustal reflection sequence beneath the Iberian Massif [74]. Top: time migrated seismic section. Bottom: line drawing.

The correlation with high conductivity would seem to suggest that fluids (partial melts?) are still present, but the lack of modern magmatism in the geological record argues against such an inference. Carbonell et al. [76] suggests instead that the conductivity correlation is due to the enhancement of connectivity between graphite deposits in the overlying Ossa-Morena Zone during the emplacement of the IRB.

The IRB underlies the Ni-Fe deposits of the Aguablanca Ni-Fe deposits [77] and terminates beneath the massive sulfide deposits of the Iberian Pyrite belt [78] near Rio Tinto. Both of these Late Variscan mineralizations imply a substantial heat source at depth, most likely from mantle-derived mafic intrusions. That the IRB represents the remains of mafic magmatism that provided both the heat and fluids to generate the overlying ore deposits may be speculative, but it is certainly plausible [76]. The tectonic origin of this magmatism is unclear, though Carbonell et al. [76] suggest that is linked to a Carboniferous mantle plume which impacted a large part of northwestern Europe.

2.2.6. Mantle Sills?

Although the mantle has proven far less heterogeneous than the crust, at least in terms of reflectivity, prominent mantle reflections have now been traced by a number of seismic reflection profiles. Perhaps the best known of these have been mapped by the BIRPS (British Institutions Reflection Profiling Syndicate) in northwestern Britain [79–81]. The mantle reflections which have received the most attention are dipping features, often interpreted as "fossil" subduction zones [82–85]. However, relative extensive subhorizontal reflections have also been observed in the mantle near Britain which resemble those in the upper crust that have been interpreted as sills (e.g., Figure 13). Some of these "flat" mantle reflections are spatially linked to the nearby dipping mantle reflections [86] and thus may be genetically related. In any case, they too may represent igneous sills, although other speculative explanations have been put forward (e.g., detachments [87]).

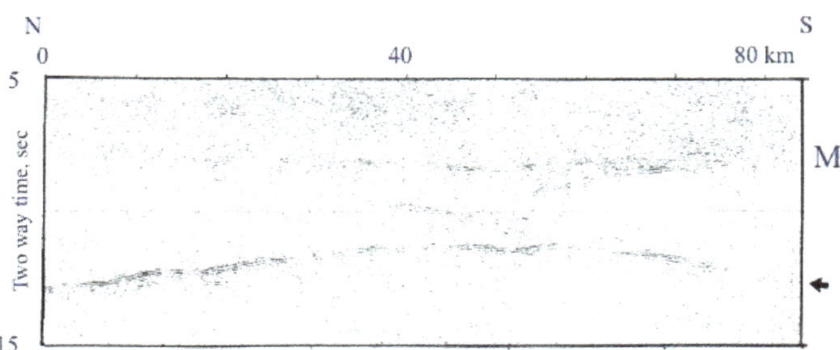

Figure 13. Strong, subhorizontal reflector in the mantle beneath northwest Great Britain from BIRPS marine data (After Snyder and Flack [86]). M indicates the position of the Moho at approximately 8 s (ca 25 km). Note that the top of this section corresponds to 5 s TWTT instead of the customary 0 s (surface).

3. Discussion

The preceding description of prominent, relative extensive reflections and reflection sequences has emphasized their interpretation as mafic sills. The basis for these interpretations ranges from unequivocal (e.g., Siljan reflections—drilled) to likely (Arizona—outcrop) to speculative (central U.S. seismic character). There are a number of caveats that must be considered in evaluating the interpretations of these and other reported images of deep intrusions. For example, the seismic sections shown in this paper are all 2D. Without 3D control, apparently subhorizontal reflections could just as well be along-strike images of features that are actually dipping at right angles to the line of the section. However, most of the examples here are based on surveys that did include 3D control, either in the form of a local grid of surveys (e.g., Siljan [54], North Texas [65], Great Britain [81]) or as formal three D seismic arrays (e.g., the Winagami reflections sequence [61], layered reflections in eastern New Mexico [69]).

Another consideration is that igneous intrusions may be manifest with different seismic characteristics than that exemplified by the Siljan sequence—i.e., narrow bands of strong reflectivity separated by larger bands of non-reflectivity. For example, Figure 14 (Upper left) shows a sample of BIRPS marine deep seismic reflection profiling data from northwest Britain [88] that indicates a lower crust that is highly layered and strongly reflective, a pattern commonly referred to as Layered Lower Crust (LLC). However, the layering in this case is more of a lamination, without the distinctive separation between reflective units that has been used to associate many of the sill examples in this paper with Siljan. This laminated appearance is characteristic of a number of seismic profiles in western Europe, particularly in areas affected by post-Variscan extension [89]. Based on the strong amplitudes of reflections making up this lamination, Warner [88] argues that they are most likely due to igneous intrusions or the juxtaposition of contrasting metamorphic compositions by pervasive shearing, although this remains a matter of debate [18]. Meissner et al. [90] argue from seismic anisotropy that the LLC develops by ductile processes (extension?) within warm, low-viscosity felsic lower crust with intercalations of mafic intrusions. Numerical modeling by Gerya and Berg [91] illustrates how crustal rheology can control the geometry of mafic intrusion, with a warm lower crust resulting in the lateral spread of magma with coeval viscous deformation. Thus, the different presentations of sills like Siljan/Wollaston vs. the LLC may be due to a contrast in crustal rheology—i.e., hot and ductile for the LLC vs. cold and brittle for the Siljan and its analogous reflectors—modulated by the stress field at the time of emplacement. However, the seismic data from the extended Norwegian continental margin [4,92] suggest that this maybe an over-simplification, as they show relatively continuous, distinct lower crustal sill reflectors (e.g., Figure 14; upper right) that resemble Siljan more than the LLC around Britain.

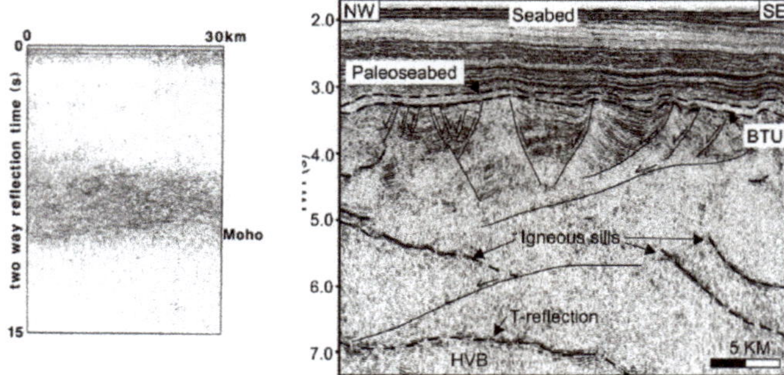

Figure 14. (**Left**) laminated reflections in the lower crust of NW Britain from the BIRPS marine reflection profiling [88]. (**Right**) distinct reflections interpreted as sills in the extended lower crust of the Norwegian continental margin [4]. T is interpreted as a reflection from the top of a high velocity body (HVB; mafic underplating?).

A number of mafic sills in both outcrop and seismic sections from basins exhibit a characteristic "saucer" shape which has fueled recent discussion about intrusion mechanics in sedimentary sequences [4,27,29,93,94]. This distinctive saucer shape has also been reported for intrabasement reflections on a several deep seismic profiles [48,95,96]. Figure 15 shows one particular distinctive example, the Surrency Bright Spot (SBS) encountered during a COCORP survey of the inferred suture zone between Laurasia and Gondwanaland buried beneath the coastal plain sedimentary rocks of southeastern Georgia. The strong amplitude of this spatially limited reflection was originally interpreted to indicate fluid involvement [95] though magma was ruled out due to its location on a long inactive passive continental margin. The saucer shape of the SBS (Figure 15) is now recognized as identical to that documented not only in oil industry 3D seismic data for various sedimentary basins but in notable outcrops of mafic intrusions [27]. The presence of saucer-shaped reflectors deep in the continental basement provides new context to evaluate mechanical models proposed to explain this geometry within sedimentary strata [4,38,93]. Thus the SBS (Surrency Bright Saucer?) is most likely a mafic intrusion, perhaps emplaced during the rifting of the Atlantic margin.

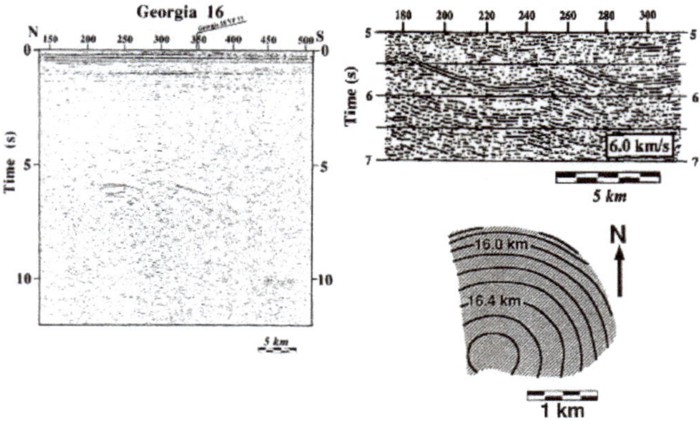

Figure 15. (**Left**) Unmigrated COCORP true amplitude section showing the Surrency Bright Spot [95]. (**Upper Right**) Blow up of migrated image of the Surrency Bright Spot [95]). (**Lower Right**) depth contour of the Surrency Bright Spot from the COCORP 3D seismic survey [97]. The "saucer" shape of this basement reflector is comparable to those associated with mafic intrusions imaged by 3D oil industry data and in outcrop [27].

Moho reflections on many deep seismic surveys also exhibit strong amplitudes (e.g., Figure 2; [98]) and/or a layered character [99]. These characteristics have been interpreted to represent "underplating", which may be another aspect of accreting/injecting mafic sills at the base of the crust [47,100].

Not every extensive prominent reflection or band of reflections is necessarily a sill. Setting aside the obvious example of sedimentary units (e.g., the finely layered, upper few seconds of the seismic sections shown in Figures 6 and 8–10). Other processes can produce distinct reflection bands traceable over large distance. Figure 16 (right) show seismic data from west central Sweden not too distant from Siljan [101]. Lacking any other constraints, it would be tempting to interpret these as sills, perhaps related to the Siljan reflections. However, the tectonic setting, outcropping geology, and—most definitively—drill holes make clear that these reflections are actually thrust faults, part of a crustal scale nappe complex emplaced during the Caledonian orogeny [101]. Regional detachment faults have also appeared on crustal reflections surveys as strong, subhorizontal reflections of regional extent. The southern Appalachian detachment traced on COCORP surveys in Tennessee and Georgia [102] and the Main Himalayan Thrust imaged by the INDEPTH surveys in southern Tibet [103] are prime examples.

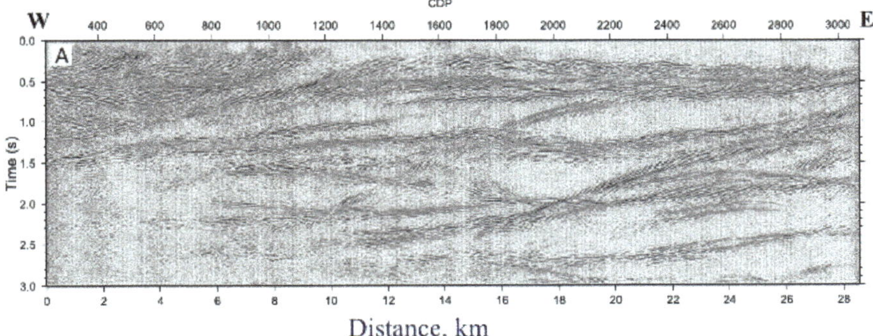

Figure 16. High amplitude, subhorizontal, and west dipping layered reflections from deep seismic surveys in west-central Sweden. Drilling and outcrop correlations identify these as thrust structures rather than intrusion [101].

Implicit in our discussion of strong reflections is the presumption that the magmas involved—whether hot or frozen—represent emplacement from below. An alternative to consider is that the reflections arise from in situ melting. This was originally one interpretation of the seismic bright spots imaged by INDEPTH in Tibet [48,104]. A fluid magma at depth, whether a planar intrusion or an in-situ melt, should give rise to a prominent reflection (e.g., Table 1) regardless of its composition. It is unlikely that reflection data, or any geophysical observations, could distinguish between these two possibilities However, if a reflector under consideration is a frozen product of in situ melt, it is unlikely to have a significant reflection contrast with its surrounding country rock (of the same composition) unless substantial fractional crystallization is involved. On the other hand, a saucer-like reflection geometry in a strong indication that mechanical intrusion is involved. Of course, intrusions may well represent crustal as well as mantle melts, though the silicic composition of the former is less likely to generate a strong reflection unless it is still molten.

The seismic character of a reflector on a seismic section depends heavily on both details of acquisition and processing. This is particularly true for reflection amplitudes. For example, the upwardly convex geometry of saucer shaped intrusions implies some degree of amplification of the associated reflection due to focusing, which in turn is ameliorated by seismic migration [105]. Therefore, caution is warranted when basing interpretation on superficial similarities in appearance between reflections on seismic sections collected and processed by different groups and individuals.

4. Tectonic Implications

The direct implications of sills within sedimentary basins have been addressed in recent papers by Magee et al. [29], Schofield [106], as well as elsewhere in this volume [107,108]. The influence of deeper intrusives may be less obvious. Haxby et al. [109] modeled the effect of a high-density intrusion into the lower crust as a driver for basin formation, though this involved massive underplating rather than intrabasement sill emplacement. The deep sills cited in this paper, which appear as relatively thin layers, may seem volumetrically small compared to other plutonic manifestations (e.g., batholiths), however the more extensive examples, such as the Winagami reflector or the COCORP midcontinent basement layering, suggest thermal perturbation over very large areas. The potential link of these two examples with the McKenzie and Keweenawan events, respectively, implies that they may serve as fingerprints of distant mantle plumes in a manner similar to the better known exposed dyke swarms [110].

The form of sill reflectivity—e.g., distinct reflections or reflection sequences vs. finely laminated zones—may be a proxy for crustal rheology and an indicator of the stress regime at the time of their emplacement [91,111,112].. The presentation of basement sills as saucers versus planar geometries is another clue to the emplacement mechanics [27]. The observation that many of these reflectors appear largely undeformed by subsequent tectonic events places constraints on the post-intrusive evolution of the crust—e.g., the lack of substantial deformation. The limited and relatively biased sampling of the crust represented by modern deep reflection profiling undermines any substantive generalization about the age distribution of observed sills, although many of the 'frozen" examples shown here either occur in Proterozoic crust or are believed to be Proterozoic in age.

5. Conclusions

Crustal seismic reflection profiles have revealed anomalously strong reflectors at numerous sites around the world. Seismic "bright spots" in tectonically active regions have been frequently attributed to fluid magma at depth, most often associated with modern extensional regimes. Some of the most distinctive reflections have been found within Precambrian basement and presumed to be "frozen" magma, most likely mafic since silicic intrusions are unlikely to provide a sufficient reflection contrast with host rock. The drilling of prominent basement reflectors near Siljan, Sweden, has confirmed them to be Proterozoic diabase sills. The correlation of similar strong reflection sequences in the southwestern US with adjacent outcrops of mafic intrusions strongly corroborates their interpretation as mafic sills as well. Extensive prominent reflections in northwestern Canada are likely to be buried manifestations of the McKenzie Dike swarm. More speculatively, an extensive layered sequence of strong reflections in the upper crust on COCORP seismic profiles that stretch from eastern New Mexico to central Ohio may be related to a distant Keweenawan mantle plume based on correlation with ultramafic rocks encountered in an oil industry borehole in west Texas. Both of these examples detail the long distance lateral transport of magma in brittle regimes. As "fingerprints" of major thermal events (plumes?), indicators of crustal rheology and stress during emplacement, and markers for post-emplacement deformation, sill reflections in the continental basement offer new constraints for models of crust and basin evolution.

Author Contributions: D.K. contributed the results for the US mid-continent and their analogy to the Canadian observations. L.D.B. provided the examples of magma bright spots and the colds sills outside North America as well as the concluding discussion. All authors have read and agreed to the published version of the manuscript.

Funding: This research received no external funding.

Conflicts of Interest: The authors declare no conflict of interest.

References

1.   Bradley, J. Intrusion of major dolerite sills. *Trans. R. Soc.* **1965**, *3*, 27–55.
2.   Hargraves, R.B. *Physics of Magmatic Processes*; Princeton University Press: Princeton, NJ, USA, 1980; p. 565.

3. Spence, D.A.; Turcotte, D.L. Magma-driven propagation of cracks. *J. Geophys. Res.* **1985**, *90*, 575–580. [CrossRef]

4. Cartwright, J.; Hansen, D.M. Magma transport through the crust via interconnected sill complexes. *Geology* **2006**, *34*, 929–932. [CrossRef]

5. Marsh, B.D. Magmatism, magma, and magma chambers. In *Treatise on Geophysics: Crust and Lithosphere Dynamics*; Watts, A.B., Ed.; Elsevier: Amsterdam, The Netherlands, 2007; Volume 6, pp. 275–333.

6. Thomson, K. Determining magma flow in sills, dykes and laccoliths and their implications for sill emplacement. *Bull. Volcanol.* **2007**, *70*, 183–201. [CrossRef]

7. Elliot, D.H.; Fleming, T.H.; Kyle, P.R.; Fland, K.A. Long-distance transport of magmas in the Jurassic Ferrar large igneous province. *Antarct. Earth Planet. Sci. Lett.* **1999**, *167*, 89–104. [CrossRef]

8. Muirhead, J.D.; van Eaton, A.R.; Re, G.; White, J.-D.; Ort, M.H. Monogenetic volcanoes fed by interconnected dikes and sills in the Hopi Buttes volcanic field, Navajo Nation, USA. *Bull. Volcanol.* **2016**, *78*, 11. [CrossRef]

9. Bott, M.H.P.; Smithson, S.B. Gravity investigations of subsurface shape and mass distributions of granite batholiths. *Geol. Soc. Am. Bull.* **1967**, *78*, 859–878. [CrossRef]

10. Oliver, H.W. Gravity and magnetic investigations of the Sierra Nevada batholith, California. *Geol. Soc. Am. Bull.* **1977**, *88*, 445–461. [CrossRef]

11. Kauahikaua, J.; Hildebrand, T.; Webring, M. Deep magmatic structures of Hawaiian volcanoes, imaged by three-dimensional gravity models. *Geology* **2000**, *28*, 883–886. [CrossRef]

12. Tizzani, P.; Battaglia, M.; Zeni, G.; Atzori, S.; Berardino, P.; Lanari, R. Uplift and magma intrusion at Long Valley caldera from INSAR and gravity measurements. *Geology* **2009**, *37*, 63–66. [CrossRef]

13. Wei, W.; Unsworth, M.; Jones, A.; Booker, J.; Tan, H.; Nelson, K.D.; Chen, L.; Li, S.; Solon, K.; Bedrosian, P.; et al. Detection of widespread fluids in the Tibetan crust by magnetotelluric studies. *Science* **2001**, *292*, 716–719. [CrossRef] [PubMed]

14. Lee, B.; Unsworth, M.; Arnason, K.; Cordell, D. Imaging the magmatic system beneath Krafla geothermal field, Iceland: A new 3-D electrical resistivity model from inversion of magnetotelluric data. *Geophys. J. Int.* **2020**, *22*, 541–567. [CrossRef]

15. Al-Chalabi, M. Some studies relating to nonuniqueness in gravity and magnetic inverse problems. *Geophysics* **1971**, *36*, 835–855. [CrossRef]

16. Beblo, M.; Björnsson, K.; Arnason, K.; Stein, B.; Wolfgram, P. Electrical conductivity beneath Iceland—Constraints imposed by magnetotelluric results on temperature, partial melt, crust and mantle structure. *J. Geophys.* **1983**, *53*, 16–23.

17. Yin, C. Inherent nonuniqueness in magnetotelluric inversion for 1D anisotropic models. *Geophysics* **2003**, *68*, 138–146. [CrossRef]

18. Thybo, H.; Artemieva, I. Moho and magmatic underplating in continental lithosphere. *Tectonophysics* **2013**, *609*, 605–619. [CrossRef]

19. Iyer, H.M.; Evans, J.R.; Zandt, G.; Stewart, R.M.; Coakley, J.M.; Roloff, J.N. A deep low-velocity body under the Yellowstone Caldera, Wyoming: Delineation using teleseismic P-wave residuals and tectonic interpretation. *Geol. Soc. Am. Bull.* **1981**, *92*, 792–798. [CrossRef]

20. Huang, H.H.; Lin, F.-C.; Schmandt, B.; Farrell, J.; Smith, R.B.; Tsai, V.C. The Yellowstone magmatic system from the upper mantle to the upper crust. *Science* **2017**, *348*, 773–776. [CrossRef]

21. Lees, J.M. Seismic tomography of magmatic systems. *J. Volcanol. Geotherm. Res.* **2007**, *167*, 37–56. [CrossRef]

22. Schuler, J.; Greenfield, T.; White, R.S.; Roecker, S.W.; Brandsdóttir, B.; Stock, J.M.; Pugh, D. Seismic imaging of the shallow crust beneath the Krafla central volcano, NE Iceland. *J. Geophys. Res. Solid Earth* **2015**, *120*, 7156–7173. [CrossRef]

23. Kiser, E.; Palomeras, I.; Levander, A.; Zelt, C.; Harder, S.; Schmandt, B.; Hansen, S.; Creager, K.; Ulberg, C. Magma reservoirs from the upper crust to the Moho inferred from high resolution Vp and Vs models beneath Mount St. Helens, Washington State, USA. *Geology* **2016**, *44*, 411–414. [CrossRef]

24. Zandt, G.; Leidig, M.; Chmielowski, J.; Baumont, D.; Yuan, X. Seismic detection and characterization of the Altiplano-Puna magma body, central Andes. *Pure Appl. Geophys.* **2003**, *160*, 789–807. [CrossRef]

25. Ward, K.M.; Zandt, G.; Beck, S.; Christensen, D.H.; McFarlin, H.M. Seismic imaging of the magmatic underpinnings beneath the Altiplano-Puna volcanic complex from the joint inversion of surface wave dispersion and receiver functions. *Earth Planet. Sci. Lett.* **2014**, *404*, 43–53. [CrossRef]

26. Waters, K. *Reflection Seismology: A Tool for Energy Resource Exploration*; Wiley: New York, NY, USA, 1981; p. 537.

27. Polteau, S.; Mazzini, A.; Galland, O.; Planke, S.; Malthe-Sorenssen, A. Saucer-shaped intrusions: Occurrences, emplacement and implications. *Earth Planet. Sci. Lett.* **2008**, *266*, 195–204. [CrossRef]

28. Thomson, K.; Schofield, N. Lithological and structural controls on the emplacement and morphology of sills in sedimentary basins. *Geol. Soc. Lond. Spec. Publ.* **2008**, *302*, 31–44. [CrossRef]

29. Magee, C.; Muirhead, J.D.; Karvelas, A.; Holford, S.P.; Jackson, C.A.; Bastow, L.D.; Schofield, N.; Stevenson, C.T.; McLean, C.; McCarthy, W.; et al. Lateral magma flow in mafic sill complexes. *Geosphere* **2016**, *12*, 809–841. [CrossRef]

30. Christensen, N.I. Poisson's ratio and crustal seismology. *J. Geophys. Res.* **1996**, *101*, 3139–3156. [CrossRef]

31. Brocher, T.M. Empirical relations between elastic wave speeds and density in the Earth's crust. *Bull. Seismol. Soc. Am.* **2005**, *95*, 2081–2092. [CrossRef]

32. Murase, T.; McBirney, A.R. Properties of some common igneous rocks and their melts at high temperatures. *Geo. Soc. Am. Bull.* **1973**, *84*, 3563–3592. [CrossRef]

33. Christensen, N.I.; Mooney, W.D. Seismic velocity structure and composition of the continental crust: A global view. *J. Geophys. Res.* **1995**, *100*, 9761–9788. [CrossRef]

34. Sanford, A.R.; Alptekin, Ö.; Toppozada, T.R. Use of reflection phases on microearthquake seismograms to map an unusual discontinuity beneath the Rio Grande rift. *Bull. Seismol. Soc. Am.* **1973**, *63*, 2021–2034.

35. Matsumoto, S.; Hasegawa, A. Distinct S wave reflector in the midcrust beneath Nikko-Shirane volcano in the northeastern Japan arc. *J. Geophys. Res. Solid Earth* **1996**, *101*, 3067–3083. [CrossRef]

36. Rinehart, E.J.; Sanford, A.R.; Ward, R.M. Geographic Extent and Shape of an Extensive Magma Body at Mid-Crustal Depths in the Rio Grande Rift Near Socorro, New Mexico. *Rio Grande Rift Tectonics Magmatism* **1979**, *14*, 237–251.

37. Brown, L.D.; Krumhansl, P.A.; Chapin, C.E.; Sanford, A.R.; Cook, F.A.; Kaufman, S.; Oliver, J.E.; Schilt, F.S. COCORP seismic reflection studies of the Rio Grande rift. *Rio Grande Rift Tectonics Magmatism* **1979**, *14*, 169–184.

38. Long, L.T.; Sanford, A.R. Microearthquake crustal reflections. *Bull. Seismol. Soc. Am.* **1965**, *55*, 579–586.

39. Brocher, T.M. Geometry and physical properties of the Socorro, New Mexico, magma bodies. *J. Geophys. Res. Solid Earth* **1981**, *86*, 9420–9432. [CrossRef]

40. Hermance, J.F.; Neumann, G.A. The Rio Grande rift: New electromagnetic constraints on the Socorro magma body. *Phys. Earth Planet. Inter.* **1991**, *66*, 101–117. [CrossRef]

41. Reilinger, R.; Oliver, J. Modern uplift associated with a proposed magma body in the vicinity of Socorro, New Mexico. *Geology* **1976**, *4*, 583–586. [CrossRef]

42. Larsen, S.; Reilinger, R.; Brown, L.L. Evidence of ongoing crustal deformation related to magmatic activity near Socorro, New Mexico. *J. Geophys. Res.* **1986**, *91*, 6283–6292. [CrossRef]

43. Fialko, Y.; Simons, M. Evidence for on-going inflation of the Socorro Magma Body, New Mexico, from interferometric synthetic aperture radar imaging. *Geophys. Res. Lett.* **2001**, *28*, 3549–3552. [CrossRef]

44. Sheetz, K.E.; Schlue, J.W. Inferences for the Socorro magma body from teleseismic receiver functions. *Geophys. Res. Lett.* **1992**, *19*, 1867–1870. [CrossRef]

45. Ross, A. Deep Seismic Bright Spots. Ph.D. Thesis, Cornell University, Ithaca, NY, USA, June 1999.

46. De Voogd, B.; Serpa, L.; Brown, L.; Hauser, E.; Kaufman, S.; Oliver, J.; Troxel, B.W.; Willemin, J.; Wright, L.A. Death Valley bright spot; a midcrustal magma body in the southern Great Basin, California? *Geology* **1986**, *14*, 64–67. [CrossRef]

47. Jarchow, C.M.; Thompson, G.A.; Catchings, R.D.; Mooney, W.D. Seismic evidence for active magmatic underplating beneath the basin and range province, western United States. *J. Geophys. Res.* **1993**, *98*, 22095–22108. [CrossRef]

48. Brown, L.D.; Zhao, W.; Nelson, K.D.; Hauck, M.; Alsdorf, D.; Ross, A.; Cogan, M.; Clark, M.; Liu, X.; Che, J. Bright spots, structure, and magmatism in southern Tibet from INDEPTH seismic reflection profiling. *Science* **1996**, *274*, 1688–1690. [CrossRef]

49. ANCORP Working Group. Seismic reflection image revealing offset of Andean subduction-zone earthquake locations into oceanic mantle. *Nature* **1999**, *397*, 341–344. [CrossRef]

50. Gase, A.C.; Van Avendonk, N.L.; Bangs, T.W.; Luckie, D.H.; Barker, S.A.; Henrys, F.G.; Fujie, G. Seismic evidence of magmatic rifting in the offshore Taupo Volcanic Zone, New Zealand. *Geophys. Res. Lett.* **2019**, *46*, 12949–12957. [CrossRef]

51. Detrick, R.S.; Buhl, P.; Vera, E.; Mutter, J.; Orcutt, J.; Madsen, J.; Brocher, T. Multi-channel seismic imaging of a crustal magma chamber along the East Pacific Rise. *Nature* **1987**, *326*, 35–41. [CrossRef]

52. Kent, G.M.; Singh, S.C.; Harding, A.J.; Sinha, M.C.; Orcutt, J.A.; Barton, P.J.; White, R.S.; Bazin, S.; Hobbs, R.W.; Tong, C.H.; et al. Evidence from three-dimensional seismic reflectivity images for enhanced melt supply beneath mid-ocean-ridge discontinuities. *Nature* **2000**, *406*, 614–618. [CrossRef]

53. Canales, J.P.; Dunn, R.A.; Arai, R.; Sohn, R.A. Seismic imaging of magma sills beneath an ultramafic hosted hydrothermal system. *Geology* **2017**, *45*, 451–454. [CrossRef]

54. Juhlin, C.; Pedersen, L.B. Reflection seismic investigations of the Siljan impact structure, Sweden. *J. Geophys. Res.* **1987**, *92*, 14113–14122. [CrossRef]

55. Juhlin, C. Interpretation of the reflections in the Siljan Ring area based on results from the Gravberg-1 borehole. *Tectonophysics* **1990**, *173*, 345–360. [CrossRef]

56. Papasikas, N.; Juhlin, C. Interpretation of reflections from the central part of the Siljan Ring impact structure based on results from the Stenberg-1 borehole. *Tectonophysics* **1997**, *269*, 237–245. [CrossRef]

57. Hauser, E.C.; Gephart, J.; Latham, T.; Oliver, J.E.; Kaufman, S.; Brown, L.D.; Lucchitta, I. COCORP Arizona Transect: Strong Crustal Reflections and Offset Moho Beneath the Transition Zone. *Geology* **1987**, *15*, 1103–1106.

58. Litak, R.K.; Marchant, R.H.; Brown, L.D.; Pfiffner, O.A.; Hauser, E.C. Correlating crustal reflections with geologic outcrops: Seismic modeling results from the southwestern USA and the Swiss Alps. *Amer. Geophys. Union Geodyn. Ser.* **1992**, *22*, 299–305.

59. Ross, G.M.; Eaton, D.W. Winagami reflection sequence: Seismic evidence for post collisional magmatism in the Proterozoic of western Canada. *Geology* **1997**, *25*, 199–202. [CrossRef]

60. Mandler, H.A.F.; Clowes, R. The HIS bright reflector: Further evidence for extensive magmatism in the Precambrian of western Canada. *Tectonophysics* **1998**, *288*, 71–81. [CrossRef]

61. Welford, J.K.; Clowes, R.M. Three-dimensional seismic reflection investigation of the crustal Winagami sill complex of northwestern Alberta, Canada. *Geophys. J. Int.* **2006**, *166*, 155–169. [CrossRef]

62. Mandler, H.A.F.; Clowes, R. Evidence for extensive tabular intrusions in the Precambrian shield of western Canada: A 160-km-long sequence of bright reflections. *Geology* **1997**, *25*, 271–274. [CrossRef]

63. Fahrig, W.F. The tectonic settings of continental mafic dyke swarms: Failed arm and early passive margin. *Spec. Pap. Geol. Assoc. Can.* **1987**, *34*, 331–348.

64. Muirhead, J.D.; Airoldi, G.; Rowland, J.V.; White, J.D.L. Interconnected sills and inclined sheet intrusions control shallow magma transport in the Ferrar large igneous province. *Antarct. Geol. Soc. Am. Bull.* **2012**, *124*, 162–180. [CrossRef]

65. Oliver, J.; Dobrin, M.; Kaufman, S.; Meyer, R.; Phinney, R. Continuous seismic reflection profiling of the deep basement, Hardeman County, Texas. *Geol. Soc. Am. Bull.* **1976**, *87*, 1537–1546. [CrossRef]

66. Brewer, J.A.; Brown, L.D.; Steiner, D.; Oliver, J.E.; Kaufman, S.; Denison, R.E. Proterozoic basin in the southern Midcontinent of the United States revealed by COCORP deep seismic reflection profiling. *Geology* **1981**, *9*, 569–575. [CrossRef]

67. Pratt, T.; Culotta, R.; Hauser, E.; Nelson, D.; Brown, L.; Kaufman, S.; Oliver, J.; Hinze, W. Major Proterozoic basement features of the eastern midcontinent of North America revealed by recent COCORP profiling. *Geology* **1989**, *17*, 505–509. [CrossRef]

68. Adams, D.C.; Miller, K.C. Evidence for late middle Proterozoic extension in the Precambrian basement beneath the Permian basin. *Tectonics* **1995**, *14*, 1263–1272. [CrossRef]

69. Kim, D.; Brown, L.D. From trash to treasure: Three-dimensional basement imaging with "excess" data from oil and gas explorations. *AAPG Bull.* **2019**, *103*, 1691–1701. [CrossRef]

70. Bickford, M.E.; Van Schmus, W.R.; Karlstrom, K.E.; Mueller, P.A.; Kamenov, G.D. Mesoproterozoic-trans-Laurentian magmatism: A synthesis of continent-wide age distributions, new SIMS U–Pb ages, zircon saturation temperatures, and Hf and Nd isotopic compositions. *Precambrian Res.* **2015**, *265*, 286–312. [CrossRef]

71. Kargi, H.; Barnes, C.G. A Grenville-age layered intrusion in the subsurface of west Texas: Petrology, petrography, and possible tectonic setting. *Can. J. Earth Sci.* **1995**, *32*, 2159–2166. [CrossRef]

72. Keller, G.R.; Hills, J.M.; Baker, M.R.; Wallin, E.T. Geophysical and geochronological constraints on the extent and age of mafic intrusions in the basement of west Texas and eastern New Mexico. *Geology* **1989**, *17*, 1049–1052. [CrossRef]

73. Ernst, R.E.; Buchan, K.L. Layered mafic intrusions: A model for their feeder systems and relationship with giant dyke swarms and mantle plume centres. *S. Afr. J. Geol.* **1997**, *100*, 319–334.

74. Simancas, J.F.; Carbonell, R.; Lodeiro, F.G.; Estaún, A.P.; Juhlin, C.; Ayarza, P.; Kashubin, A.; Azor, A.; Poyatos, D.M.; Almodóvar, G.R.; et al. Crustal structure of the transpressional Variscan orogen of SW Iberia: SW Iberia deep seismic reflection profile (IBERSEIS). *Tectonics* **2003**, *22*, 1062. [CrossRef]

75. García-Lobón, J.L.; Rey-Moral, C.; Ayala, C.; Martín-Parra, L.M.; Matas, J.; Reguera, M.I. Regional structure of the southern segment of Central Iberian Zone (Spanish Variscan Belt) interpreted from potential field images and 2.5 D modelling of Alcudia gravity transect. *Tectonophysics* **2014**, *614*, 185–202. [CrossRef]

76. Carbonell, R.; Simancas, F.; Juhlin, C.; Pous, J.; Pérez-Estaún, A.; Gonzalez-Lodero, F.; Muñoz, G.; Heise, W.; Ayarza, P. Geophysical evidence of a mantle derived intrusion in S.W. Iberia. *Geophys. Res. Lett.* **2004**, *31*. [CrossRef]

77. Tornos, F.; Casquet, C.; Galindo, C.; Velasco, F.; Canales, A. A new style of Ni-Cu mineralization related to magmatic breccia pipes in a transpressional magmatic arc. *Aguablanca. Min. Depos.* **2001**, *36*, 700–706. [CrossRef]

78. Leistel, J.M.; Marcoux, E.; Thiéblemont, D.; Quesada, C.; Sánchez, A.; Almodóvar, G.R.; Pascual, E.; Sóez, R. The volcanic-hosted massive sulphide deposits of the Iberian Pyrite Belt. *Miner. Depos.* **1998**, *33*, 2–30. [CrossRef]

79. Smythe, D.K.; Dobinson, A.; McQuillin, R.; Brewer, J.A.; Matthews, D.H.; Blundell, D.J.; Kelk, B. Deep structure of the Scottish Caledonides revealed by the MOIST reflection profile. *Nature* **1982**, *299*, 338–340. [CrossRef]

80. Brewer, J.A.; Matthews, D.H.; Warner, M.R.; Hall, J.; Smythe, D.K.; Whittington, R.J. BIRPS deep seismic reflection studies of the British Caledonides. *Nature* **1983**, *305*, 206–210. [CrossRef]

81. Flack, C.A.; Klemperer, S.L.; McGeary, S.E.; Snyder, D.B.; Warner, M.R. Reflections from mantle fault zones around the British Isles. *Geology* **1990**, *18*, 528–532. [CrossRef]

82. Warner, M.; Morgan, J.; Barton, P.; Morgan, P.; Price, C.; Jones, K. Seismic reflections from the mantle represent relict subduction zones within the continental lithosphere. *Geology* **1996**, *24*, 39–42. [CrossRef]

83. Steer, D.N.; Knapp, D.J.H.; Brown, L.D. Super-deep reflection profiling: Exploring the continental mantle lid. *Tectonophysics* **1998**, *286*, 111–121. [CrossRef]

84. Cook, F.A.; van der Velden, A.J.; Hall, K.W.; Roberts, B.J. Tectonic delamination and subcrustal imbrication of the Precambrian lithosphere in northwestern Canada mapped by LITHOPROBE. *Geology* **1998**, *26*, 839–842. [CrossRef]

85. van der Velden, A.J.; Cook, F.A. Relict subduction zones in Canada. *J. Geophys. Res.* **2005**, *110*, B08403. [CrossRef]

86. Snyder, D.B.; Flack, C.A. A Caledonian age for reflectors within the mantle lithosphere north and west of Scotland. *Tectonics* **1990**, *9*, 903–922. [CrossRef]

87. Reston, T.J. Mantle shear zones and the evolution of the North Sea basin. *Geology* **1990**, *18*, 272–275. [CrossRef]

88. Warner, M. Basalts, water, or shear zones in the lower continental crust? *Tectonophysics* **1990**, *173*, 163–174. [CrossRef]

89. Sadowiak, P.; Wever, T.; Meissner, R. Deep seismic reflectivity patterns in specific tectonic units of Western and Central Europe. *Geophys. J. Int.* **1991**, *105*, 45–54. [CrossRef]

90. Meissner, R.; Rabbel, W.; Kern, H. Seismic lamination and anisotropy of the Lower continental crust. *Tectonophysics* **2006**, *416*, 81–99. [CrossRef]

91. Gerya, T.V.; Burg, J.-P. Intrusion of ultramafic magmatic bodies into the continental crust: Numerical simulation. *Phys. Earth Planet. Inter.* **2007**, *160*, 124–142. [CrossRef]

92. Wrona, T.; Magee, C.; Fossen, H.; Gawthrope, R.L.; Bell, R.E.; Jackson, C.A.-L.; Faleide, J.I. 3-D seismic images of an extensive igneous sill in the lower crust. *Geology* **2019**, *47*, 729–733. [CrossRef]

93. Hansen, D.M.; Cartwright, J. The three-dimensional geometry and growth of forced folds above saucer-shaped igneous sills. *J. Struct. Geol.* **2006**, *28*, 1520–1535. [CrossRef]

94. Goulty, N.R.; Schofield, M. Implications of simple flexure theory for the formation of saucer-shaped sills. *J. Struct. Geol.* **2008**, *30*, 812–817. [CrossRef]

95. Pratt, T.L.; Mondary, J.F.; Brown, L.D. Crustal structure and deep reflector properties: Wide angle shear and compressional wave studies of the midcrustal Surrency bright spot beneath southeastern Georgia. *J. Geophys. Res.* **1993**, *98*, 17723. [CrossRef]

96. Ivanic, T.; Korsch, R.J.; Wyche, S.; Jones, L.E.A.; Zibra, I.; Blewett, R.S.; Jones, T.; Milligan, P.; Costelloe, R.D.; van Kranendonk, M.J.; et al. Preliminary interpretation of the 2010 Youanmi deep seismic reflection lines and magnetotelluric data for the Windimurra igneous complex. In *Proceedings, Youanmi and Southern Carnarvon Seismic and Magnetotelluric Workshop*; 2013; pp. 93–102. Available online: https://www.academia.edu/33335649/Youami_and_Southern_Carnarvon_seismic_and_magnetotelluric_MT_workshop_2013 (accessed on 22 October 2020).

97. Barnes, A.E.; Reston, T.J. A study of two mid-crustal bright spots from southeast Georgia (USA). *Geophys. J. Int.* **1992**, *108*, 683–691. [CrossRef]

98. Carbonell, R.S.; Smithson, B. The bright Moho reflection in the 1986 Nevada PASSCAL, seismic experiment. *Tectonophysics* **1995**, *243*, 255–276. [CrossRef]

99. Klemperer, S.L.; Hauge, T.A.; Hauser, E.C.; Oliver, J.E.; Potter, C.J. The Moho in the northern Basin and range province, Nevada, along the COCORP 40-N seismic reflection transect. *Geol. Soc. Amer. Bull.* **1986**, *97*, 603–618. [CrossRef]

100. Suetnova, E.; Carbonell, R.; Smithson, S.B. Magma in layering at the Moho of the basin and range of Nevada. *Geophys. Res. Lett.* **1993**, *20*, 2945–2948. [CrossRef]

101. Hedin, P.C.; Christopher, J.; Gee, D.G. Seismic imag8ng of the Scandinavian Caledonides to define ICDP drilling sites. *Tectonophysics* **2012**, *554–557*, 30–41. [CrossRef]

102. Cook, F.A.; Vasudevan, K. Reprocessing and enhanced interpretation of the initial COCORP southern Appalachians traverse. *Tectonophysics* **2006**, *420*, 161–174. [CrossRef]

103. Zhao, W.; Nelson, K.D.; Project INDEPTH Team. Deep seismic reflection evidence for continental underthrusting beneath southern Tibet. *Nature* **1993**, *366*, 557–559. [CrossRef]

104. Makovsky, U.; Klemperer, S.L.; Ratschbacher, L.; Brown, L.D.; Li, M.; Zhao, W.; Meng, F. INDEPTH wide-angle reflection observations of P-to-S conversion from crustal bright spots in Tibet. *Science* **1996**, *274*, 1690–1691. [CrossRef]

105. Yilmaz, O. *Seismic Data Analysis: Processing, Inversion and Interpretation of Seismic Data*; Society of Exploration Geophysicists: Tulsa, OK, USA, 2001; Volume 1, p. 2024.

106. Schofield, N.; Holford, S.; Millett, J.; Brown, D.; Jolley, D.; Passey, S.R.; Muirhead, D.; Grove, C.; Magee, C.; Murray, J.; et al. Regional magma plumbing and emplacement mechanisms of the Faroe-Shetland sill complex: Implications for magma transport and petroleum systems within sedimentary basins. *Basin Res.* **2017**, *29*, 41–63. [CrossRef]

107. Syndes, M.; Fjeldskaar, W.; Grunnaleite, I.; Løtveit, I.F.; Mjelde, R. The Influence of magmatic intrusions on diagenetic processes and stress accumulation. *Geosciences* **2019**, *9*, 477. [CrossRef]

108. Syndes, M.; Fjeldskaar, W.; Grunnaleite, I.; Løtveit, I.F.; Mjelde, R. Transient thermal effects in sedimentary basins with normal faults and magmatic sill intrusions—A Sensitivity Study. *Geosciences* **2019**, *9*, 160. [CrossRef]

109. Haxby, W.F.; Turcotte, D.L.; Bird, J.M. Thermal and mechanical evolution of the Michigan basin. In *Developments in Geotectonics*; Bott, M.H.P., Ed.; Elsevier: Amsterdam, The Netherlands, 1976; Volume 12, pp. 57–75.

110. Ernst, R.E.; Buchan, K.L. The use of mafic dyke swarms in identifying and locating mantle plumes. *Geol. Soc. Am. Spec. Pap.* **2001**, *352*, 247–275.

111. Gretener, P.E. On the mechanics of the intrusion of sills. *Can. J. Earth Sci.* **1969**, *6*, 1415–1419. [CrossRef]

112. Menand, T. The mechanics and dynamics of sills in layered elastic rocks and their implications for the growth of laccoliths and other igneous complexes. *Earth Planet. Sci. Lett.* **2008**, *267*, 93–99. [CrossRef]

Publisher's Note: MDPI stays neutral with regard to jurisdictional claims in published maps and institutional affiliations.

Article

Transient Thermal Effects in Sedimentary Basins with Normal Faults and Magmatic Sill Intrusions—A Sensitivity Study

Magnhild Sydnes [1,2,*], Willy Fjeldskaar [1], Ivar Grunnaleite [1], Ingrid Fjeldskaar Løtveit [1] and Rolf Mjelde [2]

1   Tectonor AS, P.O. Box 8034, NO-4068 Stavanger, Norway; wf@tectonor.com (W.F.); ig@tectonor.com (I.G.); ifl@tectonor.com (I.F.L.)
2   Department of Earth Science, University of Bergen, Box 7803, 5020 Bergen, Norway; Rolf.Mjelde@uib.no
*   Correspondence: ms@tectonor.com

Received: 13 March 2019; Accepted: 2 April 2019; Published: 5 April 2019

Abstract: Magmatic intrusions affect the basin temperature in their vicinity. Faulting and physical properties of the basin may influence the magnitudes of their thermal effects and the potential source rock maturation. We present results from a sensitivity study of the most important factors affecting the thermal history in structurally complex sedimentary basins with magmatic sill intrusions. These factors are related to faulting, physical properties, and restoration methods: (1) fault displacement, (2) time span of faulting and deposition, (3) fault angle, (4) thermal conductivity and specific heat capacity, (5) basal heat flow and (6) restoration method. All modeling is performed on the same constructed clastic sedimentary profile containing one normal listric fault with one faulting event. Sills are modeled to intrude into either side of the fault zone with a temperature of 1000 °C. The results show that transient thermal effects may last up to several million years after fault slip. Thermal differences up to 40 °C could occur for sills intruding at time of fault slip, to sills intruding 10 million years later. We have shown that omitting the transient thermal effects of structural development in basins with magmatic intrusions may lead to over- or underestimation of the thermal effects of magmatic intrusions and ultimately the estimated maturation.

Keywords: normal faulting; sill intrusions; transient thermal effects; steady state; basin modeling; volcanic basins

---

1. Introduction

Hydrocarbon discoveries associated with magmatic intrusions are common in many sedimentary basins throughout the world [1–3], and these intrusions may potentially affect all parts of the petroleum system [4]. The impact of magmatic intrusions has been studied in several basins worldwide, e.g., Vøring Basin, Norway (e.g., [5–7]), Karoo Basin, South Africa (e.g., [8–10]), Gunnedah Basin, Australia [11], Neuquén Basin, Argentina (e.g., [12,13]), Bohai Bay Basin, China (e.g., [14,15]). All these studies conclude that magmatic intrusions significantly influence the basin thermal history and thus the maturation of organic material in their vicinity.

Several studies have identified the pre-intrusion temperature of the host rock as an important variable for the ultimate thermal and maturation levels of sedimentary basins with magmatic intrusions [8,10,13,16–18]. However, this variable is one of the most difficult parameters to constrain, as the traces of pre-intrusion temperature and maturation are erased instantly in the thermal aureole of magmatic intrusions when magmatic emplacement occurs [19] and the basin has been subject to later geological development. As magmatic intrusions often are emplaced into structurally complex sedimentary basins, it is crucial to understand how the basin's structural evolution affects the temperature development. With such knowledge it should be possible to discern temperature effects of the structural development from temperature effects of magmatic intrusions. This is important as the pre-intrusion temperatures have implications for the magnitude of the thermal effect of sills, i.e., size and temperatures of the thermal aureole, which eventually will have implications for the estimated maturation of the basin.

Magmatic intrusions and their effect on temperature and maturation in the surrounding host rocks have been subject of several studies (e.g., [8,13,19–24]). Only a few studies model sills intruding into structurally complex basins under development, while taking the temporary host-rock temperature and maturation into consideration (e.g., [7,25]). Structurally complex basins are characterized by numerous faults, changing lithologies with different physical properties and area specific heat flow. To estimate the basin's thermal development as a function of time, the geohistory of the basin must first be established by reconstruction. In standard basin modeling, the impact of the structural development on the temperature history is seldom taken into account and this may give incorrect thermal and maturation predictions [26].

Several studies exist on temperature modeling in fault zones, but many of these focus on the effects related to uplift and erosion observed in the footwall part of the fault zone and not on the thermal transients in the hanging wall section (e.g., [27–34]). To our knowledge no studies includes magmatism in a fault setting. In our study the main focus is on the transient thermal effects of normal faults with syn- and post-rift deposits, and the influence such structural development has on the thermal effects of sills. Possible transient thermal effects due to erosion have not been pursued in this study.

The calculated pre-intrusion temperatures are dependent on the representation of the basin's structural development [13,25,26]. The purpose of this study is to quantify the thermal effects of the most important factors affecting the thermal history in structurally complex sedimentary basins with magmatic sill intrusions. We do this by running a series of simple models. A more realistic model would require higher structural resolution with variation of lithological and geometrical properties in space and time. However, many simultaneously varying factors affecting the thermal evolution would obscure the magnitude and impact of the individual factors. We therefore first study the thermal effects on structural complex basin on a simple, synthetic profile followed by analysis of the subsequent sensitivity sets: (1) fault displacement, (2) time span of faulting and deposition, (3) fault angle, (4) thermal conductivity and specific heat capacity, (5) basal heat flow and (6) restoration method. The overall goal is to test the sensitivity of faulting, physical properties of the lithologies and choice of fault restoration method on the thermal effect of magmatic intrusions in complex sedimentary basins. We aim for some general conclusions that are applicable to sedimentary basins with normal faults and magmatic sill intrusions.

2. Methods

Our study investigates the relative effect faulting, different physical properties and choice of restoration method have on the resulting thermal and maturation history in structurally complex sedimentary basins intruded by sills. A simple, synthetic profile containing one listric fault and a number of horizons, including syn- and post-rift deposits, is used as the basis for this study (Figure 1). Throughout the simulations, one parameter is changed at a time which allows for evaluating the impact each parameter has on the modeling results. Table 1 shows the different modeling sets, the tested parameter values and default values of non-changing parameters. Shale and sandstone are the applied lithologies as these are common in sedimentary basins worldwide.

Table 1. Input parameters for the modeling. w = with, sh = shale, sst = sandstone, ant = antithetic, inc = inclined, synth = synthetic, kyr = thousand years.

Simulation Set	Tested Parameter	Tested Values	Fault Displacement	Time Span of Faulting and Deposition	Fault Angle	Thermal Conductivity/ Heat Capacity	Basal Heat Flow	Restoration Method
Set 1	Fault displacement	1200 m 500 m 1000 m 2000 m 3000 m	X	10 kyr	Original	All shale	47 mW·m^{-2}	Vertical shear
Set 2	Time span of faulting and deposition	10 kyr 1 Myr 5 Myr 10 Myr 20 Myr	1200 m	X	Original	All shale	47 mW·m^{-2}	Vertical shear
Set 3	Fault angle	Original Steepest Less steep Least steep	1200 m	10 kyr	X	All shale	47 mW·m^{-2}	Vertical shear
Set 4	Thermal conductivity/ Heat capacity	All shale All sandstone Sh basin w/sst layer Sst basin w/sh layer	1200 m	10 kyr	Original	X	47 mW·m^{-2}	Vertical shear
Set 5	Basal heat flow	47 mW·m^{-2} 40 mW·m^{-2} 60 mW·m^{-2} 80 mW·m^{-2}	1200 m	10 kyr	Original	All shale	X	Vertical shear
Set 6	Restoration method	Vertical shear No fault restoration 10° ant. inc. shear 20° ant. inc. shear 30° ant. inc. shear 10° synth. inc. shear	1200 m	10 kyr	Original	All shale	47 mW·m^{-2}	X

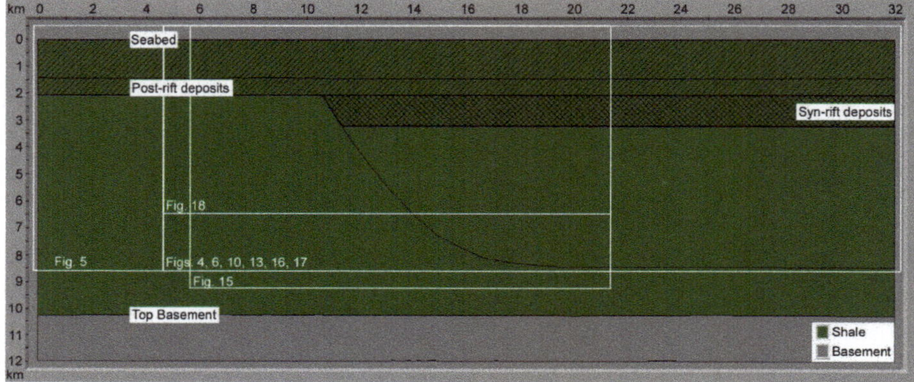

Figure 1. The synthetic profile with one active, listric fault used in the modeling. The green color (in this case) represents shale lithology. White squares indicate the blow-up area of following figures.

2.1. Thermal and Maturation Modeling

The geological, structural, thermal, and maturation history of the studied synthetic profile is performed with BMT (Basin Modelling Toolbox, Tectonor AS), a high-resolution 2D basin modeling software [26,35,36]. All thermal and maturation modeling simulations starts with present day geometry, where every horizon is given a specific age and all polygons are assigned a lithology with related porosity/depth trend, thermal conductivity and specific heat capacity (Table 2, upper part).

Table 2. Lithological parameters used in the modeling, based on standard values published in the literature (e.g., [37,38]).

Lithology	Porosity-Depth Trend		Conductivity (kv) (W·m^{-1}·K^{-1})		Heat Capacity
	Surface Porosity	Exponential Constant (km^{-1})	Low Porosity	High Porosity	(J·kg^{-1}·K^{-1})
Shale (Default)	0.63	0.51	3.00 (6%)	2.80 (60%)	1190
Sandstone (Default)	0.45	0.27	3.30 (6%)	1.50 (40%)	1080
Basement, metamorphic			3.10	3.10	1100
Magmatic intrusions			3.10	3.10	1100
Asthenosphere			3.50	3.50	1100
Shale, average conductivity	0.63	0.51	1.98 (6%)	1.19 (60%)	1190
Shale, max. conductivity	0.63	0.51	4.08 (6%)	2.08 (60%)	1190
Sandstone, average conductivity	0.45	0.27	2.36 (6%)	1.72 (40%)	1080
Sandstone, max. conductivity	0.45	0.27	6.24 (6%)	4.20 (40%)	1080
Shale, min. heat capacity	0.63	0.51	3.00 (6%)	2.80 (60%)	840
Shale, max. heat capacity	0.63	0.51	3.00 (6%)	2.80 (60%)	1420
Sandstone, min. heat capacity	0.45	0.27	3.30 (6%)	1.50 (40%)	760
Sandstone, max. heat capacity	0.45	0.27	3.30 (6%)	1.50 (40%)	3350

The first modeling step is a backstripping process, where one horizon at a time is removed, faults are restored and underlying deposits are decompacted. This process is repeated all the way down to top basement and in this way the section's geohistory is built. All elements that characterize the basin, such as faults, horizons, and lithologies, are parts of the geohistory process, and must be carefully defined. For the geohistory reconstruction a special type of grid was developed in BMT [26]. The grids are vertical line segments that are connected to the base of a polygon. A grid is always created at each digitized point in the present-day polygon. Additional grid columns are added to the section automatically. The number of inserted grid points can be controlled by the user (the default is 70). The foundation for thermal and maturation modeling is established during the geohistory process, and it is therefore important that the geological reconstruction of the basin is accurate.

The next step in the modeling process is the thermal development of the basin. BMT utilizes finite difference calculations by conduction with a rectangular finite difference grid of varying sizes (cf., [7,26]). For every reconstructed timestep in the geohistory, BMT builds a new high-resolution thermal modeling grid. Where needed, grid lines are automatically inserted so that the geometry is accurately represented. Around small features, like sills, the grid size is especially fine to ensure realistic calculations. The finite difference grid in this study consists of minimum 400×400 cells of varying sizes with an average size of 80 m $\times$ 30 m (width and height). The spatial variation in rock properties and possible differences from one timestep to the next are adjusted for so that appropriate finite difference calculations are maintained.

The finite difference calculation by conduction is controlled by the temperatures from the previous timestep, thermal conductivity (vertical and horizontal) and specific heat capacity of the basin's lithology/lithologies (see Appendix A for details on the numerical temperature model from Fjeldskaar et al. [7]). Temperature-dependent thermal conductivity is used, which commonly leads to reduction of the conductivity with increasing temperature. However, compared to conduction variations derived from differences in porosity and lithology, the temperature dependent variations are considered to be modest [35]. The lower boundary condition of the temperature calculations is the basal heat flow from the mantle, and the upper boundary condition is the paleo-surface temperature. The surface temperature is kept constant at 7 °C, and the heat flow is constant over the profile. Fjeldskaar et al. [7] tested BMT's numerical model versus an analytical model on the temperature effect of sill emplacement, and documented good performance for high-resolution modeling, both spatially and temporally.

Maturity modeling is completed in BMT and all calculations in this study assume kerogen type II, the most common in marine shales [39]. Classical first-order kinetics for the decomposition reactions is the basis for the maturation model in BMT (see [7,26]). In this study the whole basin is set as source rock in order to study the potential maturation effect of magmatic sill intrusions. However, for a case study, only the potential source rock would be defined as such a sequence.

2.2. Restoration of Faults

Several algorithms exist for restoring the structural evolution of basins. The different algorithms result in different basin and fault geometries, which again impact the calculated thermal history. Commonly, basin modeling simplifies the structural reconstruction of basins, e.g., by not reconstructing the faults through time, regardless the fact that hydrocarbons often accumulate in complex geological structures associated with faults [26]. According to Dula [40] reconstruction with simple inclined antithetic shear of 20° is one of the methods that best represents the observed fault shape, but all tested reconstruction methods in that study gave adequate results.

The present study utilizes vertical shear as a standard fault reconstruction method. However, different segments of simple shear, including inclined antithetic and synthetic shear are explored (Figure 2). Furthermore, we investigate how the different resulting geometries impact the thermal history. Because BMT restore faults solely by vertical shear, and the software Move (Petex Ltd.) does not have the ability to perform thermal and maturation modeling, both Move and BMT were used to model the structural development of the section. In the restoration process BMT and Move both use the backstripping process, which removes layers one by one and decompact the underlying sequences based on the given porosity-depth trend for the assigned lithology/lithologies. The porosity calculations are based on exponential functions (cf., [37]) and the same values are used in both softwares (Table 2). Move is a commercial software restoring geological cross sections by kinematic algorithms. Six algorithms are available for fault restoration, these are: simple shear, fault parallel flow, fault bend fold, fault propagation fold, trishear, and detachment fold. BMT is a non-commercial basin modeling software with the ability to perform fault restoration (by vertical simple shear) so that structural effects on temperature estimates are accounted for (cf., [26]).

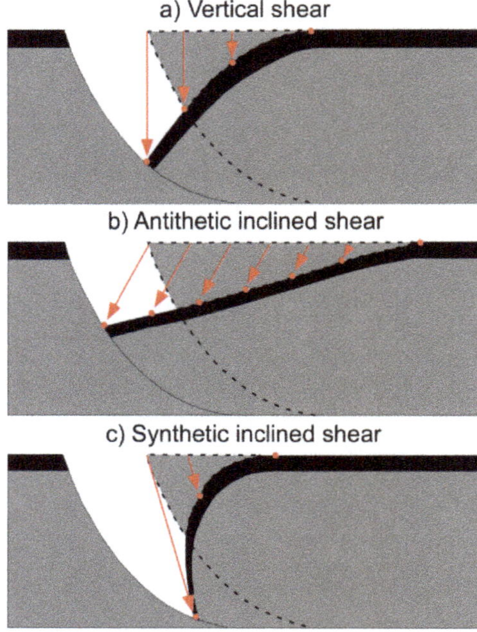

Figure 2. Illustration of the technique of (**a**) vertical simple shear, (**b**) antithetic inclined simple shear and (**c**) synthetic inclined simple shear. Antithetic simple shear results in a broader basin compared to for instance synthetic shear. Modified after Fossen [41].

In our study Move was used to restore the fault by antithetic and synthetic inclined shear. The tested angles for inclined shear are summarized in Table 1. The resulting reconstructions from Move are replicated in BMT by utilizing a "volume editing" function (cf., [26]). This method enables calculations of the thermal histories for the different restoration methods. All models have the same starting geometry. However, through restoration by different algorithms the paleo geometries will differ and so will the temperature histories.

2.3. Thermal Effects of Sills

On reflection seismic data, upper and lower boundaries of sills can be difficult to discern due to limited vertical resolution (cf., [42–45]). Many sills may therefore be left out in seismic interpretations due to sill thicknesses below the detection limit of the data (e.g., [44,45]). Sill thickness, both when sills intrude individually or as clusters, has a large impact on the temperature and maturation in sedimentary basins (e.g., [8,25]). The relative timing between sills emplaced as clusters is especially important if potential source rocks are located between sills [25]. Reported sill thicknesses from the literature vary from ~10 cm to >400 m (e.g., [44,46–55]). In our study a modest sill thickness of ~50 m has been chosen for the modeled sills and we assume the sills to intrude during one pulse with a magma temperature of 1000 °C. In BMT this is done by changing the lithology in sill polygon from shale to sill lithology, with related physical properties at time of intrusion. The sills have been modeled to intrude at various times relative to fault slip. This is done to study the interaction of two transient thermal effects simultaneously. As the largest thermal and maturation impact of sill intrusions are found at 3–5 km depths [25], the sills are modeled to intrude within this depth range. Commonly magmatic sills observed in the field and on seismic images (e.g., [7,52]) are layer concordant. Therefore, the sills in the study are modeled accordingly.

Latent crystallization heat introduces an extra source of heat when magma cools and starts to crystallize [56,57]. This extra heat source is not taken into account, as it is the relative differences between scenarios that are of interest in our study. Heat advection by fluids in relation to magmatic intrusions is common (e.g., [5,58–61]). However, convection is not accounted for in our modeling, as convection is considered insignificant in low permeability rocks (e.g., [7,8,13,25,62]) and is beyond the scope of this study.

3. Modeling Strategy and Results

A basin not subjected to tectonic deformation undergoes a slow rate of erosion or sedimentation and has a gradual temperature increase with burial depth. Such a basin is in steady state [63]. On the other hand, structurally active basins may undergo sudden erosion or high influx of sediments, e.g., due to increased accommodation space during faulting, leading to abrupt changes in the geothermal gradient. When magmatic intrusions are emplaced into such active basins, the temporary, local change in temperature is not solely an effect of the hot magma intruding the system, but also due to transient thermal effects caused by structural development. Figure 3 shows the studied synthetic profile assuming thermal steady state with input parameters as given in Table 2.

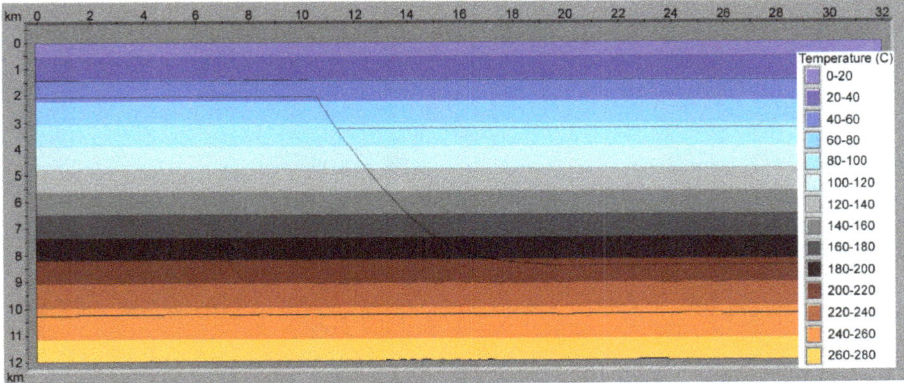

Figure 3. Temperature regime for the modeled basin in steady state solution.

Emplacement of sills and significant structural development may occur within the same timeframe (e.g., [30,64,65]). Therefore, we want to investigate the combined transient thermal contributions from the structural development and magmatic intrusions. The results should be considered as trends, not absolute values, as the study aim to give some general conclusions. The models are considered to have arrived at steady state when the whole basin show transient temperatures <1 °C from the steady state temperatures. However, for some cases exact steady state conditions are obtained right after reaching this limit, for other cases several million years (Myr) are required before actual steady state is obtained.

3.1. Thermal Effect of Slip along a Single, Normal Fault

Normal faulting causes downward displacement of sediments and generally deposition of syn-rift sediments (ref., Figure 1) in the hanging wall section of the fault zone. These down faulted sediments in the hanging wall are initially colder than the sedimentary rocks at the same depth in the footwall part, which result in transient thermal effects mostly in the hanging wall part (Figure 4). The basin is thermally unstable and results in a temperature difference up to 40 °C from one side of the fault zone to the other at the time of fault slip (Figure 4). Post-rift sediments deposited over the whole basin leads to thermal transient effects also in the foot wall section. However, these temperature differences will vary depending on physical properties and geometry of the basin. In the event of extension and normal faulting, the footwall temperatures are hardly influenced at time of fault slip. At 1000 years after the fault movement the isotherms in the footwall show a small downward bend towards the fault zone, indicating a small thermal influence by the neighboring colder sediments in the hanging wall (Figure 4).

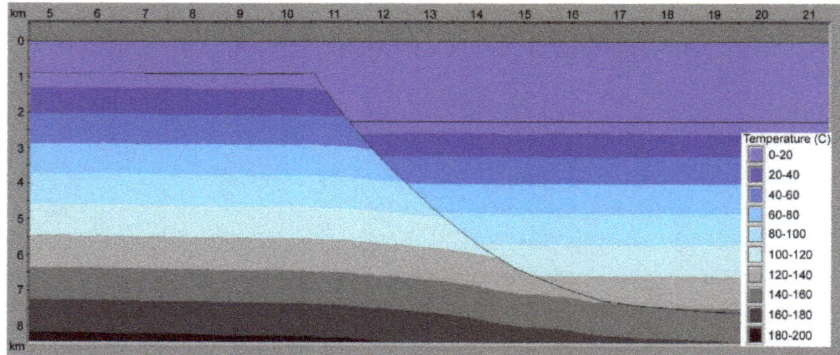

Figure 4. Temperature regime in the uppermost 8 km of the basin 1000 years after fault movement and sediment deposition. Fault slip is 1200 m.

The main temperature effect of magmatic sills lasts only some thousand years (see e.g., [25]). In accordance with this, we have here assumed that the syn- and post-rift sediments (Figure 1) of the timestep with fault movement are deposited over a period of only some thousand years. Deposition of these sediments leads to thermal instability in the basin. Figure 5 shows the temperature difference between steady state and transient models for deposition of these sediments at 10,000 years (10 kyr) and 500 kyr after fault slip. This causes larger parts of the basin to have transient temperatures up to 20 °C lower than under steady state conditions even up to 500 kyr after fault slip (Figure 5). Around 10 Myr is required for the basin to reach steady state temperatures.

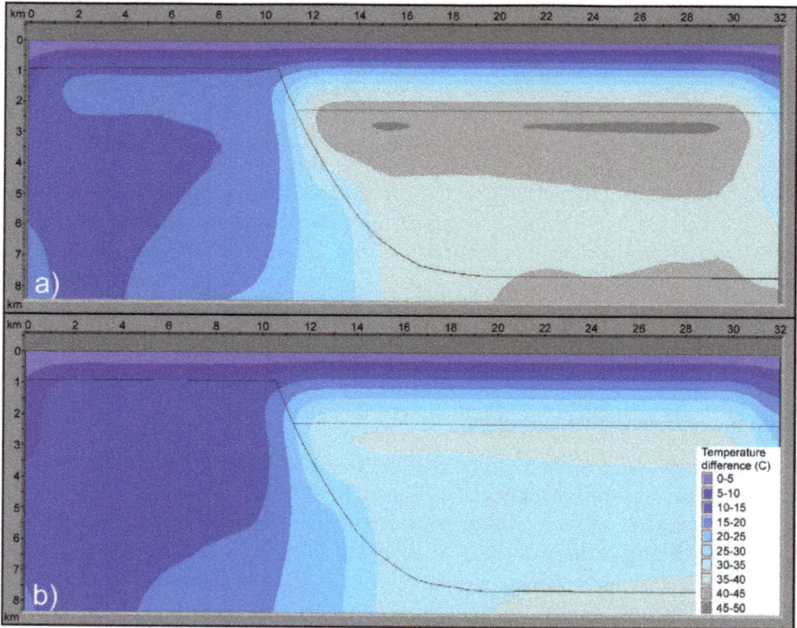

Figure 5. Temperature difference between steady state and transient model of the studied synthetic profile at 10 kyr (**a**) and 500 kyr (**b**) after fault displacement.

3.1.1. Fault Displacement

The original model has a fault displacement of 1200 m with syn-rift deposits of the same thickness and about 600 m of post-rift deposits, taking place over one timestep (Figure1). These post-rift deposits are kept unchanged in all tested scenarios. Additional four fault displacement models were tested; 500 m, 1000 m, 2000 m, and 3000 m (Table 1). As expected, temperature differences across the fault zone increases with the fault displacement. With 3000 m fault slip there is a temperature difference of more than 50 °C across the fault zone immediately after fault slip (Figure 6a). In the footwall part, the isotherms make a gentle downward bend towards the fault zone. As time passes, a gradual heating of the sediments in the hanging wall section occurs and the bend of the isotherms from the footwall side ties with those of the hanging wall (~10 kyr after faulting) creating continuous isotherms (Figure 6b,c). There is also a slight temperature change in the footwall part due to the deposition of cold post-rift sediments over that area.

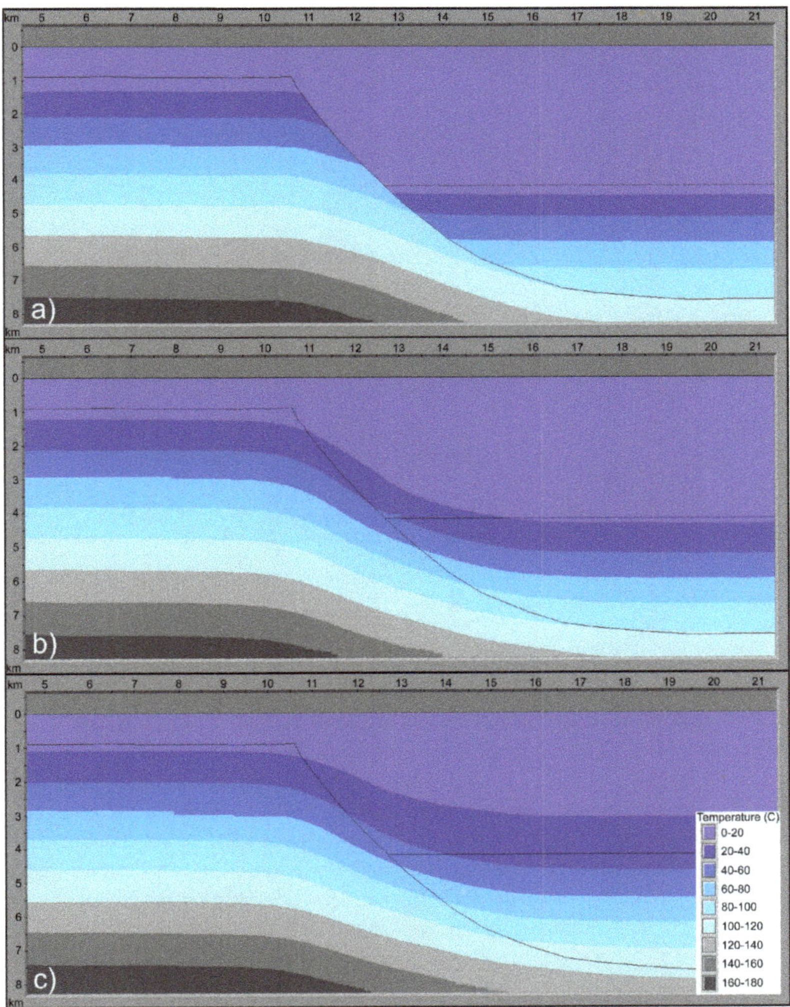

Figure 6. Temperature development in a basin with 3000 m fault slip: temperatures at 1 kyr (**a**), 500 kyr (**b**), and 5 Myr (**c**) after fault slip.

With increasing fault slip but within the same timeframe, the volume of deposited sediments and time needed for basin to regain steady state increases. For the modeled basin with 500 m of fault slip, large parts of the basin differ ~20 °C from the steady state basin 10 kyr after displacement (Figure 7a). Around 9 Myr later the basin obtains steady state. When fault displacement is 3000 m, large parts of the hanging wall section differ more than 50 °C from the steady state basin at 10 kyr after fault slip (Figure 7b). Steady state is achieved approximately 11 Myr after fault slip. As expected a basin with larger fault slip and higher influx of sediments requires more time to regain steady state, compared to a basin with smaller fault slip and less sediment deposits.

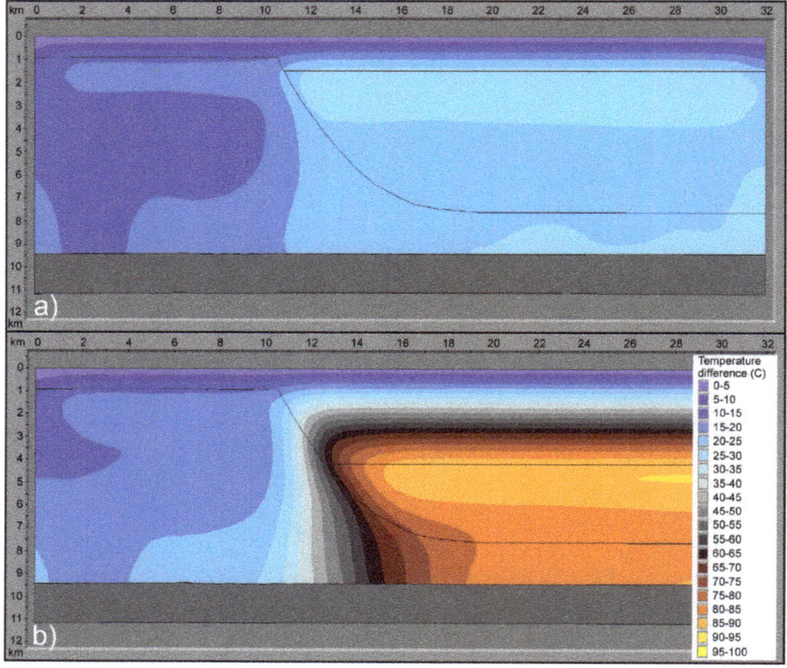

Figure 7. Temperature difference between steady state and transient temperatures for fault slip of 500 m (**a**) and 3000 m (**b**) at 10 kyr after fault slip.

3.1.2. Time Span of Faulting and Deposition

So far the modeling assumes the fault slip to be almost instantaneous (taking place over 10 kyr). The next set of models explores four different time spans of faulting and deposition of the same amount of sediments: 1, 5, 10, and 20 Myr. The results show that slower faulting and deposition rates, as expected, keep the basin closer to a steady-state condition compared to high rates (Figure 8a,b). Adding 10 Myr to the faulting and deposition time results in a basin with transient temperatures much closer to those of steady state throughout the whole period. For a faulting and deposition rate over 20 Myr, the basin is never in a state of thermal instability. The resulting trend show that basins with the same amount of sediment input and faulting and deposition time up to 10 Myr regains steady state approximately 10 Myr after faulting initiated (Figure 8c). This means that from time of faulting and deposition to the process is finalized, more time is required for a basin with rapid fault slip and deposition to regain steady state compared to a basin with slow fault slip and deposition. It emphasizes how crucial the time relation is for faulting and deposition and basin's thermal development after fault slip.

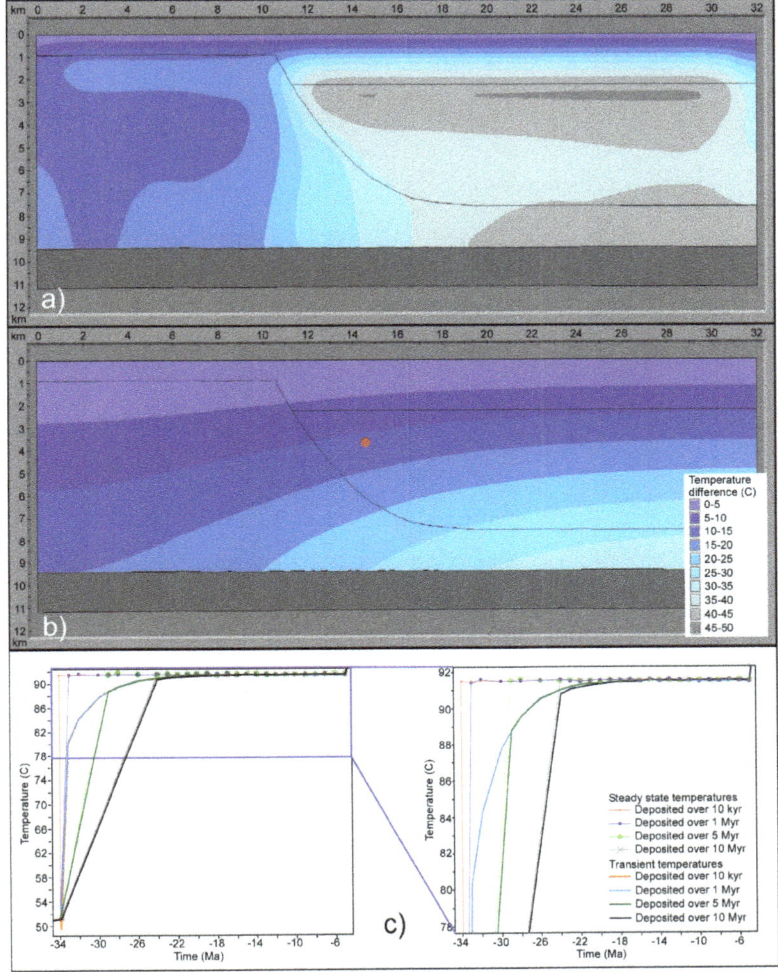

Figure 8. (**a**) Temperature difference between steady state and transient temperatures with fault slip and deposition over 10 kyr. (**b**) Temperature difference between steady state and transient temperature with fault slip and deposition over 1 Myr. Temperatures in (**a**,**b**) are from the same timestep, 10 kyr and 1.01 Myr respectively after faulting started. The red point in (**b**) indicates location of point plot in (**c**). (**c**) Resulting temperatures for four tested time spans of deposition of syn- and post-rift sediments.

3.1.3. Fault Angle

The previous results show that the largest thermal differences between steady state and transient temperatures are found in the hanging wall part of the basin. For all the tested fault angles the results are broadly similar, but the temperature effect will change with the fault angle. As the fault angle is changed, so is the affected area of the hanging wall. With a steeper fault angle a smaller part of the basin experiences the largest thermal instabilities (Figure 9a). The opposite is the case for lower fault angles, a larger part of the basin will then experience higher thermal instabilities before the basin regain steady state (Figure 9b). Five different fault angles have been tested (Figure 9a,b). All scenarios show temperature differences up to 45 °C between transient and steady state models. Around 10 Myr after fault slip all scenarios regain steady state temperatures.

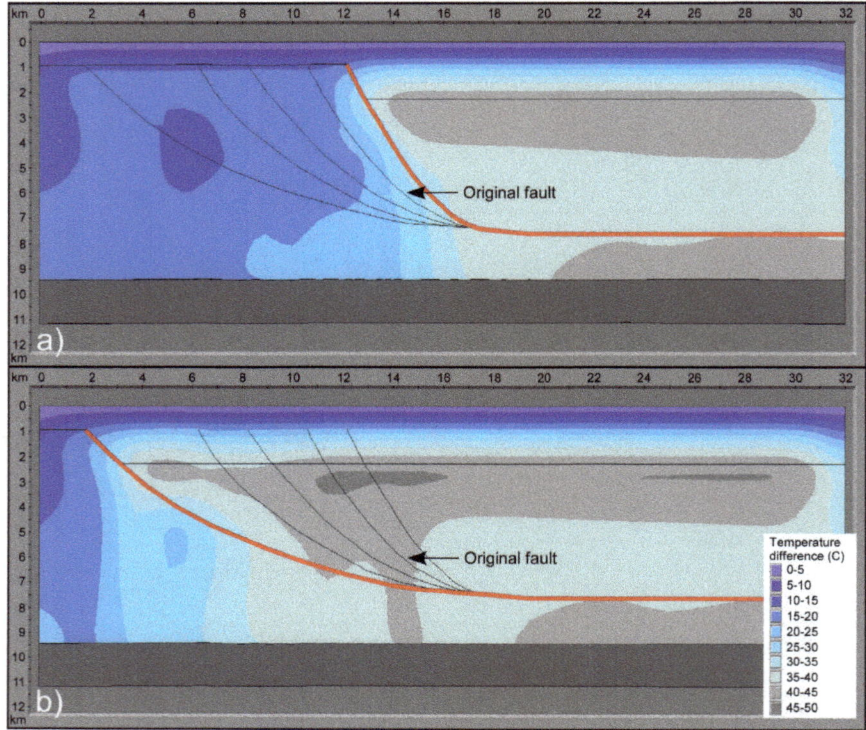

Figure 9. Temperature difference between steady state and transient model 10 kyr after fault slip for the steepest (**a**) and the least steep (**b**) studied listric fault.

3.1.4. Thermal Conductivity and Specific Heat Capacity

The thermal conductivity ($\mathrm{W \cdot m^{-1} \cdot K^{-1}}$) of a lithology relates to the rock's ability to transfer heat. Rocks with low thermal conductivity result in warm basins, while high thermal conductivity rocks give colder basins. A lithology's specific heat capacity ($\mathrm{J \cdot kg^{-1} \cdot K^{-1}}$) is the physical property crucial for the time frame needed to transfer heat through the stratum and thus for the time needed for basins to regain steady state after sediment deposition or erosion. So far the section has been modeled as all shale with properties as listed in Table 2. Commonly, basins consist of altering layers of different lithology types, often shale and sandstone. Therefore, the following four scenarios have been compared: all shale, all sandstone, all shale with one sandstone layer, and all sandstone with one shale layer.

Thermal conductivities and specific heat capacities of shale and sandstone show a large variation (e.g., values in Table 2). The default values are given in Table 2 and to illustrate the span of values within the same lithology segment, we refer to Čermác and Rybach [38]. Their minimum and maximum specific heat capacities and the average and maximum thermal conductivities of sandstone and shale are used in this study (Table 2). In order to determine the thermal conductivity values based on the rock's porosity, the mixing law arithmetic mean model [66] has been used:

$$k = \Phi \cdot k_f + (1 - \Phi)k_s, \tag{1}$$

where the thermal conductivity (k) is obtained on the basis of the rock porosity (Φ) by combining the thermal conductivity of rock matrix (k_s) with that of the pore fluid (k_f).

Generally, sandstones have higher thermal conductivities than shales, resulting in colder basins where they are abundant. The results from the homogeneous basins, either all shale or sandstone, coincide with this, showing high temperatures in the basin when the thermal conductivity is low, and lower temperatures when the thermal conductivities are high. A comparison between two basins with the same fixed specific heat capacity, but with one basin modeled with average shale conductivity, to another basin modeled with default shale conductivity, shows up to 140 °C difference between the two models shortly after fault slip (Figure 10a). With fixed specific heat and changing conductivity the results show that the basin needs longer time to regain steady state in basins with low thermal conductivity values due to the higher temperatures obtained in these basins (Figure 10b). This applies for both scenarios with all shale and all sandstone of varying thermal conductivity. Steady state is regained from 3–22 Myr after fault slip and the quickest scenario to regain steady state is the sandstone basin with maximum thermal conductivity.

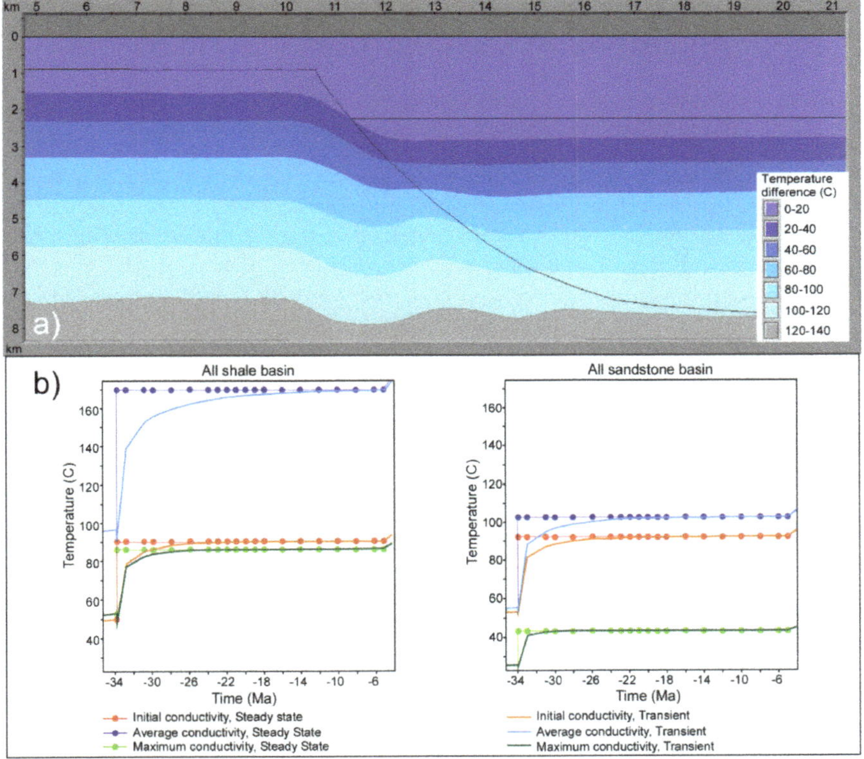

Figure 10. (**a**) Thermal difference between transient temperatures of basin (10 kyr after fault slip) modeled with average conductivity to basin modeled with default conductivity. (**b**) Point plot results from location of point given in Figure 8b, showing the steady state and transient temperatures for the three tested conductivities for shale and sandstone. Conductivity values given in Table 2.

In the case where the thermal conductivity is fixed and the specific heat capacities are changed in the different modeling scenarios (Table 2, lower part), the steady state temperatures for all the scenarios are, as expected, the same (Figure 11). However, with increasing specific heat, the time needed for the basin to obtain steady state increases (Figure 11). This is especially visible in the point plot for sandstone basin in Figure 11, because of large difference in the published heat capacity values of sandstone, but the effect is also visible for the shale basin. For all tested scenarios steady state is obtained between 8–18 Myr after fault slip.

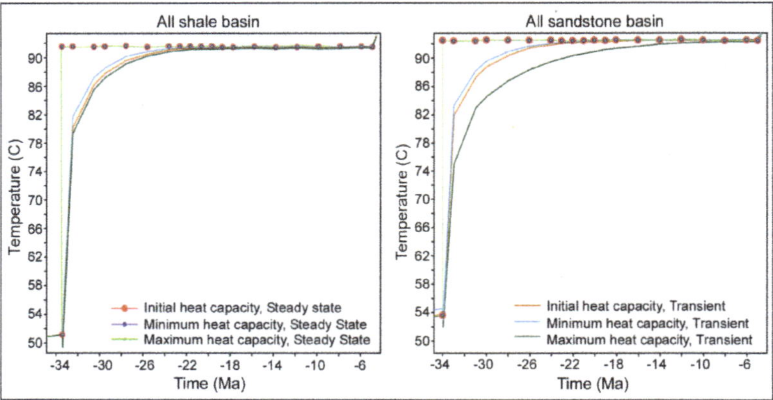

Figure 11. Resulting temperatures for the three tested specific heat capacities for shale and sandstone. Location of studied point indicated in Figure 8b. Heat capacity values given in Table 2.

To accommodate the fact that basins commonly consists of different lithologies, the polygon with syn- and post-rift deposits have been changed in the following simulation (Figure 12a). The thermal conductivity is fixed with values as for the default shale or sandstone (Table 2). However, the specific heat capacity is changed using default, minimum, and maximum values for the sandstone and shale (Table 2). With the presence of a sandstone layer in an otherwise shale basin, the results show that the temperatures in the basin increases. For the opposite case, a layer of shale in an otherwise sandstone basin, the temperatures in the basin decreases (Figure 12b).

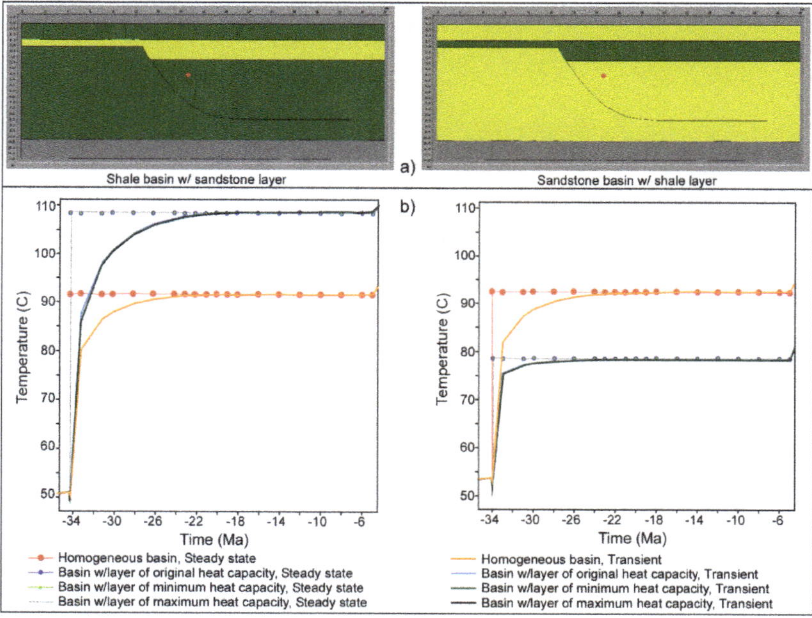

Figure 12. (**a**) Shale and sandstone basins with sandstone layer and shale layer respectively. The red point indicates location for point plot in (**b**). (**b**) Results show the temperatures for the three tested heat capacities for shale basin with one sandstone layer and sandstone basin with one shale layer. For comparison, temperatures for homogeneous basin are also plotted. For heat capacity values see Table 2.

The variation of specific heat capacity in a homogeneous basin resulted in quite a time gap (8–18 Myr) for the sandstone basin to regain steady state (Figure 11). However, the shale basin required less time to arrive at steady state for the tested specific heat capacities (8–10 Myr). For homogeneous basins, shale or sandstone, steady state was regained around 10 Myr after fault slip (Figure 12b). The presence of another lithology influences the basin temperatures and thus the time needed for the basins to regain steady state. For a shale basin with a sandstone layer, steady state was obtained around 13 Myr after fault slip for all the specific heat capacities. Sandstone basin with a shale layer regains steady state around 6 Myr after fault slip regardless of the specific heat capacities (Figure 12b). These results are related to the porosities at the depth at which the different lithology layers are modeled and the contrasting thermal conductivity for shale and sandstone.

3.1.5. Basal Heat Flow

The basal heat flow, which is the heat sourced from the mantle, transferred through the basin and constitutes the lower boundary condition in the models, is so far kept constant at 47 mW·m^{-2}. This basal heat flow value is slightly above one heat flow unit and is typical for some continental shelves, e.g., parts of the Norwegian Continental Shelf. However, to study the influence of the basal heat flow on the temperature development in the basin, three additional values were tested: 40 mW·m^{-2}, 60 mW·m^{-2}, and 80 mW·m^{-2} (Table 1). Figure 13b shows the results for 40 and 80 mW·m^{-2} together with the default heat flow (47 mW·m^{-2}). Increased heat flow leads to higher steady-state temperatures and longer time is therefore needed for the basin to regain steady state (Figure 13b). The resulting transient temperatures of basins modeled with 80 mW·m^{-2} and 40 mW·m^{-2} show thermal differences up to 140 °C (Figure 13a). For the tested values, the models regain steady state approximately between 9 and 12 Myr after the fault slip. Increasing basal heat flow gives higher temperatures in the basin, and as a consequence, a longer time is needed for the basin to arrive at steady state after fault slip.

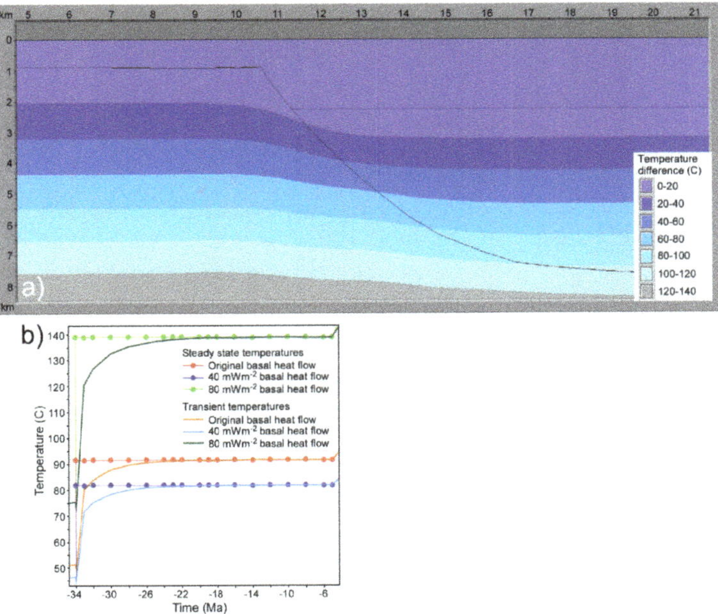

Figure 13. (**a**) Thermal difference between transient temperatures in basin with basal heat flow of 80 mW·m^{-2}, to basin with basal heat flow of 40 mW·m^{-2} at 10 kyr after fault slip. (**b**) Results show the temperatures for the default and the two extreme values for heat flow: 47 mW·m^{-2}, 40 mW·m^{-2}, and 80 mW·m^{-2}. Location of studied point is indicated in Figure 8b.

3.2. Restoration Methods

Different fault restoration methods cause different basin geometries. It is of interest to explore what effect these different geometries have on the resulting basin temperatures. In addition to vertical shear, we have therefore tested the basin reconstruction for 10° antithetic and synthetic shear, 20° and 30° antithetic shear and a case without fault reconstruction (Table 1). Basin reconstruction without fault restoration was done by moving the hanging wall section vertically up such that the timeline split by the fault zone is horizontally connected. This process gives no lateral mass movements of the basin. Vertical simple shear modeling was, as mentioned in the method section, done by BMT, while the alternative fault reconstructions were performed with Move. Antithetic simple shear results in a wider basin (Figure 2) and the larger the angle, the wider the basin. Synthetic simple shear results in a narrower basin (Figures 2 and 14) relative to the other methods. Vertical simple shear falls in between these two, while basin reconstruction without fault restoration results in a geometry which is quite different from the other restored scenarios. Figure 14 shows the resulting geometries of the fault plane and top and bottom basement after fault slip of the six tested reconstruction methods. The basins with non-restored fault and restored by 10° synthetic inclined shear (purple and green line respectively, Figure 14) result in the shallowest hanging wall and top basement.

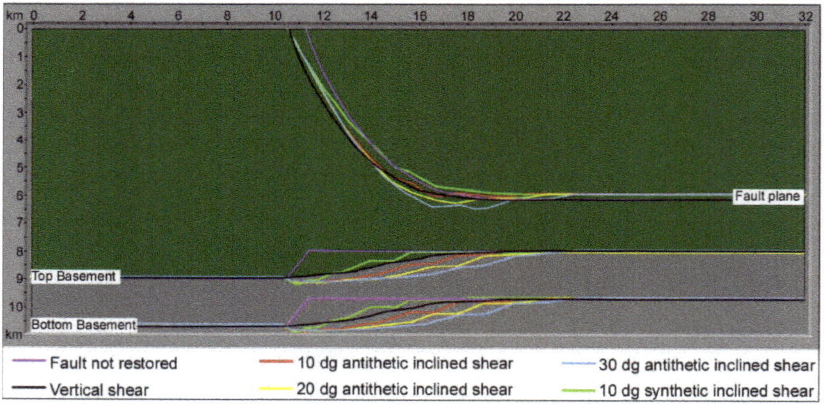

Figure 14. Resulting basin geometries after fault slip with the tested fault restoration methods. The different colored lines represent the resulting geometry of fault plane and Top and Bottom Basement after fault is restored by the six tested restoration methods. dg = degrees.

The geometry variations resulting from the tested restoration methods involve varying footwall and hanging wall area and burial depth, which result in differences in the thermal calculations, especially in the footwall section. Therefore, the thermal differences between them are concentrated in the footwall part of the basin at the time of fault slip (Figure 15). Figure 15c shows that the temperatures prior to fault slip show a larger difference on the footwall side compared to the hanging wall side. The comparison between the non-restored basin to basin restored by 30° antithetic shear, shows the largest area with thermal differences (Figure 15a), while the thermal differences between vertically- and non-restored basins are the most pronounced (Figure 15b). Although the thermal differences do not last long, ~1 Myr, such temperature differences might play a role concerning timing of generation and ultimately migration of hydrocarbons. For this particular case, the modeled thermal differences result in up to 40% maturation difference of the potential organic matter in a limited time and area of the footwall section (not shown here). All the tested restoration methods require approximately 10 Myr to achieve a steady state after fault slip. We therefore conclude that the tested restoration methods do not lead to large differences in time needed for the basins to achieve steady state. However, temperature differences due to different restoration methods may lead to temperature variations that might influence the maturation calculations.

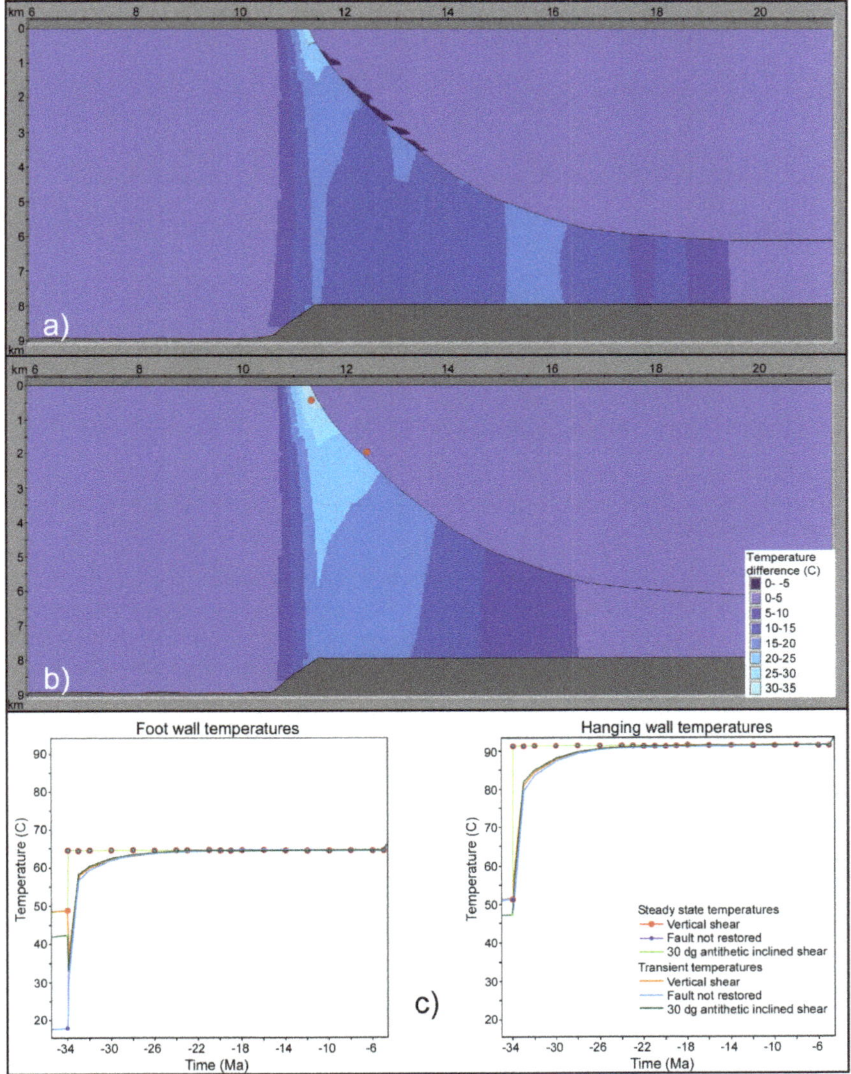

Figure 15. (**a**) Temperature difference between a scenario restored with 30 degrees antithetic inclined shear, compared to a scenario with non-restored fault. (**b**) Temperature difference for a scenario restored with vertical shear compared to a scenario with non-restored fault. (**c**) Temperature point plot from foot wall (left) and hanging wall (right) in the two points indicated in (**b**).

3.3. Sill Intrusions in the Basin

As mentioned above, earlier studies have pointed out the importance of host-rock temperatures at time of emplacement for the resulting thermal effects of intrusions [8,13,16–18]. In our study two sills, ~50 m thick, one in the hanging wall and one in the footwall, have been modeled to intrude the basin with increasing time lapse after fault slip. Figure 16 shows the results for sills intruding at time of fault slip, 1 Myr and 10 Myr after fault slip. There is a clear thermal instability on both sides of the fault zone. However, due to the masses of faulted rock and larger deposits of sediments in the hanging wall section, the thermal effects of the fault displacement are more pronounced here. Therefore, the largest differences in the thermal effects of the sills as time passes are expected to be found here.

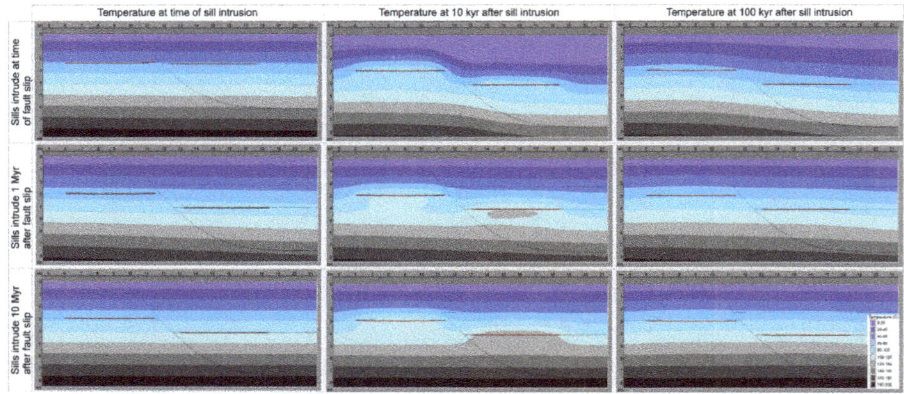

Figure 16. Temperature results for sills intruding with different timing in relation to fault slip. Temperatures to the left are at time of intrusion, in the middle, 10 kyr after intrusion, to the right, 100 kyr after intrusion.

At the time of intrusion, the temperatures in the close proximity of the sills increases dramatically (cf., [25]). As time passes the area with increased temperatures grow and consequently the temperatures starts to fall. The thermal effect of sills has a quick rise and fall within the first 1 Myr after intrusion.

Modeling results at 1000 years after fault slip show that there is a temperature difference up to 40 °C on either side of the fault zone (Figure 4). Results also show that as time passes the basin regains steady state (Figure 6a–c), which for most scenarios in this study occur somewhere between 3 and 22 Myr after fault slip. For sills intruding at time of fault slip the host-rock temperature effects are lower compared to sills intruding into a basin with a time lapse after fault slip, which have temperatures closer to that of a basin in steady state. A consequence is that sills intruding with a time lapse in relation to fault slip have higher background temperatures and the thermal effects will be more prominent (Figure 16). These thermal differences are still present 100 kyr after sill intrusion. In the hanging wall there is a temperature difference around 40 °C between the cases where the sills intrude at time of fault slip and 10 Myr after fault slip (Figure 16). The highest potential host-rock temperature effect of intruding sills will be in a steady state basin.

The largest thermal differences between basins with faults restored in different ways, was found between basin with fault restored by vertical shear to basin with non-restored fault (Figure 15b). Therefore, we have tested the possible influence of a non-restored fault basin with sills on the calculated thermal effects. As the thermal differences between a basin restored by vertical shear to a basin with a non-restored fault was found in the footwall part of the fault zone (Figure 15b), this is also the area where it is expected that possible differences in thermal effects of sills will be found (Figure 17a). However, the thermal results of the basin with sills and non-restored fault show very small thermal differences from basin with sills and fault restored by vertical shear (Figures 16 and 17a). The most pronounced thermal difference is found between the scenarios where sills intrude 1 Myr after fault slip. These small temperature differences will not significantly influence the maturation calculations for the two scenarios.

Sedimentary basins worldwide normally consist of changing layers of different lithologies with contrasting properties. A homogeneous shale basin with a lower thermal conductivity, such as the average conductivity used in this study (Table 2), result in a warmer basin (Figure 10a,b) and consequently gives a higher host-rock temperature effect for intruding sills. This emphasizes that the basin lithologies and their thermal conductivities have large influence on the temperatures in the basin. We have here tested a case where the sills intrude into a shale basin with a sandstone layer (Figure 12a). The results show no differences for the sills intruding into the basin at time of fault slip (Figure 17b). However, as time passes (1 Myr and 10 Myr after fault slip) the presence of the sandstone layer makes a difference in the resulting thermal effects of the sill intrusions. Sills intruding 10 Myr after fault slip cause thermal differences in some areas up to 20 °C in both the footwall and hanging wall 10 kyr after sill intrusion (Figures 16 and 17b).

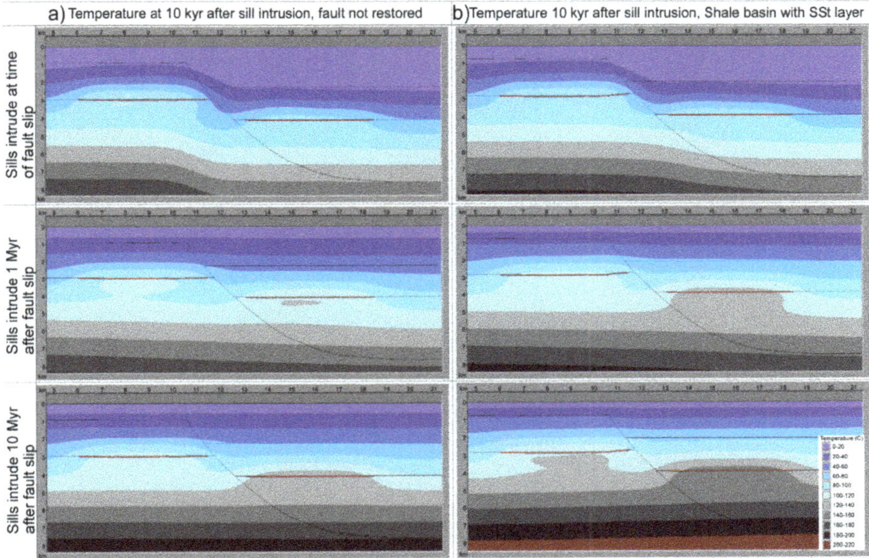

Figure 17. (**a**) Calculated temperatures 10 kyr after sills intrude into a basin with non-restored faults. (**b**) Calculated temperatures 10 kyr after sills intrude into a shale basin with a sandstone layer as shown in Figure 12a left.

As the results in Figures 16 and 17 show, the thermal impact of sill intrusions are more pronounced when emplaced into a basin closer to thermal steady state. This more pronounced thermal impact also influences to a small degree the maturation levels around the sills, especially the area above the sills in the hanging wall (Figure 18a). Reports from onshore and offshore sedimentary basins with magmatic intrusions show that sills often occur in clusters (e.g., [6,9,52,67]). It is considered that multiple sills intruding at different levels within a certain time frame thermally impacts a larger rock volume of the basin compared to one single intrusion (e.g., [8,21,68]). To replicate a basin with clusters of sills, two more sills have been added to the model, one on either side of the fault zone. The thermal effect of sill swarms intruding with different timing relative to fault slip have a large impact on the matured organic material, particularly in the area between the sills on both sides of the fault zone (Figure 18b). In the heterogeneous basin, the warmer host-rock temperatures compared to the homogeneous basin, result in more transformed organic material on both sides of the fault zone for sills intruding at time of fault slip, as well as for sills intruding 10 Myr after fault slip. However, the maturation differences between the homogeneous and heterogeneous basins are more prominent when the sills intrude 10 Myr after fault slip (Figure 18b).

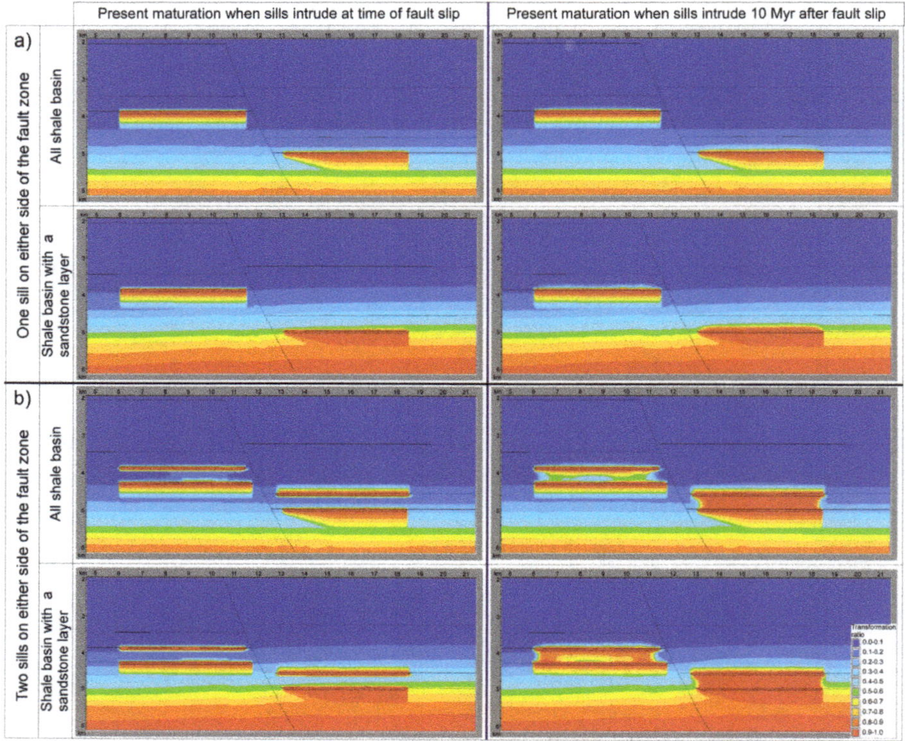

Figure 18. (**a**) Calculated maturation for one sill intruding on either side of the fault zone. (**b**) Calculated maturation for two sills intruding on both sides of the fault zone. Left side show results for sills intruding at time of fault slip. Right side show results for sills intruding 10 Myr after fault slip.

4. Discussion

The purpose of this work was to study the transient thermal effects in structurally complex basins with normal faults and sill intrusions. To uncover trends of several parameters transient thermal effects in sedimentary basins, both extreme and realistic values were tested. Three specific cases were addressed: the transient thermal effects of normal faulting, the ultimate effect of fault restoration method on thermal modeling, and the impact of sill intrusions in structurally complex basins. Interesting effects were discovered and are discussed separately in the following.

4.1. Transient Thermal Effects in Relation to Normal Fault Slip

Figure 4 shows that the geothermal gradient, especially in the hanging wall, changes abruptly after fault displacement and with time the basin arrives at a thermal steady state (Figure 6). The isotherms in the footwall are somewhat affected by down-faulted colder rocks and deposition of colder sediments in the hanging wall, which is in accordance with the work by ter Voorde and Bertotti [29]. By preserving a constant heat flow from the basement the model has been simplified which is acceptable as it does not change the relative differences between compared scenarios. A constructed synthetic profile has been used to uncover the thermal effects of normal faulting in structurally complex basins. Steady state is regained 3–22 Myr after the fault slip, but the time elapsed for the tested parameters and model runs to achieve this varies considerably: for fault displacement 9–11 Myr, thermal conductivity 3–22 Myr, specific heat capacity 8–18 Myr and basal heat flow 9–12 Myr. For the remaining tested parameters (Table 1) the variation of time needed to restore steady state is much smaller between the tested values, only some 100 kyr.

The syn- and post-rift deposits (Figure 1) in the studied section are 1800 m thick. These sediments are modeled to be deposited over five different time spans (Table 1); 20, 10, 5, and 1 Myr and 10 kyr, which correspond to a sedimentation rate of 0.09–180 mm·year^{-1}. The sedimentation rate roughly overlaps the reported sedimentation rates (0.4–100 mm·year^{-1} for the referenced papers) (e.g., [69–75]). ter Voorde and Bertotti [29] suggest that the extension rate somewhat controls the thermal effect of sediment deposition. They argue that high sedimentation rates will temporarily cool the basin while low sedimentation rates lead to heating. According to their study the "turning points" are at sedimentation rates of 2.0, 2.2, and 1.4 mm·year^{-1} for shale, siltstone, and sandstone, respectively. Our results show that sedimentation rates > 0.18 mm·year^{-1} for shale result in basins with transient thermal effects, in agreement with results reported by Ehlers et al. [34]. The numerical temperature model is different in the two studies. Additionally, the resolution of the finite difference grid is much finer in our study as the focus is on a smaller part of the basin compared to the work by ter Voorde and Bertotti [29]. These elements may possibly be the reason for the different results.

All the tested deposition time spans ≤10 Myr for the syn- and post-rift sediments regain steady state approximately 10 Myr after fault slip. However, there is a difference of 600 kyr for regaining steady state between the modeled scenarios. When faulting and deposition occur over a time span of 10 Myr, steady state is restored 200 kyr after the fault slip ceased. Basins with faulting and deposition over a time span of 10 kyr and 1 Myr, regains steady state after 9.6 and 9.7 Myr, respectively. Although the compared models keep the basin in an unstable thermal condition over approximately the same time span, the slow rate of faulting and deposition makes the thermal instability less prominent as opposed to basins with a high rate of faulting and deposition (Figure 8). This coincides with the result for the models were the fault displacements are varied. The larger the fault slip, the larger the thermal differences on either side of the fault zone (Figure 7), which results in longer time required for the basin to regain steady state. For basins with faulting and deposition occurring over >10 Myr, the basin is close to, or in, steady state throughout the whole deposition process; this is in accordance with Bertotti and ter Voorde [30] and ter Voorde and Bertotti [29].

In addition, our calculations show that there is a 40 °C difference after fault slip from one side of the fault zone to the other. Larger temperature differences have been reported in the literature, e.g., by Graseman and Mancktelow [76] (and references therein) who observed differences above 100 °C on either side of the Simplon Fault Zone in Switzerland. Although such a high difference may not be enough to cause regional disturbances in the geothermal field [30], it can lead to significant local maturation differences of deposited organic material if such thermal differences occurred in basins of sedimentary rocks. Since the oil window typically lies between 80 and 150 °C [77], some tens of degrees difference in temperature may thus lead to different stages of organic maturation on either side of a fault zone. The maturation calculations in BMT follow the Arrhenius equation, which ties the reaction rates of kerogen to the temperature differences and the activation energies of the kerogen type. For given activations energies and temperatures, the reaction rates of kerogen are roughly doubled for every 10 °C rise (e.g., [39]). This indicates that a 10 °C difference will have implications for the estimated maturation in the basin.

Figure 10 demonstrates the importance of the thermal conductivity for the obtained temperature levels in the basin. Worldwide basins show a variety of alternating lithologies, and consequently differences in thermal conductivities [38,63]). Our modeling shows that thermal conductivity, as well as the fault movement, influence both the steady state and the transient temperatures in the basin (Figure 12), and therefore the time needed to regain steady state. This is in keeping with results reported by Fjeldskaar et al. [78] who demonstrated the need for an accurate description of the thermal-conductivity variation to obtain a correct thermal history, on which the prediction of maturation is based.

The thermal conductivity of lithologies, amount of fault slip, time span of faulting and deposition, angle of fault zone, and basal heat flow will influence both the size of the affected area and level of the thermal imbalance. Specific heat capacity will determine the time needed for a basin to regain steady state (Figure 11) [26,35].

4.2. Fault Restoration and Its Effect on Thermal Modeling

Hydrocarbons are often trapped in structurally complex basins, yet structural reconstruction is often over-simplified in traditional basin modeling [26]. We have shown that the tested restoration methods do not lead to substantial differences in the thermal estimates. Dula [40] concludes that 20 degrees antithetic simple shear result in a geometry resembling the natural basin the most. However, resulting thermal calculations for vertical shear and 20 degrees antithetic shear show negligible differences. We therefore conclude that thermal estimates of transects restored by vertical shear give reasonable results.

The largest thermal differences are found between basins with the greatest deviation in restored geometry and, consequently, deviation in burial depth. Thermal differences up to 35 °C are found when the temperatures obtained for basin restored by vertical shear are compared to basin with non-restored fault. Fjeldskaar et al. [26] demonstrated the importance of fault reconstruction for the temperature modeling on a 2D section from the Gulf of Mexico. Up to 70 °C in temperature differences were observed in the footwall part of the basin between model with reconstructed faults versus model with no fault reconstruction. This strongly suggests that accounting for the structural evolution in sedimentary basins is essential for a reliable thermal and maturation modeling. The results from this study along with those of Fjeldskaar et al. [26] emphasize the importance of proper structural reconstruction of sedimentary basins and highlights the need to study the transient thermal effects that occur in a basin proceeding a fault slip in order to make good thermal and maturation estimates.

4.3. Fault Slip and Magmatic Intrusions

As mentioned above, the pre-intrusion host-rock temperatures are acknowledged to be important for the thermal effect of magmatic intrusions and the size of the thermal aureole (e.g., [8]). Fault slip and magmatic intrusions may occur within the same timeframe (e.g., [30,64,65]). Results in this study show that the thermal effects of magmatic intrusions are sensitive to the timing of the fault slip relative to the timing of the sill emplacement (Figures 16–18). The basin with no sills shows that millions of years are required to restore steady state after fault movement (Figures 5 and 6). Before fault slip, sediments in the hanging wall are buried at shallower depths in the basin compared to after fault slip. The cold sediments in the hanging wall will gradually experience a temperature increase up to 40 °C over a time span of ~10 Myr. With warmer host-rock temperatures the thermal effect of intrusions is higher. Figures 8 and 10–12 show temperature histories as a function of various parameters with time to equilibrium of several Myr. However, temperatures rebound within 1 Myr or less to only a few degrees from steady state temperatures. Sedimentary basins in general would possibly not show variation in maturation due to transient thermal effects over ~1 Myr. Unless, the thermal differences were of a certain magnitude, as the reaction rates of kerogen is doubled for every 10 °C rise (e.g., [39]) as mentioned above. Sydnes et al. [25] show that the main thermal effect of magmatic sill intrusions occurs within the first 1 Myr after intrusion. This indicates that if faulting and sill intrusion occurs within the same timeframe, two transient thermal effects operate simultaneously in the basin. Disregarding these transient thermal effects could possibly lead to over- or underestimation of the basin maturation.

Sills intruding around the time of fault slip are emplaced into a basin with lower background temperatures than sills intruding 10 Myr after fault displacement. However, these differences in host-rock temperatures lead to larger thermal effects (Figure 16), as mentioned above, but do not result in large differences in maturation of organic material when one sill intrudes on both sides of the fault zone (Figure 18a). This is in contrast to the results of Aarnes et al. [8] who concluded that the host-rock temperature is an important parameter in relation to aureole thickness and thus the amount of generated gas after intrusion of one sill. They found that increasing the host-rock temperature by 50 °C has a larger impact on the aureole thickness than both the option of increasing the temperature of the intruding magma by 50 °C or by increasing the sill thickness by 50 m. They also found that thinner sills have smaller thermal aureoles than thicker sills [8]. In this study ~50 m thick sills have been used as opposed to the work by Aarnes et al. [8], who studied a 100 m thick sill. This difference in sill thickness might possibly be the reason for the different conclusions in these two studies.

As for sill clusters, a 40 °C difference in background temperature results in a larger area with matured organic material around the intruded sills (Figure 18b). This is in agreement with Aarnes et al. [21] and Sydnes et al. [25] who showed that the thermal aureole of sill clusters at multiple levels thermally impacts a larger volume of surrounding rocks than single sills. Both these studies also show that sill thickness plays a role in the thermal and maturation effects. A 100 m thick sill will mature organic material in a larger rock volume than a thinner sill [21,25].

Figure 10 shows clearly that the thermal conductivity plays a significant role for the obtained basin temperatures. The results of sills intruding into a heterogeneous shale basin with a sandstone layer has a greater thermal effect on the surrounding host rocks, and is especially visible for sills intruding 10 Myr after fault slip (Figure 17). With the presence of higher thermal conductivity lithologies the host-rock temperatures increases and the thermal and maturation effect of magmatic intrusions are enhanced (Figure 18a,b), a conclusion supported by others (e.g., [8,16,21]).

4.4. Limitations of the Study

4.4.1. Assumptions

A simple model with one listric fault and one faulting event has been used in this study. The layers end horizontally in the fault zone and most of the tested scenarios assume a homogeneous lithology. Deposited sediments are modeled to reach the surface in every timestep, meaning that there are no constraints on the sediment supply and sea level is not accounted for. The magmatic sill intrusions are modeled to be horizontal and concordant to the layering with a constant sill thickness. Natural basins commonly consist of several faults, several faulting events, and different fault dips possibly dipping in various directions. Furthermore, layers often show a bending downwards towards the fault zone of a listric fault and they commonly show a large lithological heterogeneity, which might lead to a variation in physical properties. All these factors may result in variations in the estimated transient thermal effects and the time needed for the basin to regain steady state after fault slip, and contribute to contrasting host-rock temperatures. The occurrence of salt in the vicinity of sills will in particular influence the thermal situation in basins. Including all the above mentioned factors is possible and will turn the model into a more realistic example. However, doing so will limit the general conclusions that can be drawn from our study. By keeping the model simple the study is focused on fundamental questions related to transient thermal effects in basins with normal faults and magmatic sills. Revealed trends in this study are therefore applicable for basins in general with such features, as the relative differences between the tested parameters will remain more or less the same.

Sills vary normally in thickness and shape, e.g., saucer-shaped, v-shaped, and transgressive sills (cf., [67,79,80]). Sill thickness influences the temperature effects in the surrounding area; a thick magmatic sill has a wider thermal influence on the surroundings than a thin sill. It influences the size of the thermal aureole surrounding it (e.g., [8,25]) and an uneven sill thickness could possibly lead to an uneven thermal aureole. In a 2D model, the intrusion is extended infinitely in the 3rd direction. This is a good approximation as sills commonly have much larger horizontal dimensions compared to the vertical dimension. The sills are often several km in lengths and widths, while possibly only 50–100 m thick. For typical sill dimensions, the 2D modeling is sufficient for realistic temperature calculations. Fjeldskaar et al. [7] studied the 3D effect of magmatic sill intrusions and investigated what length/width ratio of the sill that requires a 3D calculation to give realistic temperature results.

In the temperature modeling the initial temperature of the intruding magma is set to 1000 °C and latent crystallization heat is not accounted for. Most magmas have temperatures ranging from 700 to 1300 °C (cf., [81] and references therein) making the chosen temperature in this study a mean value. Higher magma temperatures will increase the thermal effects of the sills, whilst a lower magma temperature will have the opposite effect. However, the relative differences between the compared models in this study will stay the same, as the intruding magma initial temperature is constant in all models.

Magmatic sills are often fed by dykes in magmatic upwelling zones. These near vertical features will contribute with additional heat to the thermal history of basins with sills. However, magmatic sills can also be fed by other sills over relatively large distances which has been observed in the field (e.g., [52]).

Latent heat of crystallization is thermal energy released when the sill undergoes a phase transition from liquid to solid. The composition of the magma may differ, which may lead to different timing of the emissions of latent heat as components have different crystallization temperatures. Peace et al. [62] have estimated the additional latent heat to be 488 °C, which is a substantial thermal contribution. Heat due to solidification of the magma will be released as long as the sill contains liquids and until the sill has solidified completely. However, introducing this additional heat as one pulse, either by increasing the starting temperature of the magma or introducing it as an additional heat when the sills start to cool, may result in an overestimation of the effect of sills [20].

4.4.2. Temperature Model

The temperature calculations are done by a finite difference scheme. The grid resolution is very high around magmatic sills and in areas of irregular structures. Fjeldskaar et al. [7] showed that the numerical temperature results for magmatic sills match the analytical solutions quite well, both spatially and temporally. On this basis, we argue that the uncertainty of the conduction-temperature calculations is of minor importance.

In our model we assume that the sills cool by conduction, but it is possible that the process could be accelerated by convection. Water convecting in the vicinity of sill intrusions could significantly modify the temperature and maturity effect of intrusions. Convection of hot fluids from the magma, the host-rock water, and decomposition products of kerogen is dependent on the permeability of porous host rocks or hydrofracturing in less porous host rocks (cf., [60,61,82–84]. How the fluids from magma and host rock affect the fluid pressure and flow is determined by the host-rock porosity, permeability, and amount of fluids present [85]. Wang and Manga [83] showed that for rocks with low permeability (<10 mD) symmetrical contact aureoles are produced, implying that conduction is the favorable cooling method for sills and host rocks after emplacement. On the other hand, rocks of high permeability (>50 mD) show maturation asymmetry above the sill, implying occurrence of convection-influenced maturation [83]. Annen [84] studied the maximum temperatures of three rocks of different water saturations: dry, hydrated and total water saturated, and found that in close proximity to an intrusion (~10 m) there were no differences in the temperatures, whereas at a distance of 50 m there was a ~50 °C maximum temperature difference. Dry rock has the highest temperature and water saturated rock the lowest. This means that convection of water lowered the temperature by ~50 °C. It was concluded that diffusivity contrasts between magma and host rock and incremental sill emplacement had more effect on the aureole thickness than host-rock water content [84]. For the present study this implies that for highly porous and permeable lithologies (e.g., sandstone) the predicted time for cooling might be overestimated. However, for less porous and permeable lithologies (e.g., shale) the estimated cooling rates would be adequate. This was also shown by temperature and maturity modeling on data from offshore Norway [7].

Another aspect of convection, is the possible increased permeability that occurs when sills are emplaced. Sill emplacement may potentially induce stresses high enough on the surrounding host rocks to reactivate or produce new shear fractures or open up tensile fractures (e.g., [86,87]), allowing for increased permeability and possibly increased convection. However, accumulation of stresses and reactivation and/or production of faults and fractures are highly area specific and dependent on, e.g., the lithology, deformation features and existing weakness zones. Although increased convection due to emplacement of sills have been identified on seismic images (e.g., [67]) it is too area specific to be included in a study aiming for some general conclusions.

Radioactive decay within the sedimentary section itself can generate heat, but is not accounted for in this study. Rybach [88] found, however, that the contribution of radioactivity in the sedimentary section was generally no more than a few percent of the overall heat budget, which support that the errors by not accounting for radioactive heat, is insignificant. Others have concluded for the opposite and advocate for including radiogenic heat production in the thermal history modeling (e.g., [89,90]), but also emphasize the large area-variations in radiogenic heat production for the same lithologies [90].

Frictional heating from fault displacement may potentially have implications for the thermal history in a basin. ter Voorde and Bertotti [29] calculated for shear stress of 100 MPa and an extension rate of 2 mm·year^{-1} that the thermal contribution from frictional heat would be 2.5 °C. They concluded that this could not influence the thermal situation in the basin to a significant degree. However, with higher extensional rates and larger shear stress the frictional heating may become important.

4.4.3. Kerogen Type

Simulations of kerogen maturation can only be as good as the kinetic parameters used in the model. The parameters are derived from laboratory experiments, and are assumed to be applicable to geological processes although the scales are different by orders of magnitude. Thus the uncertainty in the calculated maturity can be significant.

As mentioned above, the maturation modeling in our study assumes kerogen type II, which is considered to be the most abundant among marine shales [39]. Kerogen degradation is temperature dependent following the Arrhenius law (e.g., [39,63]). Due to kerogen types having different activation energies, reaction rates of other kerogen types in the models presented here, could result in different maturity effects in the modeled scenarios.

5. Concluding Remarks

We have studied the transient thermal effects in a constructed basin with magmatic sill intrusions and one normal faulting event. Several factors related to faulting, physical properties, and fault restoration methods have been tested to study possible differences in resulting calculated temperature regimes over time. Sills of modest sill thickness (50 m) have been modeled to intrude before the basin has regained steady state after fault slip, enabling the study of two transient thermal effects acting simultaneously in the basin.

From the results in this study, the following conclusions can be drawn:

- After fault slip, the basin is thermally unstable and is influenced by transient thermal effects that may last up to several million years. This implies that transient thermal effects should be accounted for if sills are emplaced after the structural events, as they might affect the pre-intrusion host-rock temperatures.
- With increasing fault displacement, the temperature effects of fault slip on either side of the fault zone increases, as does the time the basin is thermally unstable.
- For faulting and deposition occurring over a time span of more than 10 Myr, the basin is in, or close to, a steady state throughout the entire period. However, for the same basin, but with faulting and deposition occurring over less than 10 Myr, the basin is thermally unstable for ~10 Myr. This means that the shorter the time used on faulting and deposition, the longer is the time the basin is thermally unstable.
- Different fault angles barely influence the time the basin is in a transient state. All tested angles lead to steady state ~10 Myr after fault slip. However, different fault angles cause changes in

the foot wall and hanging wall areas and thus affect the host-rock temperature and therefore the temperature effect of potential sills.

- Thermal conductivity is the parameter influencing pre-intrusion host-rock temperatures in the basin the most. As temperatures increase, so does the time needed for the basin to regain steady state after fault slip. For basins with identical temperature regimes, the specific heat capacity is the important property determining the time needed for the basin to regain steady state.
- The obtained temperatures in the basin increase with increasing basal heat flow, thereby increasing the time needed to arrive at a steady state after fault slip.
- Different restoration methods result in basins of different geometries, leading to temporary thermal differences mainly in the footwall part of the basin. The largest thermal differences are found between basins with fault restored to basin with non-restored fault.
- Disregarding transient thermal effects proceeding normal fault slip may lead to both under- or overestimation of pre-intrusion host-rock temperature. This will influence the calculated effects of intruded sills and has implications for the estimate of how magmatic intrusions influence hydrocarbon maturation and is particularly the case for sills intruding as clusters at multiple levels.
- A basin that has regained steady state after normal faulting has the highest potential host-rock temperature.

Emplacement of magmatic intrusions may possibly affect the reservoir quality and the fault and fracture permeability in their vicinity. Future work should evolve around the effects magmatic intrusions have on the diagenetic processes in its surroundings.

Author Contributions: Conceptualization, M.S., W.F., I.G. and I.F.L.; methodology, M.S., W.F. and I.G.; validation, W.F.; investigation, M.S..; writing—original draft preparation, M.S..; writing—review and editing, W.F., R.M., I.G. and I.F.L.; visualization, M.S. and I.G..; supervision, W.F. and R.M..; project administration, I.F.L..; funding acquisition, I.F.L.

Funding: This research was partly funded by The Research Council of Norway and Tectonor AS as a part of the PhD project 'Effects of magmatic intrusions on temperature history and diagenesis in sedimentary basins—and the impact on petroleum systems', RCN project number 257492.

Acknowledgments: We want to express gratitude for the support. The authors acknowledge the use of the Move Software Suite granted by Petroleum Experts Limited. Nestor Cardozo is thanked for fruitful discussions throughout the course of designing the study. Craig Magee is thanked for providing helpful comments and three anonymous referees are thanked for constructive reviews.

Appendix A

Numerical temperature model,
The following equation is discretizised,

$$\frac{\partial}{\partial x} Kh \frac{\partial T}{\partial x} + \frac{\partial}{\partial z} Kv \frac{\partial T}{\partial z} = \frac{\partial}{\partial t}(cT) \tag{A1}$$

where T is the temperature, Kh is the horizontal conductivity, and Kv is the vertical conductivity. Finite differences and a cell-centered grid are used. In the block with indices (ij) the expression

$$\frac{\partial}{\partial z} Kh \frac{\partial T}{\partial z} \tag{A2}$$

is evaluated by the following formula [91]:

$$\left[\frac{\partial}{\partial x} Kh \frac{\partial T}{\partial x}\right] ij = \frac{1}{\delta x_i}\left[Kh_{i+\frac{1}{2},j}\left(\frac{2(T_{i+1,j} - T_{i,j})}{\delta x_i + \delta x_{i+1}}\right) - Kh_{i-\frac{1}{2},j}\left(\frac{2(T_{i,j} - T_{i-1,j})}{\delta x_{i-1} + \delta x_i}\right)\right] \tag{A3}$$

$Kh_{i+\frac{1}{2},j}$ is the value of *Kh* at the boundary between the blocks (*i,j*) and (*i* + 1,*j*). It is computed as the harmonic mean of $Kh_{i,j}$ and $Kh_{i+1,j}$.

The expression $\frac{\partial}{\partial z} Kv \frac{\partial T}{\partial z}$ is treated analogously.

This gives M · N equations to find the $T_{i,j}$, unknowns, where *i* = 1,2, … ,M, *j* =1,2, … ,N. Here, M and N are the number of blocks in x-direction and z-direction, respectively.

We use both Dirichlet and Neumann boundary conditions for the temperature model. For Dirichlet boundary conditions the temperature, T, at the boundary is given whereas for Neumann conditions the heat flux, $Kh\frac{\partial T}{\partial x}$ and $Kv\frac{\partial T}{\partial z}$, is given. A Neumann boundary condition with a heat flux of zero is used for the basin edges.

An iterative method is used to solve the linear system. Conjugate gradients are used as an acceleration method [92,93]. The conjugate gradient method is preconditioned by nested factorization [94].

References

1. Schutter, S.R. Occurrences of hydrocarbons in and around igneous rocks. In *Hydrocarbons in Crystalline Rocks*; Petford, N., McCaffrey, K.J.W., Eds.; Geological Society, Special Publications: London, UK, 2003; Volume 214, pp. 35–68.
2. Zou, C.; Zhang, G.; Zhu, R.; Yuan, X.; Zhao, X.; Hou, L.; Wen, B.; Wu, X. *Volcanic Reservoirs in Petroleum Exploration: Petroleum Industry Press*; Elsevier Inc.: Amsterdam, The Netherlands, 2013.
3. Senger, K.; Planke, S.; Polteau, S.; Ogata, K.; Svensen, H. Sill emplacement and contact metamorphism in a siliciclastic reservoir on Svalbard, Arctic Norway. *Nor. J. Geol.* **2014**, *94*, 155–169.
4. Schutter, S.R. Hydrocarbon occurrence and exploration in and around igneous rocks. In *Hydrocarbons in Crystalline Rocks*; Petford, N., McCaffrey, K.J.W., Eds.; Geological Society, Special Publications: London, UK, 2003; Volume 214, pp. 7–33.
5. Svensen, H.; Planke, S.; Malthe-Sørenssen, A.; Jamtveit, B.; Myklebust, R.; Eidem, T.R.; Rey, S.S. Release of methane from a volcanic basin as a mechanism for initial Eocene global warming. *Nature* **2004**, *429*, 542–545. [CrossRef] [PubMed]
6. Svensen, H.; Planke, S.; Corfu, F. Zircon dating ties NE Atlantic sill emplacement to initial Eocene global warming. *J. Geol. Soc.* **2010**, *167*, 433–436. [CrossRef]
7. Fjeldskaar, W.; Helset, H.M.; Johansen, H.; Grunnaleite, I.; Horstad, I. Thermal modelling of magmatic intrusions in the Gjallar Ridge, Norwegian Sea: Implications for vitrinite reflectance and hydrocarbon maturation. *Basin Res.* **2008**, *20*, 143–159. [CrossRef]
8. Aarnes, I.; Svensen, H.; Connolly, J.A.D.; Podladchikov, Y.Y. How contact metamorphism can trigger global climate changes: Modeling gas generation around igneous sills in sedimentary basins. *Geochim. Et Cosmochim. Acta* **2010**, *74*, 7179–7195. [CrossRef]
9. Svensen, H.; Corfu, F.; Polteau, S.; Hammer, Ø.; Planke, S. Rapid magma emplacement in the Karoo Large Igneous Province. *Earth Planet. Sci. Lett.* **2012**, *325–326*, 1–9. [CrossRef]
10. Moorcroft, D.; Tonnelier, N. Contact Metamorphism of Black shales in the Thermal Aureole of a dolerite sill within the Karoo Basin. In *Origin and Evolution of the Cape Mountains and Karoo Basin, Regional Geology Reviews*; Linol, B., de Wit, M.J., Eds.; Springer International Publishing: Cham, Switzerland, 2016; pp. 75–84.
11. Othman, R.; Arouri, K.R.; Ward, C.R.; McKirdy, D.M. Oil generation by igneous intrusions in the northern Gunnedah Basin, Australia. *Org. Geochem.* **2001**, *32*, 1219–1232. [CrossRef]
12. Monreal, F.R.; Villar, H.; Baudino, R.; Delpino, D.; Zencich, S. Modeling an atypical petroleum system: A case study of hydrocarbon generation, migration and accumulation related to igneous intrusions in the Neuquen Basin, Argentina. *Mar. Pet. Geol.* **2009**, *26*, 590–605. [CrossRef]
13. Spacapan, J.B.; Palma, J.O.; Galland, O.; Manceda, R.; Rocha, E.; D'Odorico, A.; Leanza, H.A. Thermal impact of igneous sill-complexes on organic-rich formations and implications for petroleum systems: A case study in the northern Neuquén Basin, Argentina. *Mar. Pet. Geol.* **2018**, *91*, 519–531. [CrossRef]
14. Wang, D.; Song, Y. Influence of different boiling points of pore water around an igneous sill on the thermal evolution of the contact aureole. *Int. J. Coal Geol.* **2012**, *104*, 1–8. [CrossRef]

15. Wang, D.; Zhao, M.; Qi, T. Heat-transfer-model analysis of the thermal effect of intrusive sills on organic-rich host rocks in sedimentary basins. In *Earth Sciences*; Dar, I.A., Ed.; InTech: London, UK, 2012.

16. Dow, W.G. Kerogen studies and geological interpretations. *J. Geochem. Explor.* **1977**, *7*, 79–99. [CrossRef]

17. Bostick, N.H.; Pawlewicz, M.J. Paleotemperatures based on vitrinite reflectance of shales and limestones in igneous dike aureoles in the Upper Cretaceous Pierre Shale, Walsenburg, Colorado. In *Hydrocarbon Source Rocks of the Greater Rocky Mountain Region*; Woodward, J.G., Meissner, F.F., Clayton, C.J., Eds.; Rocky Mountain Association of Geologists Symposium: Denver, CO, USA, 1984; pp. 387–392.

18. Raymond, A.C.; Murchison, D.G. Development of organic maturation in the thermal aureoles of sills and its relation to sediment compaction. *FUEL* **1988**, *67*, 1599–1608. [CrossRef]

19. Aarnes, I.; Planke, S.; Trulsvik, M.; Svensen, H. Contact metamorphism and thermogenic gas generation in the Vøring and Møre basins, offshore Norway, during the Paleocene-Eocene thermal maximum. *J. Geol. Soc.* **2015**, *172*, 588–598. [CrossRef]

20. Galushkin, Y.I. Thermal effects of igneous intrusions on maturity of organic matter: A possible mechanism of intrusion. *Org. Geochem.* **1997**, *26*, 645–658. [CrossRef]

21. Aarnes, I.; Svensen, H.; Polteau, S.; Planke, S. Contact metamorphic devolatilization of shales in the Karoo Basin, South Africa, and the effects of multiple sill intrusions. *Chem. Geol.* **2011**, *281*, 181–194. [CrossRef]

22. Wang, D. Comparable study on the effect of errors and uncertainties of heat transfer models on quantitative evaluation of thermal alteration in contact metamorphic aureoles: Thermophysical parameters, intrusion mechanism, pore-water volatilization and mathematical equations. *Int. J. Coal Geol.* **2012**, *95*, 12–19.

23. Wang, D.; Song, Y.; Xu, H.; Ma, X.; Zhao, M. Numerical modeling of thermal evolution in the contact aureole of a 0.9 m thick dolerite dike in the Jurassic siltstone section from Isle of Skye, Scotland. *J. Appl. Geophys.* **2013**, *89*, 134–140. [CrossRef]

24. Liu, E.; Wang, H.; Uysal, I.T.; Zhao, J.-X.; Wang, X.-C.; Feng, Y.; Pan, S. Paleogene igneous intrusion and its effect on thermal maturity of organic-rich mudstones in the Beibuwan Basin, South China Sea. *Mar. Pet. Geol.* **2017**, *86*, 733–750. [CrossRef]

25. Sydnes, M.; Fjeldskaar, W.; Løtveit, I.F.; Grunnaleite, I.; Cardozo, N. The importance of sill thickness and timing of sill emplacement on hydrocarbon maturation. *Mar. Pet. Geol.* **2018**, *89*, 500–514. [CrossRef]

26. Fjeldskaar, W.; Andersen, Å.; Johansen, H.; Lander, R.; Blomvik, V.; Skurve, O.; Michelsen, J.K.; Grunnaleite, I.; Mykkeltveit, J. Bridging the gap between basin modelling and structural geology. *Reg. Geol. Metallog.* **2017**, *72*, 65–77.

27. Benfield, A.E. The effect of uplift and denudation on underground temperatures. *J. Appl. Phys.* **1949**, *20*, 66–70. [CrossRef]

28. Birch, F. Flow of heat in the Front Range, Colorado. *Geol. Soc. Am. Bull.* **1950**, *61*, 567–630. [CrossRef]

29. ter Voorde, M.; Bertotti, G. Thermal effects of normal faulting during rifted basin formation, 1. A finite difference model. *Tectonophysics* **1994**, *240*, 133–144. [CrossRef]

30. Bertotti, G.; ter Voorde, M. Thermal effects of normal faulting during rifted basin formation, 2. The Lugano-Val Grande normal fault and the role of pre-existing thermal anomalies. *Tectonophysics* **1994**, *240*, 145–157. [CrossRef]

31. Johnson, C.; Harbury, N.; Hurford, A.J. The role of extension in the Miocene denudation of the Nevado-Filábride Complex, Betic Cordillera (SE spain). *Tectonics* **1997**, *16*, 189–204. [CrossRef]

32. Bertotti, G.; Seward, D.; Wijbrans, J.; ter Voorde, M.; Hurford, A.J. Crustal thermal regime prior to, during, and after rifting: A geochronological and modeling study of the Mesozoic South Alpine rifted margin. *Tectonics* **1999**, *18*, 185–200. [CrossRef]

33. Ehlers, T.A.; Chapman, D.S. Normal fault thermal regimes: Conductive and hydrothermal heat transfer surrounding the Wasatch fault, Utah. *Tectonophysics* **1999**, *312*, 217–234. [CrossRef]

34. Ehlers, T.A.; Armstrong, P.A.; Chapman, D.S. Normal fault thermal regimes and the interpretation of low-temperature thermochronometers. *Phys. Earth Planet. Inter.* **2001**, *126*, 179–194. [CrossRef]

35. Lander, R.H.; Langfeldt, M.; Bonnell, L.; Fjeldskaar, W. *BMT User's Guide*; Tectonor AS Proprietary Publication: Stavanger, Norway, 1994.

36. Fjeldskaar, W. BMT^{TM}—Exploration tool combining tectonic and temperature modeling: Business Briefing: Exploration & Production. *Oil Gas Rev.* **2003**, 1–4.

37. Sclater, J.G.; Christie, P.A.F. Continental stretching: An explanation of the post-mid-cretaceous subsidence of the central North Sea basin. *J. Geophys. Res.* **1980**, *85*, 3711–3739. [CrossRef]

38. Čermác, V.; Rybach, L. Thermal properties: Thermal conductivity and specific heat of minerals and rocks. In *Landolt-Börnstein Zahlenwerte und Functionen aus Naturwissenschaften und Technik, Neue Serie, Physikalische Eigenschaften der Gesteine*; Angeneister, G., Ed.; Springer Verlag: Berlin/Heidelberg, Germany; New York, NY, USA, 1982; pp. 305–343.

39. Tissot, B.P.; Welte, D.H. *Petroleum Formation and Occurrence*, 2nd ed.; Springer-Verlag: Berlin/Heidelberg, Germany, 1984.

40. Dula, W.F. Geometric Models of Listric normal Faults and Rollover Folds. *Am. Assoc. Pet. Geol. Bull.* **1991**, *75*, 1609–1625.

41. Fossen, H. *Structural Geology*; Cambridge University Press: Cambridge, UK, 2010.

42. Osagiede, E.E.; Duffy, O.B.; Jackson, C.A.-L.; Wrona, T. Quantifying the growth history of seismically imaged normal faults. *J. Struct. Geol.* **2014**, *66*, 382–399. [CrossRef]

43. Magee, C.; Maharaj, S.M.; Wrona, T.; Jackson, C.A.-L. Controls on the expression of igneous intrusions in seismic reflection data. *Geosphere* **2015**, *11*, 1024–1041. [CrossRef]

44. Schofield, N.; Holford, S.; Millett, J.; Brown, D.; Jolley, D.; Passey, S.R.; Muirhead, D.; Grove, C.; Magee, C.; Murray, J.; et al. Regional Magma Plumbing and emplacement mechanisms of the Faroe-Shetland sill Complex: Implications for magma transport and petroleum systems within sedimentary basins. *Basin Res.* **2017**, *29*, 41–63. [CrossRef]

45. Eide, C.H.; Schofield, N.; Lecomte, I.; Buckley, S.J.; Howell, J.A. Seismic interpretation of sill complexes in sedimentary basins: Implications for the sub-sill imaging problem. *J. Geol. Soc.* **2017**, *175*, 193–209. [CrossRef]

46. Francis, E.A. Magma and sediment-I: Emplacement mechanism of late Carboniferous tholeiite sills in northern Britain. *J. Geol. Soc.* **1982**, *139*, 1–20. [CrossRef]

47. Chevallier, L.; Woodford, A. Morpho-tectonics and mechanism of emplacement of the dolerite rings and sills of the western Karoo, South Africa. *S. Afr. J. Geol.* **1999**, *102*, 43–54.

48. Galerne, C.Y.; Neumann, E.-R.; Planke, S. Emplacement mechanisms of sill complexes: Information from the geochemical architecture of the Golden Valley Sill Complex, South Africa. *J. Volcanol. Geotherm. Res.* **2008**, *177*, 425–440. [CrossRef]

49. Galerne, C.Y.; Galland, O.; Neumann, E.-R.; Planke, S. 3D relationships between sills and their feeders: Evidence from the Golden Valley Sill Complex (Karoo Basin) and experimental modelling. *J. Volcanol. Geotherm. Res.* **2011**, *202*, 189–199. [CrossRef]

50. Hansen, J.; Jerram, D.A.; McCaffrey, K.; Passey, S.R. Early Cenozoic saucer-shaped sills of the Faroe Islands: An example of intrusive styles in basaltic lava piles. *J. Geol. Soc.* **2011**, *168*, 159–178. [CrossRef]

51. Richardson, J.A.; Connor, C.B.; Wetmore, P.H.; Connor, L.J.; Gallant, E.A. Role of sills in the development of volcanic fields: Insights from lidar mapping surveys of the San Rafael Swell, Utah. *Geology* **2015**, *43*, 1023–1026. [CrossRef]

52. Eide, C.H.; Schofield, N.; Jerram, D.A.; Howell, J.A. Basin-scale architecture of deeply emplaced sill complexes: Jameson Land, East Greenland. *J. Geol. Soc.* **2016**, *174*, 23–40. [CrossRef]

53. Walker, R.J.; Healy, D.; Kawanzaruwa, T.M.; Wright, K.A.; England, R.W.; McCaffrey, K.J.W.; Bubeck, A.A.; Stephens, T.L.; Farrell, N.J.C.; Blenkinsop, T.G. Igneous sills as a record of horizontal shortening: The San Rafael sub-volcanic field, Utah. *Geol. Soc. Am. Bull.* **2017**, *129*, 1052–1070. [CrossRef]

54. Svensen, H.H.; Frolov, S.; Akhmanov, G.G.; Polozov, A.G.; Jerram, D.A.; Shiganova, O.V.; Melnikov, N.V.; Iyer, K.; Planke, S. Sills and gas generation in the Siberian Traps. *Philos. Trans. R. Soc. A Math. Phys. Eng. Sci.* **2018**, *376*, 1–18. [CrossRef]

55. Svensen, H.H.; Polteu, S.; Cawthron, G.; Planke, S. Sub-volcanic Intrusions in the Karoo Basin, South Africa. In *Physical Geology of Shallow Magmatic Systems: Dykes, Sills and Laccoliths (Advances in Volcanology)*; Breitkreuz, C., Rocchi, S., Eds.; Springer International Publishing AG: Cham, Switzerland, 2018; pp. 349–362.

56. Lange, R.A.; Cashman, K.V.; Natrosky, A. Direct measurements of latent heat during crystallization and melting of a ugandite and an olivine basalt. *Contrib. Mineral. Petrol.* **1994**, *118*, 169–181. [CrossRef]

57. Atkins, P.; de Paula, J.; Keller, J. *Atkins' Physical Chemistry*; Oxford University Press: Oxford, UK, 2017.

58. Svensen, H.; Planke, S.; Jamtveit, B.; Pedersen, T. Seep carbonate formation controlled by hydrothermal vent complexes: A case study from the Vøring Basin, the Norwegian Sea. *Geo-Mar. Lett.* **2003**, *23*, 351–358. [CrossRef]

59. Svensen, H.; Planke, S.; Polozov, A.G.; Schmidbauer, N.; Corfu, F.; Podladchikov, Y.Y.; Jamtveit, B. Siberian gas venting and the end-Permian environmental crisis. *Earth Planet. Sci. Lett.* **2009**, *277*, 490–500. [CrossRef]

60. Iyer, K.; Rüpke, L.; Galerne, C.Y. Modeling fluid flow in sedimentary basins with sill intrusions: Implications for hydrothermal venting and climate change. *Geochem. Geophys. Geosyst.* **2013**, *14*, 5244–5262. [CrossRef]

61. Iyer, K.; Schmid, D.W.; Planke, S.; Millett, J. Modelling hydrothermal venting in volcanic sedimentary basins: Impact on hydrocarbon maturation and paleoclimate. *Earth Planet. Sci. Lett.* **2017**, *467*, 30–42. [CrossRef]

62. Peace, A.; McCaffrey, K.; Imber, J.; Hobbs, R.; van Hunen, J.; Gerdes, K. Quantifying the influence of sill intrusion on the thermal evolution of organic-rich sedimentary rocks in nonvolcanic passive margins: An example from ODP 210–1276, offshore Newfoundland, Canada. *Basin Res.* **2017**, *29*, 249–265. [CrossRef]

63. Allen, P.A.; Allen, J.R. *Basin Analysis: Principles and Application to Petroleum Play Assessment*, 3rd ed.; Wiley-Blackwell: Chichester, West Sussex, UK, 2014.

64. Hendrie, D.B.; Kusznir, N.J.; Hunter, R.H. Jurassic extension estimates for the North Sea 'triple junction' from flexural backstripping: Implications for decompression melting models. *Earth Planet. Sci. Lett.* **1993**, *116*, 113–127. [CrossRef]

65. Trommsdorf, V.; Piccardo, G.B.; Montrasio, A. From magmatism through metamorphism to sea floor emplacement of subcontinental Adria lithosphere during pre-Alpine rifting (Malenco, Italy). *Schweiz. Mineral. Und Petrogr. Mitt.* **1993**, *73*, 191–203.

66. Somerton, W.H. Thermal properties and temperature-related behavior of rock/fluid systems. In *Developments in Petroleum Sciences*; Elsevier: Amsterdam, The Netherlands, 1992.

67. Planke, S.; Rasmussen, T.; Rey, S.; Myklebust, R. Seismic characteristics and distribution of volcanic intrusions and hydrothermal vent complexes in the Vøring and Møre basins. In *Petroleum Geology: North-West Europe and Global Perspectives—Proceedings of the 6th Petroleum Geology Conference*; Doré, A.G., Vining, B.A., Eds.; Geological Society: London, UK, 2005; pp. 833–844.

68. Hanson, R.B.; Barton, M.D. Thermal development of Low-Pressure Metamorphic Belts: Results from two-dimensional numerical models. *J. Geophys. Res.* **1989**, *94*, 10363–10377. [CrossRef]

69. Kennett, J.P.; Watkins, N.D.; Vella, P. Paleomagnetic Chronology of Pliocene-Early Pleistocene Climates and the Plio-Pleistocene Boundary in New Zealand. *Science* **1971**, *171*, 276–279. [CrossRef] [PubMed]

70. Kennet, J.P.; Watkins, N.D. Late Miocene-Early Pliocene Paleomagnetic Stratigraphy, Paleoclimatology, and Biostratigraphy in New Zealand. *Geol. Soc. Am. Bull.* **1974**, *85*, 1385–1398. [CrossRef]

71. Karlin, R.; Levi, S. Geochemical and sedimentological control of the magnetic properties of hemipelagic sediments. *J. Geophys. Res.* **1985**, *90*, 10373–10392. [CrossRef]

72. Turner, G.M.; Roberts, A.P.; Laj, C.; Kissel, C.; Mazaud, A.; Guitton, S.; Christoffel, D.A. New paleomagnetic results from Blind River: Revised magnetostratigraphy and tectonic rotation of the Marlborough region, South Island, New Zealand. *N. Z. J. Geol. Geophys.* **1989**, *32*, 191–196. [CrossRef]

73. Tric, E.; Laj, C.; Jéhanno, C.; Valet, J.-P.; Kissel, C.; Mazaud, A.; Iaccarino, S. High-resolution record of the Upper Olduvai transition from Po Valley (Italy) sediments: Support for dipolar transition geometry? *Phys. Earth Planet. Inter.* **1991**, *65*, 319–336. [CrossRef]

74. Szmytkiewicz, A.; Zalewska, T. Sediment deposition and accumulation rates determined by sediment trap and ^{210}Pb isotope methods in the Outer Puck Bay (Baltic Sea). *Oceanologia* **2014**, *56*, 85–106. [CrossRef]

75. Peketi, A.; Mazumdar, A.; Joao, H.M.; Patil, D.J.; Usapkar, A.; Dewangan, P. Coupled C-S-Fe geochemistry in a rapidly accumulating marine sedimentary system: Diagenetic and depositional implications. *Geochem. Geophys. Geosyst.* **2015**, *16*, 2865–2883. [CrossRef]

76. Graseman, B.; Mancktelow, N.S. Two-dimensional thermal modelling of normal faulting: The Simplon Fault zone, Central Alps, Switzerland. *Tectonophysics* **1993**, *225*, 155–165. [CrossRef]

77. Gluyas, J.; Swarbrick, R. *Petroleum Geosciences*; Blackwell Publishing: Oxford, UK, 2015.

78. Fjeldskaar, W.; Christie, O.H.J.; Midttømme, K.; Virnovsky, G.; Jensen, N.B.; Lohne, A.; Eide, G.I.; Balling, N. On the determination of thermal conductivity of sedimentary rocks and the significance for basin temperature history. *Pet. Geosci.* **2009**, *15*, 367–380. [CrossRef]

79. Jackson, C.A.-L.; Schofield, N.; Golenkov, B. Geometry and controls on the development of igneous sill-related forced folds: A 2-D seismic reflection case study from offshore southern Australia. *Geol. Soc. Am. Bull.* **2013**, *125*, 1874–1890. [CrossRef]

80. Galland, O.; Bertelsen, H.S.; Eide, C.H.; Guldstrand, F.; Haug, Ø.T.; Leanza, H.A.; Mair, K.; Palma, O.; Planke, S.; Rabbel, O.; et al. Storage and transport of magma in the layered crust—Formation of sills and related flat-lying intrusions. In *Volcanic and Igneous Plumbing Systems, Understanding Magma Transport, Storage and Evolution in the Earth's Crust*; Burchardt, S., Ed.; Elsevier: Amsterdam, The Netherlands, 2018; pp. 113–138.

81. Jain, S. *Fundamentals of Physical Geology*; Springer Geology: New Dehli, India, 2014.

82. Podladchikov, Y.Y.; Wickham, S.M. Crystallization of Hydrous Magmas: Calculation of Associated Thermal Effects, Volatile Fluxes, and Isotopic Alteration. *J. Geol.* **1994**, *102*, 25–45. [CrossRef]

83. Wang, D.; Manga, M. Organic matter maturation in the contact aureole of an igneous sill as a tracer of hydrothermal convection. *J. Geophys. Res. Solid Earth* **2015**, *120*, 4102–4112. [CrossRef]

84. Annen, C. Factors affecting the thickness of thermal aureoles. *Front. Earth Sci.* **2017**, *5*, 1–13. [CrossRef]

85. Hanson, R.B. The hydrodynamics of contact metamorphism. *Geol. Soc. Am. Bull.* **1995**, *107*, 595–611. [CrossRef]

86. Gudmundsson, A.; Løtveit, I.F. *Sills as Fractured Hydrocarbon Reservoir: Examples and Models*; Geological Society London Special Publications: London, UK, 2012; Volume 374, pp. 252–271.

87. Montanari, D.; Bonini, M.; Cori, G.; Agostini, A.; Ventisette, C.D. Forced folding above shallow magma intrusions: Insights on supercritical fluid flow from analogue modelling. *J. Volcanol. Geotherm. Res.* **2017**, *345*, 67–80. [CrossRef]

88. Rybach, L. Amount and significance of radioactive heat sources in sediments. In *Thermal Modelling in Sedimentary Basins*; Burrus, J., Ed.; Technip: Paris, France, 1986; pp. 311–323.

89. Keen, C.E.; Lewis, T. Measured radiogenic heat production in sediments from continental margin of eastern North America: Implications for petroleum generation. *AAPG Bull.* **1982**, *66*, 1402–1407.

90. McKenna, T.E.; Sharp, J.M. Radiogenic heat production in sedimentary rocks of the Gulf of Mexico Basin, South Texas. *AAPG Bull.* **1998**, *82*, 484–496.

91. Thomas, G.W. *Principles of Hydrocarbon Reservoir Simulation*, 2nd ed.; International Human Resources Development Corporation: Boston, MA, USA, 1982.

92. Hageman, L.A.; Young, D.M. *Applied Iterative Methods*; New York Academic Press: New York, NY, USA, 1981.

93. Young, D.M.; Jea, K.C. *Generalized Conjugate Acceleration of Iterative Methods, Part 2: The Nonsymmetrizable Case*; Report CNA-163; Center for Numerical Analyses, University of Texas at Austin: Austin, TX, USA, 1981.

94. Appleyard, J.R.; Chesire, I.M. *Nested factorization: Proceedings of the Seventh Symposium on Reservoir Simulation*; SPE Paper 12264; Society of Petroleum Engineers of AIME: San Francisco, CA, USA, 1983.

Article

The Influence of Magmatic Intrusions on Diagenetic Processes and Stress Accumulation

Magnhild Sydnes [1,2,*], **Willy Fjeldskaar** [1], **Ivar Grunnaleite** [1], **Ingrid Fjeldskaar Løtveit** [1] and **Rolf Mjelde** [2]

[1]   Tectonor AS, P.O. Box 8034, NO-4068 Stavanger, Norway; wf@tectonor.com (W.F.); ig@tectonor.com (I.G.); ifl@tectonor.com (I.F.L.)
[2]   Department of Earth Science, University of Bergen, Box 7803, 5020 Bergen, Norway; Rolf.Mjelde@uib.no
*   Correspondence: ms@tectonor.com

Received: 27 September 2019; Accepted: 11 November 2019; Published: 13 November 2019

Abstract: Diagenetic changes in sedimentary basins may alter hydrocarbon reservoir quality with respect to porosity and permeability. Basins with magmatic intrusions have specific thermal histories that at time of emplacement and in the aftermath have the ability to enhance diagenetic processes. Through diagenesis the thermal conductivity of rocks may change significantly, and the transformations are able to create hydrocarbon traps. The present numerical study quantified the effect of magmatic intrusions on the transitions of opal A to opal CT to quartz, smectite to illite and quartz diagenesis. We also studied how these chemical alterations and the sills themselves have affected the way the subsurface responds to stresses. The modeling shows that the area in the vicinity of magmatic sills has enhanced porosity loss caused by diagenesis compared to remote areas not intruded. Particularly areas located between clusters of sills are prone to increased diagenetic changes. Furthermore, areas influenced by diagenesis have, due to altered physical properties, increased stress accumulations, which might lead to opening of fractures and activation/reactivation of faults, thus influencing the permeability and possible hydrocarbon migration in the subsurface. This study emphasizes the influence magmatic intrusions may have on the reservoir quality and illustrates how magmatic intrusions and diagenetic changes and their thermal and stress consequences can be included in basin models.

Keywords: magmatic intrusions; diagenesis; stress; porosity; permeability; basin modeling; stress modeling

---

1. Introduction

Magmatic intrusions are commonly emplaced with much higher temperatures compared to their host rocks. Therefore, when emplaced into sedimentary basins, they may influence all parts of the petroleum system [1]. Several studies have shown how such intrusions influence the temperatures and hydrocarbon maturation in sedimentary basins (e.g., [2–11]). Other studies focus on the stress induced by the sills on the host rocks as they intrude (e.g., [12,13]). However, only a few studies have reported on the effect magmatic sill intrusions have on the diagenetic processes in sedimentary basins. Haile et al. [14] concluded in their work at Edgeøya (Svalbard) that conductive heat from intrusions did not seem to have affected the diagenetic products in the area. However, at Wilhelmøya (Svalbard) there is evidence suggesting that hydrothermal fluid flow originating from sills has affected the chemical transformations [15]. In a study of sandstones at Traill Ø (East Greenland) it was concluded that a combination of conductive and convective heat from magmatism enhanced the diagenetic process [16].

In essence, porosity and permeability determine the quality of petroleum reservoirs [17,18]. At the time of sediment deposition, the process of sediment lithification starts [19], driven by mechanical and chemical diagenesis. Mechanical alterations are related to compaction of the sediments by burial, while chemical alteration is the compaction occurring when chemical compounds are dissolved and re-deposited or new components are precipitated. While the mechanical changes are strictly a result of increasing burial depth and vertical loading, chemical changes are less predictable and highly dependent on the chemical compounds and the temperature (e.g., [20]). Diagenesis results in porosity loss, increased rock densities and seismic velocities (e.g., [21–25]). Due to the sensitivity of chemical diagenesis to temperature, basins subjected to magmatic intrusions are particularly prone to abrupt and sudden changes of physical properties. Laboratory measurements show an increase of physical rock strength (>100%) on the transition of opal A to opal CT [23], which imply that diagenetic processes may also affect the way the rocks respond to subsurface stresses.

Diagenetic transformations are reported from sedimentary basins worldwide (e.g., [26–34]). The Monterey Formation in California, USA, has been extensively studied with regard to the transformation of opal A to opal CT (e.g., [35–39]), but diagenetic alterations have also been observed several places in wells and on seismic data from the Norwegian Continental Shelf (e.g., [24,40–47]). All of the latter studies report observed transitions of either opal A to opal CT or smectite to illite in the Vøring Basin, offshore mid-Norway. We used a 2D section from the Vøring Basin with numerous sills as the basis for the modeling, as it represents the structures of a magmatic basin better as opposed to a synthetic profile. Modeling of a real 2D section will thus give more realistic results, even though the parameters are of global nature. A detailed case study of the effects of magmatic intrusions on diagenesis in the Vøring Basin is beyond the scope of this work.

The main goal of the study is to quantify the effect of magmatic intrusions on transitions of opal A to opal CT to quartz, smectite to illite, and quartz diagenesis which was done with basin modeling software (BMT^{TM}, [48]). A second goal is to assess the influence of diagenetic processes and of the sills themselves on the stress field in a sedimentary basin and the potential impact on fracture and fault permeability, which is of significant importance for the petroleum systems. The results show that magmatic sills and related thermal effects might have notable implications for the porosity loss due to diagenesis in their vicinity. Diagenetic alterations and the sills themselves influence the location and magnitude of stress accumulations in the basin and thereby have implications for the fault and fracture development, and implicitly for the migration of fluids.

2. An Example of the Evolution of a Volcanic Basin—The Vøring Basin

The Vøring Basin is located offshore mid-Norway and is bounded by the Bivrost Lineament to the NE and the Vøring Transform Margin to the SW (Figure 1). The area consists of grabens, basins, and structural highs developed over three main rifting episodes from Carboniferous to Eocene times (e.g., [49] and references therein); (1) Carboniferous-Permian; coincided with the onset of rifting in the North Atlantic [40,50–55]; (2) Late-Mid Jurassic to Early Cretaceous; led to subsidence and development of accommodation space for the thick Cretaceous sedimentary sequence [56,57]; (3) Late Cretaceous to Early Eocene; coincided with the opening of the North Atlantic, development of the Vøring Marginal High, and intrusion of numerous sills in the Cretaceous basin fill [49,58]. Subsequent events were dominated by seafloor spreading and accretion of oceanic crust in the expanding Norwegian-Greenland/North Atlantic Sea [59,60]. In the post-rift phase, the Vøring area experienced localized tectonic uplift, erosion, sediment deposition, subsidence, flexure, and isostatic uplift partly due to numerous glaciations and deglaciations of the Fennoscandian landmasses [61,62].

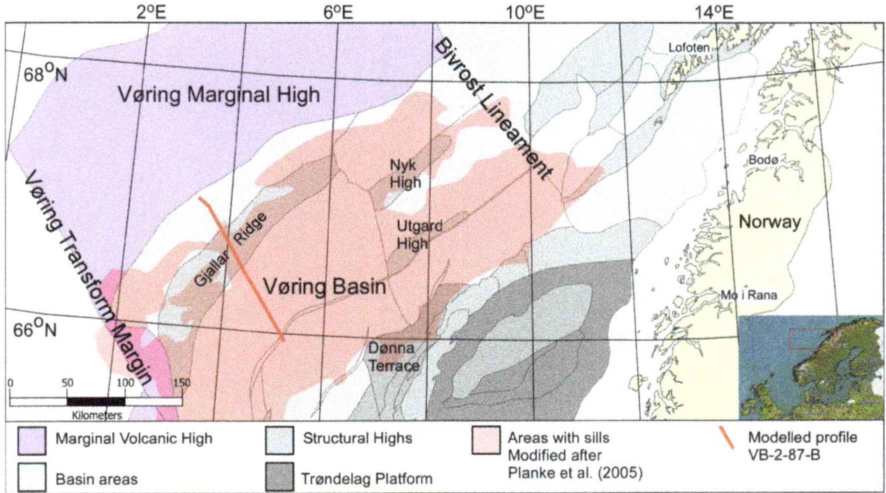

Figure 1. Location of the studied profile (VB-2-87-B) in the Vøring Basin offshore mid Norway. Modified after Sydnes et al. [10].

The sedimentary deposits of the Vøring Basin mainly consist of marine to deep marine sediments (mainly shale) with intercalated shallow marine sediments (mainly sandstone) in the Upper Jurassic and Cretaceous [49]. The Paleocene consists of shale with minor sandstone and limestone deposits [63], and the Lower Eocene consists of shales. The magmatic intrusions were emplaced in lower Eocene time within the entire sedimentary sequence. Oligocene, Neogene, and Quaternary deposits are mainly shales with some ice rafted debris in the Plio-Pleistocene. The outer Vøring Basin was in the Miocene and early Pliocene dominated by deep-water deposits consisting of biosiliceous hemipelagic sediments [46,64].

The utilized 2D transect, VB-2-87-B (Figure 1), is about 100 km long and holds some 20 interpreted sills located at ~3–5 km depth with lateral extension of the sills varying from about 3 to 30 km (Figure 2). A velocity model by Hjelstuen et al. [64] was used for depth conversion of the section. The time of intrusion activity is estimated at ~56 Ma, which is supported by ages derived from zircon dating of magmatic sills in the Vøring area (55.6 ± 0.3 Ma and 56.3 ± 0.4 Ma, 6607/5-2 and Utgard wells; [65]).

From Carboniferous and up to the beginning of Eocene, the Vøring Basin was part of an active rift phase [49]. As the active rifting ceased the Vøring area subsequently experienced thermal subsidence and ridge push due to continental break up and opening of the North Atlantic. Repeated glaciations, mainly since 2.6 Ma, have led to sediment depositions in the Vøring Basin. Currently the regional stress field of coastal and mainland Norway is somewhat affected by post-glacial uplift and erosional unloading/loading [62,66–70] in addition to compressive ridge-push. To make our results generally applicable, the section from the Vøring Basin is solely used as a basis for the modeling, and site-specific parameters such as lithology and rock properties, are substituted by global values.

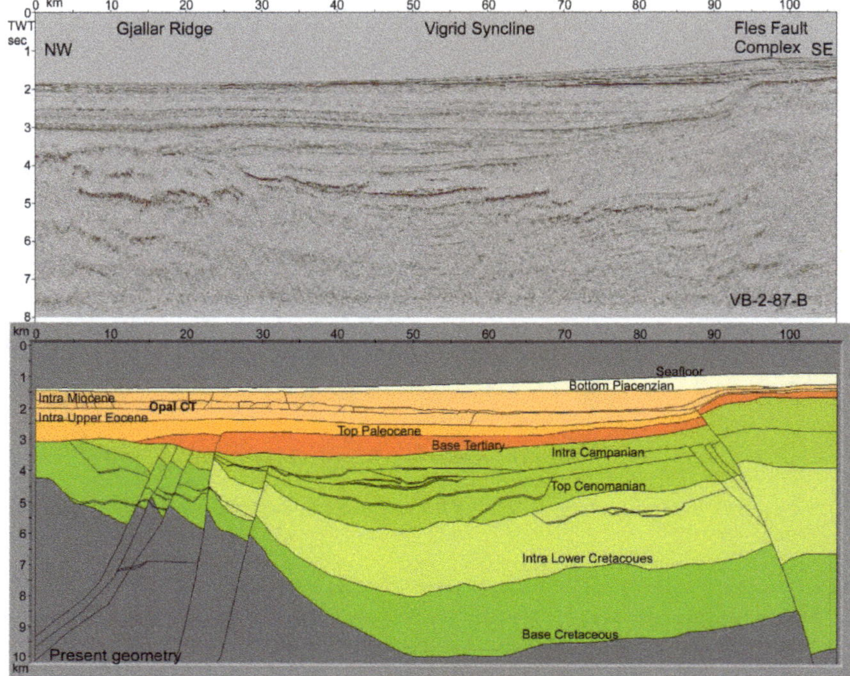

Figure 2. Seismic section and the interpreted, depth converted section of VB-2-87-B in the Vøring Basin offshore mid-Norway. Interpretation is obtained from Blystad et al., [40].

3. Methods and Results for Magmatic Intrusion's Influence on Diagenetic Processes

To perform the modeling of structural evolution, geohistory, thermal development, maturation, and diagenesis, BMT^{TM} (Tectonor AS) was used. All models are in 2D, which infer the third dimension to be infinite. This is a good approximation when sills and faults are modeled, as the third dimension commonly is much larger than the two others (e.g., [12]).

3.1. Geohistory, Temperature, Maturation, and Diagenesis

BMT is a high-resolution 2D basin modeling software utilizing a backstripping process, starting with present day geometry, in order to reconstruct the geohistory of an interpreted seismic section (Figure 2) [48,71,72]. One horizon at a time is stripped from the section and fault restoration and decompaction (based on the given porosity-depth trends assigned for the lithologies) of underlying sequences follow. This process is repeated all the way to the top basement, thus building the geohistory. The fault restoration in BMT is performed by vertical simple shear, a method that leaves no gaps in the geometry and allows for thermal and maturation calculations to be performed (cf. [48]).

BMT bases the thermal calculations on the finite difference method by conduction. The resulting temperature history depends on the thermal conductivity and specific heat capacity of the sediments as well as the basal heat flow from the mantle [48,71]. The surface temperature is the upper boundary condition in the calculations and is kept at 7 °C, whereas the lower boundary condition is the basal heat flow which is set to 47 mW·m^{-2} over the profile and over time. With these settings, the BMT calculations generate the basis for the maturation and diagenetic modeling. Kerogen type II is assumed for the maturation modeling, and maturation is calculated using classical first-order kinetics [48].

3.1.1. Silica Diagenesis—Quartz Diagenesis

Quartz diagenesis occurs over three connected steps involving dissolution of quartz grains, transport of dissolved silica over short distances, and precipitation of quartz cement on clastic quartz grains [19,73]. BMT performs quantitative estimates of quartz cementation caused by dissolution of silica as described in Walderhaug [74,75] and Lander and Walderhaug [76]. The method is based on an empirically derived quantification of quartz precipitation as a function of temperature, and the precipitation is considered rate limiting for the whole cementation process (in [75] and references therein). This is a different approach than several other quantitative models where quartz dissolution is considered the limiting factor (e.g., [77–81]). As the diagenesis is thermally dependent (e.g., [74,75]), quartz cementation is an indicator of the maximum reached temperature and the burial depth.

Diagenesis is a complex process consisting of a number of transformations occurring simultaneously. According to Walderhaug [74,75], the rate limiting step is the precipitation, which is described by first-order kinetics and is proportional to the rate constant (k) given by the Arrhenius equation (Equation (1)).

$$k = A_0 e^{\left(\frac{-E_a}{RT}\right)}, \tag{1}$$

where k is the rate constant of mineral precipitation per unit surface area (mol/cm^{2}s), A_0 is a pre-exponential constant/frequency factor (mol/cm^{2}s), E_a is the activation energy (J/mol), R is the real gas constant (8.314 J mol^{-1} K^{-1}), and T is the temperature. A_0 is the number that describes how often molecules collide within the system under consideration. E_a is the energy needed for a specific chemical reaction to occur and both parameters are unique for any reaction [17]. The kinetic equation is integrated over temperature and time for a given timestep. By multiplying this with the available grain surface area obtained from the porosity history of the reconstructed geohistory, the mineral cement precipitated is calculated. However, increasing grain coating on quartz minerals reduces the possibility for precipitation of diagenetic minerals. The porosity loss calculations in this study assume zero grain coating, the porosity loss due to cementation might therefore be overestimated.

3.1.2. Silica Diagenesis—Opal A to Quartz Via Opal CT

Diatoms and radiolarian are two examples of siliceous plankton that after death enrich deep water sediments with dissolved non-crystalline silica known as opal A [82]. Such fragments are associated with low-energy deposition environments and amorphous silica is commonly found in deep water sediments amongst clay and silt deposits. Through burial and increased temperatures opal A converts to quartz via opal CT; this is observed in several basins worldwide, as documented in the introduction. The conversion of biogenetic/amorphous silica, opal A, to diagenetic/microcrystalline silica, opal CT, and subsequently to stable/crystalline silica, quartz, occurs through several dissolutions and precipitations as the burial depth increases (e.g., [39,83,84]). This reorganization from non-crystalline to crystalline silica causes changes in rock structures, mechanical properties, porosity, and permeability and may potentially generate traps for hydrocarbons (e.g., [23,38,39,84,85]).

One approach to determine the depth of these diagenetic processes is an empirical derived schematic presentation of the relation between temperature, rock composition, and silica-phase changes [86,87], which can be used in areas of known temperatures. Another approach is to determine experimentally the reaction kinetics under given chemical conditions which can be used for basins worldwide (e.g., [22,38,88]. A third approach was discovered by Neagu et al. [25] who found a connection between rotated fault planes and amount of porosity loss due to the conversion of opal A to opal CT.

Generally, each chemical reaction has its specific pre-exponential constant and activation energy which can be derived experimentally [17]. Additionally, all chemical reaction rates are temperature dependent [89]. Therefore, the method described above for quartz diagenesis was used to predict the precipitation rate for the transition of opal A to opal CT to quartz. Laboratory studies show that the activation energy for the opal A to opal CT transition is lower than for the transition of opal CT to quartz and this indicates that reorganization of amorphous silica to microcrystalline silica require lower temperature than the conversion of microcrystalline silica to crystalline silica [22]. This implies that the opal A/opal CT transition is expected to be located at shallower depths than the opal CT/quartz boundary.

In this study the values reported by Dralus et al. [39] for the transition of opal A→ opal CT→ quartz was used (Table 1), as their study results in improved predictions of the opal A to opal CT boundary compared to other published kinetics [38]. They also found that high content of organic matter slows the transitioning of opal A to opal CT and that different kinetic parameters are required for rocks with high and low organic content.

3.1.3. Clay Diagenesis—Smectite to Illite

During burial diagenesis, illitization of smectite is a commonly occurring mineralogical reaction of clay rich sediments and shales. The conversion is a gradual process leading to mixed layers with different illite/smectite ratios [90] and is accompanied by volume loss and reduction in permeability [91]. With increasing burial depth and temperatures, the percentage of illitic beds in the illite/smectite mixed layers increases (e.g., [24,92]). The temperature is an important factor in the smectite to illite diagenetic process (e.g., [91]), but the presence of pore water and potassium (K^+) are also essential for this transition to take place [42,93–95]. Smectite reacts with potassium, possibly sourced by K-feldspar, and the outcome is illite, silica, and water (e.g., [42]) from a complex dissolution/precipitation process where the precipitation is rate limiting (e.g., [96–99]). A common approach for modeling the conversion of smectite to illite is by using the Arrhenius equation and regarding the transformation as either a one step reaction with one reaction rate (e.g., [94,100]) or as several parallel reactions with rates spread over a defined range (e.g., [101]).

Table 1. Kinetic data used for the opal A/CT/quartz diagenesis and smectite to illite modeling. *Values obtained from Sachsenhofer et al. [103].

Diagenetic Process		E_a (kJ mol^{-1})	A_0 (mol/cm^{2}s)
Opal A—opal CT—quartz	Dralus et al. [39]	36.99 (low TOC) 33.44 (high TOC) 32.52 (Quartz)	1.04×10^{14} 3.01×10^{10} 1.96×10^9
Smectite to illite	Roaldset et al. [95]	33	1.02×10^8 2.72×10^9
	Huang et al. [94]	28	8.08×10^4
	Hillier et al. [102]*	31	1.81×10^3

The method described above for quartz diagenesis is also applied for the estimation of smectite to illite diagenesis. However, it is important to note that this method does not take into account the presence of chemical components that can act as catalysts or quenchers. As for quartz diagenesis, precipitation is considered rate limiting for the smectite to illite transition and it is assumed that critical chemical components, e.g., potassium, are present for the reaction to occur. Kinetic data from three different studies were tested: (1) Roaldset et al. [95], (2) Huang et al. [94], and (3) Hillier et al. [102]. Their preferred parameters are given in Table 1.

3.2. Opal A to Opal CT—Testing of the Method

To test the quantification methods described above for the diagenetic processes, 1D modeling was performed on a well in the North Sea. Roaldset and Wei [23] observed the opal A to opal CT transition to be at 1740 m depth in well 30/9-B-50H (Figure 3) in the Oseberg area (North Sea). This area has experienced several millions of years of Cenozoic tectonic quiescence [104–107] and is presently close to maximum burial depth [95], which makes this particular well excellent for testing of the method. The stratigraphy was obtained from a well close by (well 30/6-2; Figure 3) consisting mainly of sandstones and claystones (see Table 2 for parameters).

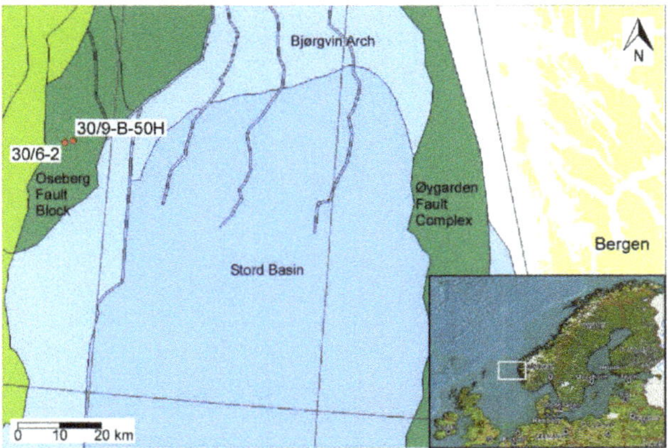

Figure 3. Location of well 30/6-2 and 30/9-B-50H on the Oseberg fault block in the North Sea. The wells are located approximately 2 km apart. Modified from map on www.npd.no.

Table 2. Lithological parameters used in the modeling, based on standard values published in the literature (e.g., [108,109]).

Lithology	Porosity—Depth Trend		Conductivity (kv) $(Wm^{-1} K^{-1})$		Heat Capacity
	Surface Porosity	Exponential Constant (km^{-1})	Low Porosity	High Porosity	$(Jkg^{-1}K^{-1})$
Sandstone	0.45	0.27	3.30 (6%)	1.50 (45%)	1080
Claystone	0.63	0.51	1.70 (6%)	1.00 (60%)	940
Shale	0.63	0.51	2.00 (6%)	1.40 (60%)	1190
Basement, metamorphic			3.10	3.10	1100
Magmatic intrusions			3.10	3.10	1100
Asthenosphere			3.50	3.50	1100

Roaldset and Wei [23] showed that the opal A to opal CT transition zone coincided with a porosity reduction of around 20%. Results from the 1D model of well 30/9-B-50H shows that 20% porosity loss starts around 1740 m below seafloor (Figure 4). This coincides with the observations by Roaldset and Wei [23], which gives a strong indication that the method gives realistic results.

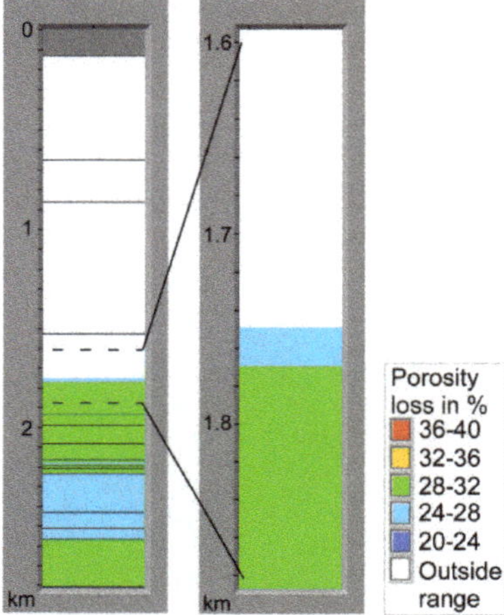

Figure 4. The modeled porosity loss in well 30/6-2 in the North Sea. The result coincides well with observations by Roaldset and Wei [23] which shows the porosity loss starting approximately around 1740 m depth.

3.3. Example from the Vøring Basin, Section VB-2-87-B

To study the effect emplaced sills have on the diagenesis in basins, the methods described in Section 3.1 were used on the seismic line VB-2-87-B (Figure 2) in the Vøring Basin (Figure 1). This transect contains approximately 20 interpreted magmatic sills, as well as a horizon denoted opal CT (Figure 2). The sills were modeled to intrude in one pulse at 56 Ma, with a magma temperature of 1000 °C which is within normal magma temperature range [12]. Table 2 summarizes the parameters used in the modeling. The thermal effect of sills and the effect on the diagenetic processes are largest the first million years after intrusion [5,10]. The figures below show results for 1 Myr after the intrusion of the sills.

3.3.1. Quartz Cement

The resulting diagenetic effects (Figure 5) show large differences between models where sills are accounted for and when they are disregarded. It is well known that sill thickness plays a role in the size of the thermal aureole, as well as the amount of matured organic matter in the vicinity of sills (e.g., [5,10]). This is also clearly illustrated in Figure 5 where results for alternatives with no sills, 50 m and 100 m thick sills are shown. The entire basin is assumed to have sandstone lithology (see Table 2 for properties). Some areas show up to 40% difference in porosity loss between the models (Figure 5). The areas where the largest differences are found are located close to the sills and in the areas between clusters of sills. With increasing sill thickness a larger area between the sill clusters have increased porosity loss (Figure 5). This is in accordance with Sydnes et al. [10] who found that the vitrinite reflectance in a well in the Barents Sea is a function of the spacing and the possible heat exchange between neighboring sills. For single, relatively shallow lying sills, the sill thickness does not impact the size of the area of porosity loss to a significant degree. However, for deeper lying single sills, the sill thickness will impact the size of the diagenetically altered area (Figure 5).

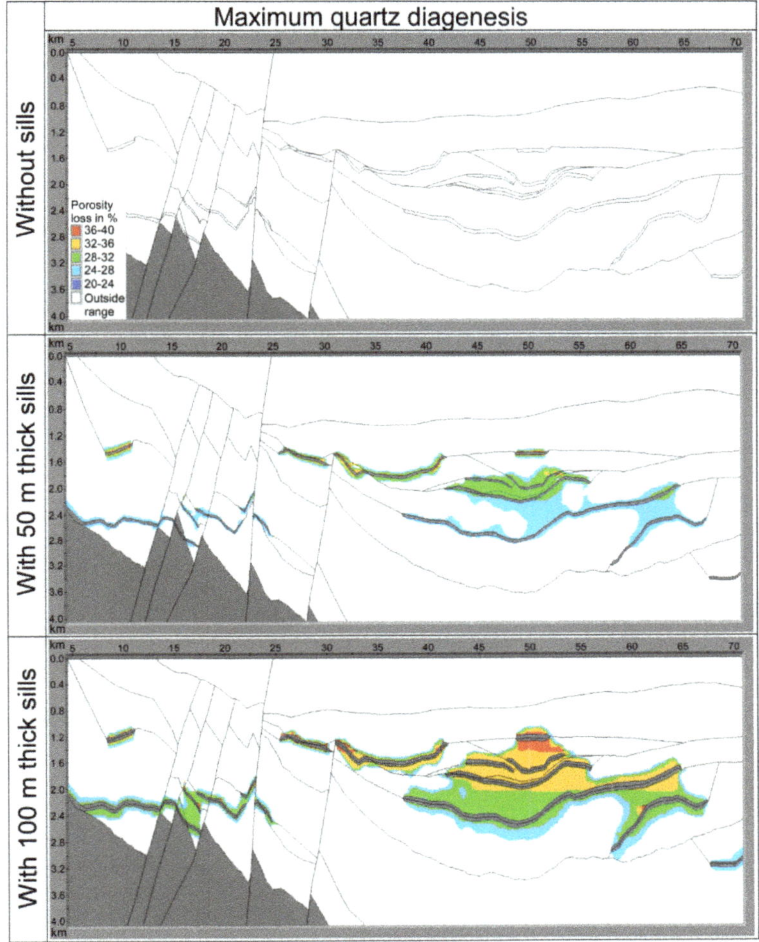

Figure 5. Results for porosity loss due to quartz diagenesis for basin holding no sills (upper), 50 m thick sills (middle), and 100 m thick sills (lower).

Quartz diagenesis due to sill emplacement will have implications for the reservoir quality in the vicinity of sills, in particular for reservoirs lying between clusters of sills. For our basin with this particular geohistory and thermal development, sills intruding at any depth in the sedimentary rocks will enhance the porosity loss due to quartz diagenesis. The generated diagenetic changes as a result of intruding sills (as shown in Figure 5) should still be visible today, as the surrounding host rocks at the same depth have not yet reached the same amount of quartz diagenesis.

3.3.2. Opal A to Opal CT to Quartz

The transition from opal A to opal CT is considered to occur over some tens of meters and up to 200 m [41]. Roaldset and Wei [23] found the transition zone of opal A to opal CT to start where the porosity reduction is around 20%. Therefore, we assume that the transition zone starts where the modeling result show 20% porosity loss and ends around 200 m deeper down. The basin is now set as all shale and assumes presence of amorphous silica. Figure 6 shows the modeled opal CT transition zone for the studied transect in the Vøring Basin with and without sills included in the calculations. The largest contrast between the two models are found in the shallower areas and for sills situated at depths <1500 m below seafloor (Figure 6). For the scenario where sills are not included, the transition zone results in a more or less horizontal zone around 1200 m depth. When the thermal contribution of sills is included in the calculations, the opal A/CT boundary is found as shallow as approximately 700 m depth. This implies that the additional heat from the sills moved the transition zone up to 500 m shallower depths. The influence of sills on the further transition of opal CT to quartz is more or less the same as for opal A to opal CT and therefore not shown here.

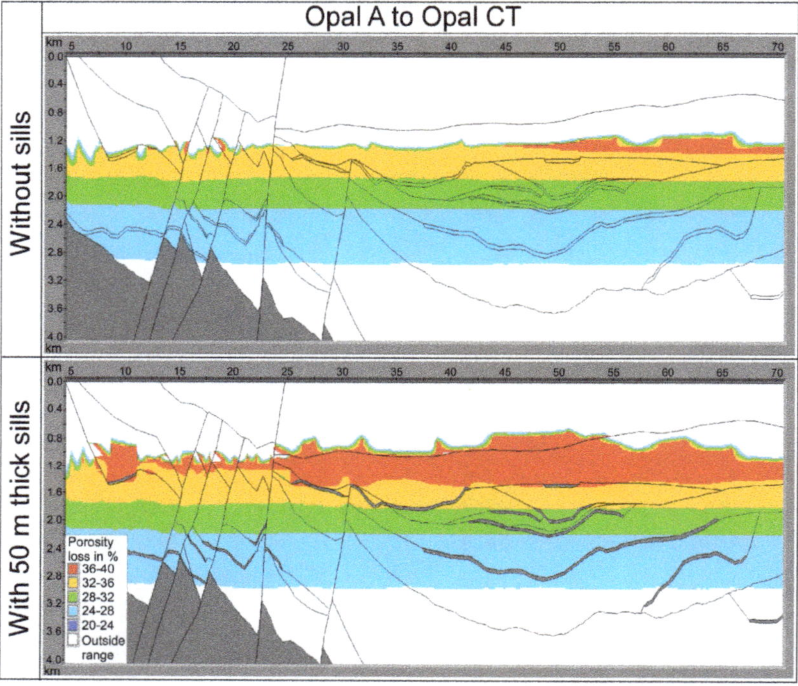

Figure 6. The results for the opal A/CT transition zone when sills are disregarded (upper) and when they are included (lower). The transition zone is assumed to start where the porosity loss exceeds 20%.

3.3.3. Smectite to Illite

Transformation of smectite to illite may occur in sediments and rocks containing clay minerals, and the basin is therefore assumed to have only shale lithology. The results for all the tested kinetic values show that magmatic intrusions and their heat contribution influence the diagenesis and result in transition of smectite to illite at shallower depths than without sills (Figure 7). However, the published and used A_0 and E_a values give significantly different results (Figure 7). The values proposed by Roaldset et al. [95] with high frequency factor (Table 1; Figure 7) give results similar to the result for the transition of opal A to opal CT (Figure 6). Without sills the transition from smectite to illite occurs around 1200 m depth and with sills this boundary lies about 500 m shallower. Results for the parameter values proposed by Hillier et al. (Table 1; [102]) show a porosity loss of 10% and increasing with depth, starting at >2000 m depths when sills are disregarded (Figure 7 pink, stippled lines lower left). For models where sills are included a porosity loss up to 40% is obtained in areas close to the sills (Figure 7). The results are quite similar to results displayed for quartz diagenesis (Figure 5). For the Vøring Basin, a transition in seismic velocity, porosity, and density, believed to resemble the smectite to illite transition, has been observed at depths corresponding to 80–90 °C [24]. The modeled isotherms of 80 and 90 °C for the model without sills are shown in Figure 7 (left). According to Peltonen et al. [24], the kinetic parameter values preferred by Roaldset et al [95] result in a transition of smectite to illite at a shallower depth than 80–90 °C. The kinetic values of Hillier et al. [102] result in a transition zone that lies deeper than the 80–90 °C isotherm.

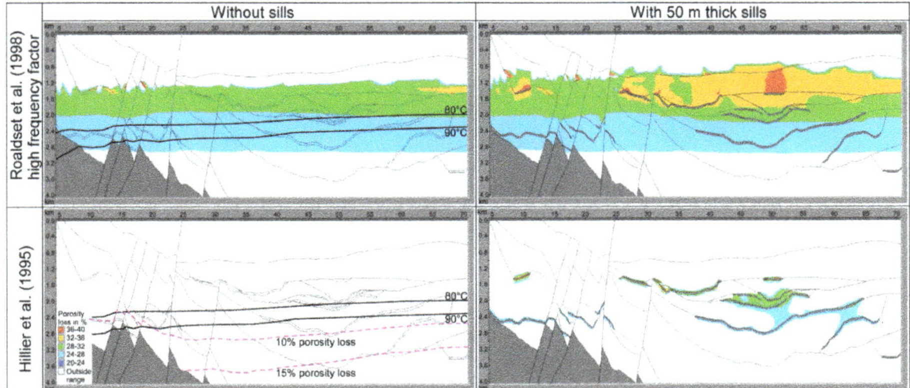

Figure 7. Modeling results of smectite to illite diagnesis when sills are disregarded (left) and when they are included (right). The kinetic parameters of Roaldset et al. [95] and Hillier et al., [102] are used (cf. Table 1). The modeled isotherm of 80–90 °C is marked for scenario without sills (left). Lower left, pink, stippled lines indicate location of 10% and 15% porosity loss.

From field observations it has been reported concurrent onset of oil generation and the conversion of smectite to illite (e.g., [110]). When comparing the resulting maturation assuming kerogen type II in the section from the Vøring Basin (Figure 8) with and without sills, there is a good correlation for both scenarios when compared to the smectite to illite conversion, when kinetic values by Hillier et al. [102] were used. Without sills the maturation starts at depths >2900 m at lateral location 50 km on the transect (Figure 8). At the same point in the transect without sills and Hillier et al. parameters ([102]; Figure 7), a porosity loss of 10% due to conversion of smectite to illite starts around 2700 m depth (Figure 7 pink, stippled lines lower left), which gives about 200 m difference in the location of the onset of hydrocarbon generation and smectite to illite conversion. For the modeled hydrocarbon maturation with sills (Figure 8) there is also a good correlation with the modeled porosity loss due to conversion to illite with the use of Hillier et al. [102] kinetic values (Figure 7).

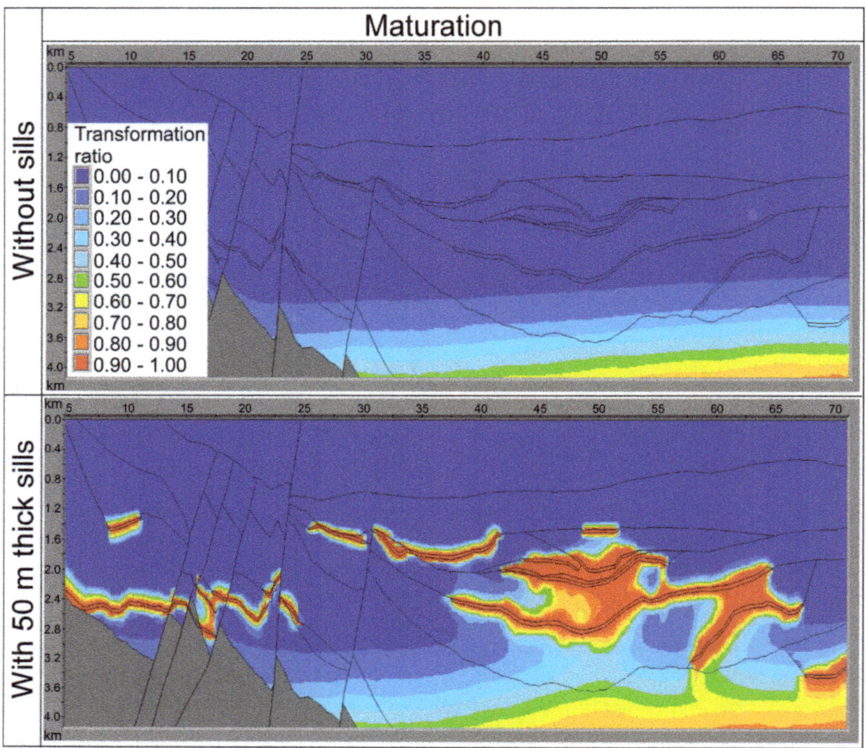

Figure 8. Hydrocarbon maturation for the studied transect without (upper) and with (lower) sills. The onset of maturation of organic material and the onset of porosity loss due to transition of smectite to illite is almost concurrent (compare to Figure 7).

4. Methods and Results for Stress Accumulation Influenced by Diagenetic Processes and Sills Themselves

As stated in the introduction, the physical properties of rocks subject to diagenesis will be altered (e.g., [22,23,95]. The increase in rock stiffness, as a consequence of diagenesis is of particular interest when studying how diagenesis may change the way the rocks respond to stresses. Additionally, the sills themselves are much stiffer than the host rocks in sedimentary basins and will accumulate stresses when subject to compressional or extensional loading [12,111].

4.1. Stress Modeling

To investigate how diagenesis and sills might affect stress accumulations in the subsurface, we used Comsol Multiphysics 5.2.a (www.comsol.com), a commercial finite element method (FEM) software that solves problems based on partial differential equations. The stress effects were studied on two synthetic cases and on the section from the Vøring Basin (Figure 2). All the models are in 2D and have fault zones divided into two separate units, consisting of a fault core which is surrounded by a damage zone with stiffnesses of 0.1 and 1 GPa, respectively (Figure 9). A typical Poisson's ratio of 0.25 [112] was used in all the models and they were all subject to either horizontal compressional or extensional stress of 5 MPa to resemble horizontal regional or local stresses. All sills were assumed to be solidified and have Young's modulus values resembling such a situation. Properties are as summarized in Table 3. The lower boundary was fastened in all models to avoid translation and rotation of the rock body.

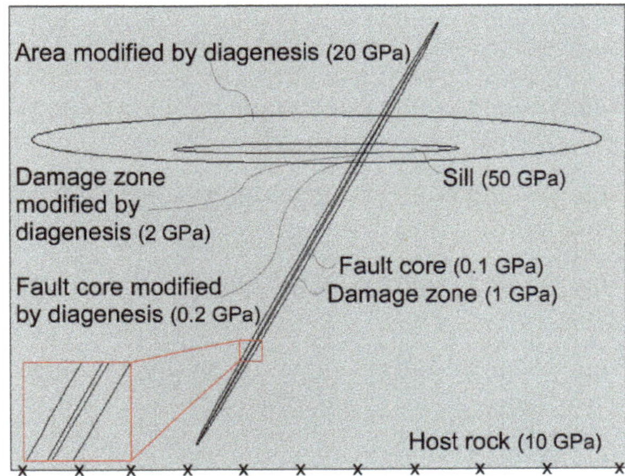

Figure 9. Schematic overview of the synthetic models with an enlargement of the fault zone and definition of the units and Young's modulus values used in the models. The X's indicate the boundary fixated to avoid translation and rotation of the rock body during loading.

Table 3. Physical properties used in the stress modeling, based on standard values for rock properties published in the literature (e.g., [111]).

			Young's Modules		
Host Rock	**Area Modified by Diagenesis**	**Basement**	**Fault Core/ Damage Zone**	**Fault Core/Damage Zone Modified by Diagenesis**	**Sills**
10 GPa	20 GPa	50 GPa	0.1 GPa/1 GPa	0.2 GPa/2 GPa	50 GPa

4.2. Stress Effects of Sills and Diagenesis

In general, stress accumulations indicate areas prone to either opening of fractures or activation/reactivation of faults. Areas where tensile stresses accumulate may lead to opening of fractures or keep existing fractures open, whereas activation/reactivation of faults may occur in areas where shear stresses accumulate. Active faults and open fractures increase the permeability in the area where they are located. Typical global properties of rocks were used in the models (Table 3) and the results show the location of stress concentrations, and how sills and diagenesis affect stress accumulations. The models were based on the assumption that rocks behave as linear elastic up to 1%–3% strain at low temperature and pressure [113,114] depending on the rock strength. Strain exceeding 1%–3% commonly results in failure. Normally the shear strength of rocks lies between 1 and 12 MPa and the tensile strength is half of the shear strength, normally between 0.5 and 6 MPa [115–117]. Stresses exceeding these limits will commonly initiate fault movement and open fractures, respectively.

When comparing models where sills and sill-initiated diagenesis are accounted for versus when it is disregarded, it is possible to determine if and how these units may have influenced fracture and fault related permeability. Laboratory studies show that the physical strength of rocks increases by more than 100% on the transition of opal A to opal CT [23]. Therefore, the stiffness of the diagenetically altered area is doubled compared to the Young's modulus value of the rocks not altered by diagenesis.

Synthetic Case

A synthetic case of arbitrary dimensions shows that in a homogeneous basin, stresses build up at and around the fault tips when subject to compression and extension (Figure 10). Red and yellow colors in the models represent areas where stresses concentrate, while blue colors are areas of small to no stress accumulations. Fractures may open and increase the permeability close to the sides of the fault tips where the tensile stresses concentrate (Figure 10). The direction of the fractures will follow the direction of σ_1 (maximum principal compressive stress) and open in the direction of σ_3 (minimum principal compressive stress or maximum principal tensile stress) [111]. Initiation of fault movement may occur at the fault tips where shear stresses accumulate, and as the shear stresses are larger at the upper tip, this is the area most likely for fault slip initiation (Figure 10).

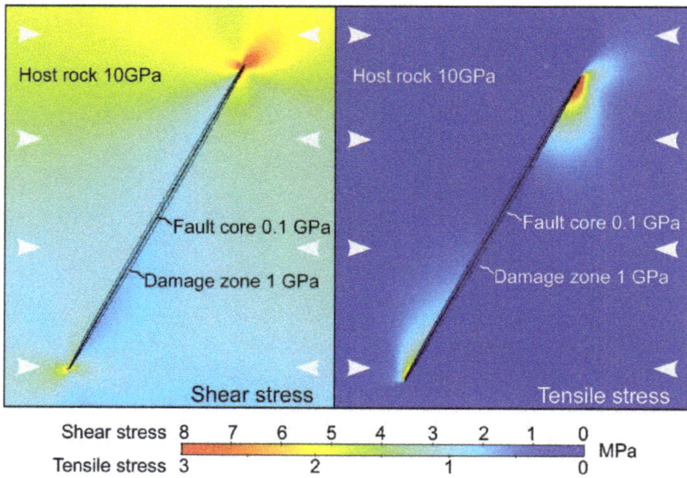

Figure 10. Stress results for the synthetic case of a fault zone with arbitrary dimensions in a homogeneous basin. The tensile and shear stress developed in a homogeneous basin with one fault subject to horizontal compressive stress of 5 MPa (indicated by horizontal arrows). The host rock has a Young's modulus value of 10 GPa, the damage zone is 1 GPa, and fault core 0.1 GPa.

As shown in Sections 3.3.1–3.3.3, magmatic sills will influence the diagenetic processes in their proximity (see Figures 5–7). In order to include this impact in the synthetic model, a second model featuring the same fault, but with one sill surrounded by a diagenetically altered area was added. The results (Figure 11) show that in addition to accumulating shear and tensile stresses at the fault tips, due to compression, stresses mainly build up in the sill and in the area modified by diagenesis surrounding the sill. This occurs as stresses tend to build up in layers/zones of stiffer rocks in contrast to softer rocks in heterogeneous basins (e.g., [111,118,119]). These shear stress accumulations are also present in the fault zone, particularly where the sill crosses the fault, and consequently, potential fault reactivation may start in this area. The accumulated shear stresses in the sill and diagenetically modified area may result in linking of present weaknesses and potentially initiate growth of new faults within this area, if the shear strength of the rock is exceeded [111]. Tensile stresses concentrating in the area modified by diagenesis along the fault zone and within the diagenetically altered damage zone may, if the tensile strength of the rock is exceeded, open fractures [111]. Similar results are obtained for the same model subject to extension, however, the resulting tensile stresses will be much larger compared to when subject to compression (not shown here) and accordingly the chance of opening fractures will increase.

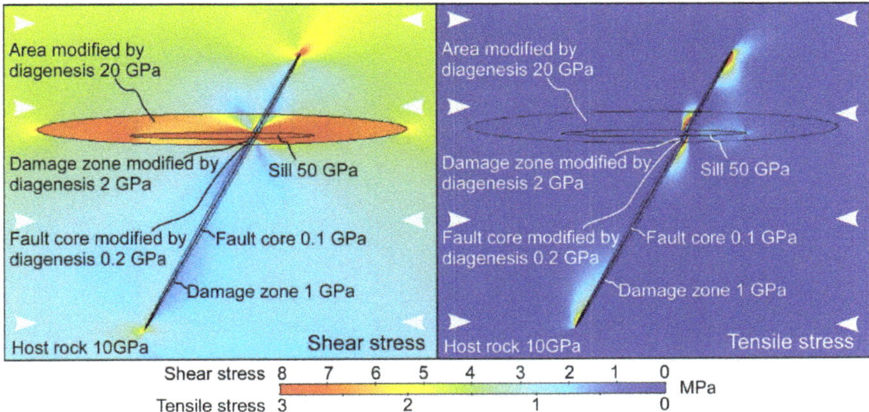

Figure 11. The resulting tensile and shear stresses of a fault with sill intruded through the fault zone with a surrounding area of diagenetic alteration when subject to horizontal compressive stress of 5 MPa (indicated by horizontal arrows). Dimensions are arbitrary in this synthetic model.

4.3. Example from the Vøring Area

For the transect VB-2-87-B in the Vøring Basin, the influence of sills and diagenetically altered area (as given in Sections 3.3.1–3.3.3) was included to investigate its effect on the stress accumulations. The mid-Norwegian margin is presently experiencing compression. However, this is not a specific study of the Vøring area, therefore, to study the response to compressive and tensile horizontal stresses the model was subjected to both. The area to the left (Figure 12a) with several faults and sills, is of particular interest as it displays the interaction between faults, sills, and areas modified by diagenesis. As the smectite to illite transition zones (Figure 7) show boundaries much like the one of opal A to opal CT (Figure 6) and quartz diagenesis (Figure 5), this transition was not included in the study, as the results are expected to be more or less the same as for the opal CT transition zone and quartz diagenesis.

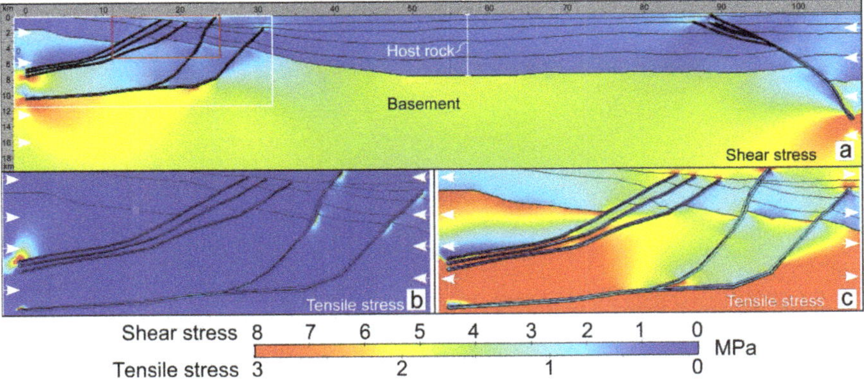

Figure 12. (**a**) Stress results for the Vøring profile when the sills are excluded, and diagenesis is disregarded. The results show the shear stresses due to 5 MPa compressive horizontal stress (indicated by horizontal arrows). The white box indicates the area enlarged in (**b**) and (**c**), the red box indicates the area enlarged in Figures 13 and 14. (**b**) The same scenario as in (**a**) but with the tensile stresses due to compression. (**c**) The same scenario as in (**a**) and (**b**), but with tensile stresses due to 5 MPa extension (indicated by horizontal arrows).

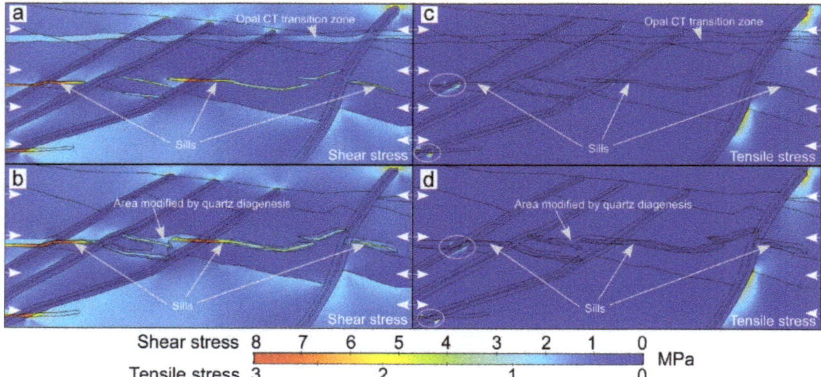

Figure 13. Stress results for the basin with 50 m thick sills subject to horizontal compressive stress of 5 MPa. The area and scale are shown by the red box in Figure 12. (**a,b**) The results for the basin when the sills and the transition zone of opal CT are accounted for. (**c,d**) The results when sills and quartz diagenesis are accounted for. White circles indicate areas where the sill does not penetrate the fault zone, however, parts of the fault zone have area altered by diagenesis.

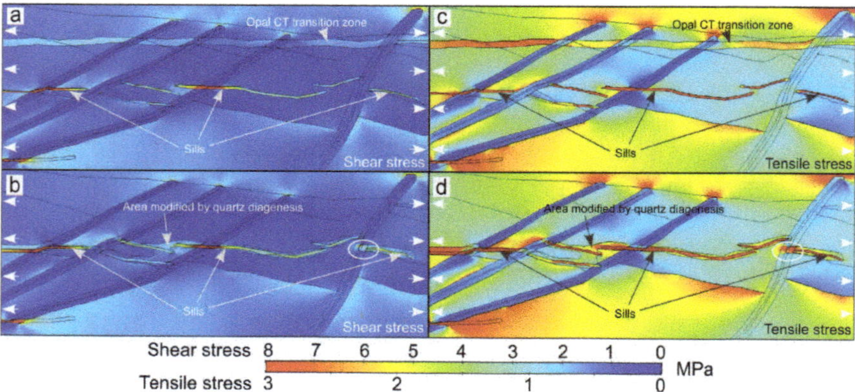

Figure 14. Stress results for the basin with 50 m thick sills subject to horizontal tensile stress of 5 MPa. The area and scale is shown by the red box in Figure 12. (**a,b**) Results for basin where sills and the transition zone of opal CT is accounted for. (**c,d**) Results for when sills and quartz diagenesis is accounted for. The white circle to the right indicates the area where the sill does not penetrate the fault zone, however, parts of the fault zone have area altered by diagenesis. This area concentrates stresses, and when high enough, may contribute to fault reactivation and fracture development. The sills are 50 m thick.

The transformation zone of opal A to opal CT starts when the porosity reduction is around 20% (cf. Figure 6; [23]) and is modeled to end around 200 m deeper (cf. [41]). Five different scenarios were tested and subject to horizontal compressive and tensile stresses of 5 MPa: (1) the basin with both sills and diagenesis disregarded (Figure 12), (2) the basin with sills but diagenesis disregarded, (3) the basin without sills but with opal A to opal CT transition zone included, (4) the basin with sills and opal A to opal CT transition zone included (Figures 13 and 14) and, (5) the basin with sills and quartz diagenesis (Figures 13 and 14).

Common results for all the models, whether subject to compressive or tensile stress, is that stresses build up at the fault tips (like the synthetic models), and in particular at the lower tips (Figure 12a–c). This is due to the larger contrast in rock stiffness between the fault zones and the basement compared to the contrast between the fault zones and the host rock (cf. Table 3). Shear stresses are more or less the same with regard to location and strength of accumulations, whether they are subject to compression or extension. However, the tensile stresses are much lower for the models subject to compression (Figure 12b) than those subject to extension (Figure 12c).

When subject to both compression and extension, the synthetic models show that both shear and tensile stresses mainly concentrate in the sill and area modified by diagenesis, also where these features cross the fault zone (Figure 11). Figure 13 shows the results for the Vøring basin subject to compression and when sills and the opal CT transition zone are included (Figure 13a,c), as well as when sills and the quartz diagenetically modified area are accounted for (Figure 13b,d). For all cases, areas of increased tensile stresses are few and small, but potential fractures may open along the fault zone to the right and increase the permeability if the stresses become high enough (Figure 13c,d). In addition, areas indicated by white circles in Figure 13c,d, show elevated tensile stresses related to where the sill and diagenetically modified area cross the fault zone. Fractures may open or remain open if stress concentrations grow high enough, thereby contributing to increased permeability in these areas. The shear stresses tend to build up in the areas modified by diagenesis and in the sills. Where the sills cross the faults, stress concentrations are, without exceptions, particularly high (Figure 13a,b) and may reactivate the fault if they exceed the rock's shear strength of 1–12 MPa. Elevated shear stresses in the opal CT transition zone and area modified by quartz diagenesis, may potentially initiate growth of new faults if the stresses become high enough, and thereby contribute to increased permeability.

When the basin is subject to extension the shear stresses concentrate in the same location with the same strength (Figure 14a,b), as when subject to compression (Figure 13a,b), for all scenarios. If the shear stresses exceed the shear strength of the rocks, fault slip will be initiated in some areas (colored red and yellow). However, the tensile stresses are much larger for the basin subject to extension (Figure 14c,d) as opposed to compression (Figure 13c,d). The entire sills tend to concentrate tensile stresses, not just when crossing fault zones (Figure 14c,d). Additionally, the diagenetically altered area of both opal CT and quartz accumulate higher tensile stresses compared to their surroundings. The high concentrations of tensile stresses show that the basin is prone to open fractures and increase the permeability in several areas. Due to the particularly high stress concentrations in the sills and the areas that have undergone diagenetic transitions, fractures are expected to first open in these places.

5. Discussion

Magmatic intrusions may influence the location and amount of porosity loss due to diagenesis, given that the necessary physical, biogenic, and/or chemical conditions are present. Two specific topics are addressed in this study: (1) quantification of the effect of magmatic intrusions on several diagenetic processes (opal A to opal CT to quartz, smectite to illite, and quartz diagenesis) and (2) the effect of sills and diagenetically altered areas on the fracture and fault permeability in basins with emphasis on the effect of these factors on petroleum systems.

5.1. Effects of Diagenesis on Petroleum Systems

Diagenesis contributes to loss of porosity and permeability in petroleum reservoirs and thus essentially harm the reservoir quality [90]. However, the same processes may also enhance the porosity and permeability through development of secondary porosity by dissolution, grain coating, and fracturing of layers that have become brittle [90]. All diagenetic products can aid in calibrating the thermal history of basins, as the transformation of each diagenetic product commonly occurs over a specific temperature range. Thereby, they are markers for maximum temperatures and burial depths and may uncover areas exposed to unusual environments, i.e., uncommon heat flow or uplift and erosion, and they can be used together with for instance vitrinite reflectance and fluid inclusion data to calibrate thermal predictions.

The thermal conductivity of sediments could rise 4–5 times when opal CT transitions to quartz [120], as a result of the porosity loss. In the Bjørnøya Basin (Barents Sea) this difference in thermal conductivity results in about 10 °C higher temperatures in the Paleocene/Eocene shales for opal CT rich shale compared to quartz rich shale [118]. This temperature difference is caused by the nature of clay minerals that have a flattened structure favoring horizontal over vertical heat transport [120]. This is an example of the importance of including diagenesis in basin modeling. Modeling of diagenesis will ensure good thermal models when assessing the petroleum potential of sedimentary basins [120].

When the effects of diagenetic processes on petroleum reservoirs are considered, it is important to realize that chemical compounds in subsurface fluids can be catalysts or quenchers for the processes. First of all, the transition of opal A to opal CT to quartz requires presence of amorphous silica in the sedimentary rocks. Kastner et al. [83] studied the chemical controls on the opal A to opal CT transition and found that the diagenetic process is increased in carbonate rocks and reduced in clay-rich rocks. The presence of magnesium and hydroxide ions was found to enhance the conversion of opal A to opal CT. Subsequent transformation of opal CT to quartz released magnesium and hydroxide ions [83] and fluids enriched in magnesium and hydroxide ions were expelled upwards along faults increasing the transformation of opal A to opal CT at shallower depths [121]. Conversion of smectite to illite requires the presence of clay minerals in the sedimentary rocks. Furthermore, the presence of potassium has proved to enhance this conversion process. The transformation rates of smectite to illite increases with high potassium concentrations and lower temperatures are needed to start the conversion [95]. The potassium is commonly sourced by k-feldspars, which requires such minerals to be present for diagenesis to occur [91]. On the other hand, laboratory studies indicate that the presence of magnesium ions retards the conversion process in the early phases of the smectite to illite conversion [94]. Therefore, although thermal requirements are met for diagenetic alterations to occur, it is ultimately the biogenic and/or chemical compounds in the rocks and pore fluids that determine the possible diagenetic processes.

5.2. Effect of Magmatic Sills on the Diagenetic Process in Reservoir Rocks

Earlier studies of magmatic intrusions and their effect on porosity evolution and diagenesis of reservoir rocks have come to contrasting conclusions (e.g., [14,15,122–124]). The diagenetic processes of sandstone reservoirs are likely to be controlled by the initial composition of the sand, original pore water composition, content of neighboring lithologies, burial and thermal history, and timing of cementation relative to accumulation of petroleum [17,47,82]. These criteria can largely explain the contrasting conclusions that are put forward in the referenced literature.

Haile et al. [14] conclude in their study at Edgeøya (Svalbard) that quartz cementation is unaffected by the short-lived heating of magmatic sills in reservoir rocks. They suggest that slow quartz cementation rates are in contrast to the relatively short-lived magmatic heating and will therefore not leave fingerprints on the diagenetic process. This conclusion contrasts from the results in our study which clearly show that magmatic sill intrusions affect the diagenesis in their proximity (Figures 5–7). Nejbert et al. [125] state that most sills in the Svalbard area are about 10–30 m thick. The sills are relatively thin and the distances between the sills are unknown in the study by Haile et al. [14], which could explain the contrasting conclusions of their study and ours. The magmatic sill thickness, distance from sills, and clusters of sills are pointed out in several studies to have significant importance on the thermal effect of magmatic intrusions (e.g., [3,5,6,9–11]). Our study shows that thermal effects will also affect diagenetic transformations (i.e., Figure 5). The location of sampling in the study by Haile et al. [14] might also be too far from sills to be thermally affected by them.

Another aspect is the temperature of the intruding magma. As the thermal effect of sills is a function of the temperature of the intruding magma, a lower magma temperature than 1000 °C (used in this study), would result in smaller thermal aureole surrounding the sills and consequently decrease the area affected by diagenesis. This could also to some degree explain the contrasting conclusions. The host rock pre-intrusion temperature is identified to be crucial for the thermal effect of magmatic intrusions (e.g., [5,8,11]), and ultimately the effect of magmatic intrusions on diagenetic processes. Factors such as fault displacement, time span of faulting and sediment deposition, fault angle, rock thermal conductivity and specific heat capacity, basal heat flow, and restoration method used to simulate the fault movement through time, are all factors that influence the host rock pre-intrusion temperatures. For a specific, detailed area study, all these factors are crucial and require thorough assessment [11].

5.3. Silica Diagenesis

Silica diagenesis embraces two distinct types of chemical compaction processes, both of which are thermally dependent. The first studied here is quartz diagenesis and is related to pressure dissolution of quartz minerals, diffusion of dissolved silica and precipitation of quartz cement. Quartz cementation is the number one porosity reduction process in sandstones below 2000 m depth (e.g., [17,126,127]). Rapidly increasing porosity loss due to quartz cementation occurs when temperatures exceed 100–110 °C [75] and hydrocarbons do not seem to influence the porosity reduction process [91]. Grain coatings of clays have been found to inhibit quartz cementation and contribute to porosity preservation (e.g., [128–130]). Furthermore, at depths for peak mineral dissolution, such as quartz dissolution, the porosity is found to increase through secondary porosity production before it decreases again at deeper burial depth [130]. For basins without magmatic intrusions the porosity loss in petroleum reservoirs due to quartz diagenesis are expected to be found at depths corresponding to the mentioned temperature range (100–110 °C). As the results in this study show, magmatic intrusions in sedimentary basins contribute to abrupt changes in the diagenetic processes (Figure 5) and reservoirs with reduced porosity may thus be found at shallower depths than expected. Reservoirs located between sills in a cluster are particularly prone to porosity reduction due to quartz diagenesis.

The transition of siliceous sediments from opal A through opal CT to quartz, starts at lower temperatures than the silica transition discussed previously. Opal A dissolves and subsequently precipitates to opal CT at about 50–60 °C, and further transition into crystalline quartz starts around 80–90 °C [131]. These transformations are known to create hydrocarbon traps. In Onnagawa Formation, Yurihara oil and gas field (Japan) and Monterey Formation in California (USA), the opal CT/quartz boundary acts as a seal due to its low permeability, while the quartz rich layers underneath, with higher permeability, act as a reservoir [38,132]. For basins following the thermal development without particular interruptions, the transition zone of opal A to opal CT will normally be in the temperature range of 50–60 °C, and the opal CT to quartz transition at depths corresponding to 80–90 °C. Results from this study show that sills intruding at depths <1500 m moves the opal A/CT transition zone to significantly shallower depths. For basins with magmatic sill intrusions emplaced at shallow depths, such a hydrocarbon trap model would lie at shallower depths than basins without sills, and could possibly go undiscovered if sills are not included in the thermal and diagenetic calculations during basin modeling. The opal A/CT boundary has been mistaken for a flat spot as oil/water contact and drilled (e.g., [22]). When incorporating the diagenetic transition of opal A to opal CT in basin modeling, such errors may be avoided.

Opal CT Boundary in the Vøring Basin

The opal A to opal CT transition zone is found at surprisingly shallow depths in some areas of the Vøring Basin. Several reasons for this fossilized opal A/CT transition have been suggested. The transition zone is proposed to have developed during Miocene and died out in Early Pliocene (e.g., [44,49,133]). As the transition of amorphous silica to opal CT is a temperature dependent reaction (e.g., [22,24,73–75]) the fossilized transition is a record of peak paleo temperatures which may reflect max burial depth of the basin, uplift and erosion of the area, higher thermal gradient in the past, or another unknown thermal event [22,44,49,131]. The Eocene magmatic sills are, however, not the cause of the observed shallow opal A/CT zone because the extra heat provided by the intrusions did not affect the diagenesis as shallow as the Miocene/E Pliocene formations.

5.4. Smectite to Illite

The conversion of smectite to illite is one of many different clay diagenetic processes occurring in the subsurface. Illite has a fibrous texture and small amounts of this clay may dominate the pore space and largely reduce permeability [91]. The smectite to illite transition is of particular interest for the oil and gas industry as it may coincide with the onset of oil generation [90]. For the Vøring Basin, Peltonen et al. [24] found that major smectite to illite conversion occurred at depths corresponding to a temperature interval of 80–90 °C. For shales in the Gulf Coast (USA), the major rate of smectite to illite conversion was found at temperatures of 90–120 °C [134]. Other studies conclude that the illitization process starts at 50 °C and ceases around 100 °C [135]. These diverse temperature ranges for the smectite to illite conversion is also reflected in the modeling results in this study, when different published kinetic parameters are used (Figure 7). This emphasizes the complexity of the smectite to illite conversion, and possibly clay diagenesis in general. An area influenced by magmatic activity in western Pannonian Basin (Hungary), show a divergent effect of magmatic heating on clay minerals and the smectite to illite transformation [103]. Such results emphasize that, although thermal requirements are met for diagenetic processes to take place, other factors must also be fulfilled for chemical reactions to occur (cf., [136]). For the smectite to illite diagenetic functionality described here, caution must be exercised in the interpretation of the results. This means that the described method can be used to locate areas that fulfill the required kinetics for alterations to occur. This is a good first approximation in pin pointing areas prone to have undergone diagenetic changes in the basin. However, subsurface chemical knowledge is needed in order to examine if other necessary criteria are met for diagenetic alterations.

5.5. Influence of Sills and Diagenesis on Stress Accumulations

When sills have solidified and are subject to compressive and tensile stresses, this study shows that the sills and the diagenetically altered areas accumulate stress (Figures 10–14). When stresses exceed the shear and/or tensile strengths of the rock, the rocks will fail and fault movement will be initiated or fractures will develop, respectively. This occurs because stresses tend to build up in the stiffer layers as opposed to softer rocks (e.g., [111,116,117]), which will be decisive for where fractures develop and faults could be reactivated. From the models (Figures 10–14) it can be deduced that the higher the contrast in rock stiffness, the higher the stress concentration in the stiff layers. For diagenetically modified areas, with higher increase in rock stiffness than studied here (doubled the stiffness of the host rock), the potential stress accumulation in the area modified by diagenesis will be higher, and the likelihood of fault initiation and fracture opening is increased. On the other hand, in an opposite situation, the potential stress concentration will be lower and so will the likelihood of fault movement and fracture opening.

It has been proposed that diagenetic processes may affect the fluid flow properties in the subsurface (e.g., [14,47]), a statement which is confirmed by the results in this study. This is the case, not just due to how diagenesis modifies rocks porosity and permeability, but also to how these diagenetic alterations change the rock's physical properties. An active fault may increase the temporary permeability many times, and faults and fractures may dominate the fluid transport in the rock masses if they are interconnected [111]. The results in this study show that sills or diagenetically altered areas (Figures 11–14) could ultimately lead to opening of fractures or initiate fault movements if the tensile or shear stresses of the rocks are exceeded. For a petroleum system, such an outcome can be crucial, as opening of fractures and fault movement increase the permeability and supports fluid flow, possibly to new locations. Thus, we conclude that sills and their related diagenetically modified areas may influence the subsurface fluid migration pathways through time and space and increase the permeability, as opening of fractures and reactivation of faults may act as fluid conduits.

5.6. Limitations of the Calculations

All basins have their particular thermal and structural histories concerning e.g., sediment source, lithology, fluid circulation, pore fluid chemistry, and structural development. These variations all play a part in possible chemical alterations in the subsurface and lead to unique diagenetic development. Therefore, information regarding these elements must be included for more detailed local studies.

In numerical modeling there is a direct link between the quality of the input data and the quality of the output from the models. When studying a particular area, caution must be taken regarding various parameters. For instance, we only studied conductive heat transfer in the subsurface and heat convection by fluids is not accounted for. Convection by fluids could influence the resulting thermal and diagenetic calculations to various degrees (cf., [137–141]). The Vøring area shows proof of fluid convection (e.g., [142]), our results therefore show maximum porosity loss, at least for some areas. However, for basins that show little to no signs of convecting fluids, the presented estimations are adequate. We argue that the method presented here is a good approach to reveal areas needing further investigation of possible diagenetic alterations and thus possible alterations of reservoir properties.

6. Concluding Remarks

This study quantified the porosity loss due to the transition of opal A to opal CT to quartz, smectite to illite and quartz diagenesis and how magmatic intrusions may affect the diagenetic process. As the diagenetic alterations change the physical properties of the rock, it was shown how these alterations and the sills themselves influence location of stress accumulations in a basin and thus may contribute to changes in fracture and fault permeability. All these factors have implications for the petroleum system and the results are summarized as follows:

- Conductive thermal effects of sills significantly influence the diagenetic processes in sedimentary basins.
- The effect of magmatic intrusions on the diagenetic processes depends on the depth at which the magmatic sills intrude. Maximum diagenetic changes occur at different temperatures for the different processes. For sills to influence the transition of opal A to opal CT they must intrude at depths >1500 m.
- Sill thickness influences the size of the diagenetically altered area, particularly where clusters of sills are closely spaced.
- Sills and diagenetically modified areas influence location of stress accumulations and may contribute to initiation of fault movement and opening of fractures. As fractures and faults can act as conduits for fluid flow, sills and areas modified by diagenesis may therefore contribute to increased permeability.
- A thorough case study is required to determine the sill's specific effect on diagenetic processes and stress accumulations in the Vøring Basin and other volcanic basins. This is now made possible with the work done in this study.

Author Contributions: Conceptualization, M.S., W.F., I.G. and I.F.L.; methodology, M.S., W.F. and I.G.; software, M.S., W.F. and I.G.; formal analysis, M.S. and I.G.; investigation, M.S.; writing—original draft preparation, M.S.; writing—review and editing, W.F., I.G., I.F.L. and R.M.; visualization, M.S.; supervision, W.F. and R.M.; project administration, I.F.L.; funding acquisition, I.F.L.

Funding: This research was partly funded by The Research Council of Norway and Tectonor AS as a part of the PhD project 'Effects of magmatic intrusions on temperature history and diagenesis in sedimentary basins—and the impact on petroleum systems', RCN project number 257492.

Acknowledgments: We want to express gratitude for the support received. Inspiring conversations with Professors Olav Eldholm and Elen Roaldset, and valuable discussions with Professor Leiv K. Sydnes are highly appreciated. Two anonymous reviewers and Academic Editor are thanked for constructive and helpful comments that improved the contents and design of the paper.

Conflicts of Interest: The authors declare no conflict of interest.

References

1. Schutter, S.R. Hydrocarbon occurrence and exploration in and around igneous rocks. In *Hydrocarbons in Crystalline Rocks, Special Publications*; Petford, N., McCaffrey, K.J.W., Eds.; Geological Society: London, UK, 2003; Volume 214, pp. 7–33.

2. Svensen, H.; Planke, S.; Malthe-Sørenssen, A.; Jamtveit, B.; Myklebust, R.; Eidem, T.R.; Rey, S.S. Release of methane from a volcanic basin as a mechanism for initial Eocene global warming. *Nature* **2004**, *429*, 542–545. [CrossRef] [PubMed]

3. Fjeldskaar, W.; Helset, H.M.; Johansen, H.; Grunnaleite, I.; Horstad, I. Thermal modelling of magmatic intrusions in the Gjallar Ridge, Norwegian Sea: Implications for vitrinite reflectance and hydrocarbon maturation. *Basin Res.* **2008**, *20*, 143–159. [CrossRef]

4. Galushkin, Y.I. Thermal effects of igneous intrusions on maturity of organic matter: A possible mechanism of intrusion. *Org. Geochem.* **1997**, *26*, 645–658. [CrossRef]

5. Aarnes, I.; Svensen, H.; Connolly, J.A.D.; Podladchikov, Y.Y. How contact metamorphism can trigger global climate changes: Modeling gas generation around igneous sills in sedimentary basins. *Geochim. Cosmochim. Acta* **2010**, *74*, 7179–7195. [CrossRef]

6. Aarnes, I.; Svensen, H.; Polteau, S.; Planke, S. Contact metamorphic devolatilization of shales in the Karoo Basin, South Africa, and the effects of multiple sill intrusions. *Chem. Geol.* **2011**, *281*, 181–194. [CrossRef]

7. Peace, A.; McCaffrey, K.; Imber, J.; Hobbs, R.; van Hunen, J.; Gerdes, K. Quantifying the influence of sill intrusion on the thermal evolution of organic-rich sedimentary rocks in nonvolcanic passive margins: An example from ODP 210–1276, offshore Newfoundland, Canada. *Basin Res.* **2017**, *29*, 249–265. [CrossRef]

8. Spacapan, J.B.; Palma, J.O.; Galland, O.; Manceda, R.; Rocha, E.; D'Odorico, A.; Leanza, H.A. Thermal impact of igneous sill-complexes on organic-rich formations and implications for petroleum systems: A case study in the northern Neuquén Basin, Argentina. *Mar. Pet. Geol.* **2018**, *91*, 519–531. [CrossRef]

9. Spacapan, J.B.; D'Odorico, A.; Palma, O.; Galland, O.; Rojas Vera, E.; Leanza, H.A.; Medialdea, A.; Manceda, R. Igneous petroleum systems in the Malargüe fold and thrust belt, Río Grande Valley area, Neuquén Basin, Argentina. *Mar. Pet. Geol.* **2020**, *111*, 309–331. [CrossRef]

10. Sydnes, M.; Fjeldskaar, W.; Løtveit, I.F.; Grunnaleite, I.; Cardozo, N. The importance of sill thickness and timing of sill emplacement on hydrocarbon maturation. *Mar. Pet. Geol.* **2018**, *89*, 500–514. [CrossRef]

11. Sydnes, M.; Fjeldskaar, W.; Grunnaleite, I.; Løtveit, I.F.; Mjelde, R. Transient thermal effects in sedimentary basins with normal faults and magmatic sill intrusions—A sensitivity study. *Geosciences* **2019**, *9*, 160. [CrossRef]

12. Gudmundsson, A.; Løtveit, I.F. Sills as hydrocarbon reservoirs: Examples and models. *Geol. Soc. Lond. Spec. Publ.* **2012**, *374*, 251–271. [CrossRef]

13. Montanari, D.; Bonini, M.; Corti, G.; Agostini, A.; Del Ventisette, C. Forced folding above shallow magma intrusions: Insights on supercritical fluid flow from analogue modelling. *J. Volcanol. Geotherm. Res.* **2017**, *345*, 67–80. [CrossRef]

14. Haile, B.G.; Klausen, T.G.; Jahren, J.; Braathen, A.; Hellevang, H. Thermal history of a Triassic sedimentary sequence verified by a multi-method approach: Edgeøya, Svalbard, Norway. *Basin Res.* **2018**, *30*, 1075–1097. [CrossRef]

15. Haile, B.G.; Czarniecka, U.; Xi, K.; Smyrak-Sikora, A.; Jahren, J.; Braathen, A.; Hellevang, H. Hydrothermally induced diagenesis: Evidence from shallow marine-deltaic sediments, Wilhelmøya, Svalbard. *Geosci. Front.* **2019**, *10*, 629–649. [CrossRef]

16. Therkelsen, J. Diagenesis and reservoir properties of Middle Jurassic sandstones, Traill Ø, East Greenland: The influence of magmatism and faulting. *Mar. Pet. Geol.* **2016**, *78*, 196–221. [CrossRef]

17. Allen, P.A.; Allen, J.R. *Basin Analysis: Principles and Application to Petroleum Play Assessment*, 3rd ed.; Wiley-Blackwell: Chichester, UK, 2014.

18. Gluyas, J.; Swarbrick, R. *Petroleum Geosciences*; Blackwell Publishing: Oxford, UK, 2015.

19. Bjørlykke, K.; Ramm, M.; Saigal, G.C. Sandstone diagenesis and porosity modification during basin evolution. *Geol. Rundsch.* **1989**, *78*, 243–268. [CrossRef]

20. Bjørlykke, K. Relationships between depositional environments, burial history and rock properties. Some principal aspects of diagenetic process in sedimentary basins. *Sediment. Geol.* **2014**, *301*, 1–14. [CrossRef]

21. Nobes, D.C.; Murray, R.W.; Kuramoto, S.; Pisciotto, K.A.; Holler, P. Impact of silica diagenesis on physical property variations. In Proceedings of the Ocean Drilling Program, Scientific Results, College Station, TX, USA; 1992; Volume 127/128. [CrossRef]

22. Roaldset, E.; Wei, H. Silica-phase transformation of opal-A to opal_CT to quartz, Part I: An experimental diagenetic approach to natural observations. Prepared for submittal to AAPG. 1997.

23. Roaldset, E.; Wei, H. Silica-phase transformation of opal-A to opal-CT to quartz, Part II: Changes of physical properties. Prepared for submittal to AAPG. 1997.

24. Peltonen, C.; Marcussen, Ø.; Bjørlykke, K.; Jahren, J. Clay mineral diagenesis and quartz cementation in mudstones: The effects of smectite to illite reaction on rock properties. *Mar. Pet. Geol.* **2009**, *26*, 887–898. [CrossRef]

25. Neagu, R.C.; Cartwright, J.; Davies, R. Measurement of diagenetic compaction strain from quantitative analysis of fault plane. *J. Struct. Geol.* **2010**, *32*, 641–655. [CrossRef]

26. Elliot, W.C.; Aronson, J.L.; Matisoff, G.; Gautier, D.L. Kinetics of the smectite to illite transformation in the Denver Basin; clay mineral, K-Ar data, and mathematical model results. *AAPG Bull.* **1991**, *75*, 436–462.

27. Buryakovski, L.A.; Djevanshir, R.D.; Chilingar, G.V. Abnormally-high formation pressures in Azerbaijan and the South Caspian Basin (as related to smectite ↔ illite transformations during diagenesis and catagenesis). *J. Pet. Sci. Eng.* **1995**, *13*, 203–218. [CrossRef]

28. Worden, R.H.; Charpentier, D.; Fisher, Q.J.; Aplin, A.C. Fabric development and the smectite to illite transition in Upper Cretaceous mudstones from the North Sea: An image Analysis Approach. In *Understanding the Micro to Macro Behaviour of Rock-Fluid Systems*; Shaw, R.P., Ed.; Geological Society, Special Publications: London, UK, 2005; Volume 249, pp. 103–114.

29. Robin, V.; Hebert, B.; Beaufort, D.; Sardini, P.; Tertre, E.; Regnault, O.; Descostes, M. Occurrence of authigenic beidellite in the Eocene transitional sandy sediments of the Chu-Saryssu basin (South-Central Kazakhstan). *Sediment. Geol.* **2015**, *321*, 39–48. [CrossRef]

30. Dou, W.; Liu, L.; Wu, K.; Xu, Z.; Feng, X. Diagenesis of tight oil sand reservoirs: Upper Triassic tight sandstones of Yanchang Formation in Ordos Basin, China. *Geol. J.* **2018**, *53*, 707–724. [CrossRef]

31. Morley, C.K.; Maczak, A.; Rungprom, T.; Ghosh, J.; Cartwright, J.A.; Bertoni, C.; Panpichityota, N. New style of honeycomb structures revealed on 3D seismic data indicate widespread diagenesis offshore Great South Basin, New Zealand. *Mar. Pet. Geol.* **2017**, *86*, 140–154. [CrossRef]

32. Higgins, J.A.; Blättler, C.L.; Lundstrom, E.A.; Santiago-Ramos, D.P.; Akhtar, A.A.; Crüger Ahm, A.-S.; Bialik, O.; Holmden, C.; Bradbury, H.; Murray, S.T.; et al. Mineralogy, early marine diagensis, and the chemistry of shallow-water carbonate sediments. *Geochim. Cosmochim. Acta* **2018**, *220*, 512–534. [CrossRef]

33. Fawad, M.; Mondol, N.H.; Baig, I.; Jahren, J. Diagenetic related flat spots within the Paleogene Sotbakken Group in the vicinity of the Senja Ridge, Barents Sea. *Pet. Geosci.* **2019**, 122. [CrossRef]

34. He, J.; Wang, H.; Zhang, C.; Yang, X.; Shangguan, Y.; Zhao, R.; Gong, Y.; Wu, Z. A comprehensive analysis of the sedimentology, petrography, diagenesis and reservoir quality of sandstones from the Oligocene Xiaganchaigou (E$_3$) Formation in the Lengdong area, Qaidam Basin, China. *J. Pet. Explor. Prod. Technol.* **2019**, *9*, 953–969. [CrossRef]

35. Kruge, M.A. Biomarker geochemistry of the Miocene Monterey Formation, West San Joaquin Basin, California: Implications for petroleum generation. *Org. Geochem.* **1986**, *10*, 517–530. [CrossRef]

36. Behl, R.J. Chert spheroids of the Monterey Formation, California (USA): Early-diagenetic structures of bedded siliceous deposits. *Sedimentology* **2011**, *58*, 325–351. [CrossRef]

37. Weller, R.; Behl, R.J. Physical and Mechanical Characteristics of the Opal-A to Opal-CT Transition Zone: Enhanced Diatomite Permeabiltiy from Heterogeneous Diagenetic Embrittlement. *Search Discov. Artic.* **2015**, #51112, adapted from poster presentation given at Pacific section AAPG, SEG and SEPM Joint Technical Conference, Oxnard, California, USA, 3–5 May, 2015.

38. Dralus, D. Chemical Interactions between Silicates and Their Pore Fluids: How They Affect Rock Physics Properties from Atomic to Reservoir Scales. Ph.D. Thesis, Stanford University, Stanford, CA, USA, August 2013.

39. Dralus, D.; Lewan, M.D.; Peters, K. Kinetics of the Opal-A to Opal-CT Phase transition in Low- and High-TOC Siliceous Shale Source Rocks. *Search Discov. Artic.* **2015**, #41708, adapted form oral presentation given at AAPG Annual Convention & Exhibition, Denver, Colorado, USA, May 31–June 3, 2015.

40. Blystad, P.; Brekke, H.; Færseth, R.B.; Larsen, B.T.; Skogseid, J.; Tørudbakken, B. *Structural Elements of the Norwegian Continental Shelf, Part II: The Norwegian Sea Region*; Technical Report for Norwegian petroleum Directorate; Norwegian Petroleum Directorate: Stavanger, Norway, 1995; Volume 8, ISBN 82-7257-452-7.

41. Davies, R.J.; Cartwright, J.A. Kilometer-scale chemical reaction boundary patters and deformation in sedimentary rocks. *Earth Planet. Sci. Lett.* **2007**, *262*, 125–137. [CrossRef]

42. Peltonen, C.; Marcussen, Ø.; Bjørlykke, K.; Jahren, J. Mineralogical control on mudstone compaction: A study of Late Cretaceous to Early Tertiary mudstones of the Vøring and Møre basins, Norwegian Sea. *Pet. Geosci.* **2008**, *14*, 127–138. [CrossRef]

43. Davies, R.J.; Ireland, M.T.; Cartwright, J.A. Differential compaction due to the irregular topology of a diagenetic reaction boundary: A new mechanism for the formation of polygonal faults. *Basin Res.* **2009**, *21*, 354–359. [CrossRef]

44. Neagu, R.C.; Cartwright, J.; Davies, R.; Jensen, L. Fossilisation of a silica diagenesis reaction front on the mid-Norwegian margin. *Mar. Pet. Geol.* **2010**, *27*, 2141–2155. [CrossRef]

45. Davies, R.J.; Ireland, M.T. Initiation and propagation of polygonal fault arrays by thermally triggered volume reduction reactions in siliceous sediment. *Mar. Geol.* **2011**, *289*, 150–158. [CrossRef]

46. Ireland, M.T.; Davies, R.J.; Goulty, N.R.; Carruthers, D. Structure of a silica diagenetic transformation zone: The Gjallar Ridge, offshore Norway. *Sedimentology* **2011**, *58*, 424–441. [CrossRef]

47. Wrona, T.; Taylor, K.G.; Jackson, C.A.-L.; Huuse, M.; Najorka, J.; Pan, I. Impact of silica diagenesis on the porosity of fine-grained strata: An analysis of Cenozoic mudstones from the North Sea. *Geochem. Geophys. Geosyst.* **2017**, *18*, 1537–1549. [CrossRef]

48. Fjeldskaar, W.; Andersen, Å.; Johansen, H.; Lander, R.; Blomvik, V.; Skurve, O.; Michelsen, J.K.; Grunnaleite, I.; Mykkeltveit, J. Bridging the gap between basin modelling and structural geology. *Reg. Geol. Metallog.* **2017**, *72*, 65–77.

49. Brekke, H. The tectonic evolution of the Norwegian Sea Continental Margin with emphasis on the Vøring and Møre Basins. In *Dynamics of the Norwegian Margin*; Special Publications; Nøttvedt, A., Ed.; Geological Society: London, UK, 2000; Volume 167, pp. 327–378.

50. Bukovics, C.; Cartier, E.G.; Shaw, N.D.; Ziegler, P.A. Structure and development of the mid-Norway Continental Margin. In *Petroleum Geology of the North. European Margin*; Spencer, A.M., Ed.; Norwegian Petroleum Society: Stavanger, Norway; Graham and Trotman: Bristol, UK, 1984; pp. 407–423.

51. Gabrielsen, R.H.; Færseth, R.; Hamar, G.; Rønnevik, H. Nomenclature of the main structural features on the Norwegian Continental Shelf north of the 62nd parallel. In *Petroleum Geology of the North European Margin*; Spencer, A.M., Ed.; Norwegian Petroleum Society: Stavanger, Norway; Graham and Trotman: Bristol, UK, 1984; pp. 41–60.

52. Price, I.; Rattey, R.P. Cretaceous tectonics off mid-Norway: Implications for the Rockall and Faeroe-Shetland troughs. *J. Geol. Soc.* **1984**, *141*, 985–992. [CrossRef]

53. Surlyk, F.; Piasecki, S.; Rolle, F.; Stemmerik, L.; Thomsen, E.; Wrang, P. The Permian base of East Greenland. In *Petroleum Geology of the North European Margin*; Spencer, A.M., Ed.; Norwegian Petroleum Society: Stavanger, Norway; Graham and Trotman: Bristol, UK, 1984; pp. 303–315.

54. Brekke, H.; Riis, F. Tectonics and basin evolution of the Norwegian shelf between 62°N and 72°N. *Nor. Geol. Tidsskr.* **1987**, *67*, 295–322.

55. Doré, A.G.; Lundin, E.R. Cenozoic compressional structures on the NE Atlantic margin: Nature, origin and potential significance for hydrocarbon exploration. *Pet. Geosci.* **1996**, *2*, 299–311. [CrossRef]

56. Swiecicki, T.; Gibbs, P.B.; Farrow, G.E.; Coward, M.P. A tectonostratigraphic framework for the Mid-Norway region. *Mar. Pet. Geol.* **1998**, *15*, 245–276. [CrossRef]

57. Scheck-Wenderoth, M.; Raum, T.; Faleide, J.I.; Mjelde, R.; Horsfield, B. The transition from the continent to the ocean: A deeper view on the Norwegian margin. *J. Geol. Soc.* **2007**, *164*, 855–868. [CrossRef]

58. Mjelde, R.; Kodaira, S.; Shimamura, H.; Kanazawa, T.; Shiobara, H.; Berg, E.W.; Riise, O. Crustal structure of the central part of the Vøring Basin, mid- Norway margin, from ocean bottom seismographs. *Tectonophysics* **1997**, *277*, 235–257. [CrossRef]

59. Talwani, M.; Eldholm, O. Evolution of the Norwegian-Greenland Sea. *GSA Bull.* **1977**, *88*, 969–999. [CrossRef]

60. Eldholm, O.; Thiede, J.; Taylor, E. The Norwegian continental margin: Tectonic, volcanic, and paleoenvironmental framework. In Proceedings of the Ocean Drilling Program, Scientific Results, College Station, TX, USA; 1989; Volume 104. [CrossRef]

61. Stuevold, L.M.; Eldholm, O. Cenozoic uplift of Fennoscandia inferred from a study of the mid-Norwegian margin. *Glob. Planet. Chang.* **1996**, *12*, 359–386. [CrossRef]

62. Grunnaleite, I.; Fjeldskaar, W.; Wilson, J.; Faleide, J.I.; Zweigel, J. Effect of local variations of vertical and horizontal stresses on the Cenozoic structuring of the mid-Norwegian shelf. *Tectonophysics* **2009**, *470*, 267–283. [CrossRef]

63. Dalland, A.; Worsley, D.; Ofstad, K. *A Lithostratigraphic Scheme for the Mesozoic and Cenozoic Succession Offshore Mid- and Northern Norway*; Technical Report for Norwegian petroleum Directorate; Norwegian Petroleum Directorate: Stavanger, Norway, 1988; Volume 4, ISBN 82-7275-241-9.

64. Hjelstuen, B.O.; Eldholm, O.; Skogseid, J. Ceonzoic evolution of the northern Vøring margin. *GSA Bull.* **1999**, *111*, 1792–1807. [CrossRef]

65. Svensen, H.; Planke, S.; Corfu, F. Zircon dating ties NE Atlantic sill emplacement to initial Eocene global warming. *J. Geol. Soc.* **2010**, *167*, 433–436. [CrossRef]

66. Bungum, H.; Alsaker, A.; Kvamme, L.B.; Hansen, R.A. Seismicity and Seismotectonics of Norway and Nearby Continental Shelf Areas. *J. Geophys. Res.* **1991**, *96*, 2249–2265. [CrossRef]

67. Byrkjeland, U.; Bungum, H.; Eldholm, O. Seismotectonics of the Norwegian continental margin. *J. Geophys. Res.* **2000**, *105*, 6221–6236. [CrossRef]

68. Fjeldskaar, W.; Lindholm, C.; Dejls, J.F.; Fjeldskaar, I. Post-glacial uplift, neotectonics and seismicity in Fennoscandia. *Quat. Sci. Rev.* **2000**, *19*, 1413–1422. [CrossRef]

69. Hicks, E.C.; Bungum, H.; Lidholm, C.D. Stress inversion of earthquake focal mechanism solutions from onshore and offshore Norway. *Nor. Geol. Tidsskr.* **2000**, *80*, 235–250. [CrossRef]

70. Bungum, H.; Olesen, O.; Pascal, C.; Gibbons, S.; Lindholm, C.; Vestøl, O. To what extent is the present seismicity of Norway driven by post-glacial rebound? *J. Geol. Soc.* **2010**, *167*, 373–384. [CrossRef]

71. Lander, R.H.; Langfeldt, M.; Bonnell, L.; Fjeldskaar, W. *BMT User's Guide*; Tectonor AS Proprietary Publication: Stavanger, Norway, 1994.

72. Fjeldskaar, W. BMT^{TM}—Exploration tool combining tectonic and temperature modeling: Business Briefing: Exploration & Production. *Oil Gas Rev.* **2003**, 1–4.

73. Walderhaug, O.; Lander, R.H.; Bjørkum, P.A.; Oelkers, E.H.; Bjørlykke, K.; Nadeau, P.H. Modelling Quartz Cementation and Porosity in Reservoir Sandstones: Examples from the Norwegian Continental Shelf. In *Quartz Cementation in Sandstones, Special Publication*; Worden, R.H., Morad, S., Eds.; International Association of Sedimentologists: Algiers, Algeria, 2000; Volume 29, pp. 39–49.

74. Walderhaug, O. Precipitation rates for quartz cement in sandstones determined by fluid-inclusion microthermometry and temperature-history modeling. *J. Sediment. Res.* **1994**, *A64*, 324–333. [CrossRef]

75. Walderhaug, O. Kinetic Modeling of Quartz Cementation and Porosity Loss in Deeply Buried Sandstone Reservoirs. *AAPG Bull.* **1996**, *80*, 731–745.

76. Lander, R.H.; Walderhaug, O. Predicting Porosity through Simulating Sandstone Compaction and Quartz Cementation. *AAPG Bull.* **1999**, *83*, 433–449.

77. Angevine, C.L.; Turcotte, D.L. Porosity reduction by pressure solution: A theoretical model for quartz arenites. *Geol. Soc. Am. Bull.* **1983**, *94*, 1129–1134. [CrossRef]

78. Dewers, T.; Ortoleva, P. A coupled reaction/transport/mechanical model for intergranular pressure solution, stylolites, and differential compaction and cementation in clean sandstones. *Geochim. Cosmochim. Acta* **1990**, *54*, 1609–1625. [CrossRef]

79. Mullis, A.M. The Role of Silica Precipitation Kinetics in Determining the Rate of Quartz Pressure Solution. *J. Geophys. Res.* **1991**, *96*, 10007–10013. [CrossRef]

80. Ramm, M. Porosity-depth trends in reservoir sandstones: Theoretical models related to Jurassic sandstones offshore Norway. *Mar. Pet. Geol.* **1992**, *9*, 553–567. [CrossRef]

81. Stephenson, L.P.; Plumley, W.J.; Palciauskas, V.V. A model for sandstone compaction by grain interpenetration. *J. Sediment. Res.* **1992**, *62*, 11–22.

82. DeMaster, D.J. The supply and accumulation of silica in the marine environment. *Geochim. Cosmochim. Acta* **1981**, *45*, 1715–1732. [CrossRef]

83. Kastner, M.; Keene, J.B.; Gieskes, J.M. Diagenesis of siliceous oozes—I. chemical controls on the rate of opal-A to opal-CT transformation—An experimental study. *Geochim. Cosmochim. Acta* **1977**, *41*, 1041–1059. [CrossRef]

84. Grau, A.; Sterling, R.; Kidney, R. Success! Using Seismic Attributes and Horizontal Drilling to Delineate and Exploit a Diagenetic Trap, Monterey Shale, San Joaquin Valley, California. In Proceedings of the AAPG Annual Convention, Salt Lake City, Utah, USA, 11–14 May 2003.

85. Kidney, R.; Sterling, R.; Grau, A. Delineation of a diagenetic trap using P-wave and converted-wave seismic data in Miocene McLure Shale, San Joaquin Basin, CA. In *SEG Technical Program Expanded Abstracts 2005*; Society of Exploration Geophysicists: Tulsa, OK, USA, 2005; pp. 2329–2332.

86. Isaacs, C.M. Influence of rock composition on kinetics of silica phase changes in the Monterey Formation, Santa Barbara area, California. *Geology* **1982**, *10*, 304–308. [CrossRef]

87. Keller, M.A.; Isaacs, C.M. An Evaluation of Temperature Scales for Silica Diagenesis in Diatomaceous Sequences Including a New Approach Based on the Miocene Formation, California. *Geo-Mar. Lett.* **1985**, *5*, 31–35. [CrossRef]

88. Ernst, W.G.; Calvert, S.E. An Experimental Study of the Recrystallization of Porcelanite and its bearing on the Origin of some bedded Cherts. *Am. J. Sci.* **1969**, *267*, 114–133.

89. Castellan, G.W. *Physical Chemistry*, 3rd ed.; Addison-Wesley Publishing Company: Reading, MA, USA, 1969.

90. Jiang, S. Clay Minerals from the Perspective of Oil and Gas Exploration. In *Clay Minerals in Nature—Their Characterization, Modification and Application*; Valaškova, M., Martynkova, G.S., Eds.; IntechOpen: London, UK, 2012. [CrossRef]

91. Bjørkum, P.A.; Nadeau, P.H. Temperature Controlled Porosity/Permeability Reduction, Fluid Migration, and Petroleum Exploration in Sedimentary Basins. *APPEA J.* **1998**, *38*, 453–465. [CrossRef]

92. Pevear, D.R. Illite and hydrocarbon exploration. *Proc. Natl. Acad. Sci. USA* **1999**, *96*, 3440–3446. [CrossRef] [PubMed]

93. Boles, J.R.; Franks, S.G. Clay Diagenesis in Wilcox sandstones of southwest Texas: Implications of smectite diagenesis on sandstone cementation. *J. Sediment. Res.* **1979**, *49*, 55–70.

94. Huang., W.-L.; Longo, J.M.; Pevear, D.R. An experimentally derived kinetic model for smectite-to-illite conversion and its use as a geothermometer. *Clays Clay Miner.* **1993**, *41*, 162–177. [CrossRef]

95. Roaldset, E.; Wei, H.; Grimstad, S. Smectite to illite conversion by hydrous pyrolysis. *Clay Miner.* **1998**, *33*, 147–158. [CrossRef]

96. Nadeau, P.H.; Wilson, M.J.; McHardy, W.J.; Tait, J.M. Interstratified Clays as Fundamental Particles. *Science* **1984**, *225*, 923–925. [CrossRef]

97. Nadeau, P.H.; Wilson, M.J.; McHardy, W.J.; Tait, J.M. The conversion of smectite to illite during diagenesis: Evidence from some illitic clays from bentonites and sandstones. *Mineral. Mag.* **1985**, *49*, 393–400. [CrossRef]

98. Nadeau, P.H.; Bain, D.C. Composition of some smectites and diagenetic illitic clays and implications for their origin. *Clays Clay Miner.* **1986**, *34*, 455–464. [CrossRef]

99. Bjørlykke, K.; Aagaard, P.; Egeberg, P.K.; Simmons, S.P. Geochemical constraints from formation water analyses from the North Sea and the Gulf Coast Basins on quartz, feldspar and illilte precipitation in reservoir rocks. In *The Geochemistry of Reservoirs*; Special Publication; Cubitt, J.M., England, W.A., Eds.; Geological Society: London, UK, 1995; Volume 86, pp. 33–50.

100. Eberl, D.; Hower, J. Kinetics of illite formation. *Geol. Soc. Am. Bull.* **1976**, *87*, 1326–1330. [CrossRef]

101. Wei, H.; Roaldset, E.; Bjorøy, M. Parallel Reaction kinetics of smectite to illite conversion. *Clay Miner.* **1996**, *31*, 365–376. [CrossRef]

102. Hillier, S.; Matyas, J.; Matter, A.; Vasseur, G. Illite/Smectite diagenesis and its variable correlation with vitrinite reflectance in the Pannonian Basin. *Clays Clay Miner.* **1995**, *43*, 174–183. [CrossRef]

103. Sachsenhofer, R.F.; Rantitcsh, G.; Hasenhüttl, C.; Russegger, B.; Jelen, B. Smectite to illite diagenesis in early Miocene sediments from the hyperthermal western Pannonian Basin. *Clay Miner.* **1998**, *33*, 523–537.

104. Anell, I.; Thybo, H.; Artemieva, I.M. Cenozoic uplift and subsidence in the North Atlantic region: Geological evidence revisited. *Tectonophysics* **2009**, *474*, 78–105. [CrossRef]

105. Anell, I.; Thybo, H.; Stratford, W. Relating Cenozoic North Sea sediments to topography in southern Norway: The interplay between tectonics and climate. *Earth Planet. Sci. Lett.* **2010**, *300*, 19–32. [CrossRef]

106. Nielsen, S.B.; Gallagher, K.; Leighton, C.; Balling, N.; Svenningsen, L.; Jacobsen, B.H.; Thomsen, E.; Nielsen, O.B.; Heilmann-Clausen, C.; Egholm, D.L.; et al. The evolution of western Scandinavian topography: A review of Neogene uplift versus the ICE (isostasy–climate–erosion) hypothesis. *J. Geodyn.* **2009**, *47*, 72–95. [CrossRef]

107. Nielsen, S.B.; Clausen, O.R.; Jacobsen, B.H.; Thomsen, E.; Huuse, M.; Gallagher, K.; Balling, N.; Egholm, D. The ICE hypothesis stands: How the dogma of late Ceonozoic tectonic uplift can no longer be sustained in the light of data and physical laws. *J. Geodyn.* **2010**, *50*, 102–111. [CrossRef]

108. Sclater, J.G.; Christie, P.A.F. Continental stretching: An explanation of the post-mid-cretaceous subsidence of the central North Sea basin. *J. Geophys. Res.* **1980**, *85*, 3711–3739. [CrossRef]

109. Čermác, V.; Rybach, L. Thermal properties: Thermal conductivity and specific heat of minerals and rocks. In *Landolt-Börnstein Zahlenwerte und Functionen aus Naturwissenschaften und Technik, Neue Serie, Physikalische Eigenschaften der Gesteine*; Angeneister, G., Ed.; Springer: Berlin/Heidelberg, Germany; New York, NY, USA, 1982; pp. 305–343.

110. Pollastro, R.M. Considerations and Applications of the Illite/Smectite Geothermometer in Hydrocarbon-Bearing Rocks of Miocene to Mississippian Age. *Clays Clay Miner.* **1993**, *41*, 119–133. [CrossRef]

111. Gudmundsson, A. *Rock Fractures in Geological Processes*, 1st ed.; Cambridge University Press: Cambridge, UK, 2011.

112. Bell, F.G. *Engineering Properties of Soils and Rocks*, 4th ed.; Blackwell: Oxford, UK, 2000.

113. Farmer, I. *Engineering Behaviour of Rocks*, 2nd ed.; Chapman and Hall: London, UK, 1983.

114. Paterson, M.S.; Wong, T.F. *Experimental Rock Deformation: The Brittle Field*, 2nd ed.; Springer: Berlin, Germany, 2005.

115. Haimson, B.C.; Rummel, F. Hydrofracturing stress measurements in the Iceland research drilling project drill hole at Reydarfjördur, Iceland. *J. Geophys. Res.* **1982**, *87*, 6631–6649. [CrossRef]

116. Schultz, R.A. Limits on strength and deformation properties of jointed basaltic rock masses. *Rock Mech. Rock Eng.* **1995**, *28*, 1–15. [CrossRef]

117. Amadei, B.; Stephansson, O. *Rock Stress and Its Measurement*; Chapman and Hall: London, UK, 1997.

118. Gudmundsson, A.; Brenner, S.L. How hydrofractures become arrested. *Terra Nova* **2001**, *13*, 456–462. [CrossRef]

119. Gudmundsson, A.; Fjeldskaar, I.; Brenner, S.L. Propagation pathways and fluid transport of hydrofractures in jointed and layered rocks in geothermal fields. *J. Volcanol. Geotherm. Res.* **2002**, *116*, 257–278. [CrossRef]

120. Fjeldskaar, W.; Prestholm, E.; Guargena, C.; Stephenson, M. Mineralogical and diagenetic control on the thermal conductivity of the sedimentary sequences in the Bjørnøya Basin, Barents Sea. In *Basin Modelling: Advances and Applications, Special Publication*; Doré, A.G., Ed.; NPF: Stavanger, Norway, 1993; Volume 3, pp. 445–453.

121. Ireland, M.; Goulty, N.R.; Davies, R.J. Influence of pore water chemistry on silica diagenesis: Evidence from the interaction of diagenetic readtion zones with polygonal fault systems. *J. Geol. Soc.* **2010**, *167*, 273–279. [CrossRef]

122. McKinley, J.M.; Worden, R.H.; Ruffell, A.H. Contact Diagenesis: The effect of an intrusion on reservoir quality in the Triassic Sherwood Sandstone group, Northern Ireland. *J. Sediment. Res.* **2001**, *71*, 484–495. [CrossRef]

123. Bernet, M.; Gaupp, R. Diagenetic history of Triassic sandstone from the Beacon Supergroup in central Victoria Land, Antarctica. *N. Z. J. Geol. Geophys.* **2005**, *48*, 447–458. [CrossRef]

124. Grove, C. Direct and Indirect Effects of Flood Basalt Volcanism on Reservoir Quality Sandstone. Ph.D. Thesis, Durham University, Durham and Stockton-on-Tees, UK, 2014.

125. Nejbert, K.; Krajewski, K.P.; Dubinska, E.; Pécskay, Z. Dolerites of Svalbard, north-west Barents Sea Shelf: Age, tectonic setting and significance for geotectonic interpretation of the High-Arctic Large Igneous Province. *Polar Res.* **2011**, *30*, 7306. [CrossRef]

126. Ehrenberg, S.N. Relationship between Diagenesis and Reservoir Quality in Sandstones of the Garn Formation, Haltenbanken, Mid-Norwegian Continental Shelf. *AAPG Bull.* **1990**, *74*, 1538–1558. [CrossRef]

127. Bjørlykke, K.; Nedkvitne, T.; Ramm, M.; Saigal, G.C. Diagenetic processes in the Brent Group (Middle Jurassic) reservoirs of the North Sea: An overview. In *Geology of the Brent Group*; Special Publication; Morton, A.C., Haszeldine, R.S., Giles, M.R., Brown, S., Eds.; Geological Society: London, UK, 1992; Volume 61, pp. 263–287.

128. Ehrenberg, S.N. Preservation of Anomalously High Porosity in Deeply Buried Sandstones by Grain-Coating Chlorite: Examples from the Norwegian Continental Shelf. *AAPG Bull.* **1993**, *77*, 1260–1286.

129. Freiburg, J.T.; Ritzi, R.W.; Kehoe, K.S. Depositional and diagenetic controls on anomalously high porosity within a deeply buried CO_2 storage reservoir—The Cambrian Mt. Simon Sandstone, Illinois Basin, USA. *Int. J. Greenh. Gas Control* **2016**, *55*, 42–54. [CrossRef]

130. Lin, W.; Chen, L.; Lu, Y.; Hu, H.; Liu, L.; Liu, X.; Wei, W. Diagenesis and its impact on reservoir quality for the Chang 8 oil group tight sandstone of the Yanchang Formation (upper Triassic) in southwestern Ordos basin, China. *J. Pet. Explor. Prod. Technol.* **2017**, *7*, 947–959. [CrossRef]

131. PetroWiki.org. Available online: https://petrowiki.org/Diatomite (accessed on 27 September 2019).

132. Tsuji, T.; Masui, Y.; Yokoi, S. New hydrocarbon trap models for the diagenetic transformation of opal-CT to quartz in Neogene siliceaous rocks. *AAPG Bull.* **2011**, *95*, 449–477. [CrossRef]

133. Davies, R.J.; Cartwright, J. A fossilized Opal A to Opal C/T transformation on the northeast Atlantic margin: Support for a significantly elevated Palaeogeothermal gradient during the Neogene? *Basin Res.* **2002**, *14*, 467–486. [CrossRef]

134. Hall, P.L.; Astill, D.M.; McConnell, J.D.C. Thermodynamic and structural aspects of the dehydration of smectites in sedimentary rocks. *Clay Miner.* **1986**, *21*, 633–648. [CrossRef]

135. Singer, A.; Müller, G. Diagenesis in argillaceous sediments. In *Diagenesis in Sediments and Sedimentary Rocks*; Larsen, G., Chilingar, G.V., Eds.; Elsevier: Amsterdam, The Netherlands, 1983.

136. Essene, E.J.; Peacor, D.R. Clay mineral thermometry—A critical perspective. *Clays Clay Miner.* **1995**, *43*, 540–553. [CrossRef]

137. Podladchikov, Y.Y.; Wickham, S.M. Crystallization of Hydrous Magmas: Calculation of Associated Thermal Effects, Volatile Fluxes, and Isotopic Alteration. *J. Geol.* **1994**, *102*, 25–45. [CrossRef]

138. Iyer, K.; Rüpke, L.; Galerne, C.Y. Modeling fluid flow in sedimentary basins with sill intrusions: Implications for hydrothermal venting and climate change. *Geochem. Geophys. Geosyst.* **2013**, *14*, 5244–5262. [CrossRef]

139. Iyer, K.; Schmid, D.W.; Planke, S.; Millett, J. Modelling hydrothermal venting in volcanic sedimentary basins: Impact on hydrocarbon maturation and paleoclimate. *Earth Planet. Sci. Lett.* **2017**, *467*, 30–42. [CrossRef]

140. Wang, D.; Manga, M. Organic matter maturation in the contact aureole of an igneous sill as a tracer of hydrothermal convection. *J. Geophys. Res. Solid Earth* **2015**, *120*, 4102–4112. [CrossRef]

141. Annen, C. Factors affecting the thickness of thermal aureoles. *Front. Earth Sci.* **2017**, *5*, 1–13. [CrossRef]

142. Planke, S.; Rasmussen, T.; Rey, S.; Myklebust, R. Seismic characteristics and distribution of volcanic intrusions and hydrothermal vent complexes in the Vøring and Møre basins. In *Petroleum Geology: North-West Europe and Global Perspectives, Proceedings of the 6th Petroleum Geology Conference*; Dorè, A.G., Vining, B.A., Eds.; Geological Society: London, UK, 1 January 2005; pp. 833–844.

Article

Observational and Critical State Physics Descriptions of Long-Range Flow Structures

Peter E. Malin [1],*, Peter C. Leary [2], Lawrence M. Cathles [3] and Christopher C. Barton [4]

[1] Helmholtz Centre Potsdam, GFZ German Research Centre for Geosciences, Telegrafenberg, 14473 Potsdam, Germany
[2] GeoFlow Imaging, Auckland 1010, New Zealand; peter@geoflowimaging.com
[3] Department of Earth and Atmospheric Sciences, Cornell University, Ithaca, NY 14853, USA; lmc19@cornell.edu
[4] Department of Earth and Environmental Sciences, Wright State University, Dayton, OH 45435, USA; chris.barton@wright.edu
* Correspondence: pem@asirseismic.com

Received: 11 November 2019; Accepted: 24 January 2020; Published: 28 January 2020

Abstract: Using Fracture Seismic methods to map fluid-conducting fracture zones makes it important to understand fracture connectivity over distances greater 10–20 m in the Earth's upper crust. The principles required for this understanding are developed here from the observations that (1) the spatial variations in crustal porosity are commonly associated with spatial variations in the magnitude of the natural logarithm of crustal permeability, and (2) many parameters, including permeability have a scale-invariant power law distribution in the crust. The first observation means that crustal permeability has a lognormal distribution that can be described as $\kappa \approx \kappa_0 \exp(\alpha(\varphi - \varphi_0))$, where α is the ratio of the standard deviation of ln permeability from its mean to the standard deviation of porosity from its mean. The scale invariance of permeability indicates that $\alpha\phi_0 = 3$ to 4 and that the natural log of permeability has a 1/k pink noise spatial distribution. Combined, these conclusions mean that channelized flow in the upper crust is expected as the distance traversed by flow increases. Locating the most permeable channels using Seismic Fracture methods, while filling in the less permeable parts of the modeled volume with the correct pink noise spatial distribution of permeability, will produce much more realistic models of subsurface flow.

Keywords: crustal well-core poroperm; crustal fluid flow; crustal flow channeling; critical state physics; well-log spectral scaling; crustal power law scaling; lognormal; pink noise; crustal fracture seismics; crustal fracture imaging

---

1. Introduction

As discussed by Sicking and Malin (2019) [1], new methods of passive seismic data acquisition and processing allow ambient subsurface fluid connectivity to be mapped. Complementary to the information on fluid pathways from microearthquakes (MEQ) generated by massive high-pressure (active) fluid injections (e.g., [2]), the new passive technique of Fracture Seismics (FS) can directly reveal both the locations and relative permeabilities of fluid-filled fractures. Further, and perhaps more importantly, passive FS mapping shows the degree of flow connectivity at distance scales greater than a few tens of meters from the point of active fluid injection.

A variety of independent evidence indicates that locations where MEQ and FS activity is most intense are where fluid flow is most likely [1]. FS maps typically reveal long connectivity structures that conduct fluid flow, as illustrated in Figure 1. The top panel shows the intensity of FS signals, and the bottom panel indicates the fracture-connectivity skeletal pathways through the cloud where FS signals are most intense and fluid flow is most likely.

The FS pattern in Figure 1 is from a fracture stimulation in the Marcellus gas shales, and the connectivity of its permeability-related signals may be considered a surprise. However, in this case the FS flow prediction was confirmed both by gas pressure communication and by a tracer experiment. These measurements showed a high degree of connectivity over ~600 m between vertical Well B and horizontal Well A, as indicated by the FS backbone in the lower panel. This backbone was observed in the FS data from Stage 3, and then later also at Stage 4, in advance of the arrival of tracer signals in Well B, which appeared at the time Stage 5 was stimulated.

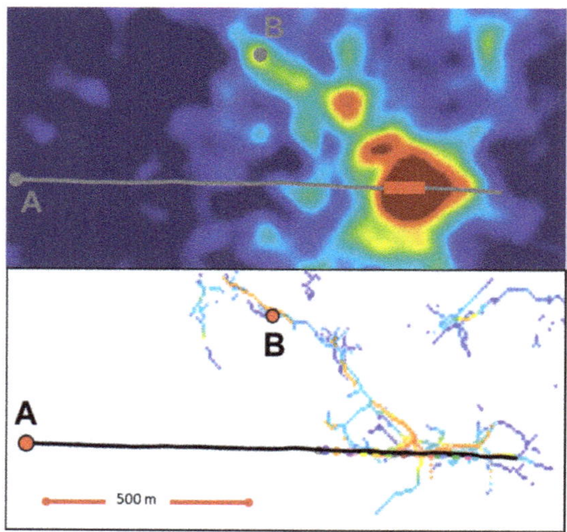

Figure 1. (**Top**) A horizontal map slice of the above background Fracture Seismic (FS) intensity (red high, blue low), activated during the third stage of a fracture stimulation treatment. (**Bottom**) The above-background FS intensity (red high, blue low) backbone of the same horizontal map slice in the FS intensity cloud for hydrofracturing Stages 3–5. Point A is the wellhead of a horizontal well with a trajectory indicted by the black line. Point B shows the location of a vertical well with pressure and chemical tracer monitors. Both pressure and tracer communicated between Well B and Well A, where the FS skeleton connects the wells over a distance of 600 m. Figure is from Lacazette et al (2013) [3].

There are good observation-based theoretical reasons to expect the kind of spatially extended crustal flow connectivity on the hundred-meter scale seen in Figure 1. The purpose of this paper is to discuss these reasons. We show that the distribution of features observed in such FS maps is expected. The expectation that crustal fluid flow over length-scales >10–20 m will be through high-connectivity flow structures is discussed below, from both (a) direct observations and physical logic and (b) the power law scale invariance in the observed self-similarity of fracture-connectivity patterns. The first approach was developed over the last three decades in a series of papers (e.g., [4–10]). The second approach has a vast and densely spaced literature (e.g., [11] and the many references therein). Our purpose here is to bring these insights together and convey an understanding of how upper-crustal fluid flow is likely to be structured, based on observations and the critical state physics of brittle rock.

2. Crustal Observations and Implications

2.1. The Observed Relationship between Crustal Porosity and Permeability

This section first shows that spatial variations in crustal porosity invariably coincide with commensurate spatial variations in the natural logarithm of permeability. It then shows that this means that the magnitude of crustal permeability commonly has a lognormal distribution, even though the magnitude of porosity that is spatially associated with permeability is normally distributed. Following Leary et al. (2017, 2018) [12,13], we derive a general expression that encapsulates this observation and deduction. The derivation shows that the combinatorial probability of crustal pore and fracture connections provides a logical explanation for why normally distributed pores are associated with a lognormal distribution of permeability. There could be a physical extension between pores and fractures [13] which we do not discuss here. With increasing strain, broken grain cements could progressively localize deformation, eventually leading to faulting, for example. This is not addressed in this paper. Here, we use observations of cores to discuss the connection conditions for permeability and learn how a normal distribution of pores and pore connections leads to the observed lognormal distribution of core permeability. We then transfer this insight to fractures.

Figures 2–4 show field data that illustrate how changes in crustal porosity are invariably associated with changes in the natural logarithm of permeability,

$$\delta\varphi \propto \delta \ln \kappa \tag{1}$$

The (possibly irregularly spaced) porosity and permeability values in these figures were measured down well-core samples according to standard oil-field procedures (e.g., [14]). The histograms on the left side of Figure 2 show that the porosity values of four North Sea cores have a normal spatial distribution, whereas the permeability values have a lognormal distribution. Similar histograms of porosity and permeability are shown in Figures 3 and 4.

On the right side of Figure 2, the porosity and permeability values are plotted as a function of sample number. In Figures 3 and 4, they are plotted as a function of sample depth. The measurements are plotted in normalized form, such that their variations have zero-mean and unit variance. This is achieved by subtracting the mean porosity (or ln permeability) and dividing by the standard deviation of porosity (or ln permeability). These profiles show that where there is a change in porosity there is a correlated change in ln permeability. Natural logs are used throughout the discussion for consistency with these observational plots and for compatibility to theory.

Whereas the number of samples with small porosity values rapidly decreases in these histograms, the same is not true for permeability histograms. The roll-off in the number of samples with small permeability values can be seen only in the plot of numbers versus the logarithm of permeability, and may be due to failure to measure variations in low permeability. The potential for undersampling suggests that the permeability distribution could be alternatively described as a power law (which has no roll-off at low permeability) instead of lognormal. Both logarithmic and power distributions are long-tailed at the high permeability end of their distributions. However, power distributions are fat-tailed, meaning this distribution has more high permeability values than a lognormal distribution.

The important point here is not these differences between power and lognormal distributions, but simply that, on simple inspection, the porosity and permeability distributions are normal and lognormal, respectively, and these relationships are general. The Figures show data from wells worldwide. They depict the most commonly observed characteristic of reservoir rock porosity and permeability. Furthermore, the normal-versus-lognormal correlations between porosity and permeability are not dependent on the orientation of the sampling well. Figure 4 shows that the same relationship is observed in cores from a horizontal well as in cores from vertical wells. Be it a vertical or horizontal well-core sampling sequence, the differences in their population distributions remain the same: normally distributed porosity and long-tail power permeability. Both these distribution and correlation relationships appear to apply just as well along rock formations as across them.

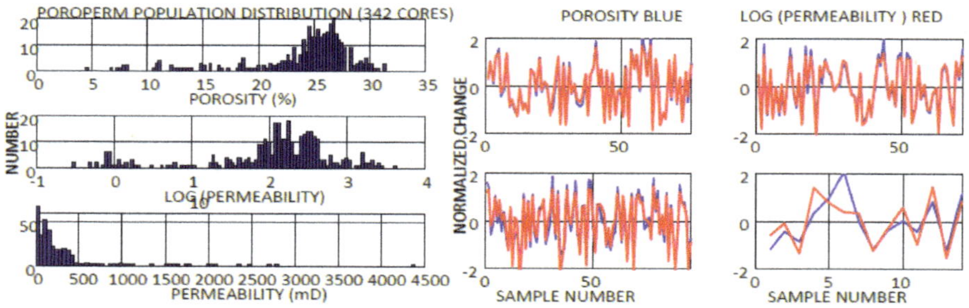

Figure 2. Porosity and permeability measurements on four sub-vertical North Sea gas field well cores. (**Left panel**) Histograms of porosity values are plotted as number of values in 1% wide bins versus percent porosity. Both log permeability and permeability data are plotted as the number of values in a bin versus millidarcies. The histograms suggest that the porosity has a normal, and the permeability a lognormal, distribution. (**Two right panels**) Porosity (blue) and natural ln permeability (red) data are plotted as functions of sample number. The porosity and permeability data are normalized to zero mean and unit variance.

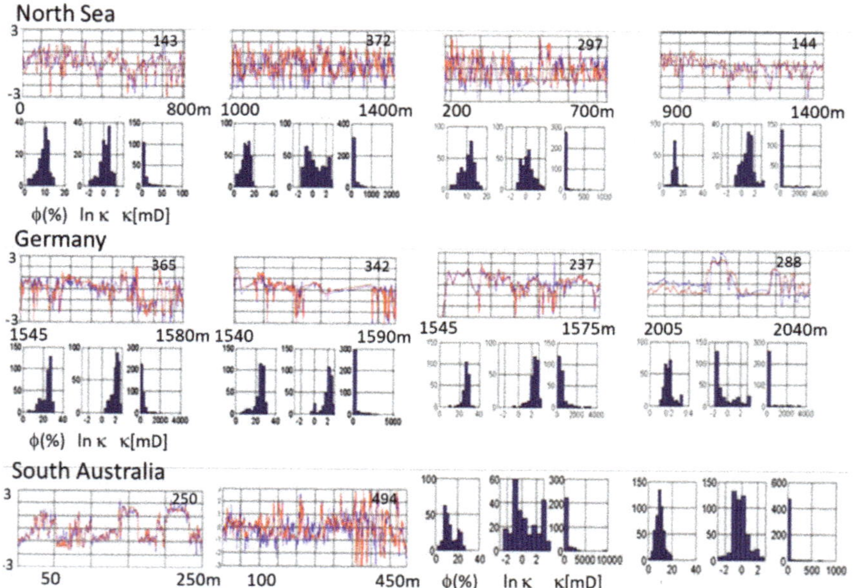

Figure 3. Porosity (φ, blue) and natural log permeability (ln κ, red) magnitudes in sub-vertical well cores from the North Sea (top), Germany (middle), and South Australia (bottom). Porosity and ln permeability are plotted against sample depth. Both are normalized to zero mean and unit variance. The number of measurements is shown on the plots. In the histograms, the porosity values are plotted as the number of values in 1% wide bins versus percent porosity. Both ln permeability and permeability data are plotted as the number of values in bins versus millidarcies (mD).

Figures 2–4 show data that establish Equation (1). Equation (1) can be integrated to yield an expression between the spatial variation of permeability, $\kappa(x,y,z)$, and the spatial variation of porosity, $\varphi\,(x,y,z)$

$$\kappa \approx \kappa_0 \exp(\alpha(\varphi - \varphi_0)), \tag{2}$$

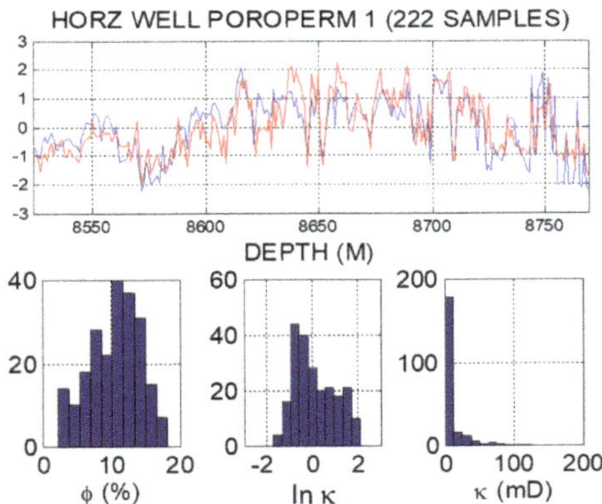

Figure 4. (**Top**) Porosity (φ, blue) and ln permeability (ln κ, red) measurements from a horizontal well from North Sea clastic oil fields, plotted with zero-mean and unit variance. (**Bottom**) Histograms of the number of measurements in bins of equal porosity, ln κ, or κ.

Here, κ_0 and φ_0 are the mean values of permeability and porosity, and α is a constant of proportionality. That (2) follows from (1) can be easily verified by differentiating (2) with respect to φ. If $\varphi = \varphi_0 \pm \sigma_\varphi$, it follows from (2) that $\ln \kappa = \ln \kappa_0 \pm \alpha \sigma_\varphi$, where α is the ratio of the standard deviation of $\ln \kappa$ to the standard deviation of φ:

$$\alpha = \sigma_{\ln \kappa} / \sigma_\varphi. \tag{3}$$

When α is very small, the series expansion of the exponential in Equation (2) reduces to a linear relationship between the departure of κ and the departure of φ from their mean, and both permeability and porosity are normally distributed. In the small-α limit, permeability equations such as the Kozeny–Carmen relationship hold. In this domain, the flow is homogeneous for a normal distribution of porosity.

When α is large, κ fits the definition of a lognormal variable, where the permeability distribution has a long permeable tail. In the large-α limit, $\ln \kappa$ has a standard deviation spread of $\sigma_{\ln \kappa}$. If this spread is large enough and, as discussed in the next section, the spatial distribution of permeability is a power distribution, the flow will be channelized. These two conditions (large α and a power spatial distribution of permeability) for channelized fracture flow are what we investigate in this paper.

2.1.1. The Connection–Condition Explanation for a Lognormal Distribution in Permeability

The reason that a normal distribution in porosity magnitude can lead to a lognormal distribution in permeability magnitude can be understood if it is recognized that permeability requires a connection condition to be met. For the current discussion, this connection condition is that fractures (or pores) must be able to pass fluids onward to other fractures (or pores).

Shockley (1957) [15], among others, illustrated how physical/statistical observables (events) that are normally distributed can interact conditionally to create other observables (events) that are lognormally distributed. His simple example asked why relatively few authors publish the bulk of scientific papers. He proposed that eight author traits are necessary to produce a publication: (1) ability to select problems; (2) competence to work on them; (3) ability to recognize a result; (4) ability to prepare a manuscript; (5) ability to present results adequately; (6) ability to profit from criticism; (7) determination to submit a manuscript; and (8) ability to respond to referees' criticism. Assuming each author trait is normally distributed through a population of scientists, the conditional nature of the final 'event', the publication of a scientific paper, depends on the product rather than the sum of the component traits. If p_1 is the probability for trait 1, p_2 for trait 2, etc., then $P = p_{1*} p_{2*} \ldots * p_8$ is the probability of publication, and only a few authors will possess all the traits needed for successful publication. Successful authors will be rare, and the distribution of author publications will have a long tail: few scientists meet the connection condition of having all eight traits, so they publish a disproportionate share of scientific papers.

Permeability requires that pores and fractures be connected. As the scale or volume of interest becomes larger, there will be more and better connections between pores, and longer and more permeable fractures within a given volume, and the probability that some subset of pores or fractures will be connected will increase with increasing volume. For example, given n independent fractures in a volume, simple counting shows there are $n! = n*(n-1)*(n-2)*$ ways to connect them. If the number of pores or fractures increases by δn, the probability of connection will increase: $\delta ln(n!) = ln([n+\delta n]!) - ln(n!) \approx \delta n \, ln(n)$. Since $ln(n)$ changes more slowly than n, it is reasonable to assume that $\delta ln(n!) \propto \delta n$ Associating n with φ, and $n!$ with κ, we find $\delta ln(\kappa) \propto \delta \varphi$, as shown in Equation (1), above. We thus see that, if permeability is viewed in terms of connections of pores or of fractures, it follows that a change in the number of pores (pore connections) or in the number of fractures will lead to a change in ln permeability.

2.1.2. Spatial Properties at the Critical State

Critical state physics principles can account for how variations in permeability become distributed spatially. This distribution is the second and final piece of information needed to determine whether the departures of ln permeability from its mean will lead to flow channeling.

One example is the application of critical state physics to the phenomenon of opalescence. Opalescence is the clouding of an otherwise clear liquid such as CO_2 at its critical temperature and pressure. Observed almost 200 years ago, this phenomenon was understood 100 years later to be related to spatial optical index fluctuations in the critical liquid which scatter light, thereby clouding the liquid. In the last 50 years, the amplitudes of these fluctuations have been found to be power law distributed as a function of their physical length scale. The longer the length scale of the fluctuation, the larger the amplitude of its density change. For example, if the density variations were decomposed into their Fourier components, the longer wavelengths would have a greater amplitude. The square of the amplitude might vary as 1/k, where k=$2\pi/\lambda$, where λ is the wavelength of the Fourier component.

Properties that vary spatially in a power law 1/k fashion are scale invariant in the sense that you cannot deduce scale in an image of the material that contains the property. In the case of opalescence in a clear liquid, the pattern of optical index variations captured in a photo frame of tenths of a millimeter on a side looks the same and is statistically the same as that captured in frames that are millimeters on a side. Furthermore, the onset of density variations and the clouding they cause is related to a "structure function" such as $\varepsilon = (T - T_c)/T_c$, where T is the system temperature, and T_c is the critical point temperature (e.g., [16]). Importantly, the opalescence is clearly related to the proximity to the critical point and not to other properties of the fluid.

The critical-state percolation model [17,18] provides additional insight. In the simplest version of a percolation model, nodes in a square grid are connected to their neighbors by bonds that are either open (and allow flow between nodes) or closed (and allow no flow). If open bonds are randomly assigned according to some probability, there is a critical fraction of open bonds at which at least one pathway of conducting bonds spans the sample. Near this critical sample-spanning fraction, percolation model properties such as permeability, electrical conductivity, and diffusion coefficient are found to be power functions, and universally depend more on spatial dimension than the morphology (internal construction) of the medium [18]. This potentially greatly simplifies understanding rock properties. The flow backbone in the sample-spanning cluster is found by deleting all the dead-end conducting bonds where no flow occurs. The flow backbone provides a basis for the depiction of backbone fracture seismic flow structures shown in the bottom section of Figure 1. Note that the seemingly dead-end backbones of flow in Figure 1 could connect to overlying or underlying depth slices through the volume, or could discharge into a broadened flow.

2.1.3. The Critical State of the Earth's Crust

A major insight of the late 20th Century in the Earth Sciences is that the earth's crust is stressed by plate tectonic forces, such that it is in a state of constant incipient failure [19]. The brittle crust is riddled with fractures that are always at or near mechanical failure. Fractures are power scaling or scale-invariant characteristics of rock [11]. Geologists must place a hammer or some other object of known scale in a picture of veins or fracture traces to indicate their scale (e.g., [11]). Leary et al. (2018) [13] have shown that the $\alpha\varphi_0$ parameter in Equation (2) above, plays the same role in the deformation of the earth's crust as the critical temperature plays in opalescence. If $\alpha\varphi_0$ is between 3 and 4, the crust is in a state of constant incipient failure. The result is that the spatial distribution of fractures is power law, and it follows that the spatial distribution of permeability is power law. Barton, Camerlo, and Bailey, 1997 [20] have shown that the distribution of fracture trace lengths is also power scaling. Leary et al. (2018) [13] show that $\alpha\varphi_0$ is between 3 and 4 for a wide variety of sedimentary rock, which means that they are in a state of incipient failure.

2.1.4. The Power Exponent of Permeability

Prior to the invention of FS methods, it was not possible to image fracture permeability directly to determine its location and spatial distribution. However, permeability proxies such as the mineral alteration caused by fluid flow through fractures and the size distribution of microearthquakes, can be used to infer the power exponent that characterizes the spatial distribution of fractures in rock [11]. The permeability power exponent can also be inferred from Equation (2) for a pink noise spatial variation in porosity or by the direct field mapping of fractures.

Figure 5 shows how the square of the amplitude (spectral density) of well-log measurements of fracture density, porosity, and fluid-flow-related mineral alteration fall off as the spatial scale of measurement (frequency) decreases. The variation in signal strength with depth, measured by well-logs, is Fourier transformed. The logarithm of the square of the amplitude (fluctuation power of the signal, $S(k)$) of each Fourier component is plotted against the logarithm of the spatial frequency, k, of the component measured in cycles/km (or any other units of measure, since the log slope is not affected by the units). The spectral density method is well described by Malamud and Turcotte (1999) [23]. All the properties related to fluid flow, plotted in Figure 5, show approximately a pink noise, scale invariant, 1/k power distribution.

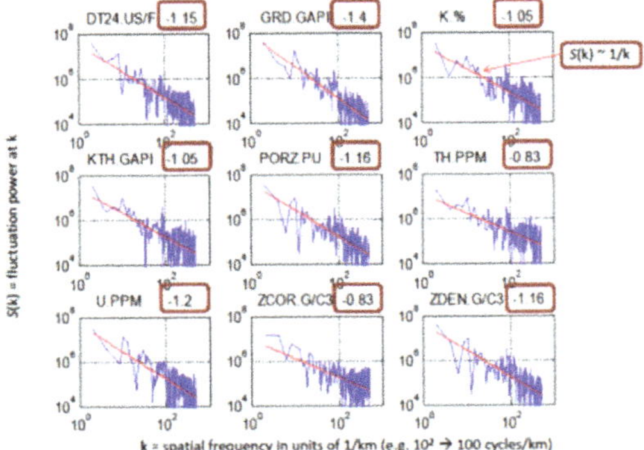

Figure 5. Time-series analysis well-log properties related to flow in fractures. From **top left to bottom right,** these properties are: acoustic velocity, gamma ray intensity, potassium abundance, potassium-thorium abundance, neutron porosity, thorium abundance, uranium abundance, mass density correlation, and mass density. In all these panels, the spectral density S(k) (y-axis) falls off with frequency as ~1/k, where k is spatial scale-length measured in number of cycles per km (x-axis). Data are from a sand-shale well-log provided to us courtesy of R Slatt [21,22]. The power-scaling exponent is shown in the red box at the upper right of each plot.

2.1.5. The Flow Significance of the Distribution of Permeability

Figure 5 shows that down hole log measurements related to flow in fractures are power-scaling with an exponent of approximately −1:

$$S(k) \propto 1/k^{\beta}, \text{ where } \beta = 1. \tag{4}$$

This value of the power exponent, β, is intermediate between zero slope of randomly distributed white noise and the slope of two of red noise (the Brownian noise, obtained by taking the running sum of randomly distributed spatial properties). A power spectrum with a slope of one is called a pink noise. Equation (4) shows that, for pink noise, when k becomes smaller (e.g., the number of Fourier cycles per km is fewer and the variation extends over a larger distance), the signal power increases. For a pink noise spatial distribution of permeability, as the volume of rock becomes larger, it is increasingly likely that a permeable channel will short-circuit flow across the entire observed volume. We show this below by first discussing how white and pink noise distributions of porosity differ, and then use Equation (2) to convert a pink noise porosity distribution to a pink noise permeability distribution and address channeling.

Figure 6 contrasts white and pink noise distributions in porosity. The plots on the left side of Figure 6 depict a hypothetical white noise porosity distribution. The top image is a depth slice map of porosity. The middle plots the porosity along an arbitrarily directed hypothetical well-log through the depth slice map shown above. The Fourier power spectrum $S(k)$ of this porosity well-log is plotted as a function of wave number k on the abscissa in the bottom image. The white noise is generated from a standard pseudo-random number generator in Matlab computational software.

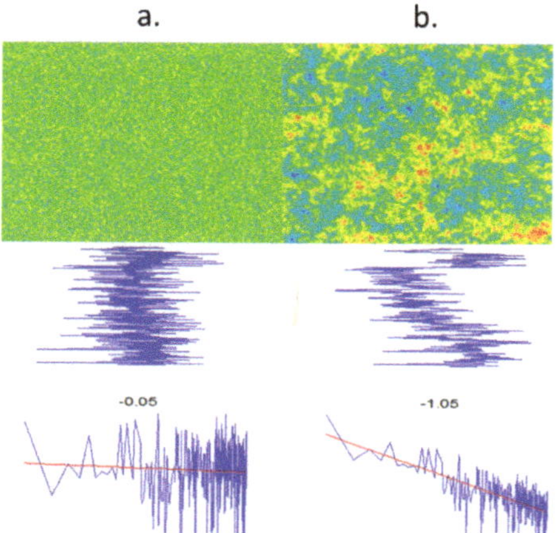

Figure 6. (**a**): Spatially uncorrelated and (**b**) correlated hypothetical porosity models. (**Top**): Map views of porosity variation indicated by color. (**Middle**) Representative synthetic porosity well-logs in an arbitrary direction through the porosity map view images above. The ordinate is distance and the abscissa is the magnitude of porosity. (**Bottom**): Power spectrum of synthetic well-log porosity, plotting the square of harmonic amplitude (ordinate) against spatial frequency (cycles per km) on abscissa. Numbers are the power exponents (-β) of the power spectra (see Equation (4)). They confirm that the left map distribution is spatially uncorrelated white noise ($\beta \sim 0$) and the right map distribution is spatially correlated $1/k$ pink noise ($\beta \sim 1$).

The working assumption behind the white noise application to crustal randomness is that, at some suitably large scale-length, whatever is physically happening at one point in the formation is independent of whatever is happening at any other point. That is, above a suitable scale length, crustal flow structure is uncorrelated. The white noise uncorrelation assumption is inherent in the representative elementary volume (REV) method of crust flow modelling (e.g., [24–26]). Whatever the pattern in the REV, it is assumed that one could make a 3D rubber stamp of the REV and replicate the pattern at a larger scale by filling in the whole volume with the stamp. The largest and smallest features remain the same, no matter how many times a pattern is repeated or how large its volume is. REV patterns cannot increase or decrease the range of properties beyond those in the REV. REV's thus have no scaling larger or smaller than that contained within the REV itself. As a consequence, the entire crustal flow volume is described by just one REV sub volume. There is no large-scale variation in permeability in the depth slice white noise map (top right).

In contrast, the *1/k* spectrum pink noise-correlated spatial distribution of porosity is illustrated on the right side of Figure 6. The pink noise distribution of porosity is generated using discrete Fourier transform methods, as described in ([23], p.35). The pink noise porosity distribution is then shown along a hypothetical well-log through the depth slice image (middle right diagram) and plotted in power spectrum form in the bottom right image. The pink noise porosity distribution reveals different intensities at every scale length contained in the volume. The plan map shows dramatic inhomogeneities which connect over large distances. A yellow to red band across the entire property map can be seen in the middle of Figure 6, for example. This porosity pattern suggests that a flow channel could pass fluid across the entire image. In contrast to the uncorrelated plan map depiction to the left, no small portion of the correlated map is representative of any other small portion. The porosity level and pattern at one location cannot be used to predict what will be found in another small portion of the image. The spatial fluctuation power spectrum has a power slope of ~1 rather than ~0. Connected flow paths are expected, but the position of the spatially correlated pore-connectivity cannot be predicted on the basis of a limited number of samples.

The tendency towards channelized flow at large scale for a *1/k* pink noise porosity distribution is demonstrated in Figure 7. Flow in this figure is calculated for a porosity distribution derived from a normal distribution of pore sizes with a *1/k* power spatially correlated distribution with a power scaling exponent of −1, in the same fashion as described above. The porosity normal distribution is converted to permeability using Equation (2), a pressure gradient is imposed across the spatial permeability distribution, and the resulting flow is computed using the Matlab finite-element partial differential equation solver. The 2D grid has 300×300 element resolution. Arrows show both the distribution of velocity magnitudes and flow directions in the model domain. The ±2 spread in the natural logarithm of the flow velocity (ln|V|) in Figure 7 approximately matches the spread of natural log permeability shown in Figures 2–4. The flow directions show that spatially correlated porosity leads to flow dramatically channeled across the model from the high-pressure boundary to the low-pressure boundary.

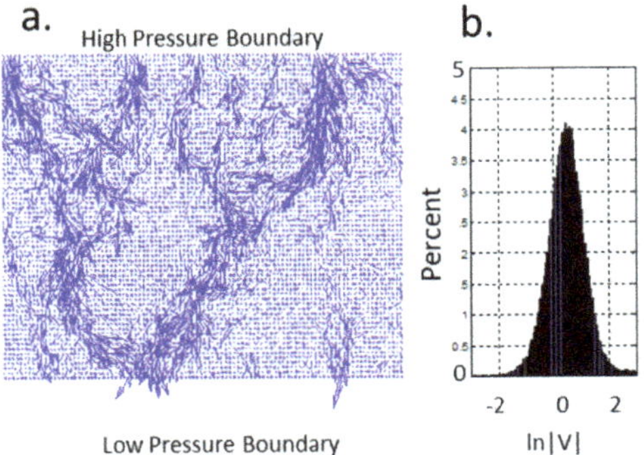

Figure 7. **a.** The flow field, computed for a medium with a permeability that varies by ±2 natural log units (see Figures 2–4 well-core data) and has a 1/k spatial distribution of ln κ, shows dramatic channeling between the high- and low-pressure boundaries. **b.** Distribution of natural log-well log-flow velocity, ln (V), where V is in arbitrary units

2.1.6. The Power Exponent of Permeability from Two-Point Analysis of Permeability

Figures 2–4 suggest a mean value of $\phi = 0.2 \pm 0.1$. Assuming a critical state crust with $\alpha\phi_0 = 4$, it follows that $\alpha = 20$, and, since $\alpha = \sigma_{\ln\kappa}/\sigma_\varphi$, $\sigma_{\ln\kappa} = 2$, then whether $\ln\kappa$ has a *1/k* crustal scaling can be tested by two-point spatial correlation methods. This analysis shows a power law separation exponent of -0.5. This corresponds to a Fourier power spectrum slope of *1/k* for the ln permeability field [10,27].

2.1.7. Power Law Exponent of Permeability from Microearthquake Distributions

Over 18,000 m^3 of water was recently injected at ~6 km depth in Finland to stimulate flow for deep geothermal heat extraction [28]. Over 54,000 microearthquakes ($-1 < M < 1.9$) resulted from this stimulation, of which 6150 were located and characterized. These events have a two-point correlation power law separation radius exponent of between -0.5 and -0.6 [10], which corresponds to a Fourier power spectrum slope of *1/k*.

2.1.8. Power Law Exponent from Field Mapping of Fractures

Barton (1995) [11] reported on outcrop fractures mapped at 17 locations at scales from centimeters to kilometers. He found that fracture length had a fractal dimension of *D*~1.5.(range 1.4 and-1.7) Since fractal dimension is related to the exponent β as $2\beta = 5\text{-}2D$ ([23], p30), Barton found $\beta \sim 1$ from outcrop mapping. A $\beta = 1$ corresponds to a *1/k* distribution, since, by convention, β is the negative of the exponent of k.

2.1.9. Power Exponent from Analysis of Fracture Seismic Images

The flow backbone shown in the bottom image of Figure 1 consists of connected linear segments with different levels of fracture seismic emission intensity and physical length. Histograms can be made of both relative fracture seismic segment length and FS emissions intensity, and their statistical distributions determined. Figure 8 shows that both of these properties have lognormal (or power law, since the roll over may simply reflect the limits of segment resolution) distributions, in which there are long fracture seismic intensity segments, and high FS emissions intensities that fall far outside of normal distributions, i.e., these distributions have long tails.

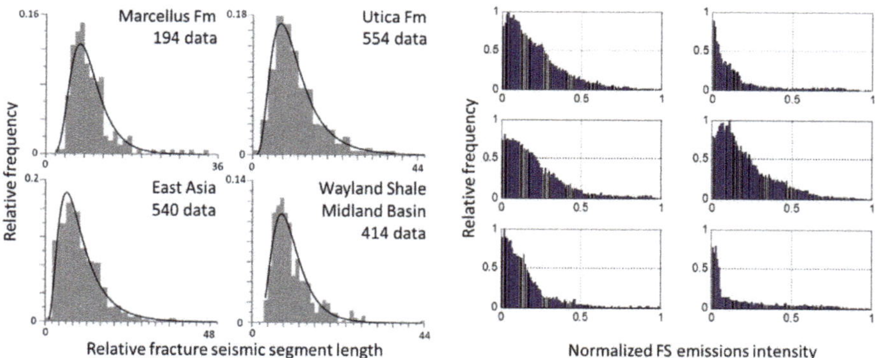

Figure 8. (**Left**) Normalized length distributions of linear fracture seismic intensity segments from four different reservoirs: (Clockwise) Marcellus, Utica, East Asia, Wayland. Linear backbone segments can be seen in the fracture seismic map slice in the bottom image in Figure 1. (**Right**) Normalized histograms of seismic fracture emission intensity recorded shale reservoir fracks, above. Data are from Lacazette et al. (2013) [29].

2.2. Flow From a Well: What Is at Stake

Figure 9 illustrates what is at stake in practical terms in the description of permeability as both lognormal and *1/k* distributed. The figure depicts flow from a well to the sides of the computation domain, where the injected fluid is allowed to freely escape the system. The figure was generated by Matlab finite-element methods for different values of β and α in Equations (2) and (4). Uncorrelated ($\beta = 0$) and correlated ($\beta = 1$) porosity distributions are assigned to 10^5 unstructured elements with the grid resolution increasing strongly toward the injection well. Corresponding permeability was derived from porosity assuming $\alpha = 1$ or 30. The mean porosity is selected such that $\alpha\phi_0 \sim$3 to 4 in all panels. The method of flow computation is the same as is used to simulate the flow vectors shown in Figure 7.

If the permeability has little variation around its mean (small $\sigma_{\ln \kappa}$), the flow proceeds radially away from the well (lower two images) regardless of whether the ln permeability (and porosity) is spatially ordered as pink ($\beta =1$, *1/k*) or white ($\beta =1$, random) noise. However, if the magnitude of permeability variation around its mean is of the order shown in Figures 2–4, and the ln permeability is spatially ordered (*1/k*), flow from the well occurs through a spatially correlated dendritic or filamentary pattern (top left image). The flow does not have this pattern if the ln permeability distribution is uncorrelated (white noise), even if the permeability has a large spread from its mean (top right image). Both substantial permeability inhomogeneity and a pink noise spatial ln permeability distribution are required to produce the channelized/filamentary flow shown in the top left panel of Figure 9.

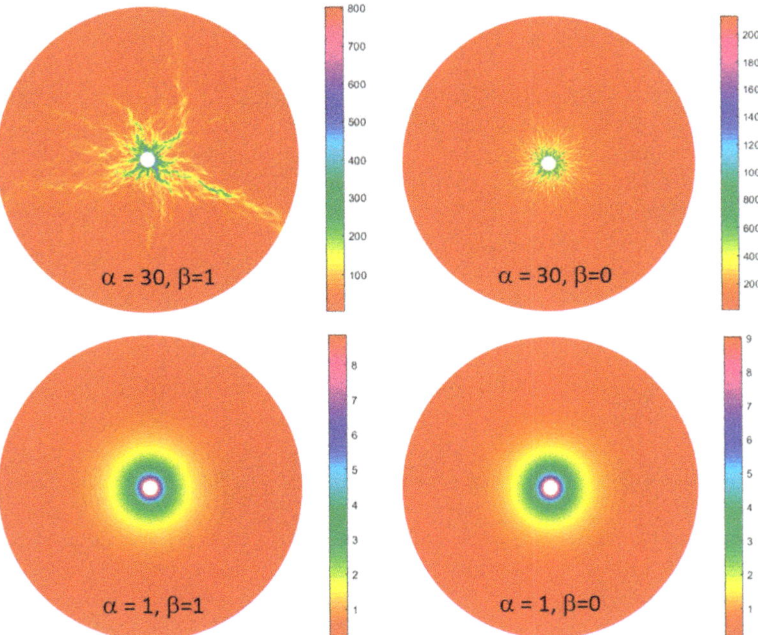

Figure 9. Flow calculated from a central well to free-flow boundaries where it escapes the system for different values of α in Equations (2) and (3) and β in Equation (4). Color indicates the relative magnitude of the fluid velocity (red low, purple high). Flow channeling (**upper left panel**) requires that both a large spread in ln κ from its mean (e.g., $\alpha =$30) and a distribution in permeability that is pink ($\beta = 1$). The variation in the magnitude of permeability from its mean can be large, but cannot produce channel flow if its spatial variation is that of spatially uncorrelated white noise (**top right panel**). The permeability distribution can be 1/k (pink noise) but cannot produce channel flow unless the variation in ln permeability is also large (**bottom left panel**). When the permeability distribution is white and the spread in ln permeability small, the flow from the well is of course uniform (**bottom right panel**).

The earth's crust meets both these criteria for channelized flow (permeability inhomogeneity and pink noise, scale invariant, spatial ln permeability distribution) as the following references in various areas attest: oil/gas (USEIA 2011), geothermal [30], mineral deposition [31–36] and trace elements [37–40]. The spatially heterogeneous, length-scale dependence of permeability is documented from lab-scale cm to reservoir-scale km [8,41–45]. The normal distributions, often assumed for modeling and analysis of flow in the earth's crust, offer no formalism for dealing with the spatial variations (channeling) observed in the actual flow system (e.g., [21,22,46–55]).

3. Summary and Discussion

Flow measurements are difficult to make at even a few meters from a drill hole. However, we can infer the characteristics of flow far away from the wellbore from the observed variation in permeability about its mean and from the observed spatial distribution of permeability. Our analysis proceeds from the observations that (1) variations in crustal porosity and pore connections are invariably associated with variations in the logarithm of crustal permeability ($\delta\varphi \propto \delta \ln \kappa$) and (2) the Fourier spectral power fluctuations in permeability magnitude (and all rock properties, micro-earthquakes, etc., associated with flow in fractures) scale inversely with spatial frequency ($1/k$).

Integrating the first observation produces an exponential relationship $\kappa \approx \kappa_0 \exp(\alpha(\varphi - \varphi_0))$, in which the variation in permeability magnitude relative to its mean is specified by a parameter $\alpha = \sigma_{\ln \kappa}/\sigma_\varphi$. The permeability in the Earth's crust has a substantial spread and the Earth's crust is critically stressed which means that $\alpha\phi_0$ = 3 to 4 and permeability has a $1/k$ (pink noise) spatial distribution. Consequently flow in the crust will become increasingly channelized as the distance of the flow increases.

The relationships described in this paper are physically fundamental. A connection condition explains why changes in porosity or the number of fractures in a volume are related to changes in log permeability magnitude. The pink noise spatial distribution of permeability is the distribution for which scale cannot be determined from observations. Geologists have long been aware that geological observations are scale invariant [56,57]. The parameter combination $\alpha\phi_0$ = 3 to 4 defines the critical point of failure for the crust, for which this scale invariance will pertain.

Critical State Physics studies show that near the point of critical failure many parameters have a pink noise spatial distribution and that their magnitudes depend more on the proximity to critical failure than to other factors [18]. This greatly simplifies understanding crustal permeability. That fractures are closely involved in the critical state of the crust may not be surprising. Fluid-filled fractures are the weakest element in the crust. The crust may be in a critical state because, in this state, fluids can move, as required by compaction and tectonic deformation, through permeable pathways at all scales (millimeters to kilometers). Fluids may control crustal properties more than presently appreciated.

The framework described in this paper is also of great practical importance. Flow channelization poses a risk to many engineering and commercial pursuits. That flow channelization must be expected in the Earth's crust as the distance of flow increases is thus significant. The fact that flow channels might be mapped by fracture seismic methods, as described by Sicking and Malin (2019) [1], means that, for the first time, the location of the major flow channels at a particular site can be determined and specified in models. The self-similar nature of permeability and the standard deviation of ln permeability from its mean allows the flow-field to be modelled in a statistically realistic way at finer scales. The combination of specific flow channel modeling with statistically realistic modeling of the finer details could produce more realistic and useful models of subsurface fluid flow for reservoir development and management, contaminant remediation, geothermal energy extraction, and many other purposes.

Author Contributions: Conceptualization, P.E.M., P.C.L.; methodology, P.C.L. and C.C.B.; writing-original draft, P.E.M. and P.C.L.; writing-review and editing, L.M.C., C.C.B. All authors have read and agreed to the published version of the manuscript

Funding: No external funding supported the content or preparation of this paper.

Acknowledgments: The authors are grateful to Peter Geiser for ongoing discussion of the flow-structure seismic imaging matters central to this paper. We also acknowledge very helpful suggestions from the reviewers.

Conflicts of Interest: The authors declare no conflicts of interest.

References

1. Sicking, C.; Malin, P. Fracture Seismic: Mapping Subsurface Connectivity. *Geosciences* **2019**, *9*, 508. [CrossRef]
2. Rutledge, J.T. Faulting Induced by Forced Fluid Injection and Fluid Flow Forced by Faulting: An Interpretation of Hydraulic-Fracture Microseismicity, Carthage Cotton Valley Gas Field, Texas. *Bull. Seism. Soc. Am.* **2004**, *94*, 1817–1830. [CrossRef]
3. Lacazette, A.; Vermilye, J.; Fereja, S.; Sicking, C. Ambient Fracture Imaging: A New Passive Seismic Method. In Proceedings of the Unconventional Resources Technology Conference, Denver, Colorado, 12–14 August 2013; pp. 2331–2340.
4. Leary, P.C. Deep borehole log evidence for fractal distribution of fractures in crystalline rock. *Geophys. J. Int.* **1991**, *107*, 615–628. [CrossRef]
5. Leary, P.C. Rock as a critical-point system and the inherent implausibility of reliable earthquake prediction. *Geophys. J. Int.* **1997**, *131*, 451–466. [CrossRef]
6. Leary, P.C. Fractures and physical heterogeneity in crustal rock. In *Heterogeneity of the Crust and Upper Mantle – Nature, Scaling and Seismic Properties*; Goff, J.A., Holliger, K., Eds.; Kluwer Academic/Plenum Publishers: New York, NY, USA, 2002; pp. 155–186.
7. Leary, P.C.; Al-Kindy, F. Power-law scaling of spatially correlated porosity and log(permeability) sequences from north-central North Sea Brae oilfield well core. *Geophys. J. Int.* **2002**, *148*, 426–442. [CrossRef]
8. Leary, P.; Malin, P.; Pogacnik, J. Computational EGS—Heat transport in 1/f-noise fractured media. In Proceedings of the 37th Stanford Geothermal Workshop, Stanford, CA, USA, 30 January–1 February 2012.
9. Leary, P.; Pogacnik, J.; Malin, P. Fractures ~ Porosity → Connectivity ~ Permeability → EGS Flow. In Proceedings of the Geothermal Resources Council 36th Annual Conference, Reno, NV, USA, 27 September 2012.
10. Leary, P.; Malin, P.; Saarno, T.; Heikkinen, P.; Diningrat, W. Coupling Crustal Seismicity to Crustal Permeability—Power-Law Spatial Correlation for EGS-Induced and Hydrothermal Seismicity. In Proceedings of the 44th Workshop on Geothermal Reservoir Engineering Stanford University, Stanford, CA, USA, 11–13 February 2019.
11. Barton, C.C. Fractal analysis of scaling and spatial clustering of fractures. In *Fractals in the Earth Sciences*; Barton, C.C., LaPointe, P.R., Eds.; Plenum Press: New York, NY, USA, 1995; pp. 141–178.
12. Leary, P.; Malin, P.; Niemi, R. Fluid Flow and Heat Transport Computation for Power-Law Scaling Poroperm Media. *Geofluids* **2017**, *2017*, 1–12. [CrossRef]
13. Leary, P.; Malin, P.; Saarno, T.; Kukkonen, I. $\alpha\phi$ ~ $\alpha\phi$crit –Basement Rock EGS as Extension of Reservoir Rock Flow Processes. In Proceedings of the 43rd Workshop on Geothermal Reservoir Engineering, Stanford University, Stanford, CA, USA, 12–14 February 2018.
14. Nelson, P.H.; Kibler, J.E. *A Catalog of Porosity and Permeability from Core Plugs in Siliciclastic Rocks*; US Geological Survey: Reston, VA, USA, 2003.
15. Shockley, W. On the Statistics of Individual Variations of Productivity in Research Laboratories. *Proc. IRE* **1957**, *45*, 279–290. [CrossRef]
16. Binney, J.J.; Dowrick, N.J.; Fisher, A.J.; Newman, M.E.J. *Theory of Critical Phenomena: An Introduction to the Renormalization Group*; Clarendon Press: Oxford, UK, 1995; p. 464.
17. Stauffer, D.; Aharony, A. *Introduction to Percolation Theory*; Taylor & Francis: London, UK, 1994; p. 192.
18. Hunt, A.G.; Sahimi, M. Flow, Transport, and Reaction in Porous Media: Percolation Scaling, Critical-Path Analysis, and Effective Medium Approximation. *Rev. Geophys.* **2017**, *55*, 993–1078. [CrossRef]
19. Bak, P. *How Nature Works: The Science of Self-Organized Criticality*; Copernicus: Göttingen, 1996; p. 229.
20. Barton, C.C.; Camerlo, R.H.; Bailey, S.W. Bedrock geologic map of the Hubbard Brook experimental forest and maps of fractures and geology in roadcuts along interstate 93, Grafton County, New Hampshire, Sheet 1, Scale 1:12,000; Sheet 2, Scale 1:200: U.S. Geological Survey Miscellaneous Investigations Series Map I-2562. 1997.

21. Slatt, R.M.; Minken, J.; Van Dyke, S.K.; Pyles, D.R.; Witten, A.J.; Young, R.A. Scales of Heterogeneity of an Outcropping Leveed-channel Deep-water System, Cretaceous Dad Sandstone Member, Lewis Shale, Wyoming, USA. In *Atlas of Deep-Water Outcrops*; Nilsen, T.H., Shew, R.D., Steffens, G.S., Stud¬lick, J.R.J., Eds.; American Association of Petroleum Geologists: Tulsa, OA, USA, 2006.

22. Slatt, R.M.; Eslinger, E.; Van Dyke, S. Acoustic and petrophysical heterogeneities in a clastic deepwater depositional system: implications for up scaling from bed to seismic scales. *Geophysics* **2009**, *74*, 35–50. [CrossRef]

23. Malamud, B.D.; Turcotte, D.L. Self-afine time series: I. Generation and Analysis. *Adv. Geophys.* **1999**, *40*, 1–90.

24. Hubbert, M.K. Darcy's law and the field equations of the flow of underground fluids. *Int. Assoc. Sci. Hydrol. Bull.* **1957**, *2*, 23–59. [CrossRef]

25. Warren, J.E.; Root, P.J. The behavior of naturally fractured reservoirs. *Soc. Pet. Eng. J.* **1963**, *3*, 245–255. [CrossRef]

26. Bear, J. *Dynamics of Fluids in Porous Media*; American Elsevier: New York, NY, USA, 1972.

27. Leary, P.C.; Malin, P.E.; Saarno, T. A physical basis for the gutenberg-richter fractal scaling. In Proceedings of the 45rd Workshop on Geothermal Reservoir Engineering, Stanford University, Stanford, CA, USA, 10–12 February 2020.

28. Kwiatek, G.; Saarno, T.; Ader, T.; Bluemle, F.; Bohnhoff, M.; Chendorain, M.; Dresen, G.; Heikkinen, P.; Kukkonen, I.; Leary, P.; et al. Controlling fluid-induced seismicity during a 6.1-km-deep geothermal stimulation in Finland. *Sci. Adv.* **2009**, *5*, eaav7224. [CrossRef] [PubMed]

29. Lacazette, A.; Geiser, P. Comment on Davies et al., 2012–Hydraulic fractures: How far can they go? *Mar. Pet. Geol.* **2013**, *43*, 516–518. [CrossRef]

30. Grant, M.A. Optimization of drilling acceptance criteria. *Geothermics* **2009**, *38*, 247–253. [CrossRef]

31. De Wijs, H.J. Statistics of ore distribution. Part I: frequency distribution of assay values. *J. R. Neth. Geol. Min. Soc. New Ser.* **1951**, *13*, 365–375.

32. De Wijs, H.J. Statistics of ore distribution Part II: theory of binomial distribution applied to sampling and engineering problems. *J. R. Neth. Geol. Min. Soc. New Ser.* **1953**, *15*, 125-24.

33. Koch, G.S.; Link, R.F. The coefficient of variation; a guide to the sampling of ore deposits. *Econ. Geol.* **1971**, *66*, 293–301. [CrossRef]

34. Clark, I.; Garnett, R.H.T. Identification of multiple mineralization phases by statistical methods. *Trans. Inst. Min. Metall.* **1974**, *83*, A43.

35. Link, R.F.; Koch, G.S., Jr. Some consequences of applying lognormal theory to pseudolognormal distributions. *Math. Geol.* **1975**, *7*, 117–128. [CrossRef]

36. Gerst, M.D. Revisiting the Cumulative Grade-Tonnage Relationship for Major Copper Ore Types. *Econ. Geol.* **2008**, *103*, 615–628. [CrossRef]

37. Ahrens, L.H. The lognormal distribution of the elements, I. *Geochim. Cosmochim Acta* **1954**, *5*, 49–73. [CrossRef]

38. Ahrens, L.H. The lognormal-type distribution of the elements, II. *Geochim. Cosmochim. Acta* **1954**, *6*, 121–131. [CrossRef]

39. Ahrens, L.H. Lognormal distributions, III. *Geochim. Cosmochim. Acta* **1957**, *11*, 205–212. [CrossRef]

40. Ahrens, L.H. Lognormal-type distributions in igneous rocks, IV. *Geochim. Cosmochim. Acta* **1963**, *27*, 333–343. [CrossRef]

41. Brace, W.F. Permeability of crystalline and argillaceous rocks. *Int. J. Rock Mech. Miner. Sci. Geomech. Abstr.* **1980**, *17*, 241–251. [CrossRef]

42. Clauser, C. Scale Effects of Permeability and Thermal Methods as Constraints for Regio-nal-Scale Averages. In *Heat and Mass Transfer in Porous Media*; Quintard, M., Todorovic, M., Eds.; Elsevier: Amsterdam, The Netherlands, 1992; pp. 446–454.

43. Neuman, S.P. Generalized scaling of permeabilities: Validation and effect of support scale. *Geophys. Res. Lett.* **1994**, *21*, 349–352. [CrossRef]

44. Neuman, S.P. On advective dispersion in fractal velocity and permeability fields. *Water Resour. Res.* **1995**, *31*, 1455–1460. [CrossRef]

45. Schulze-Makuck, D.; Cherhauer, D.S. Relation of hydraulic conductivity and dispersivity to scale of measurement in a carbonate aquifer. In *Models for Assessing and Monitoring Groundwater Quality*; Wagner, B.J., Illangasekare, T.H., Jensen, K.H., Eds.; Indigenous Allied Health Australia: Deakin, Australia, 1995.

46. Archie, G.E. Introduction to Petrophysics of Reservoir Rocks. *AAPG Bull.* **1950**, *34*, 943–961.

47. Biot, M.A. General theory of three-dimensional consolidation. *J. Appl. Phys.* **1941**, *12*, 155–164. [CrossRef]

48. Carman, P.C. Fluid flow through granular beds. *Trans. Inst. Chem. Engrs.* **1937**, *15*, 150–166. [CrossRef]

49. Amyx, J.W.; Bass, D.M.; Whiting, R.L. *Petroleum reservoir engineering: Physical properties*; McGraw-Hill: New Year, NY, USA, 1960; p. 610.

50. Timur, A. An investigation of permeability, porosity, and residual water saturation relationships. In Proceedings of the SPWLA 9th Annual Logging Symposium, New Orleans, LA, USA, 23–26 June 1968.

51. Hearst, J.R.; Nelson, P. *Well Logging for Physical Properties*; McGraw-Hill: New York, NY, USA, 1985; p. 571.

52. Nelson, P.H. Permeability-porosity relationships in sedimentary rocks. *Soc. Petrophys. Well-Log Anal.* **1994**, *35*, 38–62.

53. Jensen, J.L.; Hinkley, D.V.; Lake, L.W. A Statistical Study of Reservoir Permeability: Distributions, Correlations, and Averages. *SPE Form. Evaluation* **1987**, *2*, 461–468. [CrossRef]

54. Ingebritsen, S.; Sanford, W.; Neuzil, W. *Groundwater in Geological Processes*; Cambridge University Press: Cambridge, UK, 1999; p. 536.

55. Mavko, G.; Mukerji, T.; Dvorkin, J. *The Rock Physics Handbook: Tools for Seismic Analysis of Porous Medial*; Cambridge University Press: Cambridge, UK, 1998; p. 2009.

56. Barton, C.C.; La Pointe, P.R. *Fractals in the Earth Sciences*; Springer (Plenum Press): New York, NY, USA, 1995.

57. Barton, C.C.; La Pointe, P.R. *Fractals in Petroleum Geology and Earth Processes*; Springer (Plenum Press): New York, NY, USA, 1995.

Article

Fracture Seismic: Mapping Subsurface Connectivity

Charles Sicking [1,*] and Peter Malin [2]

1   Ambient Reservoir Monitoring, 3701 Anatole Ct., Plano, TX 75075, USA
2   Advanced Seismic Instrumentation & Research, 1311 Waterside, Dallas, TX 75218-4475, USA;
    pem@asirseismic.com
*   Correspondence: charles@ambientreservoir.com; Tel.: +1-214-763-6711

Received: 10 November 2019; Accepted: 5 December 2019; Published: 6 December 2019

Abstract: Fracture seismic is the method for recording and analyzing passive seismic data for mapping the fractures in the subsurface. Fracture seismic is able to map the fractures because of two types of mechanical actions in the fractures. First, in cohesive rock, fractures can emit short duration energy pulses when growing at their tips through opening and shearing. The industrial practice of recording and analyzing these short duration events is commonly called micro-seismic. Second, coupled rock–fracture–fluid interactions take place during earth deformations and this generates signals unique to the fracture's physical characteristics. This signal appears as harmonic resonance of the entire, fluid-filled fracture. These signals can be initiated by both external and internal changes in local pressure, e.g., a passing seismic wave, tectonic deformations, and injection during a hydraulic well treatment. Fracture seismic is used to map the location, spatial extent, and physical characteristics of fractures. The strongest fracture seismic signals come from connected fluid-pathways. Fracture seismic observations recorded before, during, and after hydraulic stimulations show that such treatments primarily open pre-existing fractures and weak zones in the rocks. Time-lapse fracture seismic methods map the flow of fluids in the rocks and reveal how the reservoir connectivity changes over time. We present examples that support these findings and suggest that the fracture seismic method should become an important exploration, reservoir management, production, and civil safety tool for the subsurface energy industry.

Keywords: fracture seismic; fracture connectivity; fracture mapping; passive seismic

1. Introduction

We use the term "Fracture Seismic Method" to refer to the method of mapping fractures using one-way depth migration applied to fracture emissions that have durations of seconds to minutes. The use of the term in this fashion distinguishes the fracture seismic method from other methods such as the reflection seismic method and the micro-seismic method. This paper presents an end-to-end description of the fracture seismic method and presents examples that map subsurface connectivity structures. The fracture seismic method extends current passive methods by making use of harmonic resonances within the fracture that are caused by interfering Krauklis waves (Krauklis, 1962) [1] initiated by dislocations on fracture tips and internal fracture fluid flows (e.g., Frehner, 2014 [2]; Tary et al., 2014) [3,4]. The fracture emissions come from short duration energy pulses and harmonic resonances of the entire fracture. The resonances are episodic, seconds to minutes long, and occur in the frequency band of 1 to 100 Hz. They are readily observed in passive, multichannel seismic recordings at both green- and brown-field sites. Two examples of fracture seismic signals are shown in the spectrograms in Figure 1.

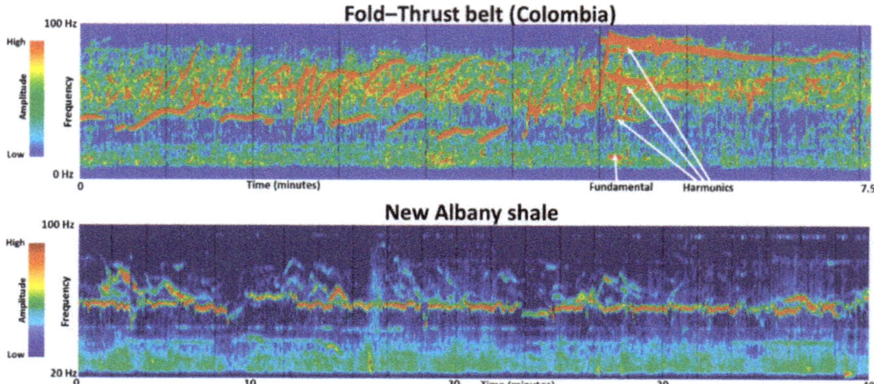

Figure 1. Spectrograms of fracture seismic data containing resonances. The top panel is from a Colombia thrust zone where the regional compressional stress is high. In the first 5 min of this panel, there are two styles of resonances. Note the harmonics at 5 min. The bottom panel is from the New Albany shale. It reveals a much simpler resonance signal where the highest intensity resonance is in the 50 to 60 Hz frequencies, with lower intensities at lower frequencies. Figure modified from Sicking et al. (2019) [5,6].

The most widely used method for monitoring of hydraulic fracturing uses geophones at reservoir depth in vertical wells that are located near the hydraulic fracturing. Maxwell et al. (2003) [7] describe this downhole method for detecting microearthquakes (MEQ) generated during stimulation operations and for imaging deformation associated with the injections.

Another method for mapping MEQ during the hydraulic fracturing uses surface or buried grid recordings. The basis of this method is Kirchhoff migration, and it is normally referred to as seismic emission tomography. Duncan et al. (2010) [8] describe the surface geophone method for detecting and mapping MEQ. The focus of these hydraulic fracture monitoring methods is to use the MEQs to infer the creation of fracture connectivity.

Kochnev et al. (2007) [9] describe a non-MEQ passive seismic imaging method for mapping the progression of hydraulic fracturing that is similar to the fracture seismic depth migration method presented here. Their method requires searching the trace data for low-energy source seismic waves that can be identified before imaging and the method is applied only to map the progression of the stimulation over time. This approach is not useful for mapping pre-existing fractures before drilling.

In work related to our fracture seismic method, Tary et al. (2012) [10] compute continuous time-frequency transforms that highlight signals that have time-varying resonance frequencies. They conclude that these signals are the result of resonance in fluid-filled fractures or, alternatively, successions of very small repetitive seismic events along the fractures. They also observe correlations between the variations in the frequency content of their recordings, the hydraulic fracturing conditions, and the occurrence of micro-seismic events. They note that there is a direct correspondence between variations in the slurry injection rate and the combined energy emitted.

Seeking to better identify these ambient emissions as opposed to MEQ events, Chorney et al. (2012) [11] present results on seismic energy sources that are associated with deformations such as tensile fracturing or slow slips. Furthermore, Bame et al. (1986) [12] note that the ambient signals they observe are unlikely to be detected by searching with seismic event triggering methods because these require sharp signal onsets.

Additional support for the origins of these episodic signals that occur over long time intervals can be found in the fracture mechanics literature. Vermilye et al. (1998) [13] and Shipton et al. (2001) [14] investigate the various release mechanics of stored elastic strain energy from rocks through field studies of fractures. This stored strain energy is not evenly distributed in the earth's crust, but it is preferentially released on fracture/fault surfaces and in the damage zones surrounding these fractures.

Fracture mechanics theory predicts that stress concentrations are associated with fractures. Accordingly, Vermilye et al. (1998) [13] and Moore et al. (1995) [15] report field and laboratory studies with clear evidence that these stress concentrations are recorded in the fracture damage zones. Vermilye et al. (1998) [13] show that damage zones consist of rock volumes with a high density of small fractures and that the density of fractures increases exponentially with their proximity to the main fracture surface.

Ziv et al. (2000) [16] show that the brittle crust is in a state of unstable frictional equilibrium. Therefore, very small changes in stress (approximately 0.01 bar or approximately 1 kPa) can cause slippage on weak fractures. Lawn et al. (1975) [17] show that failure occurs preferentially on small, optimally oriented fractures and in the zones surrounding the fractures in which crack-tip stress concentrations amplify the stress magnitudes. Hubbert et al. (1959) [18] show that this unstable equilibrium is further disturbed as additional fluid pressures reduce the normal stress on preexisting fractures. They also show that, during production, subtle movement of fluid produces similar effects.

Fracture seismic connectivity mapping started circa 2005. (Geiser et al. 2006) [19]. The end-to-end system for applying the fracture seismic method has been in practice since 2010 (Sicking et al., 2012) [20]. The main mapping step is a time-progressing depth migration of the fracture seismic resonance episodes, a process that we call streaming depth imaging (SDI; Sicking et al., 2016) [21]. Two examples of fracture information that can be computed using SDI are shown in Figure 2. For these examples, the fractures seismic intensity is summed over the time interval of a stimulation stage and the fracture surfaces are extracted from the intensity volume. The left panel shows the fracture surfaces colored by the summed intensity for the entire stage and shows that the summed intensity is highest at the perf shots and lowest at the fractures more distant from the perforations. The right panel shows the fracture surfaces colored by the clock time of the first fracture emissions. The fractures to the left of the well were stimulated first, early in the stage treatment time, and the fractures to the right of the well were stimulated progressively later in time.

Recent research on the source of fracture seismic signals has put the fracture seismic method on a solid theoretical and practical base (e.g., Tary et al., 2014 [3]; Liang et al., 2017) [22]. It has now been applied to dozens of field projects and the examples presented here come from those projects (e.g., Sicking et al., 2014; 2015; 2016; 2017 [21,23–25], Geiser et al., 2012 [26], Lacazette et al., 2013 [27]).

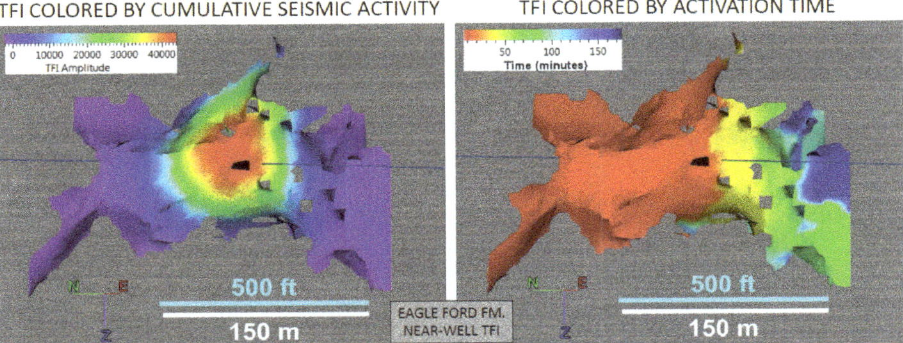

Figure 2. Fractures extracted from the local fracture seismic intensity cloud for a single stimulation stage. The left panel shows the extracted fractures colored by the fracture seismic intensity. The intensity at the perf locations (red) are the highest because they are active the longest. The right panel shows the extracted fractures colored by the time of first emission. This shows that the fractures to the left of the well stimulated much earlier during the treatment and the fractures to the right side of the well were stimulated progressively later in time. (Figure from Sicking et al., 2015) [24].

Several features distinguish fracture seismic from micro-seismic. Micro-seismic uses only slip events that are short enough to allow time separated P and S phases to be recognized on ordinary seismograms (e.g., Aki and Richards, 1980 [28]). Fracture seismic uses signals that can be viewed as the harmonic modes of fluid filled fractures embedded in the upper crust (Liang et al, 2017) [22]. After the resonances are initiated and while there is a continued input of energy, these fracture-length-and aperture-controlled modes of fluid-filled fractures can continue resonating for many seconds to even minutes (Sicking et al., 2019 [5,6]). These waves can also be initiated by the passage of earthquake waves, tectonic and tidal strains, and pressure changes caused by industrial activities. When the geometry of the fracture changes, the frequencies and intensities of the fracture's resonance modes also change.

The methods for observing fracture seismic have been greatly improved by the increase in numbers and sophistication of portable and borehole seismographs over the past few decades. The most cost-effective method is to piggyback on 3D seismic reflection surveys, the fracture seismic data being gathered during active source downtime. The offsets to be covered by the receivers is determined by the target depth of the deepest target and the noise environment. The density of receiving points needed is on the same order as used for recording reflection seismic.

The increased sophistication and speed of seismic reflection signal processing has also significantly aided fracture seismic processing. Initially, many hours of continuous fracture seismic recordings were thought to be necessary in order to build up a 3D volume of fracture intensity. Now, using spectrograms, episodes of intense resonance can be quickly identified and directed into the fracture seismic SDI workflow (Sicking et al., 2019 [5,6]). Many codes used for two-way-travel-time data analysis can be adapted for fracture seismic one-way-travel-time processing. Noise suppression methods are critical in optimal fracture seismic fracture mapping.

After creating a fracture seismic intensity volume via SDI, the local maximum energy surfaces can be tracked and mapped into a 3D image of the connectivity structure. Time-lapse versions of these structures are effective tools for resource management. Because the most intense resonances come from the most permeable fluid filled fractures, changes in relative intensity documents changes in the connectivity and fluid content.

2. Background and Methods

2.1. Fracture Seismic: Spectrograms

The presence of resonating signals in passive recordings has a long history in observational seismology. They gained prominence in Western literature in connection with volcanic activity (e.g., Dibble, 1972 [29]) and their association with fluid-filled fractures (Aki et al., 1977 [30]). Their occurrence in hydraulic fracturing was inferred during engineered geothermal system studies at Fenton Hills New Mexico (Bame and Felher, 1986 [12]). They have recently been demonstrated to be present in seismic observations of oil and gas stimulations (e.g., Tary et al., 2014 [3]). They have now also been identified in quiet time fracture seismic recordings (Sicking et al., 2019 [6]). They are most readily seen in time-verses-frequency spectrograms of multichannel seismic data.

Figure 1 shows examples of spectrograms computed from fracture seismic recordings in two very different basins. The top panel is from a thrust zone in Colombia that is under high stress from compressional tectonics, accounting for the high fracture seismic signal level. The first 5 min of this panel show a combination of chaotic and dispersive (frequency changing) resonances. The post 5 min interval shows a resonance with three harmonics.

The bottom panel of Figure 1 shows the spectrogram for data recorded in the New Albany shale during a time before the observation site was hydraulically stimulated. The type of resonances detected during this recording are typical for times when there is no industrial activity. The resonance is dominated by amplitudes in the 50 Hz to 60 Hz range, with several lower-intensity bands at lower frequencies. The difference from the Colombian thrust zone is likely due to differences in the state of stress and local geology, which is dominantly extensional.

2.2. Fracture Seismic: Signal Initiation

Fracture seismic resonance signals can be initiated in several ways, by both external and internal influences. Examples include distant earthquake strains wave, abrupt responses to accumulated earth tides, isostatic and tectonic deformations, fault creep, and hydraulic stimulation: All contribute energy that can initiate and sustain fracture seismic resonances (e.g., Gomberg, 1996 [31]; Du et al., 2003 [32]; Thomas et al., 2009 [33]; Tary et al., 2014 [3,4]; Liang et al., 2017) [22].

If a fluid-filled fracture is growing, the opening and shearing can initiate the Krauklis waves on the fracture surfaces and they are influenced by the fracture fluid and the surrounding rock. The waves travel along the fracture surfaces, quickly interfering to produce a modal/harmonic resonance of the whole fluid, fracture surface, and surrounding rock system.

Reservoir stimulation by hydraulic fracturing, flooding, or the extraction of fluids causes turbulent flow in the fractures. These flows can also initiate interfering Krauklis waves that then radiate seismic waves. These harmonic motions were first recognized in volcanic activity (e.g., Aki et al., 1977 [30]), next along plate boundaries (e.g., Zhang et al., 2010 [34]), and now in the entire fractured crust (e.g., Sicking et al., 2019 [5,6]).

Tary et al. (2014 [3,4]), divide the resonances into two end-member source types. One type involves unconnected/isolated fractures, the other connected fractures. Their resonant frequencies change with changes in the apertures and lengths of the cracks, and the forces exciting them.

In this paper, examples of resonance and turbulent flow are taken from several different areas and illustrate key features of the spectrograms that are computed for data evaluation and time window selection. Figure 2 show the system of fractures for one stage of stimulation that was computed using fracture seismic data. Both panels show the same fractures, but the left panel fractures are colored by the total local fracture seismic intensity density over the entire stimulation stage and the right panel fractures are colored by the time that each fracture first emits energy.

2.3. Fracture Seismic: Data Acquisition

Acquiring fracture seismic data requires a ground surface recording grid of geophones very similar to that used for recording 3D reflection seismic data. The area covered by the receivers should be larger than the area to be imaged. This area is selected such that the edges of the receiver grid are outside of the image area by 1.0 to 1.5 times its depth. If the receivers are laid out on the ground surface, 30 to 60 receiver points are required for each square km. The receivers used for fracture seismic recording on the surface are the same as those used for reflection seismic recording and can record a useful bandwidth that is typically 6 Hz to 1000 Hz. However, in most cases, the recording systems sample the signals at 2 milliseconds and have a Nyquist of 250 Hz. For buried grids where the receivers are placed in boreholes drilled past the local weathering layer, one to three receiver points are required per square km and the receivers may capture frequencies as low as 1 to 2 Hz.

Figure 3 shows four possible layouts for receiver grids. A uniform, face-centered hexagonal distribution is the best design and will have the smallest amplitude artifacts in the final fracture seismic intensity volume. Covering the same area but using cables in an orthogonal grid will provide very good results with only small artifacts in the amplitudes of the fracture seismic intensity volume. The star grid is widely used because it is cheaper to implement in the field and is very versatile in modifying the design to account for access to land. However, the star design is good only for the central portion of the receiver grid. As you move towards the edge of the receiver array, there will be amplitude artifacts and distortions in the locations in the fracture seismic intensity volume. The right side of Figure 3 shows a patch design. This design is sometimes used for projects where only the MEQ detections are desired. However, the patch design causes severe amplitude artifacts in the fracture seismic intensity volumes computed for fracture extractions.

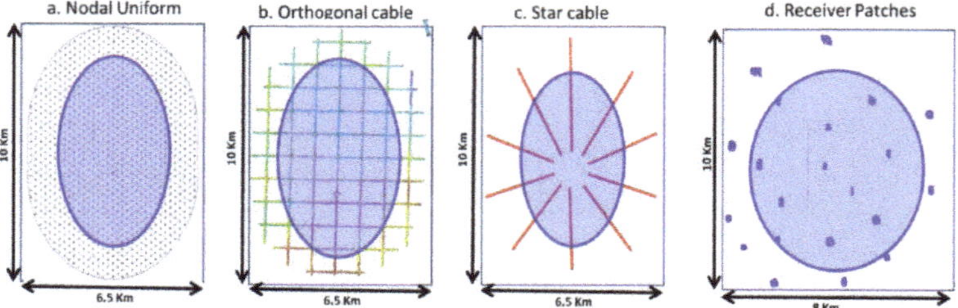

Figure 3. Surface recording grids for fracture seismic during the monitoring of stimulations. The grid should cover the desired area around the wells being monitored plus additional area to capture the required aperture for the one-way depth migration. The ideal grid is uniform distribution of geophones as shown in (**a**). The uniform layout has the minimum amount of amplitude artifacts in the fracture seismic intensity volumes. When using a cable system, the geophones are best configured in orthogonal lines (**b**). The star cable layout (**c**) provides good imaging in the center portion of the grid but the fracture seismic intensity volume suffers location distortions in the outer areas. The patch grid (**d**) provides the lowest-quality fracture seismic intensity volumes and has very high amplitude artifacts. Figure modified from Sicking et al. (2019) [5,6].

The depth of a buried grid should be sufficient to place the geophones below the seismic-signal distorting layers of the near surface. The advantage of this type is that the geophones see very little of the surface wave noise that is a major source of interference for surface geophone. For this reason, the density can be reduced to 3 or fewer per square km instead of the 30 to 60 per square km required for surface geophone arrays. Figure 4 shows a buried grid layout in which the density of receivers is 0.45 per square km.

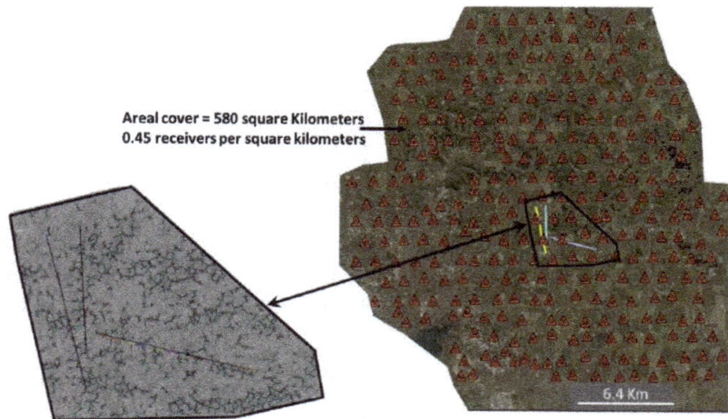

Figure 4. Buried geophone arrays are the best option if monitoring will be carried out multiple times over the same area. They are buried 30 to 100 m deep and have the advantage that they do not record the surface wave noise that is encountered on the surface arrays, so the density of geophones is reduced. Surface arrays require 30 to 60 geophones per square kilometer while buried geophone arrays require only 1 to 3 per square km. The geophones can be monitored for each new project by hooking up recorders to each geophone for the time of the project. The figure to the left is a map of the fractures extracted from the survey. Figure modified from Sicking et al. (2019) [5,6].

The cost per station is significantly higher for the buried grid than for surface recordings. However, there a many fewer geophone locations and the reduction of noise and the reuse of the same grid for stimulation and time-lapse monitoring makes up for the extra cost. If the geophone grid is reused three times, the cost of the buried grid is less than the cost of the surface geophone grid that is laid out special purpose for each observation. In addition, the quality of fracture seismic maps from a buried grid are much improved over surface grids.

For the purpose of recording fracture seismic before drilling wells, the data can be collected during the acquisition of 3D reflection seismic, whereby the recorders are switched to continuous recording mode for a few hours while the active sources are offline (Figure 5). The fracture seismic data are recorded at different locations in a roll-along mode as the reflection survey proceeds across the area. Each ground array is recorded and processed separately. The final volumes computed for each array are merged to cover a targeted area.

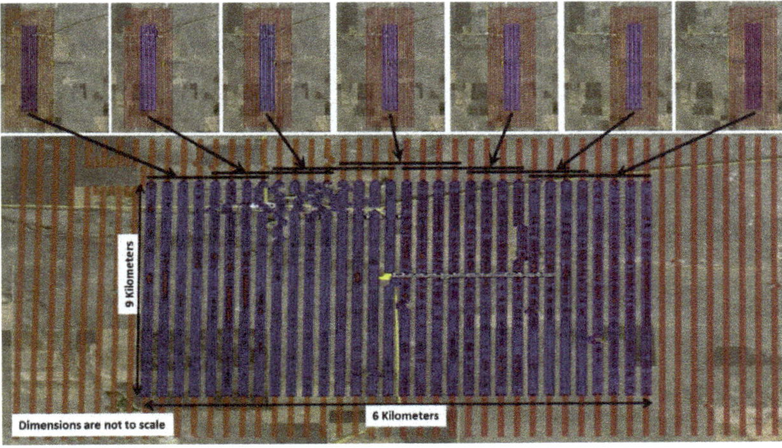

Figure 5. Passive seismic recorded using the geophones layout for the 3D reflection seismic recording. The receiver grid is rolled with the 3D acquisition and every few days the active sources are shut down for a few hours while the geophone outputs are recorded in continuous mode. In this example, the area of interest (blue) is recorded in seven separate recording times on different days. The seven fracture seismic intensity volumes have 50% overlap and are merged after the seven final intensity volumes are computed. Merging seven independently recorded and processed volumes causes artifacts at the seams. The fracture seismic intensity volume is discussed in Section 3.5.

In this method, however, because overlapping volumes are recorded at different times, the separate volumes see different amounts of fracture seismic energy and the seaming of the volumes can be problematic. The seaming problem can be avoided by laying out geophones over a large surface area and recording continuously for a few days without moving the receivers. An example of this method, discussed later in this paper, is a recording grid that had 4650 active receivers covering a 50 square km study area. In this example, the entire area is recorded simultaneously and only one fracture seismic intensity volume is computed.

The distortions in the intensity volume for different receiver layouts for fracture seismic will differ for different grid configurations. Two examples of amplitude distortions caused by recording grid layout are shown in Figures 6 and 7. Figure 6 shows the amplitude distortions for the patch grid example shown in Figure 3 that has dozens of geophones clustered in each of a few dozen patches. For fracture intensity mapping, the patch design is very poor because of the severe amplitude artifacts in the final volume. The amplitude artifacts are both short and long wavelength and significantly interfere with the interpretation of the fracture system extracted from the recordings.

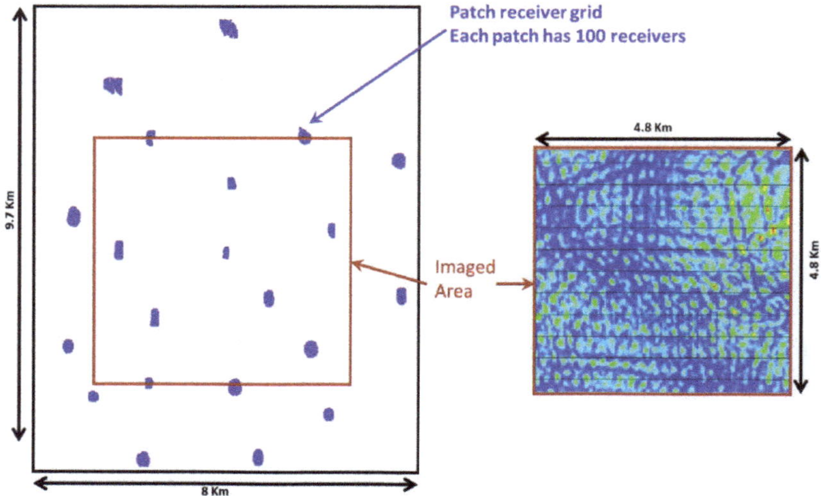

Figure 6. The patch receiver layout is designed such that there are 15 to 25 patches and within each patch there are 50 to 200 geophones. This layout allows for the suppression of surface wave noise within each patch and is focused on detecting and locating MEQs. For computing fracture seismic intensity volumes using one-way depth migration, this design is inferior. The geophone layout is shown on the left. A synthetic trace was computed for each geophone that would provide a uniform amplitude in the output fracture seismic intensity volume if a uniform grid was using the recording. When the synthetic is input to the one-way depth migration using geophone locations only at those for the patch geometry, the depth slice shown on the right is produced. The slice is for the area in the red box on the left. The patch geometry causes the short and long wavelength amplitude artifacts, and these will overprint any fracture patterns that may be imaged.

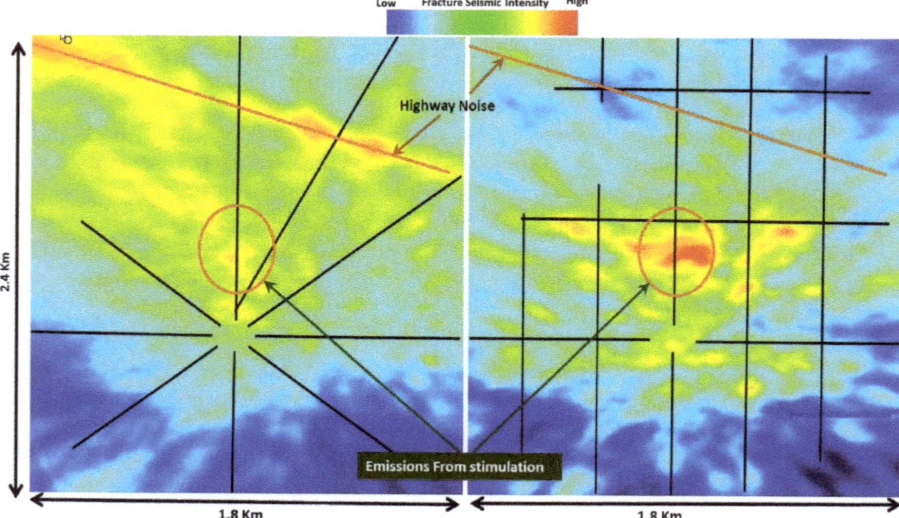

Figure 7. Highway noise in the star grid versus orthogonal grid. All data were recorded simultaneously so the signal content is the same for the star grid and the orthogonal grid. The star grid does not cancel the horizonal noise perpendicular to the star arm. The orthogonal grid suppresses the highway noise and the signal from the stimulation is enhanced.

Figure 7 shows the difference between using a star grid and a cable grid for the same study site. The fracture seismic intensity volume using a star grid does not suppress the highway noise in the final fracture seismic intensity volume while the orthogonal cable grid suppresses the highway noise. The highway is perpendicular to the cables in the star grid and the noise hitting the cable broadside cannot be suppressed. The orthogonal grid suppressed the highway noise because it has cables that are basically parallel to the highway.

2.4. Fracture Seismic: Signal Processing

The filtering and depth migration methods used for fracture seismic are based on typical reflection seismic signal processing algorithms, but are modified to deal with one-way travel times from the fracture seismic sources to the receivers. Success in using fracture seismic recordings for mapping fractures requires having high-quality non-resonant signal analysis and suppression (Sicking et al., 2016, 2017) [21,25].

The steps for processing fracture seismic can be broadly broken in to four parts: (1) Elimination of cultural and man-made seismic waveforms: (2) estimation of elevation and residual statics; (3) building the earth velocity model; and, (4) one-way travel time depth migration for a continuous signal source.

Cultural and man-made noise that is active for longer than minutes of time can be classified as stationary noise (Figure 8). This type of noise is common in industrial areas and transportation corridors and appears as noise background added to the consistent, slowly changing, harmonic character of fracture seismic signals. This long duration of noise can overwhelm fracture seismic signals, but the noise can be separated from the fracture seismic signals with cepstral filtering (Sicking, 2016) [21].

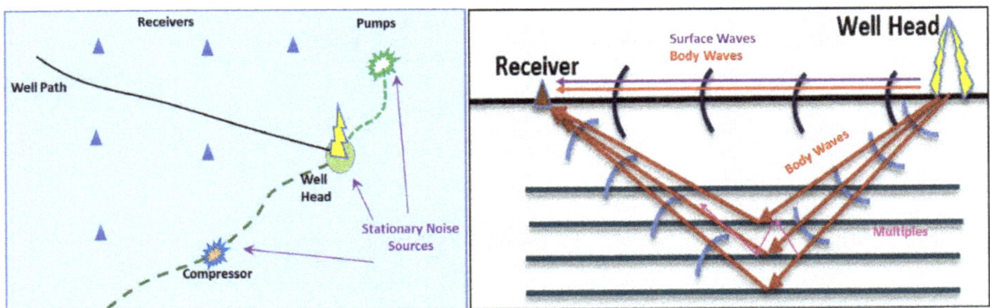

Figure 8. Continuous but erratic signals along the stationary source-receiver path can overwhelm fracture seismic signals. The left panel shows the locations for the receivers, the well head, and compressor noise sources. The right panel shows the ray paths from the well head to a single receiver for various types of seismic waves. The noise is generated at all times and the ray paths are fixed so the wave forms on the receiver trace are very repetitive. (Figure from Sicking et al., 2016) [21].

The cepstral filtering processes each fracture seismic trace independently by transforming the trace into the Cepstral domain, applying a bandpass in that domain, and inverse transforming back to time. The transform to the cepstral domain requires two forward Fourier transforms. The first Fourier transform is applied to the time data to compute the amplitude and phase as a function of frequency (a Fourier spectrum). The second forward Fourier transform uses only the amplitude versus frequency to compute the amplitude and phase as a function of quefrency (a Cepstrum). A low pass filter is applied in quefrency and the resulting amplitude and phase signal is inverse Fourier transformed to the frequency (Fourier spectral) domain. The amplitude versus frequency is combined with the original phase versus frequency before taking another inverse Fourier transform to compute the filtered trace in time.

The Fourier frequency spectrum for one trace of field data containing stationary noise is shown in the top panel of Figure 9. The erratic spectrum of the stationary noise is superimposed on the less variable, broader background of the more stable fracture seismic signal. The erratic part of the spectrum needs to be removed. Because of the large differences in their spectral character, the erratic stationary noise spectrum spreads to the full range of quefrencies while the fracture seismic signals are confined to the very lowest quefrencies. In fact, the low pass filter in quefrency needs only to keep the lowest 1% to 2% of the quefrency spectrum. The middle panel of Figure 9 shows the quefrency spectrum of the field data. A low pass filter is applied in quefrency that passes only the lowest 1% of the quefrencies. When this filtered Cepstrum is reverse-Fourier transformed back to the frequency domain (bottom panel), the stationary noise is essentially eliminated.

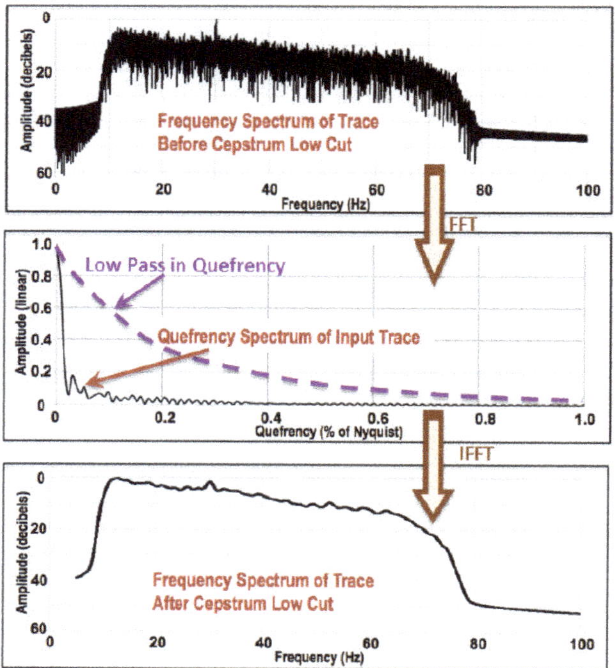

Figure 9. The cepstral filtering process to remove stationary noise. The stationary noise becomes spikes in the spectral domain (**top** panel). In the cepstral domain, these spikes are spread across all quefrencies (**middle** panel). The fracture seismic signals are in the lowest 2% of the quefrencies. After the application of the low pass filter in quefrency, the inverse transform to the spectral domain is shown in the bottom panel. (Figure from Sicking et al., 2016) [21].

The filter in the quefrency space has an amplitude of 1.0 at the smallest quefrency and an amplitude of 0.0 at the quefrency that is 3% of the Nyquist in quefrency space. More than 97% of the Quefrencies are thereby set to zero. Moreover, 95% of the trace energy is preserved in these lowest 2% of the quefrencies. Because the cepstral filter is a non-linear filter, its order of application is important: It does not commute with other linear signal processing filters. Experience shows that it should be run as the first filter in the processing flow.

The value of the cepstral filter in fracture seismic processing is illustrated in the top two panels of Figure 10. Here, the time window of noisy multichannel data includes the waveforms for a relatively large amplitude microearthquake (MEQ). The data are sorted by the azimuth direction with respect to the location of the MEQ. The traces have been shifted in time using one-way travel times from the voxel of the MEQ to each receiver on the surface. The waveforms for the MEQ should all arrive at the same time after the trace shifting. The figure shows the data before and after cepstral filtering. In the unprocessed data, the MEQ is not readily evident in the trace data. After cepstral filtering of each receiver trace, the MEQ signal emerges in the middle of the record section.

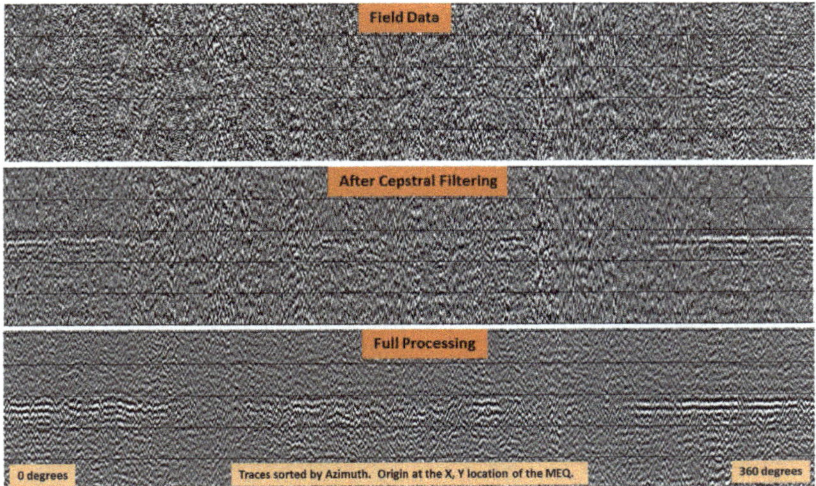

Figure 10. Cepstral filtering reveals the presence of a small microearthquake in these multichannel fracture seismic data. The top panel shows the traces as recorded in the field. The middle panel show that traces after cepstral filtering revealing the microearthquakes (MEQ). The bottom panel shows the trace data after all filtering has been applied. Figure from Sicking et al. (2016) [21].

The suppression of other types of noise also aids in obtaining clear fracture images. These include filters that help identify and clean up transient amplitude bursts, electronic line noise, and traffic noise. The trace section in the bottom of Figure 10 shows the effects of further processing to reduce such interferences. The result clearly reveals the MEQ's signals, including its azimuthally dependent radiation pattern.

Another example of stationary noise suppression is shown in Figure 11. The top panel of this Figure contains a spectrogram of fracture seismic data. A 5-min section of constant 45 Hz noise is circled. Because this signal is continuous over time, one-way depth migration spreads it out in space in the fracture seismic intensity volume, as is shown in the left and right sides of the lower panel. The key for identifying this signal as stationary noise in the fracture seismic intensity volume is the alignment of the features it produces in the final processed volume. These features are linear and point back to the surface position of the noise source, the presence of which was later identified in surface maps and images. Thus, in addition to cepstral filtering, careful selection of time windows to avoid including stationary noise greatly aids in fracture seismic intensity mapping.

Elevation and residual statics are important because the one-way travel time depth migration assumes that the traces are shifted in time to approximate a flat elevation datum. Therefore, fracture seismic traces must be shifted to account for differences in the receiver elevations and for the near surface velocity variations. The elevation statics are computed by taking the difference between the surveyed elevation of the geophone and the chosen constant elevation datum, computing the travel time for the elevation difference using the near surface velocity. The computed travel time shifts are applied to the traces before depth migration. The optimum method for analyzing the elevation statics, the residual statics, and the correct velocity model is to record the waveforms from a perforation shot that is visible on all geophones. Using the initial velocity model, the one-way travel times from the X, Y, Z location of the perforation shot to each geophone are computed and applied to the traces after correction for the elevation differences. If the first break time of the perforation shot waveform is approximately at the same time for all of the traces, the velocity model and elevation statics are accurate. The top panel of Figure 12 shows that the arrival times on average are flat in time for the full offset range for this example. The remaining variations from the same arrival time are caused by near surface velocity differences between the individual receivers. The time shifts for each trace to get them to the same arrival times are the required residual statics. The bottom panel of Figure 12 shows the traces after correction for residual statics and shows that the waveforms arrive at the same time on all traces. By using the correct velocity and good quality statics, the depth migration can be computed with very high confidence that the intensity volumes will be good quality.

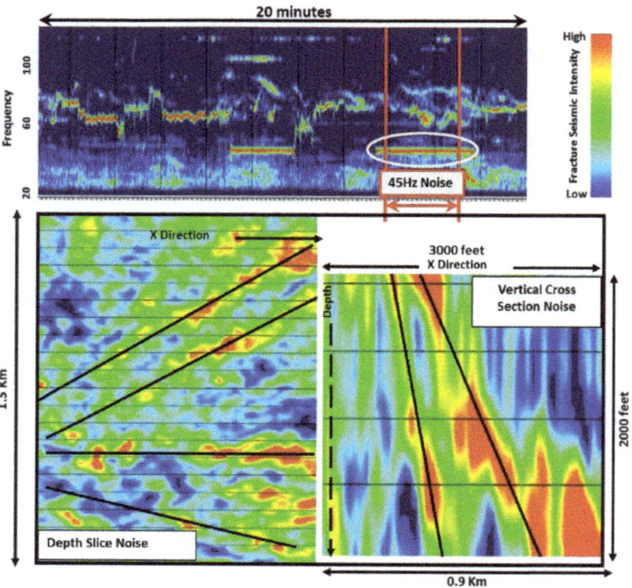

Figure 11. Surface noise in the traces appears in the spectrogram as narrow frequency band noise (**top** panel). The time window indicated by the red bars was used to compute a depth slice and a vertical slice of the fracture seismic intensity volume (**bottom** two panels). The narrow frequency band noise shown in the spectrogram causes the linear features in the fracture seismic intensity volume noted by the black lines. Tracking the black lines back to the intersections reveals the surface location of the noise source.

The velocity model must be accurate in order to obtain correct locations for the fractures in the volume. Often, a 1D velocity model is constructed from the sonic log recorded in a nearby well. This 1D velocity model is used to fill the entire 3D velocity volume such that the travel times are a function of offset only. This can work well for the small area around the well if the rock layer strata are flat lying and relatively homogeneous. For most areas of the Eagle Ford formation in Texas, there is a lateral velocity gradient such that for a constant depth the velocity decreases towards the Gulf of Mexico. When there is a lateral velocity gradient, using the same velocity for all voxels in the fracture seismic intensity volume results in a location error. Figure 13 shows an example of the velocity volume from the Eagle Ford that shows the gradient very well. When a 1D velocity model is used to focus and locate the perf shots, the gradient in the actual earth velocity causes errors in the locations of the perforation shots. The measured location errors for the perforation shots provide the information required to compute a set of statics that can be applied to the traces to force the location of the perforation shots to their known correct locations.

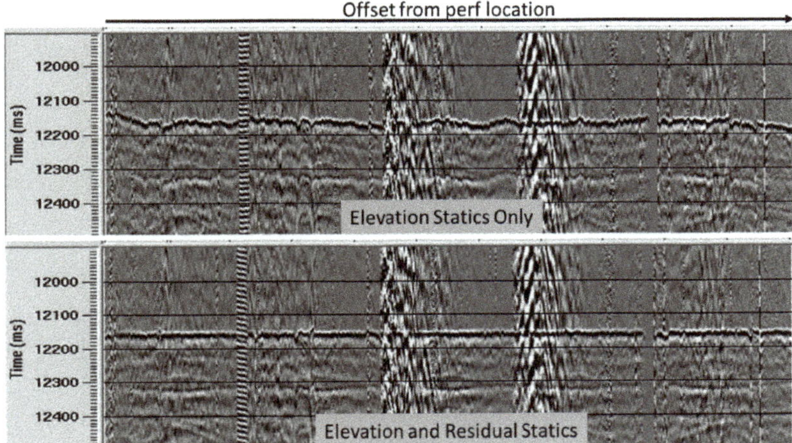

Figure 12. The top panel shows the traces sorted by offset from the X, Y location of the perf shot and with the time shifts applied using travel times from the perf location in depth to each receiver such that they should all be flattened at the same time. The traces are adjusted for elevation differences between the traces. The bottom panel shows the traces time shifted for the residual differences remaining on the traces in the top panel.

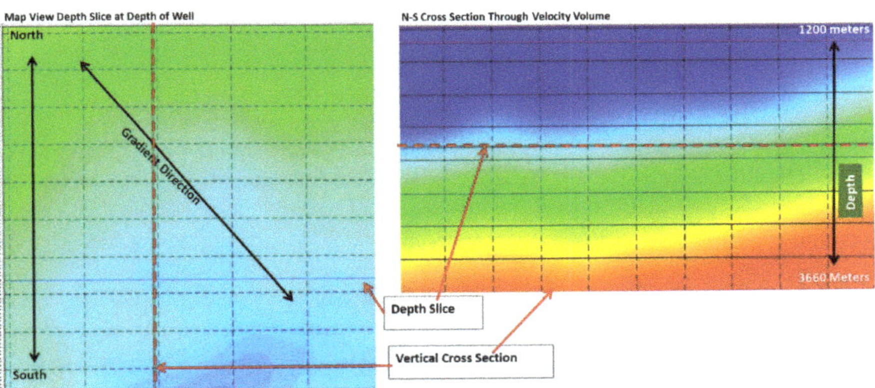

Figure 13. The velocity volume computed from the 3D reflection seismic shows a gradient. Using the 1D velocity model derived from the sonic log and this gradient, the perf shots can be positioned to match the known locations. Using this method to calibrate the location accuracy avoids the requirement to build a 3D interval velocity model.

The location correction using the gradient will not work for areas where the velocity volume has 3D complexity. For areas with 3D complexities, the full 3D interval velocity must be used for all aspects of focusing and imaging in order to obtain useful results. A 3D complex velocity model is best derived using 3D reflection data and iterative pre-stack depth migration. Figure 14 shows an example of a complex 3D velocity model in a Colombia thrust zone. For fracture seismic, the one-way travel times in a complex velocity model must be computed using a full 3D ray tracing algorithm.

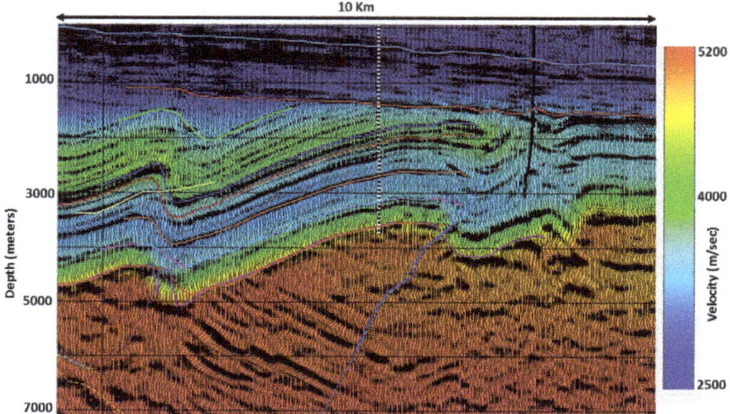

Figure 14. 3D complex velocity model in thrust zone computed using iterative depth migration. The fracture seismic intensity volumes computed with this velocity model will tie the reflection data and the fracture seismic intensity can be mapped onto the geologic structures.

2.5. Identification of Resonance and Turbulent Flow in Fracture Seismic Trace Data

Resonance and turbulent flow signals are identified in spectrograms computed from the trace data. The spectrograms are naturally noisy and extra care is taken to build up the signal during the computation of the spectrograms. For a surface location of interest, several traces very close to that location are selected for computing a single spectrogram. Spectrograms are computed for each selected trace and then the spectrograms are stacked. The first step in the computation of the spectrogram for a single trace is to compute the Fourier transform (FFT) for the first second of the trace and store the amplitude versus frequency at the first time sample in a two-dimensional array of frequency–time. The 1 s window is moved up in time by one sample, the FFT is computed, and the amplitude of the FFT is stored in the second sample along time. For trace data sampled every four-milliseconds, there are 15,000 one-second windows in 1 min of trace data. After the spectrogram is computed for every selected trace, all of the frequency–time arrays are stacked to obtain a single spectrogram for the location of interest.

Spectrogram analysis facilitates the efficient identification of time windows for use in computation of fracture seismic intensity volumes. Time periods with the strongest resonances are selected from the spectrograms and used in the depth migration method to map their spatial locations (Sicking et al., 2016, 2017) [21,25].

A spectrogram computed for data recorded during the startup of the first stage stimulation for a well in the Eagle Ford is shown in Figure 15. The quiet time before the pumping is initiated shows some resonances that are episodic. They build in amplitude over time and they transition from dispersive to turbulent flow as the pumping continues and the pressure rises.

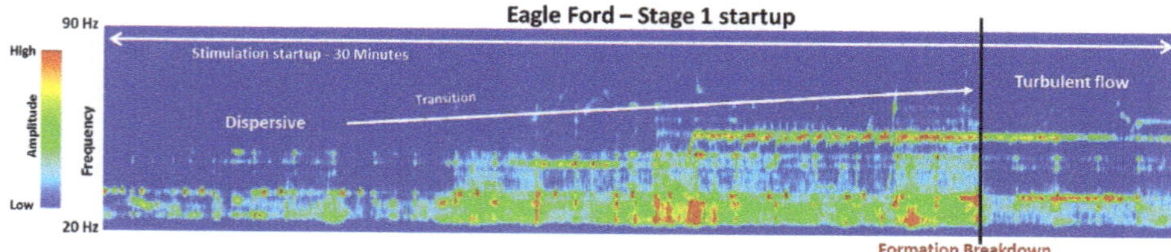

Figure 15. Resonance during stimulation. Eagle Ford during the startup of the stimulation for Stage 1. The resonances show a transition from dispersive to turbulent flow. There is a very pronounced change at the time when the formation breaks down.

Tary (2014 [3,4]) shows that fracture seismic resonances and turbulent flow are correlated with changes in pressure and slurry rate during stimulation. An example of this correlation is shown in Figure 16 where the spectrogram for 13 min of trace data recorded during stimulation is shown along with the pressure and slurry rates used for the same 13 min. The corresponding time windows between the treatment curves and the spectrogram are denoted by the vertical yellow lines. The resonance patterns in the spectrogram change at the same times that the pump curves show changes. This supports the interpretation that the resonances in the spectrograms are signal that is excited by the stimulation in the reservoir.

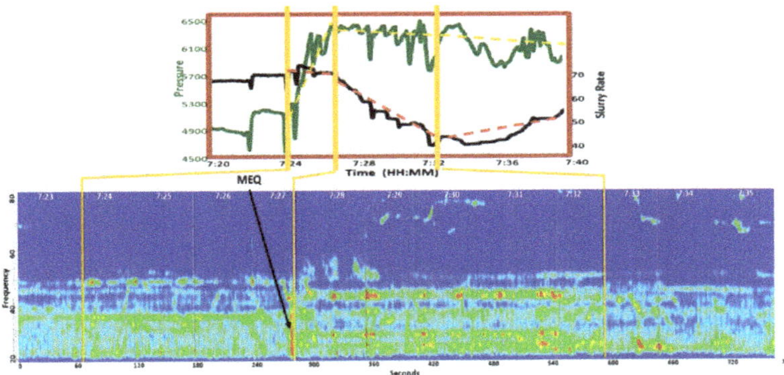

Figure 16. The spectrogram for 13 min of trace data recorded in the New Albany shale during stimulation correlates with the pressure and slurry rate curves. The pressure and slurry rate curves are shown in the top panel and the spectrogram is shown in the bottom panel. Four different time windows are denoted by the yellow lines and show that the changes in the treatment curves are correlated with changes in the spectrogram.

2.6. One-Way Travel Time Depth Migration

The fracture seismic intensity volumes are computed using Kirchhoff depth migration with one-way travel times. The one-way travel times are computed from each voxel at depth to every receiver on the surface. Kirchhoff migration is a two-step process that first applies a time shift to each trace equal to the travel time from the voxel to the surface geophone and then images across all of the time shifted traces.

An intensity volume is computed for each 200-millisecond time window of the trace data with a move up of 100 milliseconds between the intensity volumes. The intensity is computed for every voxel in the depth volume for every time window. This produces a new fracture seismic intensity depth volume at 100 millisecond steps.

Figure 17 shows a graphic of the for processing, focusing, and imaging used to compute the 3D depth fracture seismic intensity volume for each 100 milliseconds. The time interval for computation of intensity volumes can range from a few minutes to several hours. The imaging application must compute intensity volumes that can be coherently stacked over the entire time interval. Because of this stacking requirement, the fracture seismic intensity volumes are computed using semblance (defined in Figure 17) and the values in the intensity volumes are all positive. The phase of the waveforms in the trace data can change for each time window such that if the image computation method preserved the phase of the trace data, the fracture seismic intensity volumes from one depth volume to the next would not stack coherently.

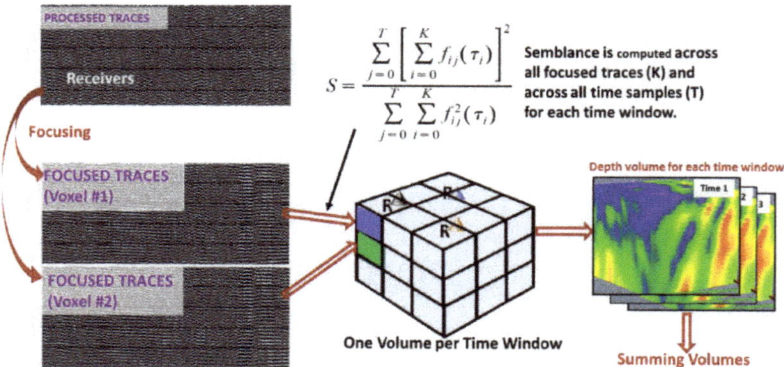

$$S = \frac{\sum\limits_{j=0}^{T}\left[\sum\limits_{i=0}^{K} f_{ij}(\tau_i)\right]^2}{\sum\limits_{j=0}^{T}\sum\limits_{i=0}^{K} f_{ij}^2(\tau_i)}$$

Figure 17. Workflow for one-way travel time depth migration. After trace processing and velocity model building and calibration, the traces are depth migrated for each time window and each depth voxel for the time interval that will be summed; f_{ij} is the trace amplitude at trace I and time j.

The differences between the method described by Kochnev (2007 [9]) and the one here are that they compute the coherency only for a subset of the depth volume and for previously identified time intervals. The streaming depth migration described here streams all of the trace data for the entire time interval of interest through the processing workflow and computes tens of thousands of depth volumes.

The post migration processing identifies the time windows in which coherent noise contaminates the fracture seismic intensity volumes and deletes them. The time windows in which large amplitude MEQ occur are detected and deleted. The final fracture seismic intensity volume is computed by summing all of the volumes that are not deleted.

2.7. 3D Fracture Extraction Methods

Fractures are extracted from the fracture seismic intensity volume by first picking all of the local maxima in the volume. Two methods that are currently employed are picking the local maxima and computing the value of the maximum negative curvature. The voxels that have local maxima or maximum curvature are connected to each other to form complex 3D surfaces that show the connectivity of the permeable fractures throughout the volume. Copeland et al. (2015) [35] describes the curvature method for tracking the maximum curvature in the intensity volume.

Petrophysical and engineering measurements support this interpretation for data recorded when there is no industrial activity (quiet times) and for data recorded during stimulation (Sicking et al., 2016, 2017) [21,25], (Geiser et al., 2012 [26]), (Lacazette et al., 2013 [27]).

2.8. Location Accuracy—Correlation with Distributed Acoustic Sensing (DAS)

The fracture systems are very complex 3D surfaces and there is always the question concerning the location accuracy of these surfaces. The location accuracy of 3D reflection seismic imaging measures the offset in three dimensions for reflections and faults from the locations determined from drilling. Fracture location accuracy can be measured using some of these same methods.

The location accuracy of images from one-way travel time depth migration is on the same order of magnitude as that obtained by reflection seismic imaging because it employs the same band pass in frequency and the same velocity model. Fracture seismic fracture locations have better location accuracy because there is integration over long time periods and the accumulation of signal over the integration time improves the location accuracy. Reflection seismic imaging does not have this advantage.

Figure 18 shows a synthetic study that demonstrates the improvements obtained from integration over time. A fracture was modeled in a 3D velocity volume and the signals emitted from the fracture are very small. However, the emissions from the fracture continue episodically for 15 min. The recording system has 2000 receivers on the surface and the fracture is located at 5000 ft. depth. The noise in the trace data has sufficient amplitude that the signal in not visible in the traces.

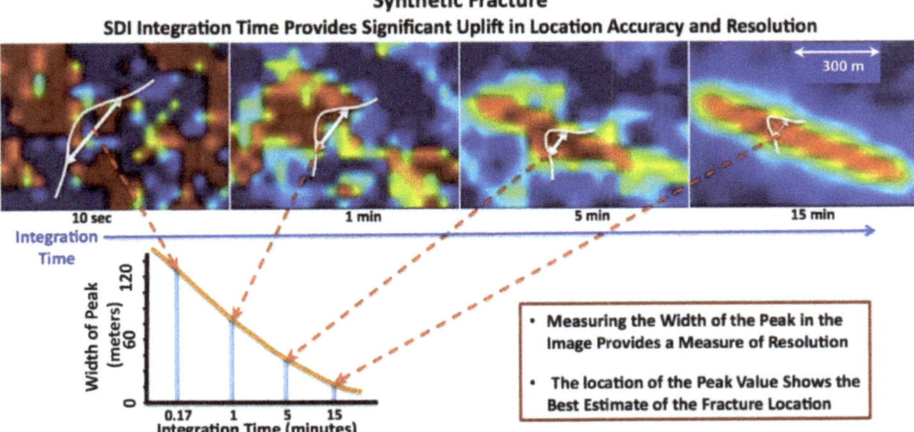

Figure 18. Integration of fracture seismic volumes increases the location accuracy, the S/N, and the resolution. For 10 s of integration, the fault is not visible. After 1 min of integration, the fault begins to be recognized. After 5 min, the fault is well defined and after 15 min, it is well resolved.

The images in Figure 18 shows a depth slice of the intensity volume through the fracture. As the integration time increases, the Signal-to-Noise ratio (S/N) increases and the resolution of the fracture improves in that the peak signal to background noise increases and the measured width of the fracture narrows. With sufficient integration time, spatial location of the fracture reaches an accuracy of 8 to 15 m.

Figure 19 shows the comparison of a fracture image from fracture seismic data integrated over the entire stage with the fiber optic cable acoustic recording for the same stage. The acoustic signal from the fiber optic log shows that most of the frack fluid came from the perf location nearest to the well head. The fracture seismic fracture image crossed the well within 3 to 8 m of that perf location. This is a direct comparison of two independent measurements and shows that the location accuracy of fracture seismic is very good.

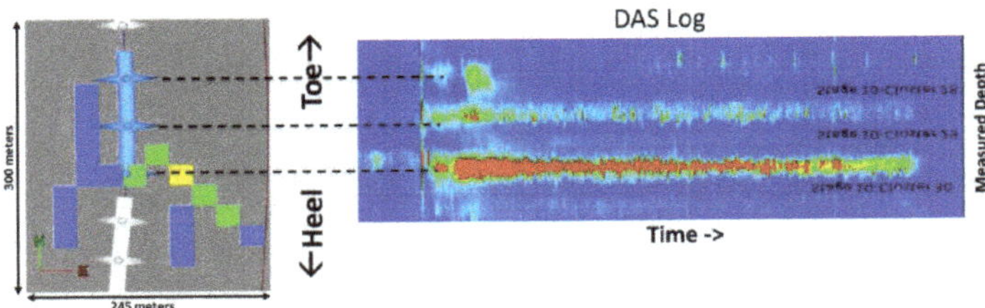

Figure 19. Comparison of the fracture image from fracture seismic traces and the distributed acoustic sensing (DAS) log from a fiber system recorded during the stimulation. The DAS plots and the fracture image show the same result, which supports the interpretation that the location accuracy of fracture imaging is on the order of 8 m.

2.9. Fracture Seismic Images and Hypocenters/MEQ

Fracture seismic was recorded using a buried grid in the Eagle Ford during the stimulation of four wells. The hypocenters were detected and located for the stimulation times for all stages of all wells and are shown in the right panel of Figure 20. The fracture seismic method was used to extract fractures for the same treatment times as was used for the hypocenter detections and they are shown in the left panel of Figure 20. In most projects where this comparison is made, it is rare that there is a direct correlation between detected hypocenters and the fracture seismic maps.

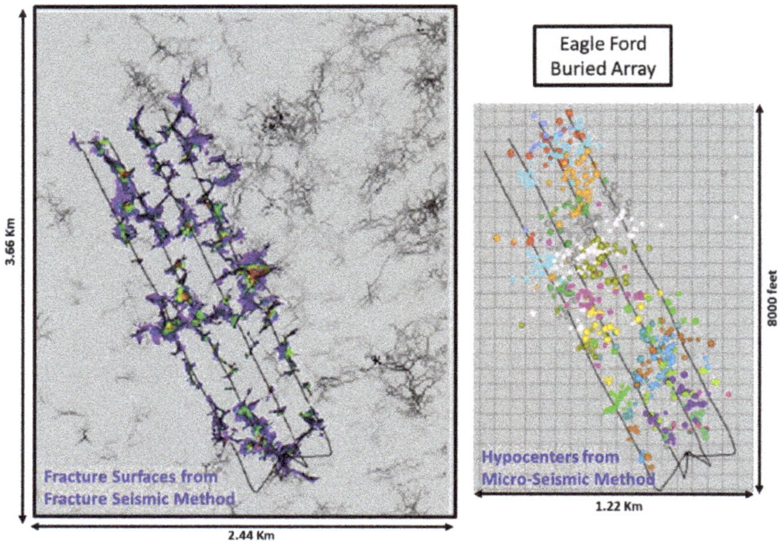

Figure 20. Comparison of fracture images from fracture seismic to the MEQ locations in the same data. The fracture seismic data were recorded using a buried array in the Eagle Ford. There is only a partial correlation between the fracture surfaces and the MEQ.

The waveforms that are emitted from the fracture tips for the hypocenters and MEQ are sourced by different mechanisms than the signals used to compute the fracture seismic intensities. The interpretation of this phenomenon is that the fracture seismic energy from fluid-filled fractures are emitted along the length of the fractures while the hypocenters occur at the tips of the fracture. Thus, the larger MEQ are not collocated with the fracture seismic image and the smaller opening mode hypocenters are at the tips of the fractures.

These differences in source mechanisms and signals explain why the fracture models computed using hypocenter locations are almost always different from the complex 3D fracture models derived using fracture seismic methods.

3. Case Studies

A large database of case histories demonstrates the capacity of fracture seismic methods to directly map fluid-filled fractures and their role in subsurface connectivity. The examples come from different basins and different geologic settings They also come from data recorded before, during, and after various kinds of industrial activities, and from both greenfield as well as brownfield sites.

Fracture seismic methods have been used to map the fracture connectivity during the stimulation of approximately 100 horizontal wells with almost 2000 stages. Fracture seismic observations before, during, and after these stimulations show that the fracture systems that produce the most fluids are the same fractures that are mapped before wells are drilled. We will also show how fracture seismic can track the fluid producing volume over the life of the well.

Figure 21 shows a subsurface rectilinear volume that contains a well for which the fracture seismic method was applied before, during, and after the stimulation of the well. The lower panel shows a depth slice of the intensity volume computed before the stimulation. The back panel show a vertical slide of the intensity volume computed during the stimulation. The 3D volumetric view in the center shows the producing volume for the well after it was being produced. The actual volume of rock that is producing fluids is quite different from the stimulated rock volume computed during the treatment.

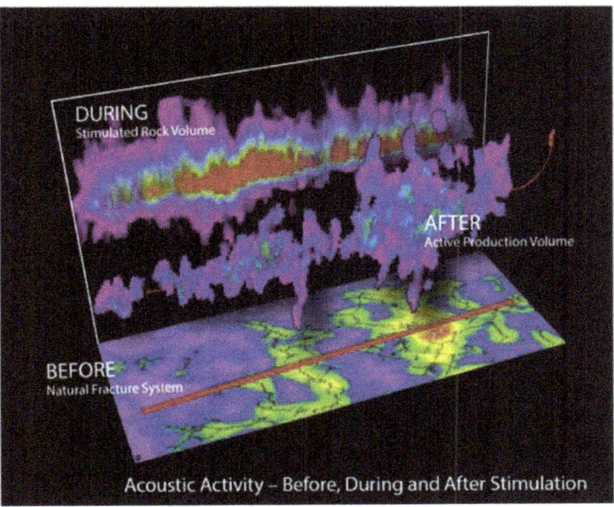

Figure 21. Fracture seismic imaging of the subsurface is applied before, during, and after hydraulic stimulation.

In addition to local hydraulic fracture stimulation projects, 15 larger-scale fracture seismic mapping studies have been completed. These include projects when the fracture seismic is recorded before drilling, stimulation, production, subsurface flooding, or other related industrial activities are underway. The correlation between the fracture systems computed for quiet times and the fracture systems computed during production indicates that the fracture systems are excited by pressure changes caused by natural earth processes such as tectonics or earth tides as well as by the stimulation treatments.

3.1. Intensity Burst during Stimulation—Texas

This example shows fracture seismic signals that are distributed along the length of a single fracture. These observations were recorded during the hydraulic stimulation of one stage in the Eagle Ford shale of Texas. It consists of a single burst of high-energy resonances that lasted for 7 s (Figure 22). The burst was recorded with a surface grid laid out with orthogonal cables that had 2100 receivers over an area of 65 square km.

The spectrogram in the left panel of Figure 22 for these 7 s shows the narrow band energy that we initially identify as resonance from turbulent flow into the fractures surrounding this stage. The seismic record sections on the right show that the amplitudes of this burst can be seen on the individual receivers with amplitude that is well above the background noise. The waveforms do not change phase with azimuth or offset, which indicates that this is not a point hypocenter or MEQ.

Fracture seismic depth migration was used to map the 7 s burst to the location of its source. The fracture seismic intensity volume shows that all this energy came from a very small area (see Figure 23). The resonance shown in the spectrogram and the vertical extent of the source location supports the interpretation that it is a resonance initiated either by flow into the fracture or by resonance along the entire fracture. The overlay of the fracture seismic intensity on the associated 3D seismic reflection section shows that it is in a small synclinal structure and is oriented in the vertical direction.

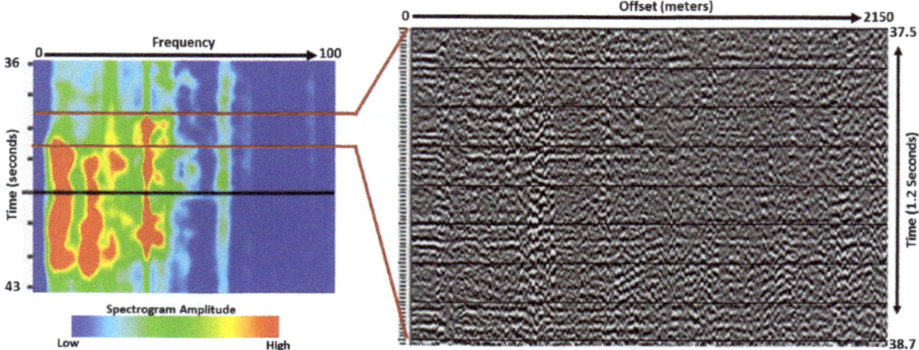

Figure 22. Intensity bursts during stimulation. A burst of higher amplitude waveforms in the trace data continued for 7 s. The spectrogram of the 7 s is shown on the left. The amplitude of the signals was sufficiently high that the waveforms are clearly visible in the individual traces as shown in the plot on the right.

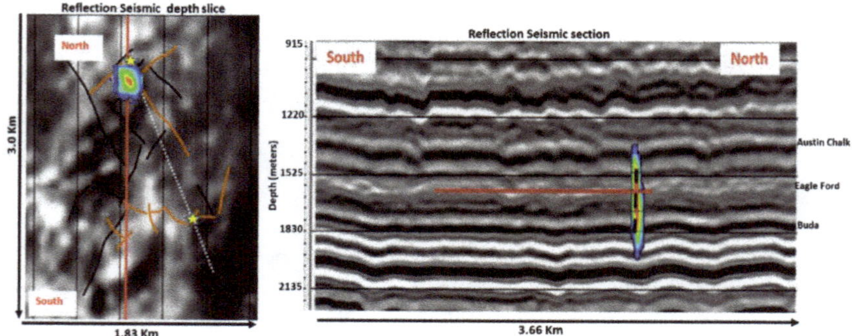

Figure 23. Computing an integrated fracture seismic intensity volume over the 7 s of data in Figure 22 and plotting on the reflection seismic shows that the energy came from a fracture in a small syncline that extends in depth from the Buda to the Austin Chalk and is focused into a very small area just to the West of the well near the stage being stimulated.

3.2. Thrust Fault Activation during Stimulation—China

This example shows that faults can have both permeable zones and zones that do not transmit pressure or fluids. A horizontal well drilled along a reservoir layer in a thrust zone is shown in Figure 24. This well was parallel to and 300 m shallower than the thrust fault mapped on reflection seismic. Given this vertical separation, it was thought that hydraulic stimulation of the well would not affect this structure. In order to evaluate the stimulated resource volume of this treatment, fracture seismic data were recorded using 1600 receivers in a surface grid.

The recording for each stage shows that there is a zone in the nearby tear fault that is permeable for only part of the length of the well. The permeable zone is active for only four of the stages during the treatment. For stimulation stages near the toe and near the heel of the well, the trust fault was seismically inactive. During the tear-fault-related four stages, the stimulation pressure was transmitted down to the thrust fault and caused many larger microearthquakes. None of the other treatment stages cause microearthquakes in the thrust. Evidently, the fracture seismic imaged permeable zone transmitted the pressure from the pumping to the thrust and a large number of MEQ are stimulated during the four stages.

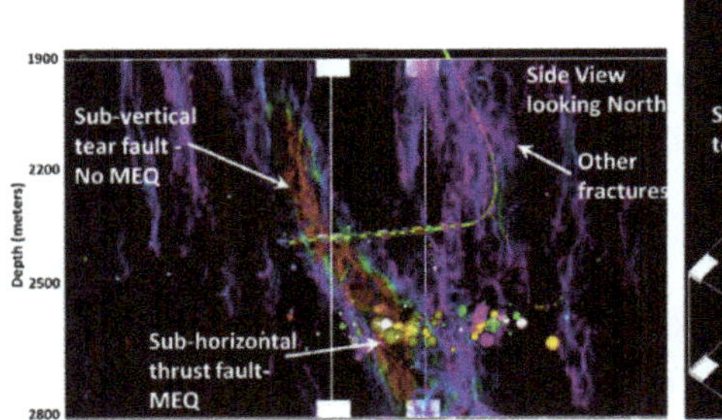
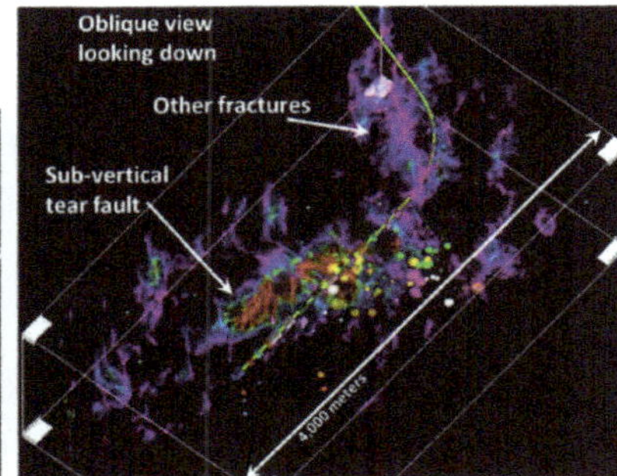

Figure 24. MEQ in the trust fault that is 300 m below the well were activated during only four middle stages of the well treatment. The pressure from the stimulation propagated along the tear fault down to the thrust and caused activation in the thrust zone. The pressure from the stimulation activated resonances in the tear fault so that it was mapped in the fracture seismic intensity volume. The side view of the fracture image volume (**left** panel) shows the tear fault, the well, and the MEQ in the trust fault. The oblique view (**right** panel) show that the tear fault is very close to the well (50 m).

The side view in Figure 24 shows the width and height of the tear fault and that is permeable. The tear fault is approximately 50 m from the well path. The fracture image volume in Figure 24 shows the tear fault width and height. The width mapped is the same width and location as the location of the four stages. The height of the tear fault goes vertically above the well and below the trust fault depth.

3.3. Large Single Grid in Thrust Zone, No Stimulation—Colombia

This example shows how fracture imaging using fracture seismic is integrated with 3D reflection seismic in a new area to select drilling location (Sicking et.al., 2017) [25]. This large area in Colombia had not been actively explored for over 60 years. The reservoir in this area is the Rosa Blanca formation that produces gas in areas where it is naturally fractured. During the decade of the 1950s, this area had been drilled with little success. There were many dry holes but one well struck a highly fractured zone and blew out. Subsequently, the area was abandoned.

In 2012, before initiation of new drilling, a modern 3D reflection seismic survey was collected and, as an independent acquisition, a 4650-channel fracture seismic dataset was recorded that covered 50 square km. The fracture seismic area was in the middle of the area covered by the reflection seismic survey. Before any wells were drilled, the reflection seismic data was processed and migrated using pre-stack depth migration and a complex detailed 3D interval velocity model was computed (Figure 14).

The fracture seismic was processed using the fracture seismic method and a fracture seismic intensity volume was computed that included data from 15 h of recordings. The depth tie between the active and fracture seismic data was ensured by the use of the same interval velocity model for both depth migrations. The intensity volume was integrated with the reflection seismic depth migration volume in order to find the structural positions of the most active fractures in the Rosa Blanca.

The fracture seismic traces were analyzed for resonance using spectrograms (Figure 1). The high fracture seismic intensity and resonance observed at this site is likely due to the high stress state caused by the compressional forces in this region of the northern Andes. The spectrograms help to identify time periods for computing the fracture seismic volumes. While the behavior of the resonances is not yet well understood, experience dictates that very active resonance time periods are the best times to use in fracture seismic depth migration to compute the intensity volume.

Figure 25a shows a map view of the horizon slice extracted along the target horizon from the fracture seismic intensity volume. It is overlaid with the structural contours interpreted from the reflection seismic volume for the same horizon. The horizon map shows that the highest intensities are in the hanging wall of the thrust fault. The black symbols are the locations of dry wells drilled in the 1950s. They are all in the low fracture seismic areas of the Rosa Blanca. The small red circle shows the location of the well that blew out in the 1950s. This well is at the top of the structure and in a higher fracture seismic area, which indicates fracturing on the Rosa Blanca. A new well shown by the large black circle was drilled into the hanging wall of the thrust fault and in an area of high fracture seismic.

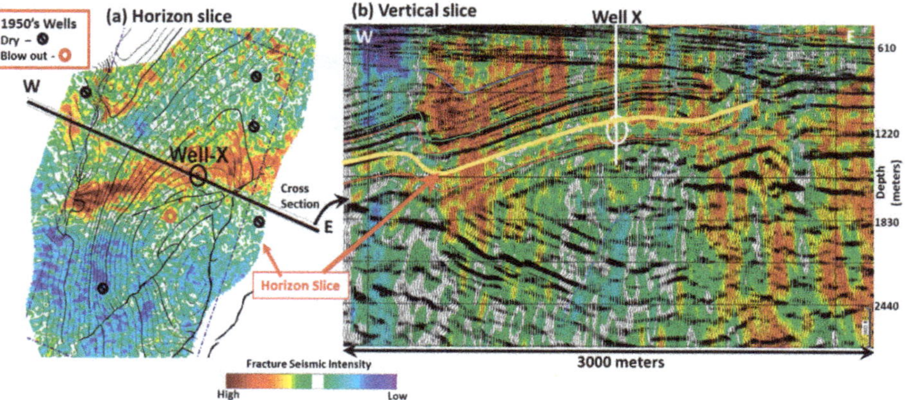

Figure 25. Identifying zones of fracturing in reservoir before drilling. Panel (**a**) shows the fracture seismic intensity extracted from the volume along the reservoir horizon with the overlay of the structural contours interpreted from the 3D reflection seismic. The traces of the thrust faults are shown. The highest fracture seismic are parallel to and out in front of the thrust fault. Panel (**b**) shows a vertical cross section through well X with the fracture seismic overlaid on the reflection seismic. The yellow line shows the horizon that was extracted to get the slice shown in (**a**). The fracture seismic intensity is higher down dip from where well X was drilled. Dry holes (small black circles) drilled before this analysis are in zones of low fracture seismic intensity. An old well that blew out (shown by the small red circle) and Well X are in zones of high fracture seismic.

Figure 25b shows the vertical cross section of the reflection seismic and the fracture seismic intensity volume through the new well. The reflection seismic data are in black and the overlay of the intensity in color. The intensities show that the hanging wall of the thrust is the most active and that the structure in the fracture seismic volume follows the structure in the 3D reflection data.

Integration of fracture seismic, together with interpretation of reflection seismic, can be used to indicate potential areas for drilling. Based on these data, a test well was drilled, and a heavily fractured reservoir was found in the zone of high intensity. From when the well was put on production and until April of 2019, this well has produced 1.7 BCF gas. Based on the results of Well X, we conclude that the high fracture seismic intensity along the flank of the thrusts does indicate areas of active fractures. By active, we mean seismically active with the implication, based on experience, that the fractures are also permeable.

3.4. Roll Along Fracture Seismic during 3D Reflection Seismic Acquisition—Texas

This example shows how fracture seismic can aide in selecting drilling locations. A 3D reflection seismic survey was collected in the Permian basin of west Texas using a nodal recording system. There were approximately 6500 simultaneously active nodes in the receiver array.

The fracture seismic data were collected concurrently with the reflection seismic by programming the nodes to record at night for 2 h when the crew was not shooting. This recording schedule meant that a very large area was covered by the fracture seismic for each day of recording. Each day of the fracture seismic recordings were processed as an independent data set. Multiple days of recording were selected such that there was substantial overlap in the independent intensity volumes. Nine hours of recordings over the area of the proposed horizontal well were selected for fracture seismic intensity computation.

Map and cross section slices of the fracture seismic intensity volume are shown in Figure 26. The well shown in Figure 26 was not drilled until two years after the seismic acquisition. The depth slice shows that the volume has areas of high intensity and areas of low activity. The proposed well path is predominantly in a zone of low activity with some higher activity at its toe and heel.

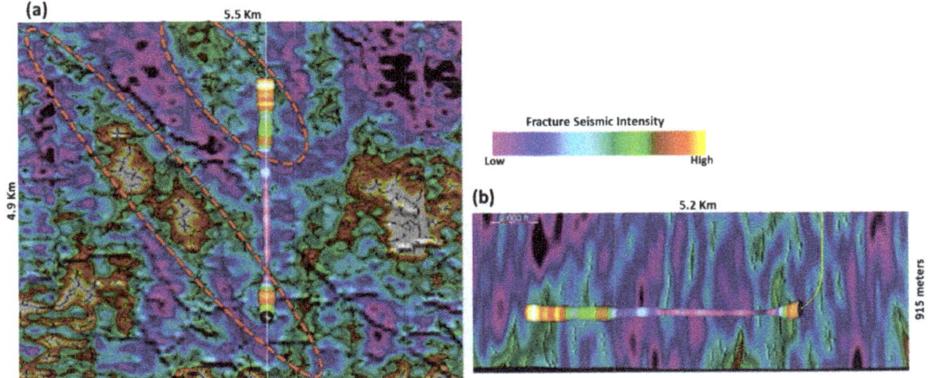

Figure 26. Slices of the fracture seismic intensity volume that was computed from data recorded during a 3D reflections seismic acquisition before the well was drilled. The depth slice (**a**) of the fracture seismic intensity volume is shown at the depth of the proposed well. The vertical slice (**b**) is along the path of the proposed well. The logs of the acid uptake show that the uptake was highest in the zones of highest fracture seismic intensity and lowest in the zones of low fracture seismic intensity. This result supports the interpretation that the fracture seismic intensity shows the zones of natural fractures.

Shown in both the depth slice and the vertical slice along the well path is the log for the volume of acid uptake during the stimulation. This log shows that the acid uptake is highest in the zones that had higher fracture seismic intensity and was lowest in the zones of low fracture seismic intensity. They are consistent with the interpretation that the zones of higher fracture seismic intensity have higher fracture density and connectivity. This well was not economic. If the fracture seismic intensity volume had been used to plan the well, it could have been relocated to a zone of high-density natural fractures.

3.5. Roll along Fracture Seismic during 3D Reflection Shoot—Wyoming

This example shows how fracture seismic is used for areal evaluation before development. A 3D reflection seismic survey was recorded in Wyoming that covered a very large area. A smaller area was selected for the fracture seismic recording. The entire receiver array of 6000 geophones was used for recording several hours on seven different days during the active seismic acquisition. The fracture seismic was recorded such that there was a 50% overlap in the fracture seismic intensity volume between recordings. Each recording time was processed as an independent project and the seven fracture seismic intensity volumes were merged later.

The depth slice at the reservoir depth from the fracture seismic intensity volumes for all seven recordings is shown in Figure 27. The slice shows a fault zone across the Northern part of the volume that is also mapped in the 3D reflection seismic. There are two large areas of high fracture seismic intensity in the SW and NE corners of the survey area. These areas are separated by a NW to SE trend of lower fracture seismic activity.

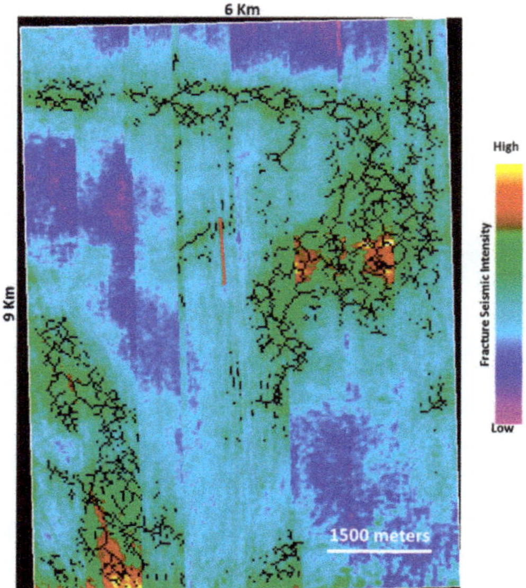

Figure 27. Depth slice of the fracture seismic intensity volume that was computed from fracture seismic recorded during a 3D reflections seismic acquisition before the well was drilled. The seams in the final volume are the result of merging the seven independent volumes. The linear feature that runs East to West in the North of the Volume is also seen in the 3D reflection volume.

The seams in the final merged volume are readily apparent in the depth slice shown in Figure 27. The fracture seismic intensity volumes are from data that were recorded a few days apart. The fracture seismic intensity during one day of recording is not the same for the other days. Differences in the fracture resonances can account for much of the differences. Processing may also account for small differences in the amplitude from one day to the next. It should be noted that there was a 50% overlap in the intensity volumes from one day of recording to the next such that every voxel in the volume has contributions from two days. Even with the presence of the seams, the fracture seismic intensity volume reveals important information on the natural fracture zones of this prospect and the optimal locations for drilling and stimulation.

3.6. Roll Fractures and Well Treatments—Illinois

This example shows how pre-existing fracture systems impact stimulation. Pre-stimulation fracture seismic was recorded over a well site in the New Albany shale after the well was drilled, but a few weeks before the well was stimulated. These data were used in the planning for the stimulation. Fracture seismic was also recorded during the stimulation. This allowed for the comparison of the fracture seismic intensity volume before, and after stimulation.

Spectrograms were computed from the fracture seismic data recorded pre-stimulation and from the fracture seismic data recorded during stimulation. Samples for both are shown in Figure 28. The top panel shows the spectrogram for a time window pre-stimulation and reveals a narrow band resonance in the 50 Hz to 60 Hz range, along with a broad distribution of signals at lower frequencies. For comparison, the spectrogram for a time window during the stimulation is shown in the lower panel of Figure 28. The resonances are mostly in the lower frequency bands but have substantial changes in amplitude, character, and frequency band. These changes correlate with the stimulation pressure and fluid rate pump curves (Figure 16).

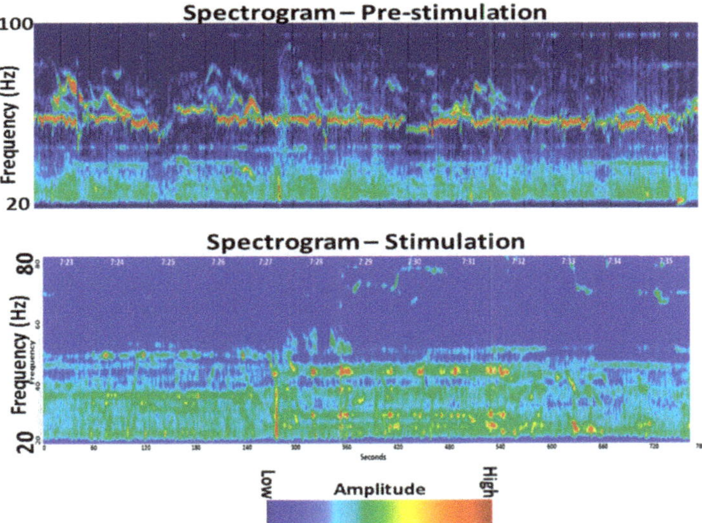

Figure 28. Spectrograms computed from pre-stimulation fracture seismic data (**top**) and data recorded during stimulation (**bottom**).

The fracture seismic intensity volume computed for this pre-stimulation time period shows that the fracture seismic activity is very high at the toe of the well and has significant zones of low activity in the middle of the well path (Figure 29, left panel). The fracture seismic volume from the pre-stimulation time period was used to plan the stimulation. During the first three stages of the stimulation, problems were encountered in getting the fluid to flow into the formation. The pre-treatment data were used to analyze the stress field to determine that the pressure used in the pumping should be reduced to solve this problem.

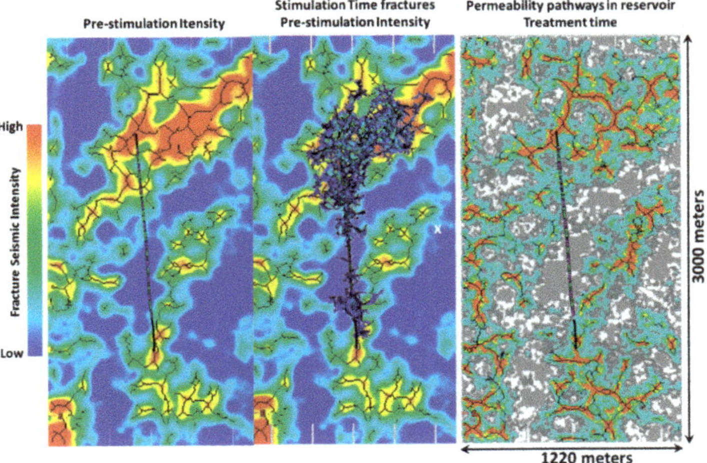

Figure 29. The fracture seismic intensity volume computed from fracture seismic data recorded before stimulation show that the stimulation activated the same fractures that were mapped before the stimulation activity. The left panel shows a depth slice of the intensity volume from the pre-stimulation data. The middle panel shows the depth slice of the pre-stimulation intensity with the overlay of the fractures active during stimulation. The right panel shows the connectivity pathways computed from the 16 intensity volumes computed for each stage.

Comparing the pre-stimulation fracture seismic (left) to the fracture volume (middle) that was computed during the stimulation shows that the fractures that are computed from the data recorded during the stimulation follow the intensity patterns in the fracture seismic volume computed pre-stimulation. The overlay of the fractures computed during stimulation on the fracture seismic intensity computed pre-stimulation show that the pre-existing fracture system impacts the performance of the stimulation with the highest intensity zone during the pre-frack time having the highest density of activated fractures during the stimulation.

Using the fracture seismic intensity volume computed from the pre-stimulation and the volume computed during stimulation we forecast the connectivity pathways (Figure 29, right panel) in the reservoir that will produce the most fluid flow during production. These were computed by first thresholding the amplitude in each of the intensity volumes for each stage of the stimulation. In each voxel for each stage, the amplitudes that were below this threshold were reset to zero. For each voxel, the number of volumes that were above the threshold for that voxel were counted and the count for that voxel was stored in the repeated activity volume.

For example, with 16 stages, each voxel has the possibility of containing a number between 0 and 16. The connectivity pathways are then seen as the highest number of threshold crossings. This attribute volume is used to model the pressure and fluid transmission through the reservoir. The connectivity pathways for this example show that zones of best connectivity in the reservoir are the most active in the pre-stimulation fracture seismic volume.

3.7. Pump Startup Time Fracture System—Texas

This example shows how the pressure from the stimulation produces fracture seismic intensity in the fracture system before formation breakdown. Formation breakdown is when the fractures near the well open and allow the initiation of fluid flow into the reservoir. The fracture seismic data recorded before formation breakdown can provide very useful details about the reservoir.

Figure 30 shows a spectrogram for 30 min of data recorded in the Eagle Ford shale during pressure buildup for the first stage of a hydraulic stimulation. During startup, the increasing pressure moves into the rocks causing resonances in the permeable fractures that are connected to the well at Stage 1. The observed resonances grow in amplitude and complexity with increasing pressure. The resonances transition from low-amplitude dispersive to high-amplitude turbulent after formation breakdown. These resonances are very different from those observed during the pre-stimulation in the New Albany shown in Figure 28.

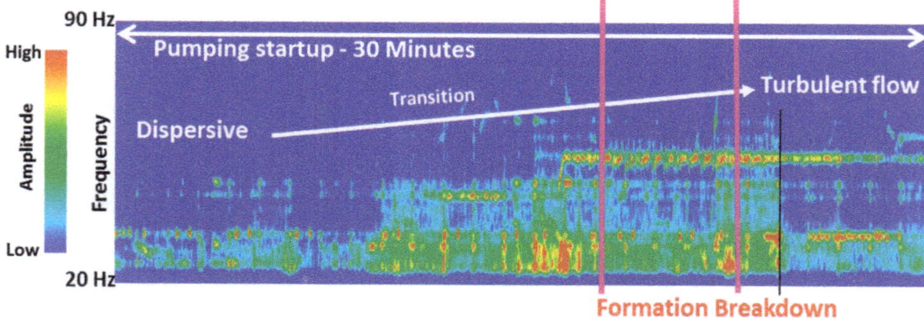

Figure 30. Spectrogram during Stage 1 startup shows the resonances transition from low amplitude to high amplitude as the pressure is increased with pumping time. Formation breakdown causes a substantial change in the resonance as the fluid begins to flow into the formation. The lavender bars mark the 5 min of data that are used to compute the fracture images.

The 5 min of trace data marked by the lavender lines in Figure 30 were selected for computation of the pre-breakdown fracture seismic volume. These minutes were selected because of the high fracture seismic signals in the spectrogram and because it is before formation breakdown. Previously, it was not expected that the fractures would emit such high-intensity fracture seismic resonances before formation breakdown.

The depth slice at the well depth of the fracture seismic intensity volume computed from these 5 min is shown in Figure 31. The feature trending from SW to NE and terminating at the well is a fault that was previously observed in the 3D reflection seismic data. The depth slice shows the fracture seismic intensity stimulated by the startup of pumping with the overlay in black lines of the fractures. The fractures map the connectivity to the perforation location for Stage 1.

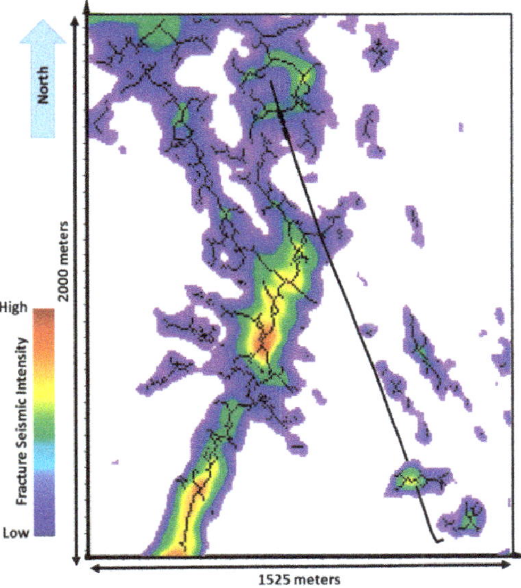

Figure 31. Depth slice of fracture intensity volume computed for 5 min fracture seismic data and the overlay of the computed fractures. The fractures show the connectivity pathways that connect to the well at the Stage 1 perforations.

This section of the Eagle Ford has a large number of fractures, as can be seen in the 3D reflection section shown in Figure 23. The interpretation is that only a few of these fractures were activated during Stage 1 startup. Most of the fractures in the volume are not activated with the increase in pressure. Only the most permeable fractures are activated during the startup time before the first stage formation breakdown. The fracture lines form a pattern that might have been in related to a previous stress direction that is different from the one in place today.

3.8. Stimulation Time—Texas

The fracture seismic fracture surfaces shown in Figure 32 are computed from data recorded during the stimulation for a well in the Permian. The fractures activated by the stimulation open out into the reservoir for approximately 15 m and then turn parallel to the well. This well was not economic because the fractures that opened did not have sufficient rock volume. The interpretation is that the well was either not drilled along the maximum horizontal stress direction or that the pumping pressures changed the local stress causing the fractures to turn parallel to the well.

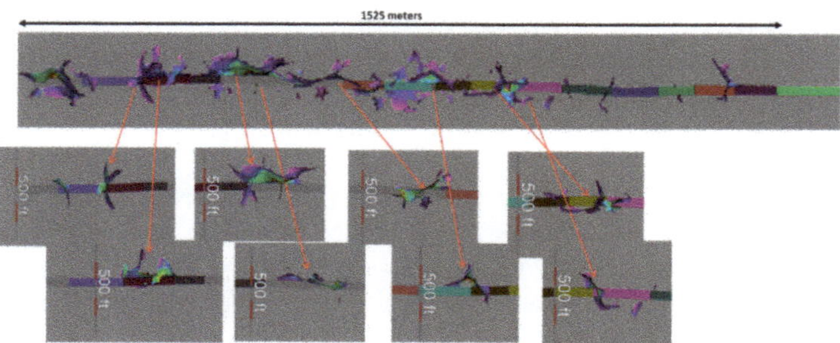

Figure 32. Fracture images computed during stimulation for the stages of a well in the Permian showing the that the stimulated fractures open perpendicular to the well and then turn parallel to the well.

3.9. Prediction of Well Interferences on Adjacent Well—Pennsylvania

The location accuracy of fracture seismic fracture imaging is demonstrated in a project in the Marcellus. One well had been drilled and seven more planned when a buried grid was installed over the pad site. Well A in Figure 33 was stimulated and put on production before Well B was drilled. The fracture seismic signals initiated by the fluid flow into Well A were used to compute a fracture seismic intensity volume before Well B was stimulated.

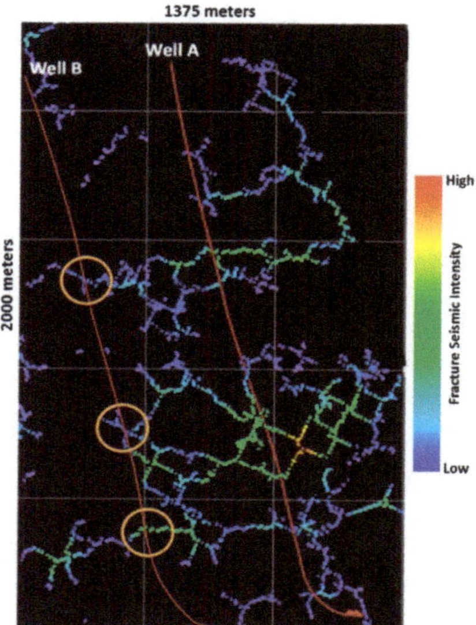

Figure 33. Predicting pressure interferences on an adjacent well. The fracture intensity computed before the stimulation of Well B but while Well A was producing show the fractures crossing the path of Well B. The locations shown by the circles mark the stage locations along Well B that were being stimulated when pressure interferences were measured at the well head of Well A.

Well A was in production during this fracture seismic recording and the fractures computed from the fracture seismic data intersect with the path for Well B. This indicates that when Well B is fracked, there could be pressure hits on Well A at three separate stages of the Well B stimulation. The locations of these three predictions are shown by the circles in Figure 33. Later, when Well B was stimulated, pressure changes in Well A were recorded by a gauge at the head of Well A. These pressure changes occurred for treatments located at fracture seismic imaged fracture crossings. These engineering data confirm the predicted connection between the wells and that the fracture map shows the fractures that transmitted the pressure from Well B to Well A.

3.10. Actively Producing Volume Before and After Pressure Hits—Pennsylvania

The fluid flow into the well during production causes turbulent resonance in the fractures and allows for the computation of the actively producing volumes. Fracture seismic intensity volumes computed from data recorded over a producing well are used to extract the active voxels that are connected to the well. An intensity threshold is first selected and applied to the intensity volume. The remaining voxels that are connected to the well are extracted from the fracture seismic intensity volume using an iterative process whereby the active voxels that are touching the well are extracted in the first iteration. The subsequent iterations detect and extract voxels that are touching the previously extracted voxels. The iterations continue until no more active voxels connected to the well are detected. The volume of extracted voxels is the actively producing volume that can be used for planning additional wells or reservoir treatments.

Two wells were drilled in this example from the Marcellus Shale shown in Figure 34. As discussed for Figure 33, fracture seismic was recorded before Well B was stimulated but while Well A was in production. The producing volume for Well A was computed using fracture seismic recorded before the stimulation of Well B (Figure 34, left) and again after the stimulation of Well B (Figure 34, right). Comparing the two producing volumes shows that the producing volume for Well A is 25% smaller after the stimulation of Well B than it was before the stimulation.

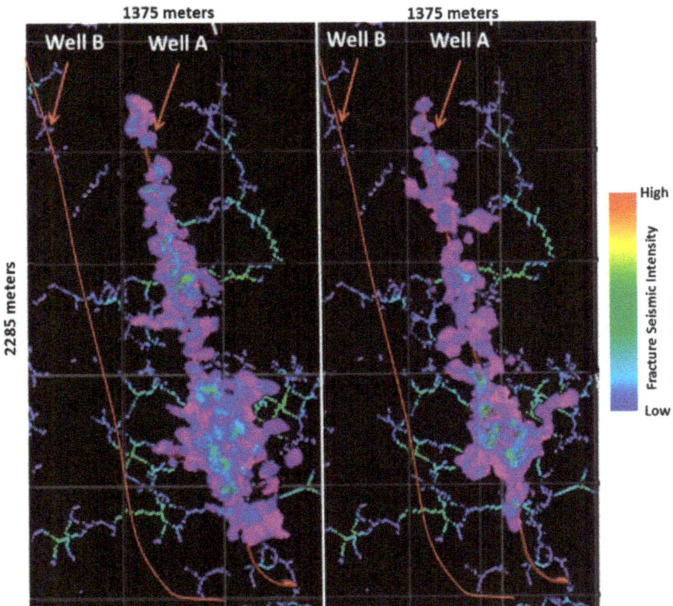

Figure 34. Producing rock volume around Well A before (**left**) and after (**right**) the stimulation of Well B. The pressure changes in Well A (Figure 33) during the stimulation of Well B caused production declines and also reduction in the producing volume.

The production data recorded at the wellhead of Well A show that the production from Well A was reduced by 30% during the stimulation of Well B. The well head production reduction agrees with the fracture seismic intensity volume reduction and supports the interpretation that the hits on Well A caused the reduction in production.

3.11. Fluid Producing Volumes over Time—Texas

The Eagle Ford producing volume images on the left in Figure 35 show the fracture seismic intensity volumes computed using fracture seismic recorded on a permanent buried grid for the stimulation time, at two years and at three years after the well was put on production. The buried grid was activated during the stimulation and again two years and three years later.

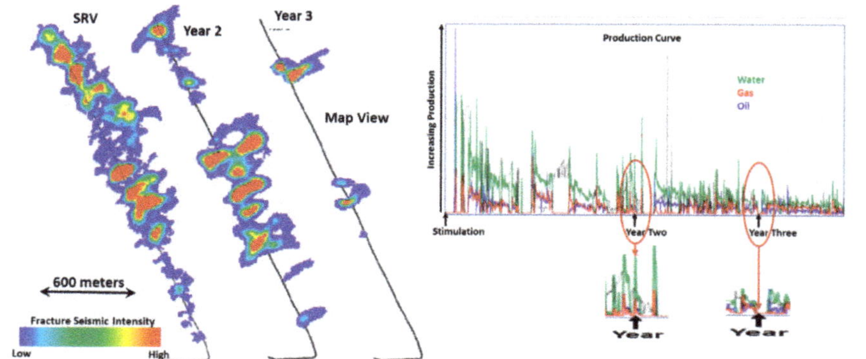

Figure 35. Fracture seismic intensity signals decline over three years. Left: Fracture seismic intensity volumes showing the rock volume activated during stimulation, the rock volume that is active after two years of production, and the rock volume that is active after three years of production. Right: Production curves recorded at the well head showing the fluids produced over time. The passive data for year three were recorded at a time when the production curves show the well was not in production.

Computing the fracture seismic intensity volumes for each of these times allows the active voxels connected to the well to be extracted for each time. The top down view of the stimulated rock volume and the actively producing volumes are shown. These three time-lapse volumes show that the stimulated volume during the treatment is much larger than the producing volume after two and three years of production. After two years, there are portions of the well that are not producing and after three years, only short segments of the well are producing.

The production curves in the right panel of Figure 35 show the volume of production measured at the well head. Zooming into the curves for the times of the active producing volumes shows that the well head production volumes correlate with the active volumes measured from the fracture seismic monitoring. The fracture seismic recording time was at a time when the well was producing and the fracture seismic actively producing volume is large. The recording at year three was at a time when the production from this well was very low and perhaps shut in and the fracture seismic actively producing volume is small.

3.12. Forecasting Production Before Drilling—Texas

This example shows an attempt to extract the reservoir connectivity volume that is connected to a well path before the well is drilled. Recording using a buried grid allowed extraction of the intensity along the planned well path before drilling and again during production 2.5 years after it was put on production. The fracture seismic intensity volumes shown in Figure 36 compare the predicted producing volume before the well was drilled to the actively producing volume after 2.5 years of production. The data were recorded twice using a buried grid. The first recording was two months before the well was drilled and the second recording was 2.5 years after the well was put on production. The planned well path for this well was used for the pre-drill active voxel extraction from the intensity volume. Figure 36 shows the forecast producing volume in black (left) and the measured producing volume 2.5 years later in blue (middle). Their overlay is shown on the right. There is good correlation between the forecast and the measured producing volumes.

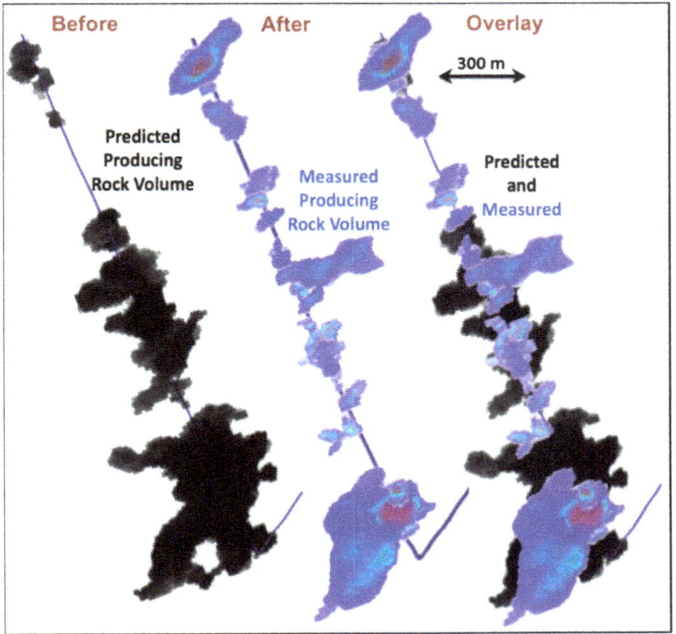

Figure 36. Forecasting production before the well is drilled. Left side: Pre-drill, fracture seismic intensity-based forecast of the producing rock volume around the well. Center: Fracture seismic-based measurement of the producing volume after 2.5 years of production. Right: Overlay of the forecast and the observed producing rock volumes.

These fracture seismic intensity volumes show that the production is coming from zones in the reservoir that were permeable before the well was drilled. This result establishes that time-lapse monitoring of the reservoir using fracture seismic, from pre-development through the production life of the reservoir, provides essential information for optimal management of the reservoir.

4. Conclusions

Based on several converging lines of evidence, we and others have concluded that fracture seismic recordings contain episodic signals generated by resonating, fluid-filled fractures. Theoretical studies by others show these resonances can be generated by interfering seismic waves originating from either dislocation at the fracture tips or turbulent fluid entries from other fractures and hydraulic stimulations. In this paper, we focus on showing how these signals can be used to map the most permeable structures in the subsurface.

Given their long durations and narrow-band frequency content, these signals can be understood as harmonic vibrations of fluid-filled cracks that are elastically coupled to the surrounding rock mass. The method for recording, processing, and imaging these signals is called fracture seismic in order to distinguish it from micro-seismic methods. Micro-seismic methods detect the impulsive dislocations that generate distinguishable P and S waves to locate fractures. Fracture seismic captures and images the signals from the entire fracture and builds a three-dimensional image of the fractures. Fractures that are interpreted from micro-seismic have only a partial correlation with the fracture systems mapped using the fracture seismic method.

Fracture seismic observations can be acquired with both high density, multi-receiver, reflection seismographic equipment and lower density buried grids. The number of receivers necessary per square km of study area is between 30 and 60 for surface-based observations and 1 to 3 for buried grids. The density of the receiver grid for any acquisition is determined by the noise environment for the project and the economics of permitting and physical access constraints. The quality of the fracture seismic map is impacted by the density. The area to be instrumented is determined by both the depth and area to be imaged. From the edge of the area to be imaged, the farthest offset receivers should be 1.2 to 1.5 times the depth of the target area.

The fracture seismic method computes fracture emission intensity volumes using one-way depth migration. These intensity volumes can be computed using modern digital signal processing of fracture seismic data recorded during the acquisition of multi-receiver reflection seismic survey data. The redundancy of such data allows for the removal of other passively recorded signals, including earthquakes and cultural and industrial generated background noise. Well-known seismic reflection processing codes such as cepstral filters, noise analysis and filtering, and depth migration, can be readily adapted to the one-way-travel-time depth migration used in the fracture seismic method for computing fracture seismic intensity.

A substantial base of fracture seismic observational case histories now exists. These examples establish that fracture seismic methods can reveal the locations of the subsurface fluid-flow pathways. Pre-drill fracture seismic mapping can be used to guide well paths, establish optimal treatment programs, and forecast well interferences. Stimulation time fracture seismic can be used to measure treatment effectiveness. Combined with pre-treatment fracture seismic maps, both potential and actual fluid production can be readily and accurately estimated. Time-lapse fracture seismic tracks the evolution of flow paths over time.

These attributes of fracture seismic permeable structure mapping establish its importance in future exploration, development, production, and management of subsurface resources. With the rapid expansion of the number of receivers that can be fielded and the speed of modern computers, fracture seismic acquisition can be integrated with 3D seismic reflection acquisition. Both reflection seismic volumes for detailed interpretation of the geologic structure and fractures seismic intensity volumes can be computed simultaneously and allow the integration of the subsurface connectivity with the geologic formations. As a consequence of these developments and the value of results of our case studies, we believe that studies of the kind we have presented here will soon become standard practices, for both commercial and social purposes.

Author Contributions: Conceptualization, C.S.; formal analysis, C.S. and P.M.; investigation, C.S. and P.M.; methodology, C.S.; project administration, C.S.; software, C.S.; supervision, C.S.; validation, P.M.; writing—original draft, C.S. and P.M.; writing—review and editing, C.S. and P.M.

Funding: This research received no external funding.

Acknowledgments: The authors wish to thank Jan Vermilye for her ideas and contributions to many of the methods presented and for her review of the manuscript. Ashley Yaner, Amanda Klaus, and Lance Bjerke processed the data and provided the data integration in the field examples.

Conflicts of Interest: The authors declare no conflict of interest.

References

1. Krauklis, P.V. About some low frequency oscillations of a liquid layer in elastic medium. *Prikl. Mat. Mek.* **1962**, *26*, 1111–1115.
2. Frehner, M. Krauklis wave initiation in fluid-filled fractures by seismic body waves. *Geophysics* **2013**, *79*, T27–T35. [CrossRef]
3. Tary, J.B.; Van der Baan, M.; Sutherland, B.; Eaton, D.W. Characteristics of fluid induced resonances observed during microseismic monitoring. *J. Geophys. Res.* **2014**, *119*, 8207–8222. [CrossRef]
4. Tary, J.B.; Van der Baan, M.; Eaton, D.W. Interpretation of resonance frequencies recorded during hydraulic fracturing treatments. *J. Geophys. Res. Solid Earth* **2014**, *119*, 1295–1315. [CrossRef]

5. Sicking, C.J.; Vermilye, J.; Malin, P.E. Resonating fluid filled fractures in passive seismic. In *SEG Technical Program Expanded Abstracts 2019*; Society of Exploration Geophysicists: Tulsa, OK, USA, 2019; pp. 3021–3025.

6. Sicking, C.J.; Vermilye, J. Resonance Frequencies in Passive Recordings Map Fracture Systems: Eagle Ford and New Albany Shale Examples. In Proceedings of the Unconventional Resources Technology Conference, Denver, CO, USA, 22–24 July 2019.

7. Maxwell, S.C.; Urbancic, T.I. Passive imaging of seismic deformation associated with injection for enhanced recovery. In Proceedings of the 66th EAGE Conference and Exhibition, Paris, France, 7–10 June 2004; pp. 458–461.

8. Duncan, P.M.; Eisner, L. Reservoir characterization using surface microseismic monitoring. *Geophysics* **2010**, *75*, 75A139–75A146. [CrossRef]

9. Kochnev, V.A.; Goz, I.V.; Polyakov, V.S.; Murtayev, I.S.; Savin, V.G.; Zommer, B.K.; Bryksin, I.V. Imaging hydraulic fracture zones from surface passive microseismic data. In Proceedings of the 2007 EAGE/SEG Research Workshop on Fractured Reservoirs "Integrating Geosciences for Fractured Reservoir Description", Perugia, Italy, 3–6 September 2007.

10. Tary, J.; van der Baan, M. Potential use of resonance frequencies in microseismic interpretation. *Lead. Edge* **2012**, *31*, 1338–1346. [CrossRef]

11. Chorney, D.; Jain, P.; Grob, M.; van der Baan, M. Geomechanical modeling of rock fracturing and associated microseismicity. *Lead. Edge* **2012**, *31*, 1348–1354. [CrossRef]

12. Bame, D.; Fehler, M. Observations of long period earthquakes accompanying hydraulic fracturing. *Geophys. Res. Lett.* **1986**, *13*, 149–152. [CrossRef]

13. Vermilye, J.M.; Scholz, C.H. The process zone: A microstructural view of fault growth. *J. Geophys. Res.* **1998**, *103*, 12223–12237. [CrossRef]

14. Shipton, Z.K.; Cowie, P.A. Damage zone and slip-surface evolution over µm to km scales in high porosity Navajo sandstone, Utah. *J. Struct. Geol.* **2001**, *23*, 1825–1844. [CrossRef]

15. Moore, D.E.; Lockner, D.A. The role of microfracturing in shear fracture propagation in granite. *J. Struct. Geol.* **1995**, *17*, 95–114. [CrossRef]

16. Ziv, A.; Rubin, A.M. Static stress transfer and earthquake triggering: No lower threshold in sight? *J. Geophys. Res.* **2000**, *105*, 13631–13642. [CrossRef]

17. Lawn, B.R.; Wilshaw, T.R. *Fracture of Brittle Solids*; Cambridge University Press: Cambridge, UK, 1975; pp. 1–378.

18. Hubbert, M.K.; Rubey, W.W. Role of fluid pressure in mechanics of overthrusting. *Bull. Geol. Soc. Am.* **1959**, *70*, 115–166. [CrossRef]

19. Geiser, P.A.; Vermilye, J.; Scammell, R.; Roecker, S. Seismic used to directly map reservoir permeability fields. *Oil Gas J.* **2006**, *104*, 37.

20. Sicking, C.J.; Vermilye, J.; Geiser, P.; Lacazette, A.; Thompson, L. Permeability Field Imaging from Microseismic. In *SEG Technical Program Expanded Abstracts*; Society of Exploration Geophysicists: Tulsa, OK, USA, 2012; pp. 1–5.

21. Sicking, C.J.; Vermilye, J. Pre-drill reservoir evaluation using passive seismic imaging. In Proceedings of the Unconventional Resources Technology Conference, San Antonio, TX, USA, 1–3 August 2016.

22. Liang, C.; O'Reilly, O.; Dunham, E.M.; Moos, D. Hydraulic fracture diagnostics from Krauklis-wave resonance and tube-wave reflections. *Geophysics* **2017**, *82*, 171–186. [CrossRef]

23. Sicking, C.J.; Vermilye, J.; Lacazette, A.; Yaner, A.; Klaus, A.; Bjerke, L. Case Study Comparing Micro-Earthquakes, Fracture Volumes, and Seismic Attributes. In Proceedings of the Unconventional Resources Technology Conference, Denver, CO, USA, 25–27 August 2014.

24. Sicking, C.J.; Vermilye, J.; Lacazette, A. Predicting Frac Performance and Active Producing Volumes Using Microseismic Data. In Proceedings of the Unconventional Resources Technology Conference, San Antonio, TX, USA, 20–22 July 2015.

25. Sicking, C.; Vermilye, J.; Yaner, A. Forecasting reservoir performance by mapping seismic fracture seismic. *Interpretation* **2017**, *5*, T437–T445. [CrossRef]

26. Geiser, P.; Lacazette, A.; Vermilye, J. Beyond "dots in a box": An Empirical View of Reservoir Permeability with Tomographic Fracture Imaging. *First Break* **2012**, *30*, 63–69.

27. Lacazette, A.; Vermilye, J.; Fereja, S.; Sicking, C. Ambient fracture imaging: A new passive seismic method. In Proceedings of the Unconventional Resources Technology Conference, Denver, CO, USA, 12–14 August 2013.

28. Aki, K.; Richards, P. *Quantitative Seismology: Theory and Methods*; WH Freeman & Co.: New York, NY, USA, 1980; pp. 1–13.

29. Dibble, R. Seismic and related phenomena at active volcanoes in New Zealand, Hawaii, and Italy. Ph.D. Thesis, Victoria University, Wellington, New Zealand, May 1972.

30. Aki, K.; Fehler, M.; Das, S. Source mechanism of volcanic tremor: Fluid-driven crack models and their application to the 1963 kilauea eruption. *J. Volcanol. Geotherm. Res.* **1977**, *2*, 259–287. [CrossRef]

31. Gomberg, J.S. Stress/strain changes and triggered seismicity following the Mw7.3 Landers, California, earthquake. *J. Geophys. Res.* **1996**, *101*, 751–764. [CrossRef]

32. Du, W.-X.; Sykes, L.; Shaw, B.; Scholz, C. Triggered aseismic fault slip from nearby earthquakes, static or dynamic effect? *J. Geophys. Res.* **2003**, *108*, 2131. [CrossRef]

33. Thomas, A.M.; Nadeau, R.M.; Bürgmann, R. Tremor-tide correlations and near-lithostatic pore pressure on the deep San Andreas fault. *Nature* **2009**, *462*, 1048–1051. [CrossRef] [PubMed]

34. Zhang, H.; Nadeau, R.M.; Toksoz, M.N. Locating nonvolcanic tremors beneath the San Andreas Fault using station-pair double-difference location method. *Geophys. Res. Lett.* **2010**, *37*, L13304. [CrossRef]

35. Copeland, D.; Lacazette, A. Fracture surface extraction and stress field estimation from three dimensional microseismic data. In Proceedings of the Unconventional Resources Technology Conference, San Antonio, TX, USA, 20–22 July 2015.

Article

Compaction and Fluid—Rock Interaction in Chalk Insight from Modelling and Data at Pore-, Core-, and Field-Scale

Mona Wetrhus Minde [1] and Aksel Hiorth [2,*]

[1] The National IOR Centre of Norway, Department of mechanical and structural engineering and materials science, University of Stavanger, 4036 Stavanger, Norway; mona.w.minde@uis.no

[2] The National IOR Centre of Norway, Department of energy resources, University of Stavanger, 4036 Stavanger, Norway

*   Correspondence: aksel.hiorth@uis.no

Received: 14 November 2019; Accepted: 16 December 2019; Published: 21 December 2019

Abstract: Water weakening is a phenomenon that is observed in high porosity chalk formations. The rock interacts with ions in injected water and additional deformation occurs. This important effect needs to be taken into account when modelling the water flooding of these reservoirs. The models used on field scale are simple and only model the effect as a change in water saturation. In this paper, we argue that the water weakening effect can to a large extend be understood as a combination of changes in water activity, surface charge and chemical dissolution. We apply the de Waal model to analyse compaction experiments, and to extract the additional deformation induced by the chemical interaction between the injected water and the rock. The chemical changes are studied on a field scale using potential flow models. On a field scale, we show that the dissolution/precipitation mechanisms studied in the lab will propagate at a much lower speed and mainly affect compaction near the well region and close to the temperature front. Changes in surface charge travel much faster in the reservoir and might be an important contributor to the observed water weakening effect. We also discuss how mineralogical variations impacts compaction.

Keywords: chalk; compaction; water weakening; rock—fluid interaction; modelling

---

1. Introduction

Many of the chalk formations in the North Sea contains large volumes of oil. In some areas, the porosity is higher than 45% [1,2]. During production, the reservoir pressure is decreased, the effective stress is increased and the reservoir subsequently compacts. Compaction is a significant driver for oil expulsion, and, as an example, for the Valhall field in the North Sea, the compaction is estimated to contribute to 50% of the total recovery [2]. For the Ekofisk field, reservoir compaction has led to seabed subsidence of approximately 9 m, where about 50% of the 9 m is due to pore pressure decline, and about 50% is due to the injected water [3]. The difference between physical and physicochemical effects is important to include when simulating the reservoir oil production and compaction.

In field-scale simulations of water-induced compaction in chalk, one uses the water saturation as an indicator of enhanced compaction [4]. In this paper, we argue that the water weakening effect is more related to changes in the water chemistry than changes in the water saturation, meaning that it is the changes in the concentration of specific ions that induce enhanced compaction, not only the ratio between oil and water. It is important to include a mechanistic understanding of fundamental processes of deformation, to model fluid flow and compaction in basins or reservoirs over observable timescales. Incorporating these insights on water weakening will produce reliable and more applicable basin models. In the next sections, we will review some of the relevant core experiments, and interpret them by simple models. Thereafter we discuss how these changes may propagate on the field scale, by the use of streamline models. The practical consequence of this is that one can, to some extent, control the fluid chemistry inside the pores, control compaction and thereby impact the oil recovery. Finally, we present discussions, summary and recommendations for the future.

2. Interpreting Triaxial Core Deformation Experiments in Terms of the de Waal Model

Hydrostatic tests are usually performed in triaxial cells, where the axial (length of the core) and, to some degree, the radial strain can be monitored during the experiment. Fluids are continuously passed through the cores, of approximately 7 cm in length, allowing for measurement of permeability and effluent fluid sampling. Back-pressure is applied to maintain a uniform pore pressure as the core compacts, and to avoid fluids boiling at elevated temperatures. A pressure difference of a few kPa is enough to drive fluids through the sample. Thus, the stress–strain development is mainly dictated by the microstructural properties of the solid (matrix) framework.

Usually, the triaxial test data are reported for two stages of testing: i) The hydrostatic loading stage, where the confining pressure and pore pressure are increased linearly, and the axial strain is measured as the axial stress is increased until the core has passed its yield point. ii) The creep stage, during which the core subsequently deforms slowly under a small constant load [5]. The yield point can be considered a measure of the "strength" or "compaction strength" of the core, while the time-dependent creep that occurs after the yield point has been exceeded might be considered as a kind of "viscosity".

Figure 1 shows an example where a chalk core from the Aalborg outcrop was heated to 130 °C, loaded up to 12 MPa, and then left to creep for 61 days at a constant stress of 12 MPa, while a solution of 0.657M NaCl was passed through at a flooding rate of 1PV (Pore Volume) per day [6]. As can be seen from Figure 1a, the axial strain is proportional to the axial stress (e.g., the core deformation is elastic) up to the yield point at ~8.2 MPa (the solid black line). The yield point is not a single point, but rather a zone of transition between 7.5 and 8.5 MPa that separates the elastic and plastic (black dotted line) phases of deformation. The elastic bulk modulus is defined as the ratio of the change in axial stress to the change in volumetric strain up to the yield point [7] (the slope of the solid line in Figure 1a divided by 3). The plastic bulk modulus is defined as the ratio of the change in axial stress to the change in volumetric strain in the phase after the core has reached the yield point (the slope of the dashed line in Figure 1a divided by 3). Assuming that the material compresses equally in all directions, the axial strain can be converted to volumetric strain by multiplying by factor of 3, and the effective viscosity can be defined as the ratio between the axial stress and axial strain rate ($\eta_{eff} = \frac{\sigma}{d\varepsilon_V/dt}$) [7]. The effective viscosity increases strongly with time in a creep experiment and as the stress is kept constant, the strain rate is reduced.

Numerous rock mechanical laboratory experiments have been carried out with a large degree of reproducibility between experiments. These experiments can be understood with reference to the grain contact model suggested by de Waal [8]. Here we describe the de Waal model and show how it provides a good basis for understanding experiments involving rock–fluid interactions and the various chemical factors that affect reservoir compaction. During the creep stage, compaction usually follows a log t trend as shown in Figure 1b. The following creep model captures this trend [8]:

$$\varepsilon_V(t) = B \log\left(1 + \frac{\dot{\varepsilon}_V(0)t}{B}\right) \tag{1}$$

B is a constant related to the friction between the grains, and $\dot{\varepsilon}_V(0)$ is the initial strain rate, which can be seen by differentiating the above equation with respect to t:

$$\frac{d\varepsilon_V(t)}{dt} = \frac{\dot{\varepsilon}_V(0)}{1 + \frac{\dot{\varepsilon}_V(0)t}{B}} \overset{t \to 0}{\rightarrow} \dot{\varepsilon}_V(0) \tag{2}$$

The model suggested by de Waal was derived from the basis of earlier work by [9] where it was shown that the friction coefficient between two blocks of rock, μ, depends on the time, t. The blocks were in stationary contact before sliding was initiated, and the friction coefficient was found to follow the empirical law:

$$\mu = \mu_0 + C \log(1 + Dt) , \tag{3}$$

where C and D are constants, and μ_0 the friction coefficient at t = 0. In [9,10] experiments were conducted over a time frame of one day, where the materials were kept in contact at a constant normal and shear stress for time t. When the static contact time was reached, the shear stress was rapidly raised to the level required to produce a slip. The reason for the dependence of friction on the stationary contact time is that the contact area between the surfaces increases as the static contact time increases [9].

De Waal argued that Equation (3) can be used to understand creep [8] if t in the equation is replaced with a time constant t_a that describes the average lifetime of grain–grain contact, and this contact time is related to the inverse strain rate. The physical explanation for the macroscopic time-dependent creep of rocks is thus, according to de Waal, that the average lifetime of contact points affects the real contact area which determines the sliding friction.

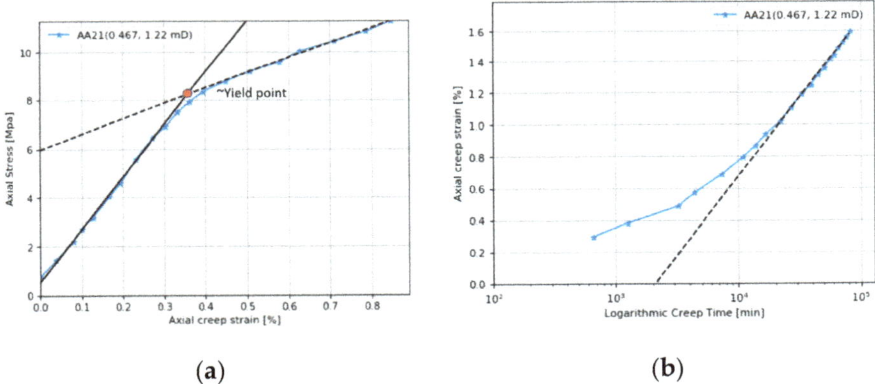

(a)   (b)

Figure 1. (**a**) Hydrostatic loading stage. The elastic bulk modulus is 720 MPa (slope of solid line divided by 3). The plastic bulk modulus is 210 MPa (slope of the dashed line divided by 3). (**b**) The creep stage deformation. The core is left to deform at a constant stress 12 MPa. The dashed line is the de Waal model (text equation 1) with parameters $\dot{\varepsilon}_V(0) = 0.012\%$/hour and $B = 0.43\%$. Data are from [6].

Westwood, Goldheim et al. [11] and Macmillan, Huntington et al. [12] extended Dieterich's [10] perspective by pointing out that the friction coefficient is also dependent on the fluid chemistry. Based on the experiments conducted, they argued that ion interactions modify the surface microhardness, and that it is greatest when the zeta potential is close to zero for, in this case, MgO. This implies that the friction is at its lowest level when the zeta potential is close to zero because grain contact is minimized by the hardness. A low value of the zeta potential should thus increase the strain rate. Westbrook and Jorgensen [13] also pointed out that water adsorption from the air affected the time dependence of microhardness observed in nonmetallic materials. They found up to a 10% decrease in the microhardness of wet calcite compared to dry, depending on the crystal plane.

The de Waal model has also been used to capture the dependence of creep strain rate on loading rate. The loading rate is known to affect the strain rate in the laboratory [14], and the laboratory strain rate is usually much larger than the strain rate in a reservoir where the loading results from the slow drawdown of fluid pressure by petroleum and water extraction. The challenge is to translate the higher strain rates observed in the laboratory to the lower field strain rates. Andersen, Foged et al. [14] extended the de Waal model from constant stress rate to constant strain rate and showed how to apply it to North Sea chalk. They found that the de Waal friction factor for the Valhall field was more or less constant over the stress range of interest, which reduced the number of tests required to estimate compaction at field depletion rates.

3. Water Activity Weakens Chalk in Loading Phase

If the friction coefficient, thus the strain rate, is dependent on the fluid chemistry, it follows that the mechanical properties of chalk also depend on the pore fluid [15–17]. The physical mechanism for water weakening has been attributed to capillary forces and differences in "wettability" between fluids [16,17], but for later publications [18], at least in part, water weakening was explained by the relations between physical and chemical properties of the chalk.

Risnes [19] found that the hydrostatic yield point of chalk (the full elastic deformation) increased by 50%, 70%, and 80% if, instead of water, the pore fluid was switched to methanol, oil and glycol, or air, respectively. The hydrostatic yield point was estimated to be 18 MPa for glycol and 10 MPa for water. These results have been confirmed by [20,21], where the chalk was saturated with supercritical CO_2, and by experiments where the activity of water was changed systematically by changing the fraction of glycol mixed into water [20]. Chalk strength increases as the activity of water is decreased.

A completely different approach by Røyne et al. [22] gives further insight into the water weakening effects. They used a double torsion method, wherein a calcite crystal with an initial crack is immersed in a fluid. The crack is extended by bending the calcite, and the crack velocity is measured. The crack velocity is considered to be a measure of surface energy, where a lower surface energy produces a greater crack velocity. By immersing the sample in a mixture of glycol and water, the surface energy of calcite was determined for different water activities. Assuming that the hydrostatic yield point is related to pore collapse and bond breaking between the individual grains in the sample, a lower surface energy should mean a weaker chalk and a lower yield point. In the Røyne et al. [22] experiment, the surface energy was 0.32 J/m^2 for glycol and 0.15 J/m^2 for water. This factor of ~2 reduction is very close to the factor of 1.8 reduction determined by Risnes [20]. The activity of water can also be changed by changing the concentration of salts. Risnes et al. [23] were able to reduce the activity of water to 0.48 by the adding of $CaCl_2$. Figure 2 plots against the activity of water, the surface energy of calcite immersed in fluids with different water activity normalized to the surface energy of calcite immersed in pure water, or the yield point of cores saturated with fluids with different water activities normalized to their yield point when saturated with pure water (activity = 1). It shows that the decrease in chalk strength with water activity is similar, regardless of how the activity of water is modified. The figure shows that not only does the yield point change by similar amounts with changes in water activity for all the experiments, but also the trend of the change with water activity is similar. Water weakens chalk, and the activity of water is a measure of the amount of weakening.

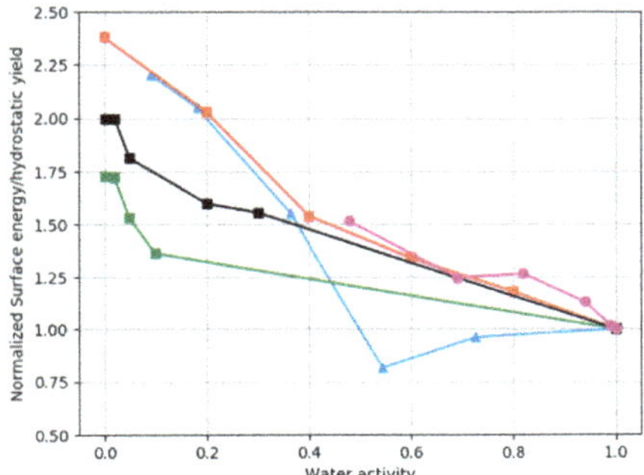

Figure 2. The normalized surface energy of calcite immersed in fluids with different water activity, or the normalized hydrostatic yield point of cores saturated with fluids with different water activity, are plotted against water activity. Normalization is achieved by dividing by the yield point or surface energy measured when the pore or emersion fluid is pure water. Blue triangular points show changes in calcite surface energy from [22]. Green, black and red square points show core tests where water activity was changed by mixing water and glycol [20]. Purple points show the changes in normalized yield point when the activity of water is modified by additions of $CaCl_2$ [23]. Experimental uncertainties are not included in the figure.

4. Water Surface Chemistry Affects Chalk Deformation and Creep

However, chalk strength depends on water *chemistry* as well as water *activity*. When comparing chalk deformed at 130 °C with seawater as the pore fluid, to chalk deformed at 130 °C with seawater without sulphate as the pore fluid, the yield point is 74% lower. The total dissolved solids concentration was kept constant by adding NaCl [24,25]. The activity of water, a_w, for a brine is [26]

$$a_w = 1 - 0.018 \sum_i m_i, \tag{4}$$

where m_i is the molality of other ions in the solution. For seawater the water activity is ~0.98; thus, the activity of water is not a factor in this comparison. Sulphate ions weaken chalk and Figure 3 shows how sulphate lowers the yield point. Notice that although the yield point is dramatically lowered (the elastic bulk modulus is strongly decreased) by the presence of sulphate, the plastic bulk modulus (the slope of the dashed lines) is not changed.

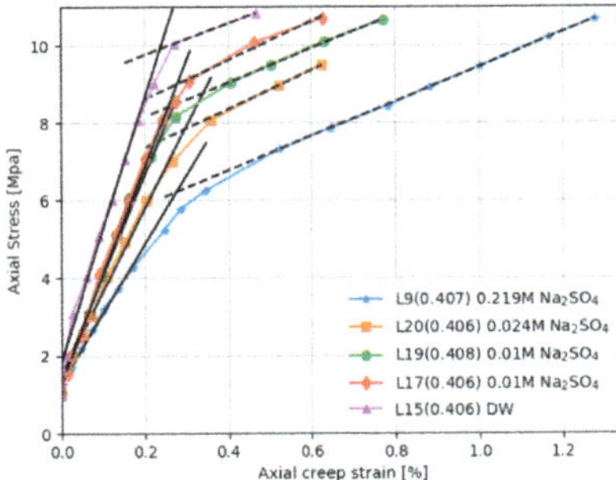

Figure 3. Loading stage deformation for Liège cores saturated with different fluids and tested at 130 °C. $NaSO_4$ concentration is varied while the amount of NaCl is adjusted to keep the ionic strength constant at 0.657 M. Pore fluids for the distilled water (DW) curve contain neither $NaSO_4$ nor NaCl. The yield point is depressed and the elastic bulk modulus decreased by sulphate (solid lines), but the creep deformation and plastic bulk modulus are not affected by sulphate.

Megawati et al. [27] hypothesized that the electrostatic interactions that give rise to an osmotic pressure (disjoining pressure) near the grain contacts could explain the enhanced weakening. Sulphate adsorbed onto the calcite grains changes the surface and zeta potentials. The surface becomes more negatively charged. The variations in hydrostatic yield for all the chalk samples was proportional to the maximum peak of the disjoining pressure. Nermoen et al. [28] performed additional experiments and suggest the electrochemical interactions that give rise to the disjoining pressure can be incorporated into an effective stress equation.

Sulphate in the brine also affects the creep rate [27]. Cores from different outcrops, which have different mineralogical compositions, have a very similar creep rate and creep rate development when the brine contains sulphate. This is in contrast to cores from the same outcrops flooded with NaCl (inert) brine, which shows greater variance in the strain rate development. This might be an indication that the grain-to-grain friction between the calcite grains is to some extend controlled by the diffusive layer counter ions, and not only the local mineralogical properties of the grains. If sulphate is introduced after the loading phase, i.e., in the creep phase, there is a sudden increase in the creep rate [29]. This effect is transient, and when the pore surfaces of the calcite grains are saturated with sulphate, the creep rate is reduced.

5. Core Deformation Related to Mineral Alteration

Additional core deformation can be the result during the creep stage when the fluid passed through the deforming cores is out of equilibrium with the minerals in the core. Figures 4 and 5 compare the axial creep strain for two sets of experiments. In one set, a wide range of chalk cores was flooded with inert (e.g., nonreactive) NaCl brine. In the second set, cores from the same outcrops are flooded with reactive $MgCl_2$. In both sets of experiments, the temperature is 130 °C. The axial creep for all the cores flooded with $MgCl_2$ is greater than those flooded with inert NaCl, and is commonly more than three times greater. The chalk cores with few impurities (>99.7% calcite, the Stevns Klint and Mons (Trivières Fm)) flooded with $MgCl_2$ show a temporary plateau in creep strain, see Figure 5. The chalk cores with lower carbonate content (the Aalborg, Liège, and Kansas) show no such plateau.

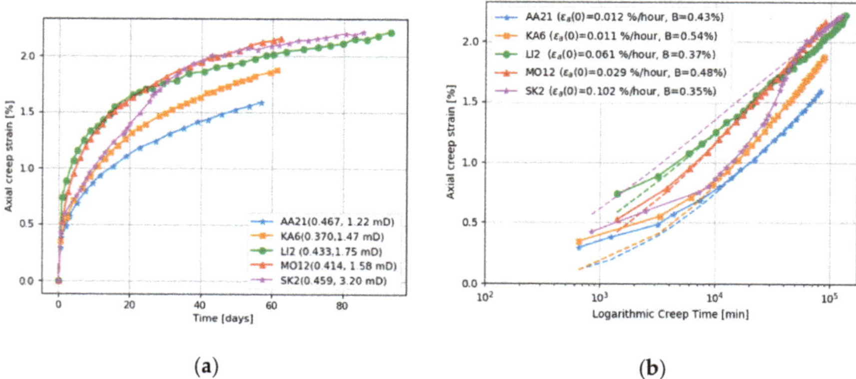

Figure 4. (**a**) Axial creep for cores flooded at 130 °C with nonreactive NaCl, and (**b**) a corresponding fit to de Waal model. AA=Aalborg, KA=Kansas, LI=Liège, MO=Mons, SK=Stevns Klint. The numbers in the parenthesis gives the porosity and permeability. All data from [6].

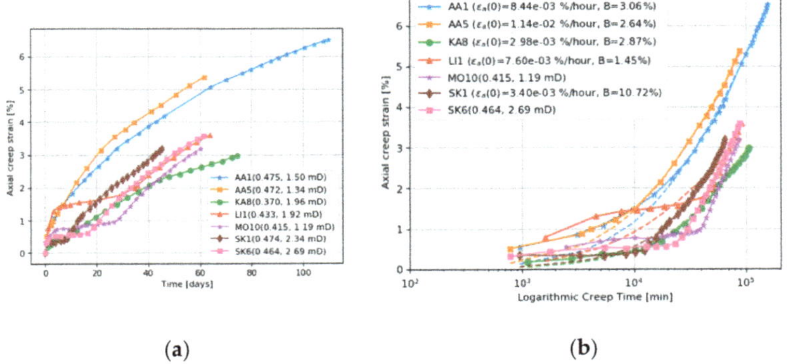

(a)   (b)

Figure 5. (**a**) Axial creep for cores flooded at 130 °C with MgCl$_2$, the numbers in the parenthesis gives the porosity and permeability, and (**b**) corresponding fit to de Waal model and model parameters in the parenthesis. AA=Aalborg, KA=Kansas, LI=Liège, MO=Mons, SK=Stevns Klint. All data from [6].

The creep strains experienced by all the cores under NaCl flooding are well described by the de Waal model (Equation (1)), as shown in Figure 4b. The de Waal model does not match well with the cores flooded with MgCl$_2$ brine, as seen in Figure 5b. We can now use the de Waal model to estimate the additional strain induced by the chemical reactions. If we simply subtract the "expected" creep from the cores flooded by MgCl$_2$, by using the results in Figure 4, Figure 6 shows the excess creep strain that results when the cores are flooded with MgCl$_2$. The excess creep strain takes a while to develop for some cores (the plateau mentioned above), but once established the excess strain rate is linear and ~0.04%/day in magnitude.

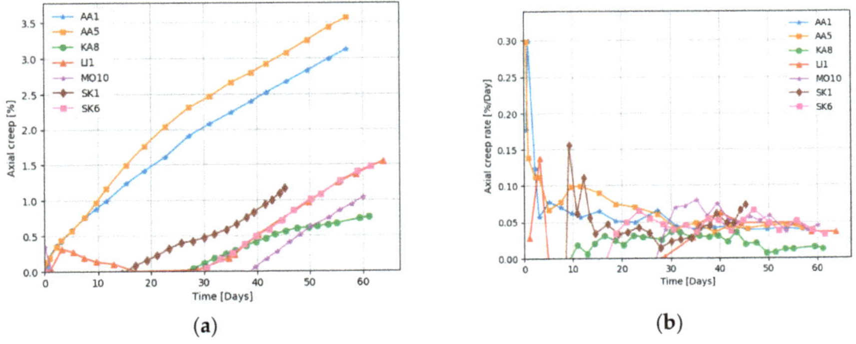

(a)   (b)

Figure 6. (**a**) Excess axial creep measured in cores flooded with reactive MgCl$_2$ obtained by subtracting the de Waal deformation for the NaCl flooding, and (**b**) the creep rate.

Post-experimental investigations by scanning electron microscopy (SEM) show mineralogical alterations already after weeks of flooding. Dissolution of calcite takes place, and is commonly observed together with precipitation of magnesite. For the chemical alteration to be fully appreciated, reactive fluids must be passed through the chalk cores at elevated temperatures for very long periods of time. Figure 7 shows the results when a Liège chalk core was flooded for ~3 years (1072 days). There is a significant flux of Ca^{2+} out of the core but no flux in, and there is an almost exactly counterbalancing flux of Mg^{2+} into the core with no Mg^{2+} flux out. The sum of the Mg^{2+} in and Ca^{2+} out always almost exactly matches the injected MgCl$_2$ concentration. This indicates that Ca^{2+} in the carbonate core is being replaced mole for mole with Mg^{2+}. Mass balance indicates that 93 to 98% of the initial calcite is replaced by magnesite over the course of this experiment [28]. The initial porosity of the core was 41%, and the final porosity 40%. The pore volume is thus conserved and the 10% axial (30% volumetric) strain resulted from the mineralogical alteration of the matrix. Mineralogical alteration can cause substantial compaction. Note that the de Waal model fits the MgCl$_2$ core flooding well in this case (see Figure 7); however, in this case, the flooding rate is changed during the experiment. At ~380 days, the observed creep is deviating from the de Waal model, but at this point, the flooding rate is decreased (less chemical reactions) and the axial creep rate is lowered. After approximately 700 days, the observed creep is below the de Waal model, but at ~800 days, the flooding rate increased and axial deformation is increased. Thus, in this case, the match with the de Waal model is a coincidence, due to changes in the flooding rate.

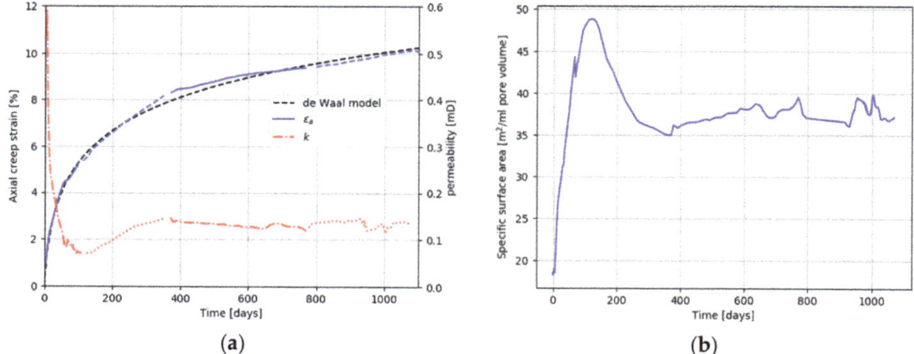

Figure 7. (a) Axial creep, permeability, and influent and effluent chemistry for a carbonate core flooded at 130 °C for ~3 years with reactive MgCl$_2$. Line style indicates changes in flooding rate from 1 PV/day to 3 PV/day. Solid black line in (b) is the specific surface area development estimated from the experimental permeability and porosity using the Kozeny–Carman equation.

As the porosity did not change in the experiment shown in Figure 7, the dramatic and immediate drop in permeability is unexpected. The drop in permeability is most likely caused by magnesite crystallization in the large, permeable pores (see Figure 8a). In other chalk types, crystallization of magnesium-bearing clay minerals may also reduce permeability (Figure 8b). Minerals crystallized in pores increases the surface area per unit pore volume. Since permeability is directly related to porosity and inversely related to the square of surface area per unit of pore volume and pore tortuosity (c.f., the Carmen-Kozeny equation), an increase in surface area per unit pore volume strongly decreased permeability. The slight recovery in permeability between 200 and 300 days (Figure 6a) could suggest either that precipitated magnesite had subsequently dissolved or that pore throats subsequently opened such that the tortuosity was reduced. Textural changes related to alteration can clearly and dramatically change permeability.

At the end of the experiment flooding with MgCl$_2$, (as shown in Figure 7) had completely replaced the original calcite with magnesite. After ~800 days, injected Mg^{2+} was no longer reacting, but passing unchanged in concentration through the core. A similar experiment that was terminated at 516 days flooded a Liège chalk core under Ekofisk field conditions with 0.219 M MgCl$_2$. Mapping by energy dispersive spectroscopy (EDS) by the use of mineral liberation analyzer (MLA) of the core from inlet to outlet after flooding, shows two regimes of mineralogical change. Toward the outlet, the core is only partly altered. The mineralogy is mainly calcite, the primary mineralogy of chalk, but precipitates of clay and magnesite are found within the calcite matrix. At the inlet, a total transformation from calcite to magnesite is observed. Only magnesite and clay are observed, and no calcite is left. For cores flooded for an even longer period, the transition between the two regimes has moved further into the core, indicating that this transformation of mineralogy moves like a front through the cores [30,31].

(a)   (b)

Figure 8. (a) Newly precipitated magnesite crystals in a large pore (edited from [30]). (b) Magnesium-bearing clay minerals crystallized in Aalborg chalk, which contains large amounts of primary opal-CT, after being flooded with $MgCl_2$.

6. Fluid–Rock Interactions in Reservoirs

The triaxial laboratory tests discussed in Section 5 cannot describe what takes place in a reservoir because they are carried out under isothermal conditions, whereas cold fluids are injected into a reservoir and reactions take place as fluids move through a thermal migration front. Figure 9 illustrates what is observed to happen in a reservoir. It plots the chemistry of fluids produced from a well 300–400 m from the nearest well into which seawater is being injected [32]. Chlorine does not react with the reservoir sediments. Seawater has a much lower chlorinity than Ekofisk formation water, and thus the production of low salinity water indicates the arrival of injected water. Following the changes in chlorinity in Figure 9, we see that during a period of 5–6 years, the chlorine concentration drops to 0.63 mol/L, which is close to the chlorine concentration in seawater of 0.525 mol/L. Thus, the water in the producer after 5–6 years is almost pure seawater.

Using the chlorine concentration as a tracer for seawater, we can predict what the concentration of all the other ions would be if no chemical reactions took place in the reservoir. The solid lines in Figure 9 represents the unreacted concentrations of calcium, magnesium, etc. The difference between the produced water chemistry (points) and these solid lines indicates the chemical exchange between pore waters and the reservoir sediments that has occurred as the injected seawater moved from the injector wells to the production well. It can be seen that a large fraction of Mg and SO_4, and a bit of K, have been lost from the fluid, presumably going into dolomite or magnesite, anhydrite, and clay, respectively.

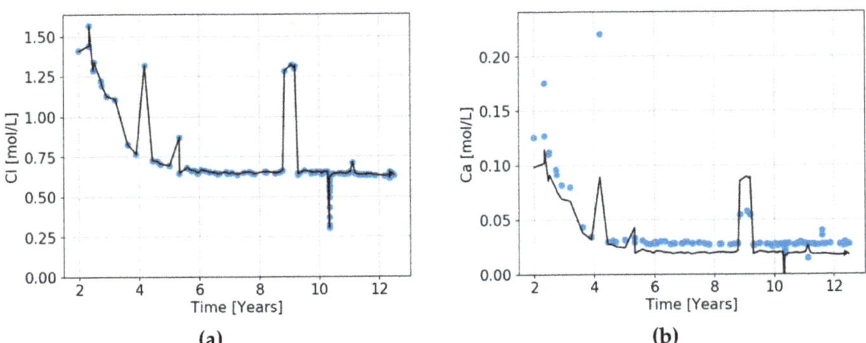

(a)   (b)

Figure 9. *Cont.*

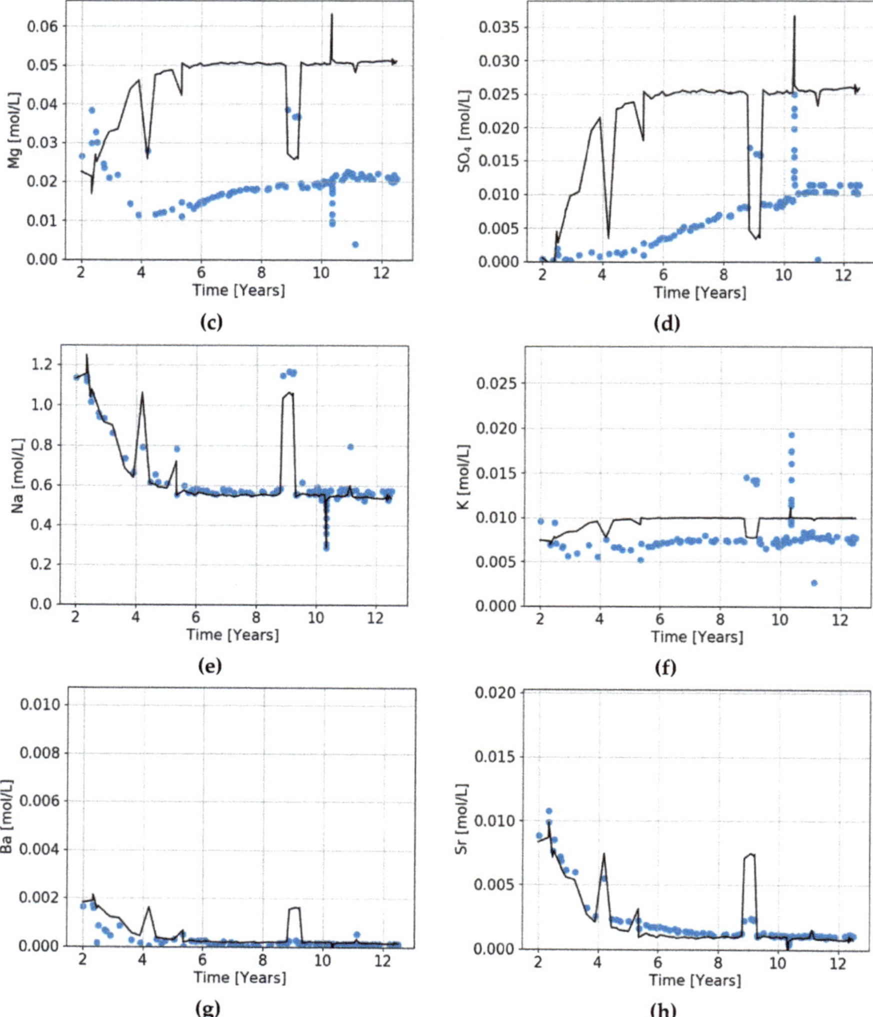

Figure 9. (a–h) Chemistry of waters produced from an Ekofisk well 400–500 m from the nearest well where seawater is being injected. The chemistry of produced water (points) is compared to the predicted initial ion concentrations (solid lines), assuming that chlorine is a conservative (ideal) tracer for unreacted seawater and that the drop in chloride concentration is solely due to mixing between Ekofisk formation water and seawater. The difference between the lines and points indicates the chemical exchange that has occurred in the reservoir. Note that the Y-axes in the various graphs have different scales.

Figures 10 and 11 show models of alteration in a reservoir for a situation similar to that shown in Figure 9. These figures show streamlines between an injection and a production well 500 m away, assuming the reservoir is homogeneous. The streamlines are calculated using potential methods [32–34], injection rate is 30 Mbbl/day and production is constant at 5Mbbl/day; the flow is two phases and through a fixed depth interval of 70 m. The modelling methods are described in more detail in [32]. The snapshots after five years of seawater injection in Figures 9 and 10 show that the temperature gradient lags behind the injected waterfront. The lag results from the fact that water only moves in the pore space, whereas the temperature moves in the total volume (matrix and pore space). Due to the temperature dependence of the mineral reaction kinetics, the chemical alteration fronts travel at speeds different form both the water and the temperature fronts. Figure 9 shows the weight fractions of some of the minerals precipitated and dissolved. Much of the magnesium carbonate alteration occurs close to the injector. After 40 years of simulated seawater injection, about 1.5 wt% of calcite is dissolved and 1.5 wt% of dolomite is precipitated near the injection well, and the calcite to magnesium carbonate transition may extends about 400 m from the injection well [32]. The sulphate bearing mineral anhydrite precipitates at high temperatures and dissolves at low temperatures. The simulations predict a sulphate wave that travels at the same speed as the temperature front, where anhydrite is precipitated and dissolved again as the thermal front passes through the reservoir.

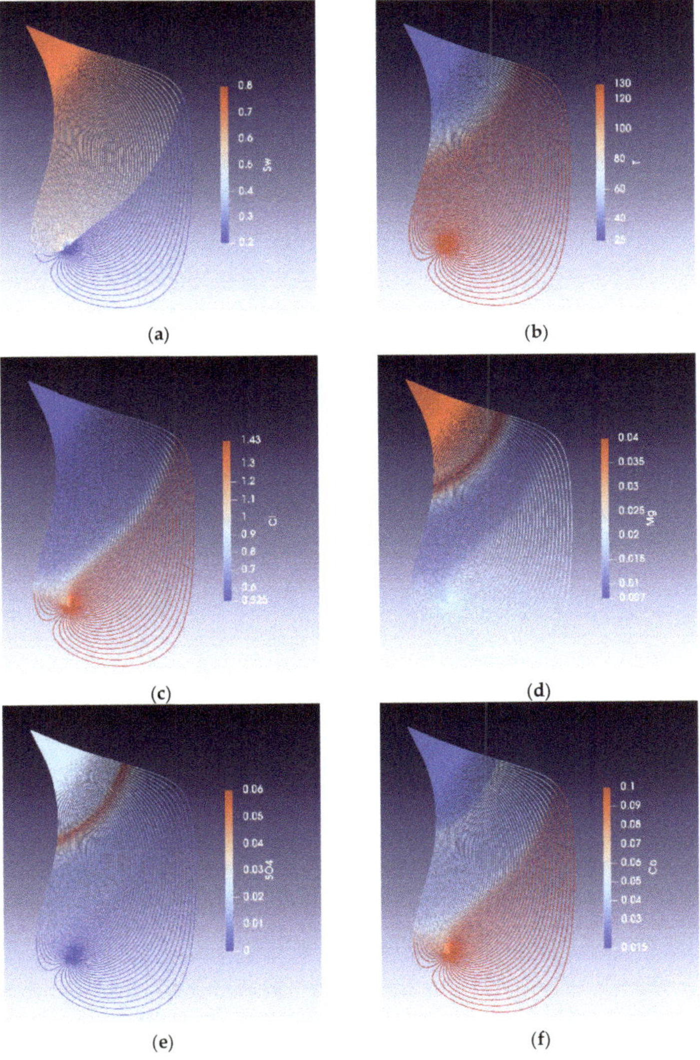

Figure 10. Streamlines approximately 5 years of seawater injection. (**a**) Water saturation (Sw), (**b**) temperature profile, (**c**) chlorine concentration, (**d**) magnesium concentration, (**e**) sulphate concentration, and (**f**) calcium concentration.

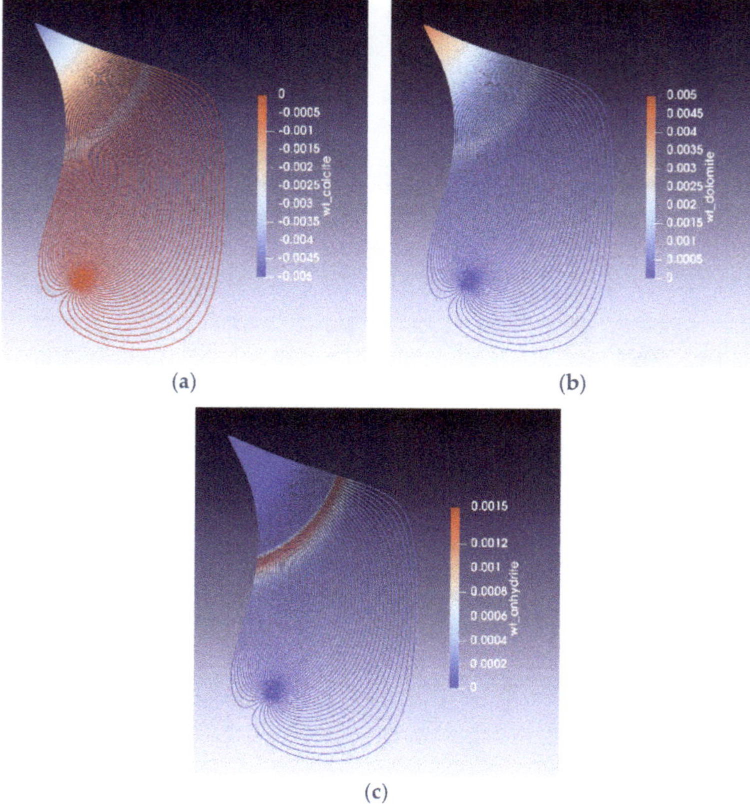

Figure 11. (a–c) Streamlines showing the weight fraction of alteration minerals in the reservoir at the time of water injection breakthrough (approximately 5 years of seawater injection). The wells are separated by 500 m.

7. Discussion and Summary

Chalk fields in the North Sea are over-pressured and during production and pressure depletion, they compact significantly. The compaction can contribute significantly to hydrocarbon recovery, and for the Valhall field, it is estimated to be responsible for about 50% [2] of the total oil recovery. For the Ekofisk field, the water weakening effect is expected to contribute to about 50% of the total compaction [3]. In the case of the Valhall field, it is believed to be much less than the Ekofisk field [35]. From the results discussed here, this can potentially be explained as a consequence of temperature differences, but another important factor is the pore pressure history. The Ekofisk, at discovery, had an initial reservoir temperature of 130 °C, whereas Valhall had a temperature of 93 °C. Both mineralogical alterations and the effect of surface charge, due to sulphate adsorption, are less significant at lower temperatures.

The field conditions are different, at the field scale the effective stress is changing due to pressure drawdown. As pointed discussed by Andersen, Foged et al. [14], it is possible to extrapolate lab results to field results by performing strain rate controlled tests. The experimental data presented and discussed in this paper have been performed at a constant hydrostatic stress, and not uniaxial constant strain rate. However, by comparing hydrostatic tests with and without chemical effects, we believe that many of our conclusions will still hold.

The effect of water activity is the most uncertain effect. The experiments discussed in Section 3, demonstrates quite nicely the transition between a phase consisting of pure fresh water, a mixture of water and glycol to pure glycol and dry samples. The experiments are also explained as a change in surface energy on the microscopic level. However, oil-bearing formations usually consist of a hydrocarbon phase that is immiscible with water. Water is believed to be trapped in small (water-wet) pores and oil resides in larger pores; over geologic time, oil adheres to the rock and change the wettability. During water flooding, the original formation water (ionic strength ~2–3 times seawater) and hydrocarbons are displaced by seawater. The change of water activity from formation water to seawater is not large. A change from a 3M brine to seawater is a change in water activity from ~0.95 to ~0.98 (see Equation (4)). However, there is likely more to this story because at low water saturations (~1–2%), the water activity is not easy to calculate because it is affected by surface charge, diffusive layers and ion interactions with the chalk surface. If most of the water in this case is bound to the chalk surface, the water activity should be much lower than 0.95 (independent on the salinity). Thus, in areas of very low water saturation, an increase in water saturation could give a sudden increase in the water activity, and weaken the chalk and induce compaction. There are some experimental data, where chalk cores have been aged with oil and water saturations about ~5% [36], in those cases when they are flooded with seawater, the creep rate does not change significantly compared to cores that are 100% water-filled. Thus, if these results are correct, the initial water saturation must be lower than 5% to see an effect of the water activity. From capillary pressure arguments, one would expect to see more compaction in areas high above the free water level, where the water saturation is low. The field data are scarce, but an observation well at the Ekofisk field shows more compaction close to the tight zone, where the water saturation is higher due to smaller pores [3]. This does not support the water activity hypothesis.

We have quantified the chemical effects in the hydrostatic creep experiments by using the de Waal model. Comparing the fitting of the de Waal model to core data, where the chemical effects are minimal and where they are present, we have shown that the effect of magnesium brines can be modelled by a linear term ~Ct. The constant C is about 0.05%/day, as shown in Figure 6.

The simulated calcite to magnesite (or dolomite) alteration in the core experiments is about 0.07 wt% per day. Thus, we find that the alteration corresponds directly to additional creep strain. The streamline simulations in Figure 11 show that close to the well, there is an alteration of 0.5 wt% change after 5 years of flooding; this should then correspond to about 0.5% additional creep (0.5 m of a 100-m reservoir interval). The chemical water weakening effects due to calcium to magnesium carbonate alteration are much smaller, but not insignificant, on the field scale compared to the lab. The main reason for the slow chemical alteration on the field scale is the cooling of the reservoir.

Sulphate, according to [27], adsorbs onto the calcite surface and causes additional elastic deformation of factor of two, compared to NaCl, see Figure 3 (seawater corresponds to 0.024 M Na2SO$_4$). If sulphate is introduced into the core after yield, in the creep phase, the core experience a sudden increase in strain rate [29]. This effect is transient, but the increase in volumetric strain due to sulphate is ~1.5%. The simulations and the field data show that sulphate breaks through to the producer in a reasonably short period of time. The sulphate front is lagging behind the water (chlorine) front, due to both precipitation of calcium sulphate and adsorption onto positively charged calcium cites, and one should therefore expect to see some delayed water weakening effect, i.e., the reservoir compaction should happen after the water front has passed a reservoir section. In the reservoir models, the grid is coarse and thus the simulation models might not be able to distinguish the sulphate front from the water saturation.

8. The Future and Final Remarks

In the future, basin models will simulate the full chemical evolution of the reservoir with all its attendant consequences. The most difficult thing to model is how the rock - fluid interactions affects compaction at the various scales. We need core experiments to test rock deformation at realistic conditions, but the drawback is that it is difficult to tease out the exact mechanisms. The mechanisms are needed in order to get the speed of chemical interactions at a larger scale. Much has been investigated regarding the chemical effects, but perhaps the most uncertain and less studied effect is how the oil saturation affects compaction. There are some results [37] showing that oil-filled cores compact when flooded with water. Later results shows less effects when flooding oil-saturated cores with water [28,36,38]. The water activity is worth pursuing, as it will travel at the speed of the waterfront, but most likely it should only be effective at very low initial water saturation (high capillary pressure).

There is still uncertainty related to the effect of initial mineralogy. Cores with low carbonate have a different strain history, than cores with non-carbonate content (~5 wt%). Impurities in chalk seem to eliminate the delay in alteration-induced creep.

Effects of water chemistry on compaction on the lab scale is important. We strongly believe that reservoir models as used by the industry today are too crude to distinguish the different field scale compaction mechanisms. By extracting streamlines from reservoir simulators and conducting chemical alteration and adsorption calculations similar to those shown here, we can see the possibility of predicting the distribution of the alteration fronts, as well as how far the sulphate adsorption front has travelled. This might provide an idea of where to expect the most compaction, both horizontally and vertically. This, in turn, can be checked against observation wells, and potentially 4D seismic data.

Author Contributions: Writing–original draft, M.W.M. and A.H. All authors have read and agreed to the published version of the manuscript.

Funding: Norges Forskningsråd: 230303.

Acknowledgments: The authors like to thank the group working with water weakening of chalk at University of Stavanger for fruitful discussions. In addition, we are grateful for the input from the editors and reviewers of the journal, and the inputs from Lawrence Cathles. The authors acknowledge the Research Council of Norway and the industry partners, ConocoPhillips Skandinavia AS, Aker BP ASA, Vår Energi AS, Equinor ASA, Neptune Energy Norge AS, Lundin Norway AS, Halliburton AS, Schlumberger Norge AS, and Wintershall DEA, of The National IOR Centre of Norway for support.

Conflicts of Interest: The authors declare no conflict of interest.

References

1. Plischke, B. Finite element analysis of compaction and subsidence—Experience gained from several chalk fields. In *Rock Mechanics in Petroleum Engineering*; Society of Petroleum Engineers: Delft, The Netherlands, 1994; p. 8.
2. Barkved, O.; Heavey, P.; Kjelstadli, R.; Kleppan, T.; Kristiansen, T.G. Valhall field-still on plateau after 20 years of production. In *Offshore Europe*; Society of Petroleum Engineers: Aberdeen, UK, 2003.
3. Doornhof, D.; Kristiansen, T.G.; Nagel, N.B.; Pattillo, P.D.; Sayers, C.J.O.R. Compaction and subsidence. *Oilfield Rev.* **2006**, *18*, 50–68.
4. Sylte, J.E.; Thomas, L.K.; Rhett, D.W.; Bruning, D.D.; Nagel, N.B. Water Induced Compaction in the Ekofisk Field. In Proceedings of the SPE Annual Technical Conference and Exhibition, Houston, TX, USA, 3–6 October 1999; Society of Petroleum Engineers: Houston, TX, USA, 1999; p. 11.
5. Griggs, D. Creep of Rocks. *J. Geol.* **1939**, *47*, 225–251. [CrossRef]
6. Andersen, P.Ø.; Wang, W.; Madland, M.V.; Zimmermann, U.; Korsnes, R.I.; Bertolino, S.R.A.; Minde, M.; Schulz, B.; Gilbricht, S. Comparative Study of Five Outcrop Chalks Flooded at Reservoir Conditions: Chemo-mechanical Behaviour and Profiles of Compositional Alteration. *Transp. Porous Media* **2018**, *121*, 135–181. [CrossRef]
7. Fjar, E.; Holt, R.M.; Raaen, A.; Risnes, R.; Horsrud, P. *Petroleum Related Rock Mechanics*; Elsevier: Amsterdam, The Netherlands, 2008.
8. De Waal, J.A. On the rate type compaction behaviour of sandstone reservoir rock. Ph.D. Thesis, Technische Hogeschool Delft, Delft, The Netherlands, May 1986.
9. Dieterich, J.H. Time-dependent friction and the mechanics of stick-slip. *Pure Appl. Geophys.* **1978**, *116*, 790–806. [CrossRef]
10. Dieterich, J.H. Time-dependent friction in rocks. *J. Geophys. Res. Solid Earth* **1972**, *77*, 3690–3697. [CrossRef]

11. Westwood, A.R.C.; Goldheim, D.L.; Lye, R.G. Rebinder effects in MgO. *Philos. Mag. J. Theor. Exp. Appl. Phys.* **1967**, *16*, 505–519. [CrossRef]

12. Macmillan, N.H.; Huntington, R.D.; Westwood, A.R.C. Westwood Chemomechanical control of sliding friction behaviour in non-metals. *J. Mater. Sci.* **1974**, *9*, 697–706. [CrossRef]

13. Westbrook, J.; Jorgensen, P. Effects of water desorption on indentation microhardness anisotropy in minerals. *Am. Mineral. J. Earth Planet. Mater.* **1968**, *53*, 1899–1909.

14. Andersen, M.A.; Foged, N.; Pedersen, H.E. The rate-type compaction of a weak North Sea chalk. In Proceedings of the 33th U.S. Symposium on Rock Mechanics (USRMS), Santa Fe, NM, USA, 3–5 June 1992; American Rock Mechanics Association: Balkema, Rotterdam, The Netherlands, 1992; p. 10.

15. Newman, G.H. The effect of water chemistry on the laboratory compression and permeability characteristics of some North Sea chalks. *J. Pet. Technol.* **1983**, *35*, 976–980. [CrossRef]

16. Homand, S.; Shao, J.F.; Schroeder, C. Plastic Modelling of Compressible Porous Chalk and Effect of Water Injection. In *SPE/ISRM Rock Mechanics in Petroleum Engineering*; Society of Petroleum Engineers: London, UK, 1998; p. 10.

17. Homand, S.; Shao, J.F. Mechanical Behaviour of a Porous Chalk and Water/Chalk Interaction. Part I: Experimental Study. *Oil Gas Sci. Technol. Rev. IFP* **2000**, *55*, 591–598. [CrossRef]

18. Homand, S.; Shao, J.F. Mechanical Behaviour of a Porous Chalk and Water/Chalk Interaction. Part II: Numerical Modelling. *Oil Gas Sci. Technol. Rev. IFP* **2000**, *55*, 599–609. [CrossRef]

19. Risnes, R. Deformation and yield in high porosity outcrop chalk. *Phys. Chem. Earth A Solid Earth Geod.* **2001**, *26*, 53–57. [CrossRef]

20. Risnes, R.; Madland, M.; Hole, M.; Kwabiah, N. Water weakening of chalk—Mechanical effects of water–glycol mixtures. *J. Pet. Sci. Eng.* **2005**, *48*, 21–36. [CrossRef]

21. Liteanu, E.; Spiers, C.; de Bresser, J. The influence of water and supercritical CO2 on the failure behavior of chalk. *Tectonophysics* **2013**, *599*, 157–169. [CrossRef]

22. Røyne, A.; Bisschop, J.; Dysthe, D.K. Experimental investigation of surface energy and subcritical crack growth in calcite. *J. Geophys. Res. Solid Earth* **2011**, *116*, B4. [CrossRef]

23. Risnes, R.; Haghighi, H.; Korsnes, R.; Natvik, O. Chalk–fluid interactions with glycol and brines. *Tectonophysics* **2003**, *370*, 213–226. [CrossRef]

24. Heggheim, T.; Madland, M.V.; Risnes, R.; Austad, T. A chemical induced enhanced weakening of chalk by seawater. *J. Pet. Sci. Eng.* **2005**, *46*, 171–184. [CrossRef]

25. Korsnes, R.; Madland, M.; Austad, T. Impact of brine composition on the mechanical strength of chalk at high temperature. In Proceedings of the International Symposium of the International Society for Rock Mechanics, Eurock, Liège, Belgium, 9–12 May 2006.

26. Garrels, R.M.; Christ, C.L. *Solutions, Minerals and Equilibria*; Freeman, Cooper & Company: San Fransisco, CA, USA, 1965.

27. Megawati, M.; Hiorth, A.; Madland, M.V. The impact of surface charge on the mechanical behavior of high-porosity chalk. *Rock Mech. Rock Eng.* **2011**. [CrossRef]

28. Nermoen, A.; Korsnes, R.I.; Hiorth, A.; Madland, M.V. Porosity and permeability development in compacting chalks during flooding of nonequilibrium brines: Insights from long-term experiment. *J. Geophys. Res. Solid Earth* **2015**, *120*, 2935–2960. [CrossRef]

29. Korsnes, R.I.; Madland, M.V. The Effect on Compaction Rates by Divalent Anion and Cations on Outcrop Chalk Tested at Reservoir Temperature and Effective Stress Conditions. *Poromechanics VI* **2017**, 706–714. [CrossRef]

30. Minde, M.W.; Zimmermann, U.; Madland, M.V.; Korsnes, R.I.; Schultz, B.; Gilbricht, S. Mineral Replacement in Long-Term Flooded Porous Carbonate Rocks. *Geochim. Cosmochim. Acta* **2019**, *268*, 485–508. [CrossRef]

31. Zimmermann, U.; Madland, M.V.; Nermoen, A.; Hildebrand-Habel, T.; Bertolino, S.A.R.; Hiorth, A.; Korsnes, R.I.; Audinot, J.-N.; Grysan, P. Evaluation of the compositional changes during flooding of reactive fluids using scanning electron microscopy, nano-secondary ion mass spectrometry, x-ray diffraction, and whole-rock geochemistry. *Am. Assoc. Pet. Geol. Bull.* **2015**, *99*, 791–805. [CrossRef]

32. Hiorth, A.; Bache, Ø.; Jettestuen, E.; Cathles, L.M.; Moe, R.W.; Omdal, E.; Korsnes, R.I.; Madland, M.V. A Simplified Approach to Translate Chemical Alteration in Core Experiments to Field Conditions. In Proceedings of the International Symposium of the Society of Core Analysts, Austin, TX, USA, 18–21 September 2011.

33. Higgins, R.; Leighton, A. Computer prediction of water drive of oil and gas mixtures through irregularly bounded porous media three-phase flow. *J. Pet. Technol.* **1962**, *14*, 1048–1054. [CrossRef]
34. Hiorth, A.; Jettestuen, E.; Vinningland, J.; Cathles, L.; Madland, M. Thermo-chemistry reservoir simulation for better EOR prediction. In Proceedings of the IEA EOR 34th Annual Symposium, Stavanger, Norway, 8–12 September 2013.
35. Kristiansen, T.G.; Plischke, B. History Matched Full Field Geomechanics Model of the Valhall Field Including Water Weakening and Re-Pressurisation. In Proceedings of the SPE EUROPEC/EAGE Annual Conference and Exhibition, Barcelona, Spain, 14–17 June 2010; Society of Petroleum Engineers: London, UK, 2010; p. 21.
36. Zangiabadi, B.; Korsnes, R.I.; Madland, M.V.; Hildebrand-Habel, T.; Hiorth, A.; Kristiansen, T.G. Mechanical Properties of High and Lower Porosity Outcrop Chalk at Various Wetting States. In Proceedings of the 43rd US Rock Mechanics Symposium and 4th U.S.-Canada Rock Mechanics Symposium, Ashville, NC, USA, 28 June–1 July 2009.
37. Andersen, M.A. Enhanced compaction of stressed North Sea chalk during waterflooding. In Proceedings of the Third European Core Analysts Symposium, Paris, France, 14–16 September 1992.
38. Sachdeva, J.S.; Nermoen, A.; Korsnes, R.I.; Madland, M.V. Impact of Initial Wettability and Injection Brine Chemistry on Mechanical Behaviour of Kansas Chalk. *Transp. Porous Media* **2019**, *128*, 755–795. [CrossRef]

Article

What Pulsating H_2 Emissions Suggest about the H_2 Resource in the Sao Francisco Basin of Brazil

Lawrence Cathles [1],* and Alain Prinzhofer [2]

[1]   Department of Earth and Atmospheric Sciences, Cornell University, Ithaca, NY 14853, USA
[2]   Geo4U Geosciences Integrated Services LTDA (Geo4U), Praia de Botafogo, 501–Botafogo, Rio de Janeiro RJ–CEP 22250-040, Brazil; prinzhofer2@gmail.com
*   Correspondence: lmc19@cornell.edu

Received: 12 February 2020; Accepted: 10 April 2020; Published: 17 April 2020

Abstract: Proterozoic sedimentary basins very often emit natural hydrogen gas that may be a valuable part of a non-carbon energy infrastructure. Vents in the Sao Francisco Basin in Brazil release hydrogen to the atmosphere mainly during the daylight half of the day. Daily temperature and the regular daily tidal atmospheric pressure variations have been suggested as possible causes of the pulsing of H_2 venting. Here, we analyze a ~550 m-diameter depression that is barren of vegetation and venting hydrogen mainly at its periphery. We show that daily temperature changes propagated only ~1/2 m into the subsurface and are thus too shallow to explain the H_2 variations measured at 1-m depth. Pressure changes could propagate deeply enough, and at the depth at which the cyclic variations are measured hydrogen concentration will have the observed phase relationship to atmospheric pressure changes provided: (1) the pressure wave is terminated by geologic barriers at about 25% of its full potential penetration distance, and (2) the volume of gas in the vents is very small compared to the volume of gas tapped by the venting. These constraints suggest that there is a shallow gas reservoir above the water table under the ~550 m-diameter barren-of-vegetation depression. The 1D-analytical and finite-element calculations presented in this paper help define the hydrogen system and suggest the further steps needed to characterize its volume, hydrogen flux and resource potential.

Keywords: hydrogen economy; natural hydrogen vents; Sao Francisco Basin; pulsing gas emission; atmospheric pressure tides

1. Introduction

Hydrogen gas may become an important part of the zero-carbon energy economy because its combustion produces only water vapor and no CO_2. Hydrogen is generally considered a vector of energy because it is an agent of energy transfer that needs to be manufactured from some other material, such as hydrocarbons or water, and thus carries energy from other sources. Several studies have shown, however, that there are natural sources of hydrogen that could be an important resource (Ward, 1933 [1]; Goebel et al., 1984 [2]; Newell et al., 2007 [3]; Sherwood-Lollar et al., 2014 [4]; Prinzhofer and Deville, 2015 [5]; Guélard et al., 2017 [6]; Prinzhofer et al., 2018 [7]). The Earth is naturally expelling native hydrogen through still-poorly determined physico-chemical processes.

As at the start of the petroleum era, we currently know only that hydrogen is seeping out of rocks and soils in many locations. Studies so far have focused on the chemistry of the seeps that have been discovered worldwide (Larin et al., 2014 [8]; Zgonnik et al., 2015 [9]; Deville and Prinzhofer, 2016 [10]; Prinzhofer et al., 2019 [11]) and the origin of the hydrogen (Larin, 1993 [12]; Gilevska, T., 2007 [13]; Milesi et al., 2015 [14]; Vacquand et al., 2018 [15]; Truche et al., 2018 [16]). The many seeps that have been found indicate the operation of an active hydrogen system, and transient accumulations of hydrogen at relatively shallow depth are known (Mali, Prinzhofer et al., 2018 [7], USA, Goebel et al.,

1984 [2]; Newell et al., 2007 [3]; Guélard et al., 2017 [5]). But we know little about the scale and economic potential of the hydrogen systems that are involved.

The most important parameter for gauging the magnitude of the hydrogen resource is the magnitude of the hydrogen (H_2) emission because it immediately suggests the probability of hydrogen accumulation under subsurface structures, and also indicates the sustainability of this source of energy. To better constrain the magnitude of the hydrogen emission, a permanent monitoring station has been installed at 16 °S in the dry, hot Sao Francisco Basin of Brazil. Venting is being monitored there with sensors that measure H_2 concentrations at ~1 m depth around the perimeter of a 550 m diameter vent (Prinzhofer et al., 2019) [11]. The data indicate daily pulses of hydrogen emission. The maximum concentration of H_2 in the venting gases usually occurs in the middle of the day, although some sensors have hydrogen concentration peaks at the beginning of the day, or even during the night. It appears that these variations are linked to complex external parameters such as biologic activity in the soils, temperature, atmospheric pressure, Earth tides, etc.

Importantly, the correlation between H_2 concentrations and atmospheric pressure is not linear, but presents a kind of hysteresis in which there is a delay between the pressure variation in free air, the hydrogen concentration measured by the sensors at 1 m depth, and the gas movements induced in the porous soil. The soil generally has very low water saturation above the water table several meters below the surface. This paper investigates whether the variation of hydrogen concentrations in soils could be caused by diurnal/nocturnal variations in temperature or atmospheric pressure.

H_2 gas is venting in the Sao Francisco Basin from slight topographic depressions that are circular, barren of vegetation, and aligned along a major fault. The water table is at 3 to 5 meters depth. The hydrogen venting from the 600×500 m study area is not continuous, but pulses on a daily cycle wherein H_2 is detected by shallow (1 m depth) sensors for only about half the day, usually centered on high noon. Prinzhofer et al. (2019) [11] have estimated the emission rate is between 7000 and 178,000 m^3 H_2 per day, based on the venting at Sensor 9 (Figure 1) where the volume of H_2 venting is greatest and the concentration of H_2 in the venting gas is sometimes above 1000 ppm. Hydrogen is thought to be generated in basins from the reduction of water attending ferrous to ferric iron oxidation (Welhan and Craig, 1979 [17]; Neal and Stranger, 1983 [18]; Abrajano et al., 1990 [19]; Sano et al., 1993 [20]; Charlou et al., 2002 [21]) and/or NH$_4$$^+$ decomposition at depths where the temperature exceeds 200 to 250 °C (Guélard et al., 2018) [22]. The H$_2$ concentration was found to be 50% to 80% (with the rest N_2 and several percent CH_4) in a deep drill hole in the Sao Francisco basin, and measurements at other sites in the basin found a source H_2 concentration ranging from 50% to ~100%. This suggests that the undiluted hydrogen source at the vent we analyze has a hydrogen concentration of at least 50%.

We investigate whether, and under what circumstances, daily changes in either surface temperature or atmospheric pressure can cause gas flow to reverse (e.g., air to flow into the ground) for half the day and thus potentially explain why H_2 is detected for only half a day. The initial method of analysis is to compute, for both temperature and pressure changes, the change in gas density as a function of depth, the variations in the rate of gas flow produced by these density changes, and the conditions under which the density-change-driven-flow could cause the flow of H_2-free air into the vent subsurface for about half the day. It is found that variations in atmospheric pressure could account for the diurnal variations in H_2 detection, but temperature variations cannot. The pressure-driven changes in hydrogen concentration at 1 m depth are then investigated quantitatively by comparing the timing and magnitude (phase relations between) of the calculated and observed modulation in hydrogen concentration. It is found that pressure variations must be blocked by a barrier of some kind when they have propagated about 25% of the distance they could potentially propagate into a gas-filled reservoir, and that the gas volume of the reservoir tapped needs to be at least 1000 times greater than the gas volume in the vent. These insights, obtained by simple 1D analytical and finite element analyses, suggest how the Sao Francisco hydrogen system is operating, how the system could be tested by future drilling, and the kind of future 3D modeling that will be required to define the operation of this hydrogen system.

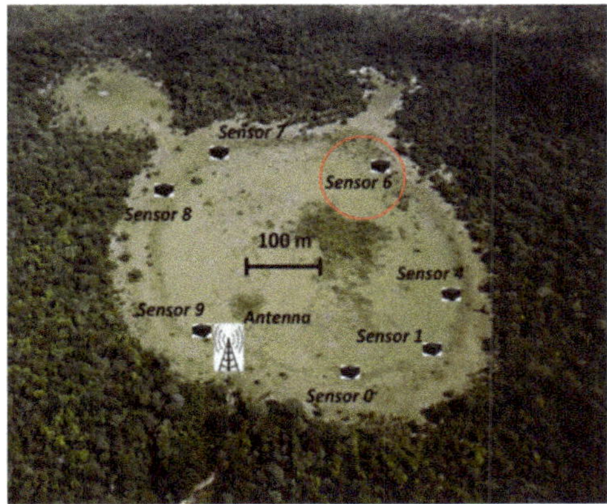

Figure 1. H_2 is venting in a pulsating fashion from the periphery of a 600×500 m topographical depression barren of vegetation in the Sao Francisco sedimentary basin of Brazil at latitude ~16 °S. Sensor 6 is circled in red. Figure is from Prinzhofer et al. (2019).

2. Analysis

2.1. Site Characterisics

2.1.1. The Study Site

As illustrated in Figure 1, H_2 is venting from a 600×500 m depression that is barren of vegetation. One possibility is that the reduced conditions caused by hydrogen venting have killed the vegetation; another is that periodic flooding of the small depressions is he cause. To be neutral, we label the area A_{barren}. The study area is the large barren area shown in Figure 1 with $A_{barren} = 300,000$ m^2. It is venting hydrogen at an estimated rate of between 7000 and 178,000 m^3 H_2 per day (Prinzhofer et al., 2019) [11]. Assuming the concentration of the source is 50% H_2, the venting rate at depth, Q_{H2}, is between 14,000 and 356,000 m^3 per day. If the flow is uniform across the vent area, the gas flux is $V_{deep} = Q_{H2}/A_{barren} = 0.05$ to 1.2 m d^{-1}. The average concentration of H_2 in the discharge measured at ~1 m depth is ~100 ppm (Prinzhofer et al., 2019 their Figure 2) [11].

2.1.2. Daily Variations in T

The temperature ranges from 15 to 35 °C in the afternoons. Because the bare surface will likely be heated above the atmospheric temperature, the range of ground surface temperature will likely be greater than the average atmospheric temperature range.

2.1.3. Daily Variations in Atmospheric Pressure

The daily variations in atmospheric pressure are quite regular at the site and are available from a local meteorological station. Pressure peaks at about 10:00 local solar time, and the steepest part of the subsequent pressure decrease occurs at about 13:00 local solar time. The pressure change over the day is between 4 and 8 mbar. Figure 2A compares the atmospheric pressure variations on 14 August 2018 to the $[H_2]$ concentrations measured at 1 m depth by Sensor 6 on the same day. Peak $[H_2]_{sensor-6}$ occurs at ~13:00, coinciding with the maximum in the rate of atmospheric pressure decline.

Figure 2B separates out the diurnal and semidiurnal sinusoidal components of the atmospheric pressure change amplitudes of 1.5 and 1.2 mb and phase lags of 16 and 14 h, respectively. When summed these components replicate the observed pressure variations very closely (compare yellow and blue curves in Figure 2A).

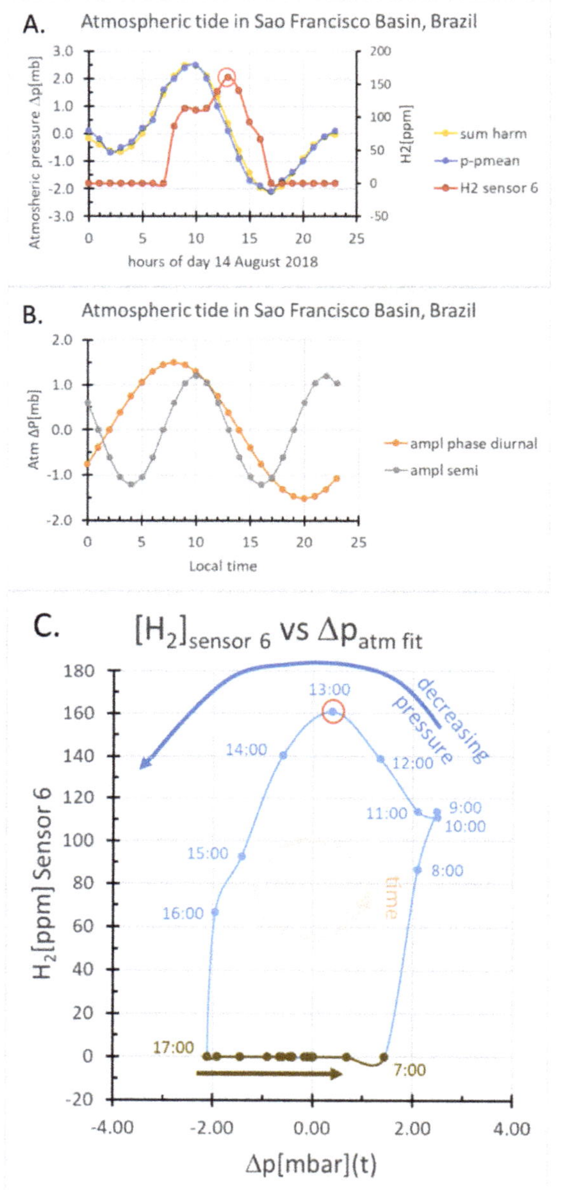

Figure 2. (**A**) Observed atmospheric pressure changes at Sao Francisco Basin H$_2$ vent, and the [H$_2$] concentrations measured at Sensor 6. (**B**) The pressure variations can be decomposed into diurnal and s semi-diurnal sinusoidal components which sum to accurately represent the observed variations (yellow curve in A). (**C**) Phase relations between atmospheric pressure and [H$_2$] measured at Sensor 6.

Figure 2C shows the phase relationships between atmospheric pressure and the hydrogen concentration measured at Sensor 6. As in Figure 2A, hydrogen is detected when the atmospheric pressure is decreasing, but the figure shows a clear hysteresis between pressure change and the hydrogen measurements that is diagnostic of pressure wave transmission into the subsurface. By hysteresis we mean that hydrogen concentrations measured in Sensor 6 when the atmospheric pressure is increasing are different from those measured when the atmospheric pressure is decreasing. In this case the contrast is dramatic since the hydrogen concentration is zero when pressure is increasing and positive when it is decreasing. The red circle in Figure 2C emphasizes that the maximum $[H_2]$ is measured at 13:00 when the surface pressure is decreasing at its maximum rate (Figure 2A).

The regular variations in atmospheric pressure are caused by atmospheric tides. At low latitudes, away from the much larger pressure variations associated with shifts of the polar front jets, the atmospheric tidal variations in pressure are typically mainly semi-diurnal with amplitudes of ~1.6 mb (Le Blancq, 2011 [23]). At our site the tidal atmospheric changes (2 mb amplitude) are similar in magnitude but have a strong diurnal component. We do not know why the daily atmospheric pressure variation at our site are more strongly diurnal than the norm.

2.2. Theoretical Analysis

This paper analyzes the hydrogen venting in the Sao Francisco basin analytically. This reveals the fundamental controls on the pulsing venting, but numerical simulations will ultimately be needed to gain a full understanding of the system. Our analysis is thus preliminary and preparatory. All symbols are defined in Table 1. Parameter values for the problem at hand are also given in the table, along with the method of their calculation or references for the values selected. Relationships given in the table indicate important physical dependencies. We consider both temperature and pressure wave transmission into the subsurface but concentrate on pressure transmission because we find that daily temperature changes penetrate only a few 10 s of centimeters and therefore cannot modulate hydrogen concentrations to a depth of one meter. Our method is to calculate the change in gas volume from either thermal expansion or compression and integrate this volume change to determine vertical (1D) gas velocity.

Thermal and Pressure Wave Propagation into the Subsurface

Surface temperature and pressure changes diffuse into the subsurface at rates and depths controlled by the subsurface thermal and hydraulic diffusivity, respectively. The diffusion equations both combine the heat and pressure flux equations with conservation of heat and mass and are identical in form and most clearly described in parallel. These equations describe how cyclic variations of temperature and pressure propagate from the surface at z = 0 into the subsurface (z negative):

The flux of heat described by Fourier's law and the flux of gas by Darcy's law are:

$$j = -K\frac{\partial T}{\partial z}\hat{z} \tag{1}$$

$$V = -\frac{k'}{\mu}\frac{\partial p'}{\partial z}\hat{z} \tag{2}$$

Conservation of mass requires:

$$\begin{aligned} \rho_T c_T \frac{\partial T}{\partial t} &= -\nabla \bullet j = K_T \frac{\partial^2 T}{\partial z^2} \\ \frac{\partial T}{\partial t} &= \kappa_T \frac{\partial^2 T}{\partial z^2} \\ \kappa_T &= \frac{K_T}{\rho_T c_T} \end{aligned} \tag{3}$$

$$\begin{aligned} \left(\beta_m + \phi\beta_f\right)\frac{\partial p'}{\partial t} &= -\nabla \bullet V = \frac{k'}{\mu_{air}}\frac{\partial^2 p'}{\partial z^2} \\ \frac{\partial p'}{\partial t} &= \kappa_p \frac{\partial^2 p'}{\partial z^2} \\ \kappa_p &= \frac{k'}{\mu_{air}(\beta_m + \phi\beta_{air})} \cong \frac{k'}{\mu_{air}\phi\beta_{air}} \end{aligned} \tag{4}$$

In (4) we have used the fact that the compressibility of air is much greater than the compressibility of the soil matrix.

The final temperature (3) and pressure (4) diffusion equations differ only in their diffusivity parameters κ. We seek the subsurface solution for $T(z,t)$ and $p'(z,t)$ for $z = 0$ to $-\infty$. Let α represent either T or p' and α_o the amplitude variation imposed at the surface. Defining $\bar{t} = t/P$, where P is the period of the harmonic variation imposed at the surface, the solution to Equations (3) and (4), (Carslaw and Jager, 1959. p.65 [24]) is:

$$\alpha(z,\bar{t}) = \alpha_o e^{z/\delta} \cos\left(2\pi\bar{t} + z/\delta\right) \tag{5}$$

where

$$\delta_\alpha = \sqrt{\frac{\kappa_\alpha P}{\pi}} \tag{6}$$

Because we want the rate of venting, we take the time derivative of (5):

$$\frac{\partial \alpha}{\partial t} = \frac{1}{P}\frac{\partial \alpha}{\partial \bar{t}} = -\frac{2\pi}{P}\alpha_o e^{z/\delta_\alpha}\sin\left(2\pi\bar{t} + z/\delta_\alpha\right) \tag{7}$$

Defining $\bar{z} = z/\delta_\alpha$, we can integrate (5) from some depth $\bar{z}_b$ to the surface (or sensor depth) to obtain the total rate of volume change from great depth to the subsurface (or up to sensor depth). By conservation of mass this must equal the gas flux, $V_{z=0}(t)$, out the top surface (or past the sensors). The sign depends on whether the forcing is by pressure or temperature change. The gas fluxes are:

$$V^{p'}_{\bar{z}=0}(t) = -\frac{2\pi}{P}\alpha_{p'o}\phi\beta_{p'}\delta_{p'}\int_{\bar{z}'=\bar{z}_b}^{0} e^{\bar{z}'}\sin\left(2\pi\bar{t} + \bar{z}'\right)d\bar{z}' \tag{8}$$

$$V^{T}_{\bar{z}=0}(t) = \frac{2\pi}{P}\alpha_{To}\phi\beta_{T}\delta_{T}\int_{\bar{z}'=\bar{z}_b}^{0} e^{\bar{z}'}\sin\left(2\pi\bar{t} + \bar{z}'\right)d\bar{z}' \tag{9}$$

Here α_o has been replaced by $\alpha_{p'o}$ or α_{To}, and the sign is such that there is volume expansion if the change in pressure is negative or the change in temperature is positive. Note the prime on $\bar{z}$ simply indicates that it is an integration variable.

If the depth of integration, $\bar{z}_b$, is sufficiently large that the underlying pressure or temperature variations are negligible, for example $\bar{z}_b = -10$, the normalized integral, $\int_{\bar{z}=-10}^{0} e^{\bar{z}}\sin(2\pi\bar{t}+\bar{z}')d\bar{z}'$ varies from -0.709 to $+0.709$ as $\bar{t}$ varies over its sinusoidal cycle. The phase shift between the temperature or pressure forcing applied at the surface ($\bar{z} = 0$) and the venting flux V are both maximum for this deep integration. If $\bar{z}_b$ is shallow and the full subsurface thermal or pressure wave is not captured in the integration, the phase shift between the pressure or temperature and the venting flux V at the $\bar{z} = 0$ is less and the rate of venting is less.

The maximum and minimum values of the harmonic gas flux at the surface are:

$$V^{\max/\min}_{\alpha}(z = 0) = \pm\frac{(0.709)2\pi}{P}\alpha_o\phi\beta_\alpha\delta_\alpha \tag{10}$$

Since the gas flux is linear, we can add the deep H_2-bearing gas flux, V_{deep}, to (6) to obtain the total gas flux, $V^{total}_{z=0}(t)$:

$$V^{total}_{z=0}(t) = V_{deep} - \frac{2\pi}{P}\alpha_o\phi\beta_\alpha\delta\int_{\bar{z}'=-10}^{\bar{z}} e^{\bar{z}'}\sin\left(2\pi\bar{t} + \bar{z}'\right)d\bar{z}' \tag{11}$$

The H_2 emissions could be pulsating if:

$$V_{deep} < \frac{2\pi}{P}\alpha_o\phi\beta_\alpha\delta_\alpha(0.709) \tag{12}$$

Table 1. Parameter values in equations derived above.

Param.	Units	Definition	Value	Method/Comment
α_{oT}	K	Amplitude harmonic $T_{z=0}(t)$	30	Ground changes > atmospheric.
$\alpha_{op'}$	Pa	Diurnal amplitude $p'_{z=0}(t)$ (lag 16 h) Semidiurnal $p'_{z=0}(t)$ (14 h phase lag)	150 120	Atmospheric tide has two harmonic components (Figure 2).
β_T	-	Thermal expansion of air	3.3×10^{-3}	$=1/T[K] = 1/298$ for 25 °C.
$\beta_{p'}$	-	Volumetric compression of ideal gas	10^{-5}	$=1/p = 1/10^5$ at atm. Pressure.
b	m	Depth of system Equation (17)	~20 m	$[H_2]$ changes only near the surface.
b_{box}	m	Depth extent box model Equation (13)	>500 m	Large b_{box} needed for sufficient V_{air}
c_s	J kg^{-1} K^{-1}	Heat capacity of solid matrix	837	
c_{air}	J kg^{-1} K^{-1}	Heat capacity of water-saturated air	4380	
δ_T	m	Thermal skin depth	ϕ δ_T 0.1 0.107 0.2 0.102 0.4 0.098	$= \sqrt{\frac{\kappa_T P}{\pi}}, \quad P = 3600 \times 24\,s$
$\delta_{p'}^{1d}$	m	Pressure skin depth for 1 day period	ϕ $\delta_{p'}$ 0.1 56.1 0.2 39.7 0.4 28.0	$= \sqrt{\frac{\kappa_{p'} P}{\pi}}, \quad P = 3600 \times 24\,s.$ Reduced by $\sqrt{2}$ for semidiurnal.
D	m^2 s^{-1}	Diffusion constant of H_2 in large excess of air	0.756×10^{-4} 6.45 m^2 d^{-1}	By comparison $D_{CO2} = 0.208 \times 10^{-4}$. Values from web.
D_E	m^2 s^{-1}	Effective diffusion constant in vent	ϕ D_E 0.1 0.323 0.2 0.645 0.4 1.29	$= \frac{D\phi}{\tau}$ values assume $\tau = 2$. Table units are m^2 per day.
ϕ	-	Porosity of vent subsurface	~0.4	
$[H_2]$	-	Volume fraction of H_2 in air		
$[H_2]_{1m}$	-	Volume fraction of H_2 in air at 1 m depth	$<10^2$–10^3 ppm	
$[H_2]_{deep}$	-	Fraction H_2 deep in the vent	0.5 to 1	Deep H_2 concentration ~1 from diverse observations (see text).
$[\overline{H}_2]$	-	H_2 as fraction of $[H_2]_{deep}$		$=[H_2]/[H_2]_{deep}$.
j	J m^{-2} s^{-1}	Heat flux	calculated	
k'	m^2	Permeability (10^{-12} m^2 = 1 Darcy)	10^{-12}	Near surface sandy soil is assumed to be very permeable.
K_T	W m^{-1} K^{-1}	Thermal conductivity of media	ϕ K_T 0.1 0.843 0.2 0.685 0.4 0.498	Calculated at 25 °C with fabric theory mix of K_s and K_{air} with fabric mixing parameter = 0.3, see Luo et al. (1994) [25].
K_s	W m^{-1} K^{-1}	Matrix thermal conductivity	4	Sandy soil with some clay at 25 °C.
K_{air}	W m^{-1} K^{-1}	Air thermal conductivity	0.0261	Air at 25 °C. T Dependence weak.
κ_T	m^2 s^{-1}	Thermal dispersivity of media	ϕ κ_T 0.1 4.2×10^{-7} 0.2 3.8×10^{-7} 0.4 3.6×10^{-7}	$= \frac{K_T}{\rho_T c_T}$.
$\kappa_{p'}$	m^2 s^{-1}	Hydrodynamic dispersivity of H_2-filled media for 1 Darcy vent permeability	ϕ $\kappa_{p'}$ 0.1 0.114 0.2 0.057 0.4 0.029	$\cong \frac{k'}{\mu_{H_2} \phi \beta_{p'}}$, $\kappa_{p'}$ for air is half H_2.
N_{pe}	-	Peclet nbr (Equation (17))	>7.72	$= \frac{Vb}{D_E} = \frac{(0.05)(100)}{0.648} = 7.72$
μ_{air}, μ_{H2}	kg m^{-1} s^{-1}	dynamic viscosity of air	$\mu_{air} = 1.78 \times 10^{-5}$ $\mu_{H2} = 8.74 \times 10^{-6}$	$\mu_{air} = \nu_{air}\, \rho_{air}$.
ν_{air}	m^2 s^{-1}	kinematic viscosity of air	1.48×10^{-5}	Regression: $\nu_{air} = -1.1555(10^{-14})T^3 + 9.5728(10^{-11})T^2 + 3.7604(10^{-8})T - 3.4484(10^{-6})$, from WWW.

Table 1. *Cont.*

Param.	Units	Definition	Value	Method/Comment
P	s	Period of P harmonic	24×3600	T surface change, tidal p' change.
$\bar{P}$	-	Period normalized to 1 day	1	P/(1 day).
p	Pa	Atmospheric pressure	10^5	
P'	kg m s^{-2}	Perturbation from atmos. pressure		
p'_o	kg m s^{-2}	Amplitude atm pressure change	200 to 400	Changes due to atmospheric tide.
ρ_T	kg m^{-3}	Density of media	$\begin{array}{cc}\phi & \rho_T \\ 0.1 & 2430 \\ 0.2 & 2160 \\ 0.4 & 1620\end{array}$	$= \rho_s(1 - \phi) + \phi\rho_{air}$.
ρ_s	kg m^{-3}	Density of solid matrix	2700	
ρ_{air}	kg m^{-3}	Density of air	1.1	Density of air at 25 °C.
$\rho_T c_T$	J kg^{-1} K^{-1}	Volumetric heat capacity air-saturated soil	$\begin{array}{cc}\phi & c_T \\ 0.1 & 2 \times 10^6 \\ 0.2 & 1.8 \times 10^6 \\ 0.4 & 1.4 \times 10^6\end{array}$	$= \rho_s c_s(1 - \phi) + \phi\rho_{air}c_{air}$.
T	K	Temperature in degrees K	~298	
T_o	K or °C	Amplitude of imposed surface T(t)	>20	Can be greater if surface T > T$_{air}$.
τ	-	Tortuosity of sediment pores	2	
τ_D	d	Time normalization in Equation (16)		$= \frac{b^2\varphi}{D_E}$.
V_{air}	m^3 m^{-2} d^{-1}	Flux of air containing H$_2$		$V_{air} = V_{cyclic} + V_{deep}$.
V_{cyclic}	m^3 m^{-2} d^{-1}	Flux of air driven only by changes p		
V_{deep}	m^3 m^{-2} d^{-1}	Deep (source) H$_2$ gas flux assuming 100% H$_2$ concentration	0.05 to 1.2	Estimated by Prinzhofer (2019) [11], assuming uniform venting in barren zone.
$V_T{}^{max}$	m d^{-1}	Maximum flux into subsurface from T(t)	$\begin{array}{cc}\phi & V_T{}^{max} \\ 0.1 & 0.005 \\ 0.2 & 0.009 \\ 0.4 & 0.018\end{array}$	$= \frac{2\pi}{(P=1d)}\alpha_{oT}\phi\beta_T\delta_T(0.709)$
$V_{p'}{}^{max}$ diurnal	m d^{-1}	Maximum flux into subsurface from diurnal p'(t), $\alpha_{op'}$ = 150 Pa. From Equation (6)a. A 1 darcy subsurface permeability is assumed.	$\begin{array}{cc}\phi & V_{p'}{}^{max} \\ 0.1 & 0.037 \\ 0.2 & 0.053 \\ 0.4 & 0.075\end{array}$	$= \frac{2\pi}{(P=1\,d)}\alpha_{op'}\phi\beta_{p'}\delta_{p'}^{1d}(0.709)$
$V_{p'}{}^{max}$ Semi-diurnal	m d^{-1}	Maximum flux into subsurface from semidiurnal p'(t), $\alpha_{op'}$ = 120 Pa. From Equation (8). A 1 darcy vent permeability is assumed.	$\begin{array}{cc}\phi & V_{p'}{}^{max} \\ 0.1 & 0.042 \\ 0.2 & 0.060 \\ 0.4 & 0.084\end{array}$	$= \frac{2\pi}{(P=0.5d)}\alpha_{op'}\varphi\beta_{p'}\frac{\delta_{p'}^{1d}}{\sqrt{2}}(0.709)$

If condition (12) is satisfied, the T-driven or p'-driven flow into the ground will exceed the base leakage rate of deep H_2-bearing gas (V_{deep}), H_2-free air could periodically surround the sensors, and the sensors could then detect low H_2 concentrations. Since V_{deep} = 0.05 to 1.2 m^3 m^{-2} d^{-1}, the question is whether the right hand side of (12) can exceed this value.

- *The box model*

It is instructive to consider a simple (but unrealistic) "box" model where, as atmospheric pressure changes, gas pressure changes instantly and uniformly within the entire box. If the subsurface where hydrogen is venting is permeable enough to some depth b_{box}, the atmospheric pressure change can be transmitted immediately, and the venting flux will be coincident with the rate of atmospheric pressure change:

$$V_{box}(\bar{z} = 0) = -\phi b_{box}\frac{\partial p}{\partial t}\beta_{p'} \tag{13}$$

For φ = 0.2, b_{box} = 1000 m, $\beta_{p'}$ = 10^{-5} Pa^{-1}, and $-\frac{\partial p}{\partial t}$ = 25 mb d^{-1} = 2500 Pa d^{-1}, the gas flux from the box V_{box} = 5 m d^{-1}. For later discussion it is important to emphasize that the assumption of very rapid pressure transmission means that there is no phase shift between surface pressure change and surface gas flux.

- *Advection-diffusion equation for [H2]*

It will take some time for gas moving into the subsurface to reach the sensors at 1 m depth, and on the way, there will be mixing with the gas in the subsurface. The advection–diffusion equation describes these phenomena. The 1D flux, j_{H2}, of H_2 (expressed as the volume fraction of H_2 in air) by diffusion and advection is:

$$j_{H2} = -D_E \frac{\partial [H2]}{\partial z} + V_{air}[H_2] \tag{14}$$

where $[H_2]$ is the volume fraction of H_2 in air. By mass conservation:

$$\frac{\partial \varphi [H_2]}{\partial t} = -\nabla \bullet j_{H2} = \frac{\partial}{\partial z} D_E \frac{\partial [H_2]}{\partial z} - V_{air} \frac{\partial [H_2]}{\partial z} \tag{15}$$

where D_E is the effective diffusion constant of H_2 in the vent, and V_{air} is the vertical flux of air (m^3 of air passing through a plan area or 1 m^2 per second) in the portion of the vent considered (in this case the uppermost part). The effective diffusion constant $D_E = \frac{D\phi}{\tau}$, where D is the diffusion constant of H_2 in air, φ is the porosity of the vent sediments, and τ is the tortuosity of the pores in the sediments, which we take to equal 2. If we make t, z and $[H_2]$ non-dimensional by defining:

$$\begin{aligned}
t &= \bar{t}\frac{b^2 \varphi}{D_E} = \bar{t}\tau_D \\
z &= \bar{z}b \\
[H_2] &= [\overline{H}_2][H_2]_{deep} \\
N_{pe} &= \frac{Vb}{D_E}
\end{aligned} \tag{16}$$

Equation (15) becomes:

$$\frac{\partial [\overline{H}_2]}{\partial \bar{t}} = \frac{\partial^2 [\overline{H}_2]}{\partial \bar{z}^2} - N_{pe} \frac{\partial [\overline{H}_2]}{\partial \bar{z}} \tag{17}$$

where N_{pe} is the Peclet number (ratio of advection to diffusion).

2.3. Site Analysis

2.3.1. The Minimum Criterion for $[H_2]_{Station-6}$ Modulation

Whether the pulsing H_2 leakage criterion (9) is met depends on the parameter values. The last 3 lines in the Table 1 show the calculated cyclic gas inflow rates for temperature and pressure variations at the Sao Fancisco site. The maximum temperature-driven gas influx is an order of magnitude less than the pressure-driven influx, and too small to reverse even Prinzhofer's lowest venting estimate. The pressure-driven influx, on the other hand, could reverse the $[H_2]$ venting if the venting were at the low end of Prinzhofer's estimated range. For convenience we show the comparison in Table 2.

Table 2. Deep H_2 flux estimated by Prinzhofer, assuming the concentration in the deep venting gas is 50% H_2 compared to air flux in and out of vent that is driven by tidal atmospheric pressure changes. The air flux exceeds the venting flux for venting fluxes at the low end of the range estimated by Prinzhofer for a subsurface permeability of 1 Darcy.

Estimation Method	H_2 or Air Flux
Prinzhofer V_{deep} for 50% H_2 concentration	0.05 to 1.2 m^3 m^{-2} d^{-1}
Calculated p-driven influx for $\varphi = 0.1$ to 0.4	0.079 to 0.159 m^3 m^{-2} d^{-1}
Calculated T-driven influx for $\varphi = 0.1$ to 0.4	0.005 to 0.018 m^3 m^{-2} d^{-1}

The calculations assume a subsurface permeability of 1 Darcy, and porosities between 0.1 and 0.4. The two harmonic components of pressure-driven gas flow in Table 1 are summed to produce the values in the Table 2. The pressure-driven gas flux is not very sensitive to porosity (factor of 2 change for a factor of 4 change in porosity). This $\sqrt{\phi}$ dependence arises because $\delta_{p'}$ depends on $1/\sqrt{\varphi}$ (see Table 1). For a subsurface permeability of 10 darcies, $\delta_{p'}^{1d}$ would be increased by $\sqrt{10} = 3.2$, and the calculated pressure-driven air flux into the ground would range between 0.25 and 0.5 m^3 m^{-2} d^{-1}.

Diurnal temperature changes penetrate so little into the subsurface that they cannot affect H_2 a meter below the surface. From Table 1, the thermal skin depth for daily temperature variations is at most 10 cm. This means that at 20 cm depth the temperature variation will be 10% of that at the surface, and at 40 cm, 1%. Temperature variations thus do not appear to be a viable way to explain $[H_2]$ variations measured at 1 m depth in the Sao Francisco basin. We therefore do not consider them further in this paper.

2.3.2. Pressure-driven Subsurface Gas Fluxes

Figure 3 shows the changes in pressure, the negative of the rate of pressure change (a proxy for decompression since gas expansion occurs when the pressure decreases), and the gas venting rate, all as a function of depth in meters into the subsurface. The calculated pressure changes take into account the resistance to gas venting offered by a 1-Darcy subsurface with 20% porosity. Note that at depth the pressure changes are out of phase with the surface pressure changes. The venting flux integrates the changes in gas volume from a depth where they are negligible to the surface. The profiles are integrated here from $\bar{z} = -10$ to 0 ($z = -394$ m to 0) using Equation (14) with $V_{deep} = 0$, but the profiles are displayed in Figure 3 only to 120 m depth. The air flux profiles in the last panel are labeled "cyclic" to emphasize that they are produced by the cyclic daily changes in atmospheric tidal pressure only.

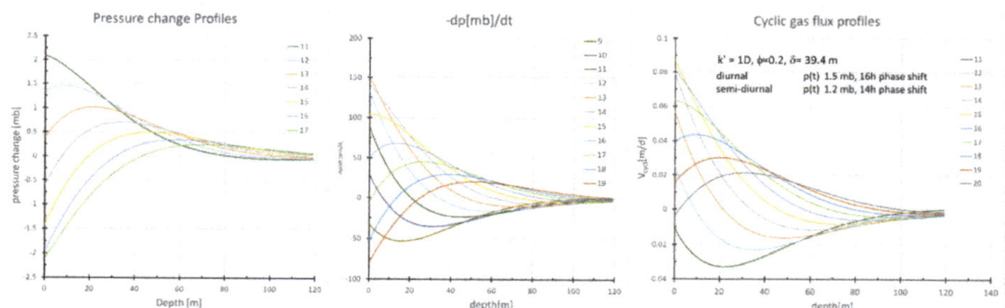

Figure 3. Calculated profiles of (**left**) pressure, (**middle**) the negative of the rate of pressure change (a proxy for gas decompression), and (**right**) the gas venting rate. Changes are caused by the changes atmospheric pressure shown in Figure 2. The subsurface permeability is 1 Darcy; the porosity 0.2 as shown in the legend in the last panel. The curves are for different hours of the day as indicated by the color key at the upper right of each diagram.

Figure 4A compares the calculated surface venting rate (black curve), the negative of the surface rate of pressure change (blue curve), and the H_2 concentration measured by Sensor 6 for 14 August 2018 (orange curve). The atmospheric pressure is decreasing at a maximum rate at 13:00 (blue curve peak). The maximum H_2 concentration measured by Sensor 6 (and most of the other sensors show in Figure 1) occurs at the same time (orange curve peak). On the other hand, the surface venting rate (black curve) peaks two hours later at 15:00.

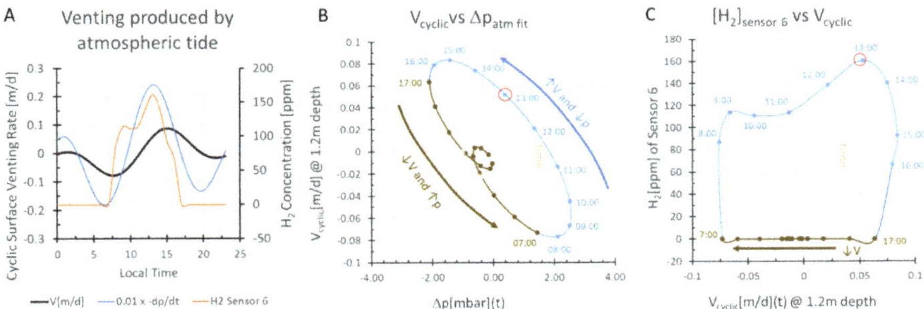

Figure 4. (**A**) The calculated gas venting rate at the surface (black curve) plotted versus local time for 14 August 2018. The blue curve shows the negative of the rate of change in atmospheric pressure at the surface. The orange curve shows the H_2 concentration measurements in Sensor 6 as a function of local time. The peaks in the H_2 concentration and –dp/dt coincide at 13:00, but the peak in surface venting rate is shifted by 2 h and occurs at 15:00. (**B**) The calculated gas flux V_{cyclic} at the surface is phase lagged relative to atmospheric pressure variations. (**C**) Calculated V_{cyclic} at 1.2 m depth is phase lagged relative to measured $[H_2]_{1m}$ at Sensor 6 is a fashion similar to atmospheric pressure variations shown in Figure 2C. The red circles indicate the time maximum $[H_2]_{1m}$ is observed in Sensor 6.

Figure 4B shows the relationship between atmospheric pressure changes and the rate of surface gas efflux (V_{cyclic}). There is considerable hysteresis between the gas flux at 1.2 m depth and the surface pressure changes that produce this flux. The changes in gas flux values are not the same when the surface pressure is increasing as they are when it is decreasing. Figure 4C shows the relationship between the measured $[H_2]$ at Sensor 6 and the calculated cyclic gas flux at 1 m depth. Again, there is substantial hysteresis, and the hysteresis is very similar to that observed and shown in Figure 2C.

Figure 5 shows that if the cyclic (compression/decompression) gas flux comes from only the shallowest parts of the full system (the first 10 m of the curves shown in Figure 3), the phase shift with respect to the peak $[H_2]_{Sensor-6}$ is reduced. If the compression/decompression gas flux derives from ~10 m (= 0.25 $\delta^{1d}_{p'}$, where $\delta^{1d}_{p'}$ = 39.4 m)of the subsurface ($\bar{z}_b$ = 0.25 in Equation (9)), the maximum in the surface venting rate (gray curve) nearly coincides in time with the maximum measured H_2 concentration. Note, however, that the magnitude of the gas flux is strongly reduced from the black curve (Figure 5A) where gas is expelled from all depths.

Figure 5B shows that the air flux profiles are nearly linear if they originate in only the upper 10 m of the profiles shown in Figure 3. Their linearity indicates that the upper part of the vent is behaving like the simple box model defined in Equation (16) where it is assumed that all the air in the subsurface is compressed equally and at the same time as the surface pressure changes. In the box case, the integrated flux will increase linearly from zero at $\bar{z}_b$, as is nearly the case in Figure 5B.

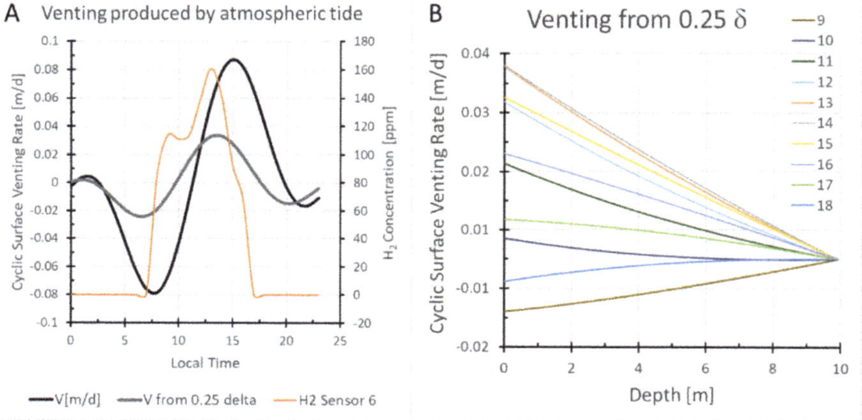

Figure 5. (**A**) The maximum cyclic gas flux calculated from a depth of 0.25δ (10 m depth for δ = 39.4 m) is more in phase with the maximum measured H$_2$. (**B**) Depth profiles of gas flux from 09:00 to 18:00. The depth profiles are almost linear and therefore approximate the flux profiles that would occur in a simple box model of the kind described in the box model section.

2.3.3. Box Model Advection/Diffusion Calculations of [H$_2$] at 1 m Depth

For a rise in atmospheric pressure to cause a drop in H_2 concentration at 1 m depth, air must penetrate at least that deep. This is a significant constraint: The air velocity v_{air} equals V_{air}/φ, so for $\varphi = 0.2$ and $V_{air} = 0.05$ m d^{-1} (the low flux estimate of Prinzhofer), the air velocity is 0.25 m d^{-1}. The atmospheric tide drives air into the ground for at most half a day, so the depth of air penetration for this low H_2 flux estimate is at most ~12 cm, which is much less than the depth of the sensors. To reach 1 m depth, the influx must be at least 8 times greater, e.g., V_{air}~0.4 m d^{-1}. Similarly, dilution of the effluent H_2 gases by influent air must be quite substantial for the measured $[H_2]_{1m}$ concentrations to be 100 to 1000 ppm. A simple stirred tank mixing model suggests reducing $[H_2]_{1m}$ by a factor of 1000 (from 50% H_2 to 500 ppm) would require mixing one volume of deep hydrogen-rich gas with ~7 volumes of air. Thus, from both these perspectives, the air influx must be about an order of magnitude greater than Prinzhofer's low estimate or ~0.05 m d^{-1}.

The advection–diffusion Equation (17) can be used to model the [H$_2$] concentration at 1 m depth. The thickness of the "box" must be ~1000 m to produce cyclic air velocities large enough to dilute the hydrogen concentration at 1 m depth by the large amounts observed. This is an unrealistically large depth but suitable for our heuristic calculations here. The calculations start with the initial steady state [H$_2$] profile in b_{box} for the assumed deep gas flux V_{deep}. This initial profile is then modified by the cyclic tidal air movements. The gas flux is constant but time varying at its 1-m-deep box value over the depth interval of analysis. The concentration changes produced by the cyclic variations in air flow are calculated to a much shallower depth, $b<<b_{box}$, using Galerkin finite element methods (Baker and Pepper, 1991 [26]) because the changes in [H$_2$] induced by the cyclic air flux do not extend very far into the subsurface (<25 m).

The calculations proceed in 1 h timesteps with 100 sub-timesteps. For plotting, [H$_2$] is normalized by dividing by the deep H_2 concentration: $[\overline{H_2}] = [H_2]/[H_2]_{deep}$. The bottom boundary condition is Neuman wherein the H_2 flux equals $V_{deep}[\overline{H_2}]_{deep}$ and $[\overline{H_2}]_{deep} = 1$ (100% H$_2$). The top boundary is Dirichlet, $[\overline{H_2}]_{z=0} = 0$. Four daily pressure cycles are computed, and the last selected for plotting. $b_{box} = 1000$ m and $V_{deep} = 0.1$ m d^{-1}. There are 100 finite elements in the 25 m interval calculated. The surface air venting rate varies from −3.4 to + 5 m d^{-1}. Calculating for 200 instead of 100 finite elements in the interval from the surface to $b = 25$ m makes no difference to the results. The 200 element curves exactly overlie the 100 element blue curves, showing the 100 element calculation has more than enough depth resolution.

Figure 6 shows the changes in the $[\overline{H_2}]$ calculated as a function of depth and time of day for the atmospheric pressure variation of 14 August 2018 in the Sao Francisco Basin of Brazil. The insert in each panel shows the gas flux at the surface over the interval plotted. It can be seen that the tidal air flux produces dramatic changes in near surface hydrogen concentrations. Periods of air inflow drive the near-surface $\left[\overline{H_2}\right]_{1m}$ concentration to lower values. Periods of air outflow increase the near-surface $\left[\overline{H_2}\right]_{1m}$. The H_2 concentration varies between the surface and ~20 m depth. Below this depth the H_2 concentration is at the deep, up-flow value.

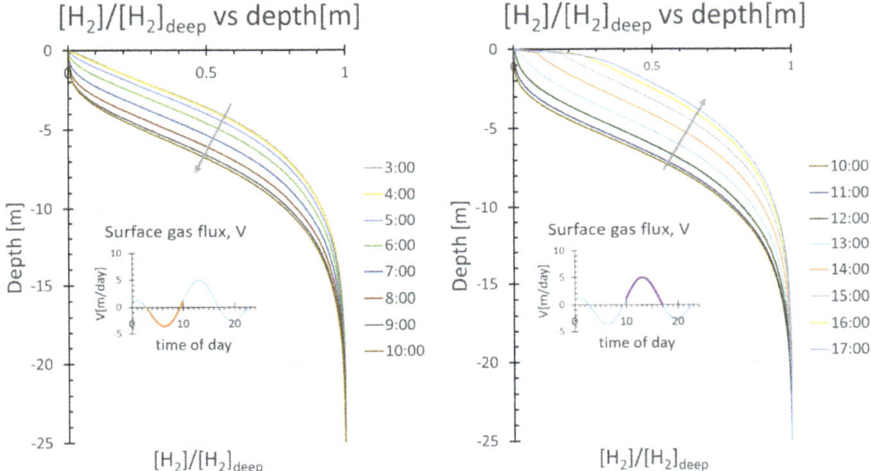

Figure 6. Calculated profiles of $\left[\overline{H_2}\right]$ for b_{box} = 1000 m and V_{deep} = 0.1 m d^{-1} for various times during 14 August 2018. The insert in each panel shows the calculated venting rate, highlighting the interval of the day when the profiles are plotted. The legend lists the time of day of each profile.

Figure 7 shows the calculated hydrogen concentration at 1 m depth, $\left[\overline{H_2}\right]_{1m}$ on 14 August 2018. Periods of air inflow drive the near-surface H_2 concentration to very low values except for a small peak at ~3 AM. Periods of venting elevate the 1 m depth concentrations to 37% of the deep input concentration. The calculated variations in $[H_2]_{1m}$ are offset from those observed (orange curve) by about ~3 h. The calculated modulation of $\left[\overline{H_2}\right]_{1m}$ is strongly dependent on the vertical extent of the high permeability "box". The dashed blue line shows the comparatively small variation in $\left[\overline{H_2}\right]_{1m}$ calculated for b_{box} = 200 m.

Finally, Figure 8 shows that, as expected, there is no substantial hysteresis between $[H_2]_{1m}$ and atmospheric pressure for the box model calculations. This is in strong contrast to observations (Figure 2C). Also, the calculated $[H_2]_{1m}$ concentration continues to increase until the gas efflux reverses (e.g., it is maximum at 17:00), rather than peaking when the rate of pressure change is greatest at 13:00 (red circle).

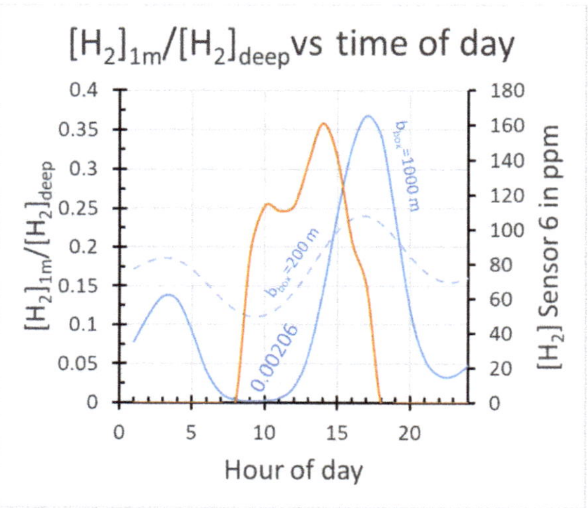

Figure 7. Computed $\left[\overline{H_2}\right]_{1m}$ for $b_{box} = 1000$ m and $V_{deep} = 0.1$ m d^{-1} (blue curve) is compared to the ppm $[H_2]_{1m}$ observed (orange curve). The dashed blue curve the H$_2$ profile computed for $b_{box} = 200$ m.

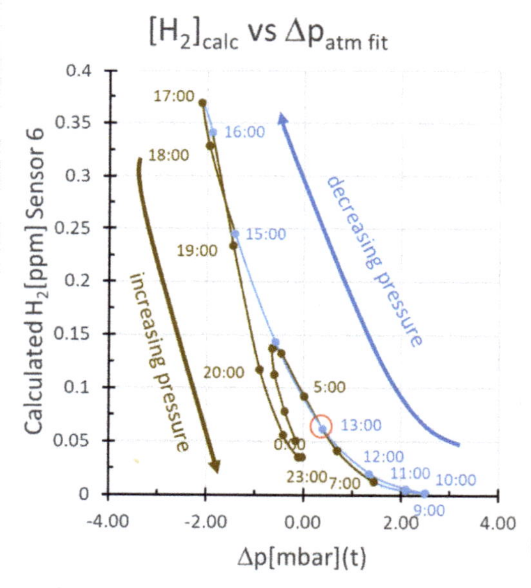

Figure 8. Phase relationship between $[H_2]_{1m}$ and the atmospheric pressure changes that drive the cyclic gas flow calculated in the box model. Unlike observations (Figure 2C), no substantial hysteresis is suggested by the box model calculations. The red circle marks the time maximum $[H_2]_{1m}$ is observed in Sensor 6.

3. Discussion

From the above analysis, it is clear that the atmospheric pressure tide could cause the pulsation in hydrogen concentration measured at 1 m depth. The main evidence supporting the hypothesis that atmospheric pressure changes are modulating the measured hydrogen concentration is the hysteresis in the observed $\Delta p'$ vs. $[H_2]_{1m}$ curve shown in Figure 2C. Figure 4C shows that the slow calculated diffusion of pressure into the subsurface produces a hysteresis between the rate of venting at 1.2 m depth, $V_{1.2m}$, and the measured $[H_2]_{1m}$ that is very similar in form to that observed between $\Delta p'$ and $[H_2]_{1m}$. (Note, the circulation is the same if $\Delta p'$ is replaced by $-\Delta p'$.) If $[H_2]_{1m}$ is proportional to $V_{1.2m}$, this hysteresis similarity strongly suggests the diffusion of pressure into the subsurface is the cause of the measured pulsing of the hydrogen venting.

A substantial reservoir of gas (compared to the volume of gas in the vents) must be compressed or expanded by the atmospheric pressure changes for hydrogen-free atmospheric air to be drawn to sensor depth. For the simple "box" model calculated above, the box must be ~1000 m thick to change $[H_2]_{1m}$ in a fashion similar to that observed (Figure 7). Instantaneous pressure transmission to 1000 m depth would require an unrealistically high subsurface permeability, so the box depth of 1000 m simply indicates that the reservoir gas volume affected by atmospheric pressure changes must be at least ~1000 times larger than the gas volume between the surface and the H_2 sensor at 1 m depth. Pressure wave calculations show that, in addition, the volume of gas accessible to the pressure wave must be about 25% of the full volume with which it could interact. This is required for the maximum venting rate, V, to coincide with the maximum $[H_2]_{1m}$ (Figure 5A). If the vent has a very low gas volume compared to the reservoir with which it interacts, there will be very little transit delay for incoming air to reach the H_2 gas sensors at 1 m depth. It is important to emphasize that the box modeling is 1D. Flow arises from vertical compression and decompression only. In reality gas would be supplied to vents laterally as well as vertically. Thus, the 25% of the potential draw should be interpreted as 25% of the 3D volume that feeds a particular vent.

It is reasonable that $[H_2]_{1m}$ should be maximum at the maximum venting rate at 1 m depth. The advection–diffusion solution shown in Figure 8 seems to suggest differently. It shows the maximum $[H_2]_{1m}$ occurs at the end of venting just before inflow begins. However, it is a 1D calculation that considers only vertical diffusion. In actuality, H_2 diffuses laterally from the vent, and, as gas rich in H_2 approaches the surface H_2 will diffuse laterally and be diluted. This dilution will be minimum when the gas efflux is maximum, and thus the maximum $[H_2]_{1m}$ should coincide with the maximum venting rate.

Observations as well as these modeling results suggest that hydrogen is venting from a shallow reservoir lying between the surface and the water table under the barren zone, and that the venting occurs mainly on the periphery, as shown in Figure 9. The barren zone is a small topographic depression that fills periodically with water. It is plausible that the top of the barren zone could be less permeable than its periphery because, due to periodic flooding, it receives more fine sediment deposition, has more evaporative salt deposition, and is more altered. Because slumping permeability might also be concentrated at the barren zone margins. If the upper layer of the central portion of the barren zone is relatively impermeable, but underlain by permeable sediments, a sealed H_2 reservoir could exist in the permeable sediments between the surface and the water table. The hypothetical reservoir could extend outside the barren zone if the sediments above the water table were as permeable outside as inside the barren zone. The reservoir in Figure 9 is shown being filled from depth by relatively pure (50 to 100%) hydrogen gas. Gas pushed into and out of the reservoir by atmospheric pressure tides through the periphery vents dilutes the reservoir near these vents as illustrated by the green hydrogen concentration contours surrounding the leftmost vent in Figure 9. Different perimeter vents will interact with the reservoir in slightly (and perhaps substantially) different ways if the permeability and porosity vary around the periphery of the reservoir. The time and concentration of the peak hydrogen concentration at different sensors could therefore differ as observed. The vents will operate as observed provided the three-to-five-meter-thick reservoir constitutes ~25% of the potential pressure wave penetration depth and the gas volume in the vents is a very small fraction of the volume compressed and decompressed by the atmospheric pressure tides impacting each vent.

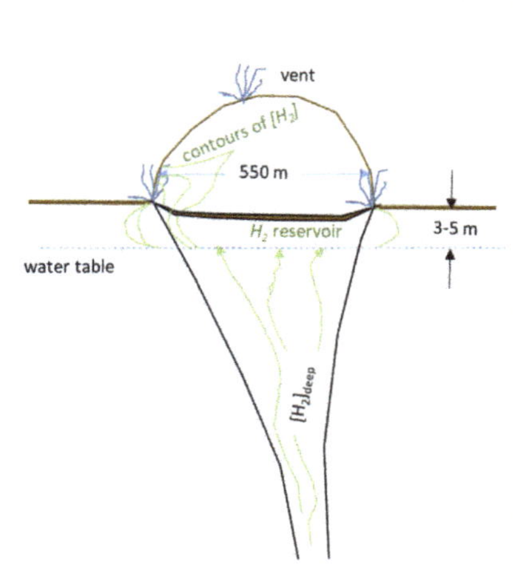

Figure 9. Schematic of H2 vent system suggested by the modeling results.

The possibility that the vent system is operating as illustrated in Figure 9 can be tested in several ways: A gas probe in the center of the barren zone would test if there is a gas reservoir between a sealed surface and the water table. The hydrogen concentration in the center of the barren zone should be >50% (or at least much greater than near the vents). Gas probes into the reservoir near sites of venting on the periphery could show how $[H_2]$ varies away from the vents. The gradient in $[H_2]$ and pressure variations at these probes could confirm the hypothesis that atmospheric-pressure variations cause the observed changes in measured H_2. Measurements of permeability and porosity would also test this hypothesis and would provide data for the kind of 3D finite element analysis that will be needed to accurately model the H_2 venting. Drill holes outside the barren zone would test the extent of the H_2 reservoir.

There is much that is not covered in our analysis. For example, as the water table at the base of our hypothetical H_2 reservoir rises and falls, accumulated H_2 will be expelled and diluted. Tracking these changes could be important to the H_2 content of the reservoir. The magnitude of H_2 venting is best provided, at least in the short term, by integrating the H_2 efflux from the periphery of the barren zone as has been done by Prinzhofer et al. (2019). We add nothing to Prinzhofer's estimates of the total H_2 venting rate in this paper. Rather, the analysis presented in this paper suggests the kind of system that could operate as observed. Ultimately 3D finite element modeling will be needed to define the hydrogen resource. For 3D modeling to contribute beyond the analysis offered here, however, more needs to be known about the shallow H_2 reservoir and its relation to the vents on the periphery of the barren zone. The needed information can be obtained by gas probe or shallow drilling.

4. Conclusions

This paper analyzes whether pressure- or temperature-driven air flow can explain the temporal variations in hydrogen concentration measured at 1 m depth along the perimeter of a 550 m diameter, largely barren depression in the Sao Francisco Basin in Brazil. Although the temporal variations could be caused by other processes such as solid earth tides or temperature-dependent bacterial H_2 generation, etc., we find:

1. The variations in hydrogen concentration measured at 1 m depth could be caused by propagation of a pressure wave into the subsurface, but not by the propagation of a thermal wave. Diurnal

temperature changes penetrate less than a meter into the subsurface and produce only weak perturbations of gas flow above the one-meter sensor depth.

2. The propagation of a pressure wave truncated at 25% of its potential penetration could produce changes in hydrogen concentration at 1 m with a phase shift relative to atmospheric tidal pressure changes similar to that most commonly observed.

3. To change $[H_2]$ concentrations at 1 m depth, the gas volume in each vent needs to be <1/1000th of the reservoir gas volume with which atmospheric pressure variations interact.

4. The venting system we infer here from observations and calculations is illustrated in Figure 9. The characteristics of this hypothetical system should be tested by measuring reservoir hydrogen concentrations with gas probes or by drilling as indicated in the discussion section.

Author Contributions: L.C. was responsible for the modeling, A.P. for field geology and framing the problem to be addressed. Both contributed equally to the writing of the paper. All authors have read and agreed to the published version of the manuscript.

Funding: The work reported here was not supported by any grant by any agency.

Acknowledgments: We thank the Scientific Direction of Engie S.A. for having launched the Natural H_2 Project. The sensors installed in Brazil that allowed monitoring measurements were developed by the Engie Laboratory CRIGEN. We specially thank Fabian Rupin for the last version of these sensors and Joao Françolin (Georisk) for his geological collaboration in the field and for data acquisition.

Conflicts of Interest: The authors declare no conflicts of interest.

References

1. Ward, L.W. Inflammable gases occluded in the pre-Palaeozoic rocks of South Australia. *Trans. R. Soc. S. Aust.* **1933**, *57*, 42–47.

2. Goebel, E.D.; Coveney, R.M.J.; Angino, E.E.; Zeller, E.J.; Dreschhoff, G.A.M. Geology, composition, isotopes of naturally occurring H_2/N_2 rich gas from wells near Junction City, Kansas. *Oil Gas J.* **1984**, *82*, 215–222.

3. Newell, K.D.; Doveton, J.H.; Merriam, D.F.; Gilevska, T.; Waggoner, W.M.; Magnuson, L.M. H2-rich and Hydrocarbon Gas Recovered in a Deep Precambrian Well in Northeastern Kansas. *Nat. Resour. Res.* **2007**, *16*, 277–292. [CrossRef]

4. Lollar, B.S.; Onstott, T.C.; Lacrampe-Couloume, G.; Ballentine, C.J. The contribution of the Precambrian continental lithosphere to global H2 production. *Nature* **2014**, *516*, 379–382. [CrossRef] [PubMed]

5. Prinzhofer, A.; Deville, E. Hydrogène Naturel. La Prochaine Révolution énergétique. Available online: https://www.belin-editeur.com/hydrogene-naturel-la-prochaine-revolution-energetique (accessed on 16 April 2020).

6. Guélard, J.; Beaumont, V.; Rouchon, V.; Guyot, F.; Pillot, D.; Jézéquel, D.; Ader, M.; Newell, K.D.; Deville, E. Natural H2in Kansas: Deep or shallow origin? *Geochem. Geophys. Geosyst.* **2017**, *18*, 1841–1865. [CrossRef]

7. Prinzhofer, A.; Cissé, C.S.T.; Diallo, A.B. Discovery of a large accumulation of natural hydrogen in Bourakebougou (Mali). *Int. J. Hydrogen Energy* **2018**, *43*, 19315–19326. [CrossRef]

8. Larin, N.; Zgonnik, V.; Rodina, S.; Deville, E.; Prinzhofer, A. Larin V.N. Evidences for natural molecular hydrogen seepages associated with rounded subsident structures. Part 1: The craton of European Russia. *Nat. Resour. Res.* **2014**, *24*, 369–383. [CrossRef]

9. Zgonnik, V.; Beaumont, V.; Deville, E.; Larin, N.; Pillot, D.; Farrell, K.M. Evidence for natural molecular hydrogen seepage associated with Carolina bays (surficial, ovoid depressions on the Atlantic Coastal Plain, Province of the USA). *Prog. Earth Planet. Sci.* **2015**, *2*, 121. [CrossRef]

10. Deville, E.; Prinzhofer, A. The origin of N2-H2-CH4-rich natural gas seepages in ophiolitic context: A major and noble gases study of fluid seepages in New Caledonia. *Chem. Geol.* **2016**, *440*, 139–147. [CrossRef]

11. Prinzhofer, A.; Moretti, I.; Françolin, J.; Pacheco, C.; D'Agostino, A.; Werly, J.; Rupin, F. Natural hydrogen continuous emission from sedimentary basins: The example of a Brazilian H2-emitting structure. *Int. J. Hydrogen Energy* **2019**, *44*, 5676–5685. [CrossRef]

12. Larin, N. *Hydridic Earth the New Geology of Our Primortidally Hydrogen-Rich Planet Translation*; Polar Publishing: Maple Ridge, BC, Canada, 1993; p. 247.

13. Gilevska, T.; Voglesonger, K.; Lin, L.-H.; Lacrampe-Couloume, G.; Telling, J.; Abrajano, T.; Onstott, T.; Pratt, L.M. Hydrogeologic Controls on Episodic H2 Release from Precambrian Fractured Rocks—Energy for Deep Subsurface Life on Earth and Mars. *Astrobiology* **2007**, *7*, 971–986. [CrossRef]

14. Milesi, V.P.; Guyot, F.; Brunet, F.; Richard, L.; Recham, N.; Benedetti, M.F.; Dairou, J.; Prinzhofer, A. Formation of CO2, H2 and condensed carbon from siderite dissolution in the 200–300 °C range and at 50MPa. *Geochim. Cosmochim. Acta* **2015**, *154*, 201–211. [CrossRef]

15. Vacquand, C.; Deville, E.; Beaumont, V.; Guyot, F.; Sissmann, O.; Pillot, D.; Arcilla, C.; Prinzhofer, A. Reduced gas seepages in ophiolitic complexes: Evidences for multiple origins of the H2-CH4-N2 gas mixtures. *Geochim. Cosmochim. Acta* **2018**, *223*, 437–461. [CrossRef]

16. Truche, L.; Joubert, G.; Dargent, M.; Martz, P.; Cathelineau, M.; Rigaudier, T.; Quirt, D. Clay minerals trap hydrogen in the Earth's crust: Evidence from the Cigar Lake uranium deposit, Athabasca. *Earth Planet. Sci. Lett.* **2018**, *493*, 186–197. [CrossRef]

17. Welhan, J.A.; Craig, H. Methane and hydrogen in East Pacific Rise hydrothermal fluids. *Geophys. Res. Lett.* **1979**, *6*, 829–831. [CrossRef]

18. Neal, C.; Stanger, G. Hydrogen generation from mantle source rocks in Oman. *Earth Planet. Sci. Lett.* **1983**, *66*, 315–320. [CrossRef]

19. Abrajano, T.A.; Sturchio, N.C.; Kennedy, B.M.; Lyon, G.L.; Muehlenbachs, K.; Bohlke, J.K. Geochemistry of reduced gas related to serpentinization of the Zambales ophiolite, Philippines. *Appl. Geochem.* **1990**, *5*, 625–630. [CrossRef]

20. Sano, Y.A.; Urabe, T.; Wakita, H.; Wushiki, H. Origin of hydrogen–nitrogen gas seeps. *Oman. Appl. Geochem.* **1993**, *8*, 1–8. [CrossRef]

21. Charlou, J.L.; Donval, J.P.; Fouquet, Y.; Jean-Baptiste, P.; Holm, N.G. Geochemistry of high H_2 and CH_4 vent fluids issuing from ultramafic rocks at the Rainbow hydrothermal field (36°14′ N, MAR). *Chem. Geol.* **2002**, *191*, 345–359. [CrossRef]

22. Guélard, J.; Martinez, I.; Sissmann, O.; Bordmann, V.; Fleury, J.M. The role of ammonium in native H2 prodiction in continental lithosphere. In Proceedings of the Goldschmidt confonference, Boston, MA, USA, 12–17 August 2018; pp. 12–178.

23. Le Blancq, F. Diurnal pressure variation: The atmospheric tide. *Weather* **2011**, *66*, 306–307. [CrossRef]

24. Carslaw, H.S.; Jager, J.C. *Conduction of Heat in Solids*; Clarendon Press: Oxford, UK, 1959; p. 510.

25. Luo, M.; Wood, J.R.; Cathles, L.M. Prediction of thermal conductivity in reservoir rocks using fabric theory. *J. Appl. Geophys.* **1994**, *32*, 321–334. [CrossRef]

26. Baker, A.J.; Pepper, D.W. *Finite Elements 1-2-3*; McGraw Hill: New York, NY, USA, 1991; 341p.

Article

Migration of Natural Hydrogen from Deep-Seated Sources in the São Francisco Basin, Brazil

Frédéric-Victor Donzé [1],*, Laurent Truche [1], Parisa Shekari Namin [1], Nicolas Lefeuvre [1] and Elena F. Bazarkina [2,3]

1   University Grenoble Alpes, University Savoie Mont Blanc, CNRS, IRD, IFSTTAR, ISTerre, 38000 Grenoble, France; laurent.truche@univ-grenoble-alpes.fr (L.T.); parisa.shekari@gmail.com (P.S.N.); nicolas.lefeuvre@univ-grenoble-alpes.fr (N.L.)
2   Institute Néel, University Grenoble Alpes, UPR 2940 CNRS, 38000 Grenoble, France; elena.bazarkina@neel.cnrs.fr
3   Institute of Geology of Ore Deposits, Mineralogy, Petrography and Geochemistry, Russian Academy of Sciences, 119017 Moscow, Russia
*   Correspondence: frederic.donze@univ-grenoble-alpes.fr

Received: 22 July 2020; Accepted: 28 August 2020; Published: 2 September 2020

Abstract: Hydrogen gas is seeping from the sedimentary basin of São Franciso, Brazil. The seepages of H_2 are accompanied by helium, whose isotopes reveal a strong crustal signature. Geophysical data indicates that this intra-cratonic basin is characterized by (i) a relatively high geothermal gradient, (ii) deep faults delineating a horst and graben structure and affecting the entire sedimentary sequence, (iii) archean to paleoproterozoïc basements enriched in radiogenic elements and displaying mafic and ultramafic units, and (iv) a possible karstic reservoir located 400 m below the surface. The high geothermal gradient could be due to a thin lithosphere enriched in radiogenic elements, which can also contribute to a massive radiolysis process of water at depth, releasing a significant amount of H_2. Alternatively, ultramafic rocks that may have generated H_2 during their serpentinization are also documented in the basement. The seismic profiles show that the faults seen at the surface are deeply rooted in the basement, and can drain deep fluids to shallow depths in a short time scale. The carbonate reservoirs within the Bambuí group which forms the main part of the sedimentary layers, are crossed by the fault system and represent good candidates for temporary H_2 accumulation zones. The formation by chemical dissolution of sinkholes located at 400 m depth might explain the presence of sub-circular depressions seen at the surface. These sinkholes might control the migration of gas from temporary storage reservoirs in the upper layer of the Bambuí formation to the surface. The fluxes of H_2 escaping out of these structures, which have been recently documented, are discussed in light of the newly developed H_2 production model in the Precambrian continental crust.

Keywords: native hydrogen; H_2 exploration; gas seeps; H_2 venting; radiolysis; serpentinization; draining faults; intra-cratonic basin

---

1. Introduction

The natural production of molecular hydrogen (hereafter hydrogen or H_2) has drawn increasing scientific attention due to the central role this molecule plays in fueling the deep subsurface biosphere or promoting the abiotic synthesis of organic molecules (Truche et al., 2020 [1]). Natural H_2 sources may also represent a new attractive primary carbon free energy resource (Smith, 2002 [2]; Smith et al., 2005 [3]; Truche and Bazarkina, 2019 [4]; Gaucher, 2020 [5]). This latter industrial perspective has motivated recent H_2 exploration studies in ophiolite, peralkaline, Precambrian shields and intra-cratonic geological settings (see review by Zgonnik, 2020 [6]).

Natural hydrogen (also known as native hydrogen) sources have been identified for several decades in seafloor hydrothermal vents, and hyperalkaline springs in ophiolite massifs. Serpentinization of ultramafic rocks is the water–rock interaction process responsible for H_2 generation in these contexts (Neal and Stranger, 1983 [7]; Coveney et al., 1987 [8]; Abrajano et al., 1990 [9]; Charlou et al., 1996 [10]; Seewald et al., 2003 [11]). However, the recent discoveries of intra-cratonic H_2 seepages and accumulations with no obvious link to an ultramafic formation challenge our current understanding of H_2 production and fate in the crust (Larin et al., 2015 [12]; Zgonnik et al., 2015 [13], Prinzhofer et al., 2018 [14]). To date, there is no in-depth understanding of the hydrogen system from source to seep in these latter geological settings. When not fortuitous, as in the Taoudeni Basin in Mali (Prinzhofer et al., 2018 [14]), the discoveries of new H_2 seepages were made thanks to the satellite detection of sub-circular soil depressions displaying vegetation anomalies, e.g., Borisoglebsk in Russia (Larin et al., 2015 [12]) and Carolina Bay in the US (Zgonnik et al., 2015 [13]). These surface features are the only evidence used to detect these H_2 seepages. This limited understanding of the H_2 systems, and this lack of robust pathfinders prevents the development of a methodic exploration strategy or resource assessment in these environments.

The São Francisco Basin belongs to this short list of intra-cratonic basins where H_2 seepages have been discovered. There, hydrogen gas vents from slight topographic depressions that are circular and barren of vegetation and, in one of them, the recorded H_2 concentrations range from 50% to 80% (Prinzhofer et al., 2019 [15]; Cathles and Prinzhofer, 2020 [16]). In the area of the São Francisco Basin, Flude et al., (2019 [17]) also, recorded up to 20% H_2, mostly accompanied with N_2 and several percent of CH_4, in the gas mixture from the head of exploration wells and natural gas seeps.

The São Francisco basin provides one of the first H_2 case studies where geological information can be collected with a sufficient level of detail to provide the primary elemental bricks that may compose the H_2 system in intra-cratonic basins. Here, we review the different layers of information that compose a supposed H_2 system in this basin and lay the foundation of a H_2 exploration guide.

2. The São Francisco Basin

Located in the Brazilian states of Minas Gerais and Bahia, the São Francisco Craton presents rocks dating back to the Paleoarchean, to the Cenozoic, and several Precambrian sedimentary successions (Heilbron, 2017 [18]) (Figure 1). The basement is mostly composed of Archean TTG (Tonalite–Trondhjemite–Granodiorite) rocks, granitoids and greenstones belts (Anhaeusser, 2014 [19]) together with Paleoproterozoic plutons and supra-crustal successions. This polycyclic substratum, assembled during late Neoarchean times under high-grade metamorphic conditions, is intruded by late tectonic K-rich granites, mafic-ultramafic units, and mafic dikes (Teixeira et al., 2017 [20]). The Southern part of the São Francisco Craton consists of several gneiss complexes and greenstone belts from the Mineiro orogeny.

The sedimentary cover is made up of units younger than 1.8 Ga: the São Francisco basin (Southern part), the Paramirin Aulacogen (Northern part) and the Recôncavo–Tucano–Jatoba rift (Northeastern part) (Heilbron et al., 2017 [18]). Besides these Proterozoic sedimentary successions, the São Francisco basin also contains Phanerozoic units (Permo-Carboniferous and Cretaceous rocks).

Along the southern edge of the São Franciso basin, the Bambuí Group fills a series of buried grabens. The Bambuí strata exposed along the area of interest in this study are generally flat lying and cover more than 300,000 km^2. The entire basin is covered by 450 to 1800 m-thick Neoproterozoic to Cambrian sedimentary successions, which are unconformably overlying the Archean-Paleoproterozoic basement (Delpomdor et al., 2020 [21]) (Figure 1b).

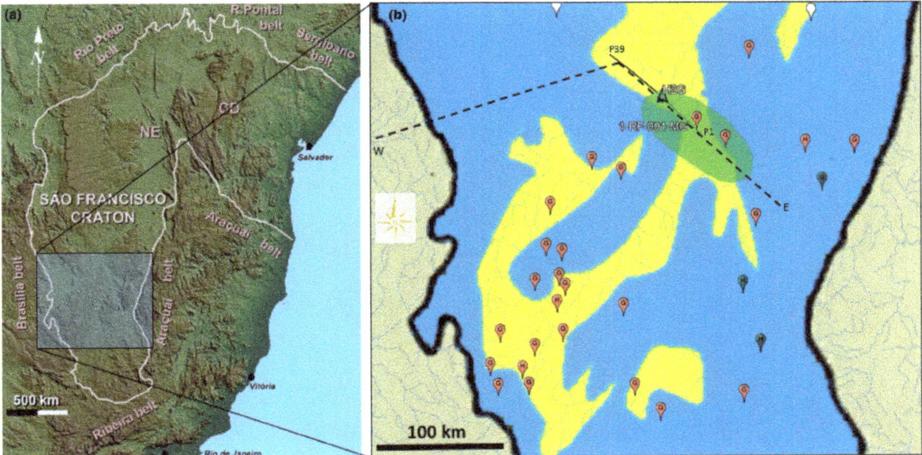

Figure 1. (**a**) Digital elevation model of southeastern Brazil, showing the topography associated with the São Francisco Craton outlined by the pink line (From Heilbron et al., 2017 [18]). (**b**) Simplified geological map (Serviço Geológico do Brasil) of the Southern part of the São Francisco Basin, outlined by the black line, with the Bambuí group in Yellow, the Phanerozoic cover in blue. Gas seepages (H_2, N_2, He, Hydrocarbons) have been observed during field investigations inside the green ellipse (Curto et al., 2012 [22]) and the triangle H2G corresponds to the gas seepage presented in Figure 2 (Prinzhofer et al.; 2019 [15]). The red markers, including the 1-RD-001-MG well, indicate locations of the exploration wells. The E–W (East–West) black dash line corresponds to the seismic sections presented in Figure 3. The solid black line linking the P1 and P39 white markers corresponds to the Magnetotelluric stations for the MT1 line set up in São Francisco Basin by Solon et al. (2015 [23]).

3. H_2 Seepages in the Bambuí Group in the Southern Part of the São Francisco Basin

To constrain the magnitude of the H_2 emission, a permanent monitoring station has been installed in a depression located 16 km North-North East of Santa Fé de Minas in the State of Mina Gerais (Prinzhofer et al., 2019 [15]) (Figure 2). The recorded emission rates range from 7000 m^3 to 178,000 m^3 of H_2 per day with H_2 concentrations in the venting gas in the order of 1000 ppm (Cathles and Prinzhofer, 2020 [16]).

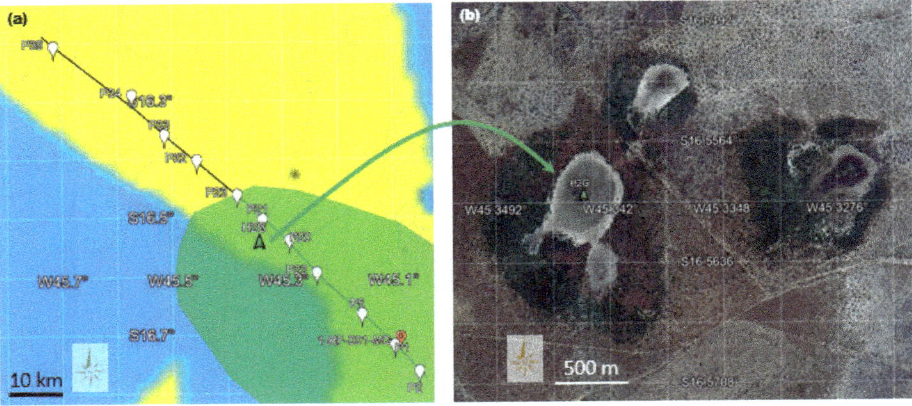

Figure 2. (**a**) Magnetotelluric stations for the MT1 line set up in São Francisco Basin by Solon et al. (2015 [23]) with the location of well 1-RF-001-MG, from Figure 1b. (**b**) H_2 seepages (H2G) observed in circular depression zones (16°33.605′ S; 45°20.620′ W) (Prinzhofer et al., 2019 [15]).

In the same area, various geophysical data acquisitions have been previously obtained from surface monitoring or exploration wells. Seismic and magnetotelluric sections show the distribution of the main stratigraphic units across the São Francisco basin (e.g., Romeiro-Silva and Zalán, 2005 [24]; Reis and Alkmim, 2015 [25]; Solon et al., 2015 [23]) (Figure 1b).

The Bambuí group includes seven stratigraphic units. Logs from the 1-RF-001-MG well, located near station P1 on line MT1 (Figure 2a), provide quantitative information regarding the thickness of each geological layer to a depth of 1848 m with the sequences listed in Table 1. The depositional age of the Bambuí Group, especially its lower part, remains controversial, e.g., the estimated date varies from 560 Ma to 762 Ma for the lowermost part of the Sete Lagoas (Delpomdor et al., 2020 [21]).

Table 1. Simplified interpretation of the well log data from the Petrobras' well 1-RF-1-MG (From Solon et al., 2015 [23]).

Depth	Composition	Lithology
surface to ~30 m	• Siltstones and sandstone.	
~30 m to ~320 m	• Mainly composed of sandstone, siltstone, mudstone and shale.	Serra de Saudade Formation
~320 m to ~480 m	• Limestone.	
~480 m to ~680 m	• Intercalations of limestone and shale.	Lagoa do Jacaré Formation
~680 m to ~980 m	• Mainly composed of siltstone.	
~980 m to ~1200 m	• Mainly consisting of limestone and dolomite.	Serra de Santa Helena Formation
~1200 m to ~1240 m	• Mainly consisting of limestone and dolomite.	Sete Lagoas Formation
~1240 m to ~1640 m	• Composed of intercalations of shales and limestone, conglomerates and diamictite.	
below ~1640 m	• Composed of shale, limestone and conglomerate.	Jequitaí Formation

4. A Possible Deep Origin for H_2

In addition to H_2 venting at location H2G (Figure 2b), He concentrations (5 ppm above atmospheric reference value) measured by Prinzhofer et al. (2019 [15]) at a depth of 1 m, suggest a possible gas migration from deep horizons, where He is generated. Other analyses of gas sampled at the surface, from the head of the exploration wells drilled in the São Francisco basin confirmed that, besides high concentrations of H_2 (up to ~20%), He (>1%) is also present, in association with methane-dominated hydrocarbons and N_2 (Flude et al., 2019 [17]). Stable isotope data also suggest an abiotic origin for the methane, while He isotopes reveal a strong crustal signature ($^3He/^4He$ < 0.02 R/Ra) (Flude et al., 2019 [17]). The nucleogenic ^3He from the decay of ^6Li could account for the $^3He/^4He$ ratios found in the head of the exploration wells drilled in the São Francisco basin, i.e., close to R/Ra = 0.01 for an average granitic crust. Moreover, Neon isotope data also suggest the presence of an Archaean crustal component in the gases, indicating that a component of the gas has likely originated from the underlying crystalline basement, or within Archaean-derived sedimentary rocks (Flude et al., 2019 [17]).

The natural production of the continental H_2 can be of various origins (Guélard et al., 2017 [26]). Studies in deep mines from the Witwatersrand basin (South Africa) and the Timmins basin (Ontario, Canada) have suggested a link between dissolved H_2 and the radiolytic dissociation of water (Lin et al., 2005a [27]). In addition to radiolysis, hydration of ultramafic rocks coupled to H_2O reduction could also be responsible for H_2 generation in Precambrian shields (Goebel et al., 1984 [28]; Sherwood Lollar et al., 2014 [29]). For example, the serpentinization of the gabbroic basement has been proposed as the process responsible for H_2 production in Kansas (Coveney et al., 1987 [8]).

Radiolysis and serpentinization both require specific environments, which can be identified from geophysical and mineralogical investigations. Regarding the São Francisco basin, we detail these two possible processes of H_2 formation in the following subsections.

4.1. Production of H_2 by Water Radiolysis

Distinct Archean gneissic–granitic complexes characterize the Southern part of the São Francisco Craton basement. They constitute a medium- to high-grade metamorphic terrain that crops out from the Quadrilátero Ferrífero towards the west, and mainly comprises TTG rocks, migmatites and K-rich granitic plutons (Teixeira et al., 2017 [20]). These rocks record deformational and metamorphic Archean episodes (from 2.55 Ga to over 3.3 Ga). It is known that the crystalline basement, rich in radiogenic elements and particularly this type of old basement Precambrian rock, represents potentially fertile deep-seated sources of H_2 (Parnell et al., 2017 [30]; Sherwood Lollar et al., 2014 [29]). Indeed, molecular hydrogen production from water radiolysis requires the presence of radiogenic elements such as U, Th or K, which split the water molecules by ionizing radiation to produce molecules of H_2. For the São Francisco Craton, the measured concentrations of uranium (U), thorium (Th) and potassium (K) are presented in Table 2. The Bambuí Group exhibits intermediate-to-high K and Th contents, while U-levels are around 2.5 ppm (Reis et al., 2012 [31]).

Table 2. Measured concentrations of Uranium, Thorium and Potassium.

Radioelement	Archean Granulitic Rocks of the Jequie Complex [1]	Brauna Kimberlite Present in the Archean Basement [2]	Bambuí Group
Uranium (U)	up to 5 ppm	up to 4.81 ppm	up to 2.5 ppm
Thorium (Th)	up to 100 ppm	up to 35.8 ppm	up to 16 ppm
Potassium (K)	up to 4.5%	NA	up to 3%

[1] Sighinolfi et al., 1982 [32]; [2] Donatti-Filho et al., 2013 [33].

In a coarse-grained rock like granite, beta-irradiation from K is more prone to affect inter-granular fluid than the shorter-range alpha irradiation from U. Since K is also more pervasively distributed than U in granite, it can contribute to a larger scale radiolysis process.

Given the rather consistent range of U, Th, and K concentrations reported in the São Francisco Basin, we could expect in this zone a production rate of radiolytic H_2 in water ranging from 10^{-8} to 10^{-7} nmol$\cdot$L^{-1}$\cdot$s^{-1} (Lin et al., 2005b [34]). The methodology proposed by Sherwood Lollar et al. (2014 [29]) to estimate the contribution of the Precambrian continental crust to H_2 production via radiolysis may then be applied to infer the regional H_2 flux. The total radiolytic H_2 production rate in water-filled fractures of the Precambrian crust was estimated to range from 0.16 to 0.47×10^{11} mol$\cdot$yr^{-1} for a corresponding surface area of 1.06×10^8 km^2. Given the surface area of the Sao Francisco Basin of 300,000 km^2, this corresponds to a H_2 diffusive flux of 0.45 to 1.34×10^8 mol$\cdot$yr^{-1}, i.e., 90 to 266 tons$\cdot$yr^{-1}.

4.2. Production of H_2 by Serpentinization or Hydration

Serpentinization occurs when meteoric or oceanic waters alter ultramafic rocks originating from the Earth's mantle, such as peridotites and volcanic rocks. These rocks undergo changes in pressure and temperature conditions, which cause them to react in the presence of water (Schlindwein and Schmid, 2016 [35]; Horning et al., 2018 [36]): They are oxidized and hydrolyzed with water into serpentine, brucite and magnetite. The anaerobic oxidation of Fe(II) by the protons of water leads to the formation of H_2 (Foustoukos et al., 2008 [37]; Proskurowski et al., 2008 [38]). In Precambrian rocks, Sherwood Lollar et al. (2014 [29]) propose that for the totality of the Precambrian crust (i.e., 1.06×10^8 km^2) around 0.2 to 1.8×10^{11} mol$\cdot$yr^{-1} of H_2 are produced by hydration. Here again, rescaling these values for the São Francisco basin (300,000 km^2), we obtain a H_2 production rate from hydration reactions of 0.56 to 5.09×10^8 mol$\cdot$yr^{-1}, i.e., 113 to 1018 tons$\cdot$yr^{-1}.

Favorable conditions to produce H_2 by a serpentinization process would imply the presence of low-silica mafic and ultramafic rocks as well as an optimum temperature.

4.3. Presence of Ultramafic Rocks

From the seismic section that crosses the São Francisco Craton from East to West (Figure 1b), the interpretations done by several authors agree on the identification of the Bambuí group (Figure 3). This unit is about 1200–1300 m deep in the area of the H2G seepage zone and lies on the Jequitaí formation which itself lies on the poorly identified older Proterozoic succession, the Macaúbas and possibly the Espinhaço formations (Solon et al., 2015 [23]).

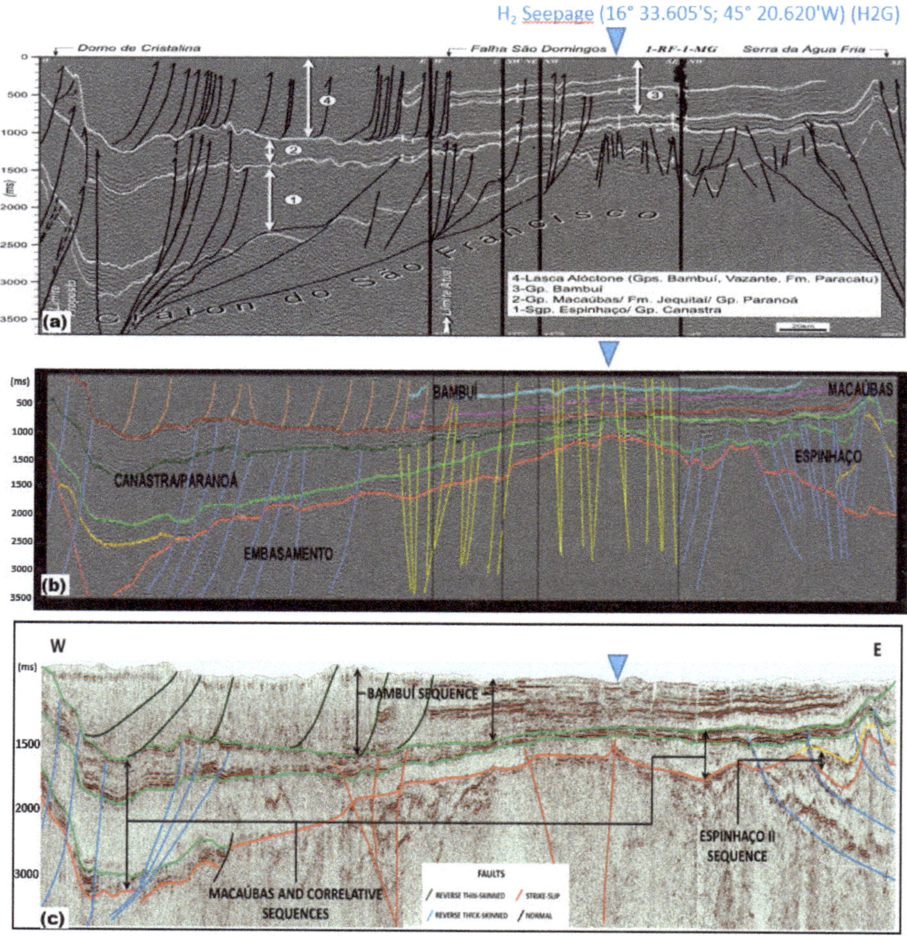

Figure 3. Three interpretations of the same reflection seismic section across the São Francisco craton (see Figure 1b for the location) from East "E" to West "W". From top to bottom: (**a**) Romeiro-Silva and Zalán (2005 [24]), (**b**) Coelho et al. (2008 [39]) and (**c**) Alkmim and Martins-Neto (2012 [40]). The locations of the exploration well A-RF-1-MG are shown in (**a**) and the gas seepage H2G (blue triangle) is reported for all cases. On all illustrations, the depth is expressed in two-way travel time (TWT).

The basal Paranoá–Upper Espinhaço sequence consists of continental sediments and volcanic rocks associated with anorogenic plutons. Mesoproterozoic anorogenic magmatism associated with multiple rifting episodes might represent a manifestation of the Columbia supercontinent breakup, which started around 1.6 Ga and ended between 1.3 and 1.2 Ga (Reis et al., 2017a, b [41,42]). The Espinhaço Supergroup is exposed on the East of the São Francisco Basin. The two basal formations of the Espinhaço sequence are composed of alluvial sandstones, conglomerates and pelites and form a ca. 300-m-thick of two coarsening-upward sequences. Despite the potential presence of K-rich alkaline volcanic and intrusives rocks (Chemal et al., 2012 [43]), this formation does not seem suitable for H_2 production.

The basement rocks of the São Francisco basin are dominated by Archaean to Palaeoproterozoic migmatites, amphibolite to granulite-grade gneisses, and granite–greenstones (Teixeira et al., 2017 [20]). For example, the Rio Itapicuru low-grade supra-crustal greenstone belt has several lithostratigraphic subdivisions, including a basal mafic volcanic unit composed of massive and pillowed basaltic flows intercalated with chert, banded iron-formation, and carbonaceous shale (Oliveira et al., 2019 [44]). The banded iron-formation is mainly composed of oxidized iron Fe(III) forming a possible mix of hematite and magnetite, which can produce a strong magnetic anomaly (Pereira and Fuck, 2005 [45]) (purple zones in Figure 4).

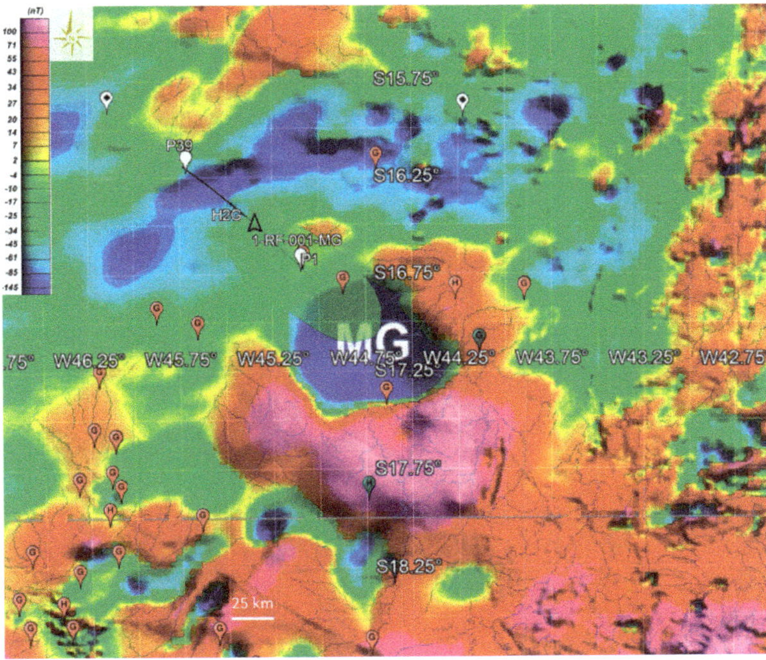

Figure 4. Magnetometric map (modified from Correa, 2019 [46]). Note that "MG" stands for "Minas Gerais".

In this area, the Bouguer anomaly map exhibits predominantly negative anomalies (Figure 5) which correlate with granitoids resulting from the crustal rejuvenation of the area, associated with partial re-fusion of the crust during past thermal events. This also suggests that major magmatic sequences affected the basement of the southern part of the São Franscico Basin, which is compatible with the magnetic anomalies (Figure 4). Since the Quadrilátero Ferrífero (a mineral-rich region with extensive deposits of iron ore) and the "greenstone" belts present the same gravimetric signature, they could have the same origin (Pinto et al., 2007 [47]). Here again, the basement rock composition presents a high potential for H_2 production.

4.4. Temperature Ranges at Depth

The potential presence of ultramafic rocks within the area of gas seepages, could suggest a serpentinization process, but to be active, this process would require a favorable range of temperatures. Crustal thermal models have been developed to examine the implications of the observed intra-cratonic variations in heat flow across the São Francisco Basin (Alexandrino et al., 2008 [48]). The thermal models take into consideration the variation of thermal conductivity with temperature. It is thus possible to get the temperature distribution calculated along a large profile that crosses the São Francisco Basin. It turns out that our zone of interest in the São Francisco Basin, exhibits an abnormally high heat flow value for a craton (Figure 6). In the gas seepage zone (green ellipse in Figure 5), the temperature gradient is about 25 °C/km (Alexandrino et al., 2008 [48]). The optimum temperature for serpentinization was found to be around 250–300 °C with a maximum production of magnetite (Klein et al., 2013 [49]). However, some experimental data suggest that below 150 °C, H_2 can still be produced through the formation of Fe^{3+} oxi-hydroxides (Mayhew et al., 2013 [50]; Miller et al., 2017 [51]), with the formation of H_2 being possibly catalyzed by the surface of spinel-structure minerals occurring in ultramafic rocks. In such thermal conditions, H_2 could still be produced at a low rate, at a depth lower than 6 km, near the gas seepage zone, and the optimum depth for its production would be at 10–12 km.

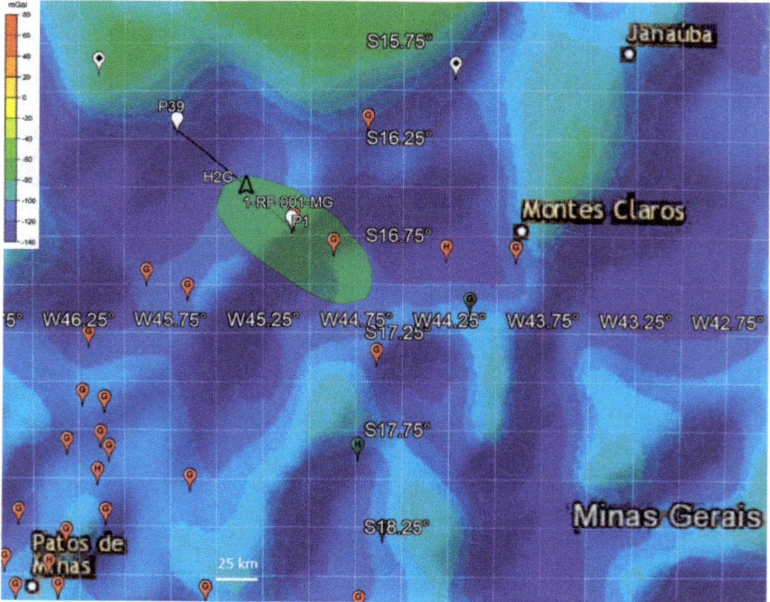

Figure 5. Bouguer anomaly map in the São Francisco Basin within the studied zone (modified from Oliveira and Andrade, 2014 [44]).

4.5. Possible H_2 Bubbling at Depth

Once produced by fluid–rock interaction processes (oxidation) at depth, H_2 can migrate as a dissolved component. The solubility of H_2 in aqueous solutions is rather low and drops when T and P decrease when approaching the surface (Figure 7d). The possible mechanism of H_2 discharge, concentrating, and transport upward to the lower T–P where H_2 is less reactive is solution boiling, i.e., formation of the vapor phase coexisting with the liquid phase. The concentrations of H_2 in vapors are many orders of magnitude higher than that in the liquid (Bazarkina et al., 2020). At the same time, rock permeability is much higher for gas-rich vapors than for salt-rich liquids. Bubble formation is a function of T, P, total salinity, and gas saturation. Thus, during fluid ascent upward to the lower T–P, gas bubble formation is favored (Figure 7d). Periodicity of H_2 emission at the surface reported by Prinzhofer et al. (2019) could be related to the kinetics of fluid–rock interaction at depth, further time-dependent bubble accumulation, and the final periodical ejections similar to those described in geysers.

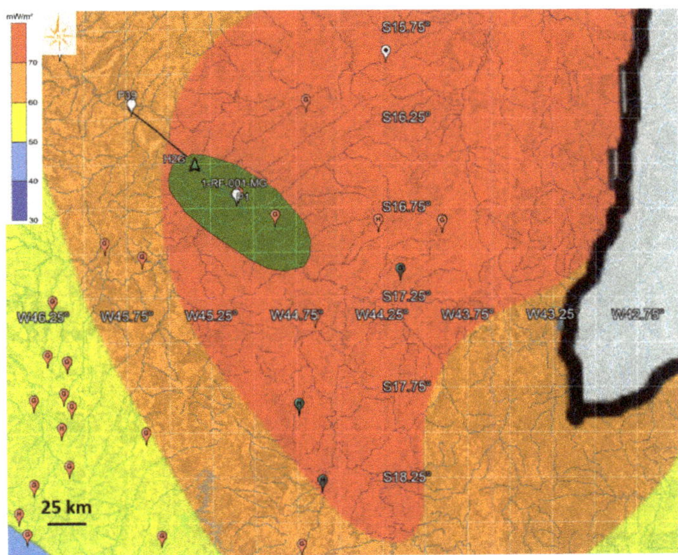

Figure 6. Heat flow map in the São Francisco Basin within the studied zone (modified from Alexandrino and Hamza, 2012 [52]).

5. Draining Fault System in the São Francisco Basin

Deep faults serve as significant channels for deep fluids to ascend into and through the crust and the $^3He/^4He$ ratio can be used to estimate the flow rate of mantle fluids through the fault zones (Kennedy et al., 1997 [53]). Since the $^3He/^4He$ signature seems to be of crustal origin in the São Francisco basin (Flude et al., 2019 [17]), the migration path followed by the H_2 mostly crosses the sedimentary cover without major changes of the ratio value. This weak interaction with the Bambuí sequences could be due to a high flow rate value along the faults. This could be possible if the faults form direct drains from the basement to the surface assuming a sufficiently high value of permeability.

Several interpretations of the available seismic data have been proposed for the fault systems (Figure 3) (e.g., Romeiro-Silva and Zalán, 2005 [24]; Coelho et al., 2008 [39] and Alkmim and Martins-Neto, 2012 [40]). Nevertheless, the São Francisco basin seems to encompass different tectonic elements such as the Proterozoic rift structures, Neoproterozoic foreland f–t-belts and Cretaceous rift structures (Reis and Alkmim, 2015 [25]). The rift structure that cuts across the central portion of the basin is characterized by a system of major NW–SE faults. One could expect that the deep-rooted faults in the graben structures (Precambrian sequence), which have been reactivated during the Neoproterozoic Macaúbas basin-cycle, could cross most of the sedimentary formation. As a major fault system, they may control the drainage at all depths and delineate some morphological features observed on satellite imagery and digital elevation models (Reis et al., 2017b [42]).

If these faults cross different geological layers, mainly shales, sandstones and limestones, their permeability values can range from 10^{-19} to 10^{-13} m^2 (Donzé et al., 2020 [54]). In terms of hydraulic conductivity for the water carrying the gas and neglecting the contribution of temperature, this could correspond to a value as high as 10^{-6} m/s. This means that in the fastest scenario, the fluid could take less than 100 years to migrate across the Bambuí sedimentary layer through a fault system.

6. Possible Temporary Shallow Zones of H_2 Accumulation

The pressure variation observed at 1 m depth in the São Francisco basin (Prinzhofer et al., 2019 [15]), with a momentary increase in H_2 pressure, could indicate that the H_2 systems are active.

There is only a small temperature window where H_2 may remain stable over a long time. This window corresponds to a T range where abiotic redox reactions such as thermochemical sulfate reduction or carbonate reduction (e.g., Fisher Tropsh type reaction) are slow (Truche et al., 2009 [55]), and where bacteria are inactive. Such a T range can be roughly approximated to be 100–200 °C. Since these temperature conditions are not met at shallow depth, H_2 will probably not survive to long residence time. Despite this fact, previous studies of H_2 seepages often indicate that the hydrogen systems are active, and transient accumulations of hydrogen at relatively shallow depth can be observed (Prinzhofer et al., 2018 [14], Goebel et al., 1984 [28]; Guélard et al., 2017 [26]). These observations may suggest a constant recharge of the aquifers by H_2 flowing from deeper levels of the basin.

The circular depression where H_2 seepage is observed (Figure 2b) could be related to a sinkhole structure resulting from a chemical dissolution process at depth. If so, these depressions will contain standing water connected with a ground-water reservoir contained in karst (De Carvalho et al., 2014 [56]). The presence of resistive carbonate and calcareous rocks was inferred from ~320 m to ~480 m followed by a layer of intercalated shales and sandstones (Solon et al., 2015 [23]). This carbonate layer, which corresponds to the Lagoa do Jacaré carbonate layer, exhibits a potential karst system according to the outcrops located East of the São Francisco Basin (Dos Santos et al., 2018 [57]).

Assuming a karst system at depth, this could imply a high level of porosity favorable for a massive storage volume of an aquifer. Since karst features are controlled by structural heterogeneities, such as faults and fractures, which influence fluid flow, they can provide preferential pathways for geofluids with the development of secondary porosity. This could agree with the fact that the circular depressions where H_2 is venting are aligned along a major fault (Cathles and Prinzhofer, 2020 [16]).

7. Putting It All Together: A Potential H_2 System within the São Francisco Basin

The first key point is related to potential source areas, e.g., the presence of both ultramafic and U, Th and K-rich rocks. The presence of Archean greenstone belts containing ultramafic rocks, TTG, migmatites and K-rich granitic plutons represent excellent H_2-producing zones either via serpentinization, or water radiolysis. Magnetic (Figure 4) and Bouguer (Figure 5) anomalies are compatible with the presence of ultramafic rock producing H_2. Temperature conditions also seem favorable for the serpentinization process: with a temperature gradient of 25 °C/km (Alexandrino et al., 2008 [48]), the optimum range of temperature would be expected at a depth of 10 km, with possible lower rate processes at a shallower depth.

The second key point is the structural/tectonic context and the presence of faults deeply rooted in the basements capable of draining a potential deep and scattered source. All interpretations of the seismic profile of the zone of interest suggest the presence of deep faults following the graben structures (Figure 3): they can be able to drain hydrogen produced at depth where the Pressure–Temperature conditions are optimal. Some interpretations suggest that some of these faults could cross the entire sedimentary sequence (Coelho et al., 2008 [39]), producing gas seepages directly at the surface. Some others predict that these faults could reach some potential shallow carbonated reservoirs (Romeiro-Silva and Zalán, 2005 [24]). These deep faults could also only reach the unconformity zone which composes the boundary between the sedimentary basin and basement (Alkmim and Martins-Neto, 2012 [40]).

The third key point concerns the storage areas (i.e., reservoirs) of H_2 at depth. As mentioned previously, the interface between the basement rocks and the sedimentary layers could represent a potential zone of accumulation. The interface is composed of the Macaúbas and the Espinhaço formations. The Macaúbas sequence is made up of sandstones, pelites, diamictites, carbonates, basic volcanic rocks, and metamorphosed banded iron formations (Alkmim et al., 2012 [40]), whereas the Espinhaço formation is a quartz–arenite dominated package. The presence of the Paranoá–Upper Espinhaço quartzite, which is tectonically uplifted, can facilitate the occurrence of sandstone reservoirs with appreciable permeability and porosity. Thus, potential reservoir rocks could be found among siliciclastics of the Macaúbas–Paranoá Megasequence (Solon et al., 2015 [23]).

An interesting characteristic of the deep topography is that the seepage zone H2G is located near the apex of the basement rock in the central part of the basin (see Figure 3). As the H_2 charged fluid reaches the Macaúbas/Espinhaço formations, it migrates along the unconformity toward this highest point before escaping to the surface in the green seepage zone (Figure 2). On its way to the surface, H_2 can also be temporarily trapped in the Sete Lagoas formation and at a shallower depth, inside the Lagoa do Jacaré formation. The permeability value of the Sete Lagoas Larst aquifer formation is estimated to range between 10^{-14} m^2 and 10^{-9} m^2 (Galvão et al., 2015 [58]). As for the Lagoa do Jacaré formation, very low permeability and porosity values were found in the Petrobras well 1-RF-1-MG. The presence of faults, possibly connecting all these reservoirs with the surface could explain the apparent structural control of the distribution of the known gas seepages (Curto et al., 2012 [22]). Nevertheless, the presence of sinkholes in the H2G seepage area suggests the existence of a shallow local karst formation, which could constitute a temporary reservoir for H_2.

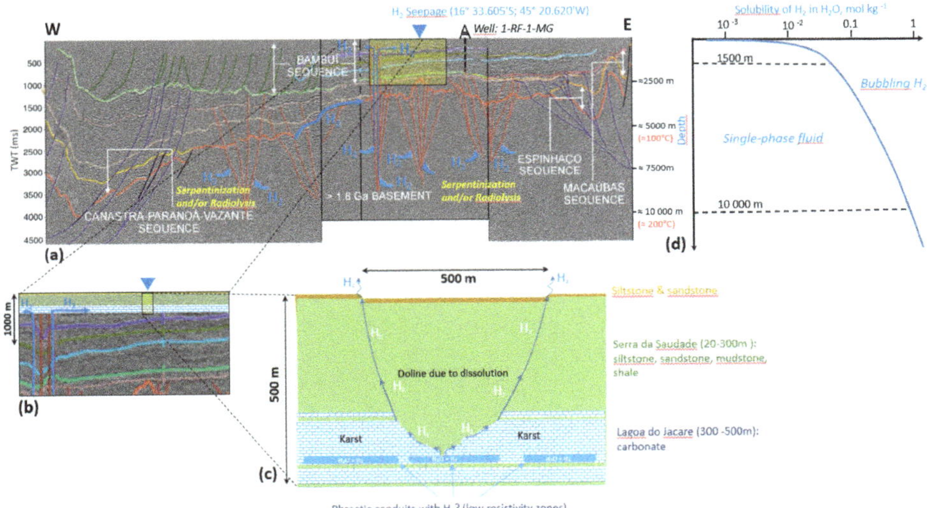

Figure 7. Conecputal model of the H_2 cycle in the Sao Francisco Basin. (**a**) Interpretated seismic section (Martins-Neto, 2009). (**b**) Zoom of the upper part of the Bambui sequence. (**c**) Possible presence of a karst structure according to the presence of sinkholes (Figure 2). (**d**) Calculated solubility of H_2 in H_2O vs. depth (Bazarkina et al., 2020 [59]).

Thus, surface seepages may be either in connection with the source rock or with intermediate leaking reservoirs since these two configurations are present in this area. A summary of H_2 migration from sources to seeps in the São Francisco basin is presented in Figure 7.

8. Discussing the H_2 Production from Radiolysis and Hydration Reactions in the São Francisco Basin

Combining the H_2 production rate from water radiolysis and hydration reactions assessed in the previous sections, we obtain an estimate of 1.01 to 6.43 × 10^8 mol·yr^{-1} H_2 production, i.e., ~200 to 1300 tons·yr^{-1} (Table 3). Cathles and Prinzhofer (2020 [23]) considered the local flux rate in the H2G seepage zone (Figure 2b) to range from 7000 to 178,000 m^3 per day. At a temperature of 21 °C and a pressure of 1 atm, these values correspond to 0.105 to 3.68 × 10^9 mol·yr^{-1}, i.e., 213 to 5400 tons·yr^{-1} (Table 3). Since the expulsion rate of H_2 is almost certainly not steady, the episodic rate measured in the H2G vent might be overestimated.

Table 3. Estimated H_2 flow rate production (tons·yr^{-1}).

System	$\times 10^9$ mol·yr^{-1}	tons·yr^{-1}	Reference
São Francisco basin (radiolysis and hydration combined)	0.101–0.643	204–1284	This study
H2G seepage zone São Francisco basin	0.105–3.68	213–5400	From Cathles and Prinzhofer, 2020 [16]
Rainbow hydrothermal field	~0.1	200	Charlou et al., 2010 [60]
Ultramafic vents along the Mid-Oceanic ridges	~10–100	20,000–200,000	Keir, 2010 [61]; Cannat et al., 2010 [62]
Global Precambrian continental lithosphere: (radiolysis and hydration combined)	36–227	72,000–454,000	Sherwood Lollar et al., 2014 [29]

In comparison, on the Mid-Atlantic Ridge (MAR), the total H_2 discharge at the Rainbow hydrothermal field is estimated to be ~10^8 moles H_2 per year, i.e., ~200 tons·yr^{-1} (Charlou et al., 2010 [60]) (Table 3). At a larger scale, the H_2 flux from all high-temperature basaltic vents along the MAR has been estimated at ~10^9–10^{10} mol·yr^{-1}, whereas the H_2 flux from high-temperature ultramafic vents along the Mid-Oceanic Ridge (MOR) has been estimated at ~10^{10}–10^{11} mol·yr^{-1} (Table 3) (Keir, 2010 [61]; Cannat et al., 2010 [62]), i.e., 20,000 to 200,000 ton.yr^{-1}.

According to our calculations, which are based on the model developed by Sherwood Lollar et al. (2014 [29]) for the Precambrian continental lithosphere, the maximum H_2 production rate from the basement rocks of the São Francisco Basin is within the same order of magnitude as the H_2 flux of one sinkhole of 500 m in diameter (H2G zone). This latter H_2 venting site would also represent from 0.047% to 7.5% of the global estimated H_2 production from the Precambrian continental Lithosphere. This large discrepancy in the results leads us to conclude that there is a need to increase the accuracy of hydrogen flux estimates through long term monitoring of soil gas migration according to different methodologies and/or to revisit the global models.

9. Conclusions

Hydrogen exploration requires a combination of the techniques and data used for both conventional petroleum and mining exploration. The first elementary bricks we provide here to evaluate the sources, migration and trapping are certainly not enough, but the following general guidelines will be extremely valuable in targeting the fertile H_2 area in intra-cratonic areas.

Explore in old provinces where basement rocks are Archean to Paleoproterozoic. Use lithologies (ultramafic rocks, U-, Th-, K-rich rocks), and He (R/Ra) as pathfinders for the H_2 generation potential. Carefully consider the local geothermal gradient as it may be of use to infer fertile zones for active serpentinization.

Identify the location of faults deeply rooted in the basement, such as horst and graben structures. Pay attention to the topography of the unconformity, which represents both a major drainage and trapping area.

Target relatively shallow traps. As for He, H_2 partitions into gas are better at a shallow depth. Surface rounded depressions, karsts and sinkholes seem to represent favorable collecting zones prior to H_2 escaping into the atmosphere. A field investigation based on electromagnetic and gravimetric prospections could help to characterize the structure of the sinkhole at depth in order to set up a geotechnical drilling at the right location. Such boreholes, carried out in the first hundreds of meters, could provide valuable information on the H_2 concentration gradient down to the upper karstic formation, which is often not possible from classical oil and gas exploration well drilling.

Dedicated exploration of boreholes is definitely required to improve this preliminary exploration guide and to strengthen the accuracy for H_2 flux measurements. Additional constraints on the

H_2–accompanying gases (He, N_2, Ne, Hydrocarbons, Rn) and on the role of H_2-consuming microbial communities at the subsurface within the emitting structure (Myagkiy et al., 2020) [63] will be extremely valuable.

Author Contributions: All authors contributed to the writing of the paper. All authors have read and agreed to the published version of the manuscript.

Funding: The work reported here was not supported by any grant by any agency.

Acknowledgments: Grateful thanks to Guest Editor Lawrence Cathles and two anonymous reviewers for their constructive reviews. The first author would like to thank Sophie-Adélaïde Magnier for helpful discussions. Laurent Truche gratefully acknowledges the support from the Institut Universitaire de France.

Conflicts of Interest: The authors declare no conflict of interest.

References

1. Truche, L.; McCollom, T.M.; Martinez, I. Hydrogen and Abiotic Hydrocarbons: Molecules that Change the World. *Elem. Int. Mag. Miner. Geochem. Pet.* **2020**, *16*, 13–18. [CrossRef]
2. Smith, N.J.P. It's time for explorationists to take hydrogen more seriously. *First Break* **2002**, *20*, 246–253.
3. Smith, N.J.P.; Shepherd, T.J.; Styles, M.T.; Williams, G.M. Hydrogen exploration: A review of global hydrogen accumulations and implications for prospective areas in NW Europe. In *Petrology Geology: North-West Europe and Global Perspectives—Proceedings of the 6th Petroleum Geology Conference*; Doré, A.G., Vining, B.A., Eds.; Petroleum Geology Conference Series; Geological Society: London, UK, 2005; Volume 6, pp. 349–358.
4. Truche, L.; Bazarkina, E.F. Natural hydrogen the fuel of the 21st century. In *E3S Web of Conferences*; EDP Sciences: Les Ulis, France, 2019; Volume 98, p. 03006.
5. Gaucher, E.C. New Perspectives in the Industrial Exploration for Native Hydrogen. *Elem. Int. Mag. Miner. Geochem. Pet.* **2020**, *16*, 8–9. [CrossRef]
6. Zgonnik, V. The occurrence and geoscience of natural hydrogen: A comprehensive review. *Earth Sci. Rev.* **2020**, *203*, 103140. [CrossRef]
7. Neal, C.; Stanger, G. Hydrogen generation from mantle source rocks in Oman. *Earth Planet. Sci. Lett.* **1983**, *66*, 315–320. [CrossRef]
8. Coveney, R.M., Jr.; Goebel, E.D.; Zeller, E.J.; Angino, E.E. Serpentinization and the origin of hydrogen gas in Kansas. *AAPG Bull.* **1987**, *71*, 39–48.
9. Abrajano, T.A.; Sturchio, N.C.; Kennedy, B.M.; Lyon, G.L.; Muehlenbachs, K.; Bohlke, J.K. Geochemistry of reduced gas related to serpentinization of the Zambales ophiolite, Philippines. *Appl. Geochem.* **1990**, *5*, 625–630. [CrossRef]
10. Charlou, J.L.; Fouquet, Y.; Donval, J.P.; Auzende, J.M.; Jean-Baptiste, P.; Stievenard, M. Mineral and gas chemistry of hydrothermal fluids on an ultrafast spreading ridge: East Pacific Rise, 17° to 19°S (Naudur cruise, 1993) phase separation processes controlled by volcanic and tectonic activity. *J. Geophys. Res.* **1996**, *101*, 899–919. [CrossRef]
11. Seewald, J.S.; Cruse, A.; Saccocia, P. Aqueous volatiles in hydrothermal fluids from the Main Endeavour Field, northern Juan de Fuca Ridge: Temporal variability following earthquake activity. *Earth Planet. Sci. Lett.* **2003**, *216*, 575–590. [CrossRef]
12. Larin, N.; Zgonnik, V.; Rodina, S.; Deville, E.; Prinzhofer, A.; Larin, V.N. Natural molecular hydrogen seepage associated with surficial, rounded depressions on the European craton in Russia. *Nat. Resour. Res.* **2015**, *24*, 369–383. [CrossRef]
13. Zgonnik, V.; Beaumont, V.; Deville, E.; Larin, N.; Pillot, D.; Farrell, K.M. Evidence for natural molecular hydrogen seepage associated with Carolina bays (surficial, ovoid depressions on the Atlantic Coastal Plain, Province of the USA). *Prog. Earth Planet. Sci.* **2015**, *2*, 31. [CrossRef]
14. Prinzhofer, A.; Cissé, C.S.T.; Diallo, A.B. Discovery of a large accumulation of natural hydrogen in Bourakebougou (Mali). *Int. J. Hydrogen Energy* **2018**, *43*, 19315–19326. [CrossRef]
15. Prinzhofer, A.; Moretti, I.; Francolin, J.; Pacheco, C.; d'Agostino, A.; Werly, J.; Rupin, F. Natural hydrogen continuous emission from sedimentary basins: The example of a Brazilian H_2-emitting structure. *Int. J. Hydrogen Energy* **2019**, *44*, 5676–5685. [CrossRef]

16. Cathles, L.; Prinzhofer, A. What Pulsating H2 Emissions Suggest about the H2 Resource in the São Francisco Basin of Brazil. *Geosciences* **2020**, *10*, 149. [CrossRef]

17. Flude, S.; Warr, O.; Magalhães, N.; Bordmann, V.; Fleury, J.M.; Reis, H.L.S.; Trindade, R.I.; Hillegonds, D.; Sherwood Lollar, B.; Ballentine, C.J. Deep crustal source for hydrogen and helium gases in the São Francisco Basin, Minas Gerais, Brazil. *AGUFM* **2019**, *2019*, EP51D-2111.

18. Heilbron, M.; Cordani, U.G.; Alkmim, F.F. The São Francisco craton and its margins. In *São Francisco Craton 2017, Eastern Brazil*; Springer: Cham, Switzerland, 2017; pp. 3–13.

19. Anhaeusser, C.R. Archaean greenstone belts and associated granitic rocks—A review. *J. Afr. Earth Sci.* **2014**, *100*, 684–732. [CrossRef]

20. Teixeira, W.; Oliveira, E.P.; Marques, L.S. Nature and evolution of the Archean crust of the São Francisco Craton. In *São Francisco Craton 2017, Eastern Brazil*; Springer: Cham, Switzerland, 2017; pp. 29–56.

21. Delpomdor, F.R.; Ilambwetsi, A.M.; Caxito, F.A.; Pedrosa-Soares, A.C. New interpretation of the basal Bambuí Group, Sete Lagoas High (Minas Gerais, SE Brazil) by sedimentological studies and regional implications for the aftermath of the Marinoan glaciation: Correlations across Brazil and Central Africa. *Geol. Belg.* **2020**, *23*, 1. [CrossRef]

22. Curto, J.B.; Pires, A.C.; Silva, A.M.; Crósta, Á.P. The role of airborne geophysics for detecting hydrocarbon microseepages and related structural features: The case of Remanso do Fogo, Brazil. *Geophysics* **2012**, *77*, B35–B41. [CrossRef]

23. Solon, F.F.; Fontes, S.L.; Meju, M.A. Magnetotelluric imaging integrated with seismic, gravity, magnetic and well-log data for basement and carbonate reservoir mapping in the São Francisco Basin, Brazil. *Pet. Geosci.* **2015**, *21*, 285–299. [CrossRef]

24. Romeiro-Silva, P.C.; Zalán, P.V. Contribuição da sísmica de reflexão na determinação do limite oeste do Cráton do São Francisco. In Proceeding of the III Simpósio Sobre o Cráton do São Francisco, Salvador, Brazil, 14–18 August 2005; pp. 44–47.

25. Reis, H.L.; Alkmim, F.F. Anatomy of a basin-controlled foreland fold-thrust belt curve: The Três Marias salient, São Francisco basin, Brazil. *Mar. Pet. Geol.* **2015**, *66*, 711–731. [CrossRef]

26. Guélard, J.; Beaumont, V.; Rouchon, V.; Guyot, F.; Pillot, D.; Jézéquel, D.; Ader, M.; Newell, K.D.; Deville, E. Natural H2 in Kansas: Deep or shallow origin? *Geochem. Geophys. Geosyst.* **2017**, *18*, 1841–1865. [CrossRef]

27. Lin, L.H.; Hall, J.; Lippmann-Pipke, J.; Ward, J.A.; Sherwood Lollar, B.; DeFlaun, M.; Rothmel, R.; Moser, D.; Gihring, T.M.; Mislowack, B.; et al. Radiolytic H2 in continental crust: Nuclear power for deep subsurface microbial communities. *Geochem. Geophys. Geosyst.* **2005**, *6*, 1–13. [CrossRef]

28. Goebel, E.D.; Coveney, R.M.J.; Angino, E.E.; Zeller, E.J.; Dreschhoff, G.A.M. Geology, composition, isotopes of naturally occurring H2/N2 rich gas from wells near Junction City, Kansas. *Oil Gas J.* **1984**, *82*, 215–222.

29. Lollar, B.S.; Onstott, T.C.; Lacrampe-Couloume, G.; Ballentine, C.J. The contribution of the Precambrian continental lithosphere to global H 2 production. *Nature* **2014**, *516*, 379–382. [CrossRef] [PubMed]

30. Parnell, J.; Blamey, N. Global hydrogen reservoirs in basement and basins. *Geochem. Trans.* **2017**, *18*, 2. [CrossRef]

31. Reis, H.L.S.; Barbosa, M.S.C.; Alkmim, F.F.D.; Soares, A.C.P. Magnetometric and gamma spectrometric expression of southwestern São Francisco Basin, Serra Selada quadrangle (1:100.000), Minas Gerais state. *Rev. Bras. Geofísica* **2012**, *30*. [CrossRef]

32. Sighinolfi, G.P.; Figueredo, M.C.H.; Fyfe, W.S.; Kronberg, B.I.; Oliveira, M.T. Geochemistry and petrology of the Jequie granulitic complex (Brazil): An Archean basement complex. *Contrib. Mineral. Petrol.* **1982**, *78*, 263–271. [CrossRef]

33. Donatti-Filho, J.P.; Tappe, S.; Oliveira, E.P.; Heaman, L.M. Age and origin of the Neoproterozoic Brauna kimberlites: Melt generation within the metasomatized base of the São Francisco craton, Brazil. *Chem. Geol.* **2013**, *353*, 19–35. [CrossRef]

34. Lin, L.H.; Slater, G.F.; Lollar, B.S.; Lacrampe-Couloume, G.; Onstott, T.C. The yield and isotopic composition of radiolytic H2, a potential energy source for the deep subsurface biosphere. *Geochim. Cosmochim. Acta* **2005**, *69*, 893–903. [CrossRef]

35. Schlindwein, V.; Schmid, F. Mid-ocean-ridge seismicity reveals extreme types of ocean lithosphere. *Nature* **2016**, *535*, 276–279. [CrossRef]

36. Horning, G.; Sohn, R.A.; Canales, J.P.; Dunn, R.A. Local seismicity of the rainbow massif on the Mid-Atlantic Ridge. *J. Geophys. Res. Solid Earth* **2018**, *123*, 1615–1630. [CrossRef]

37. Foustoukos, D.I.; Savov, I.P.; Janecky, D.R. Chemical and isotopic constraints on water/rock interactions at the Lost City hydrothermal field, 30 N Mid-Atlantic Ridge. *Geochim. Cosmochim. Acta* **2008**, *72*, 5457–5474. [CrossRef]

38. Proskurowski, G.; Lilley, M.D.; Seewald, J.S.; Früh-Green, G.L.; Olson, E.J.; Lupton, J.E.; Sylva, S.P.; Kelley, D.S. Abiogenic hydrocarbon production at Lost City hydrothermal field. *Science* **2008**, *319*, 604–607. [CrossRef]

39. Coelho, J.C.C.; Martins-Neto, M.A.; Marinho, M.S. Estilos estruturais e evolução tectônica da porção mineira da bacia proterozóica do São Francisco. *Rev. Bras. Geociências* **2008**, *38* (Suppl. S2), 149–165. [CrossRef]

40. Alkmim, F.F.; Martins-Neto, M.A. Proterozoic first-order sedimentary sequences of the São Francisco craton, eastern Brazil. *Mar. Pet. Geol.* **2012**, *33*, 127–139. [CrossRef]

41. Reis, H.L.; Suss, J.F.; Fonseca, R.C.; Alkmim, F.F. Ediacaran forebulge grabens of the southern São Francisco basin, SE Brazil: Craton interior dynamics during West Gondwana assembly. *Precambrian Res.* **2017**, *302*, 150–170. [CrossRef]

42. Reis, H.L.; Alkmim, F.F.; Fonseca, R.C.; Nascimento, T.C.; Suss, J.F.; Prevatti, L.D. The São Francisco Basin. In *São Francisco Craton, Eastern Brazil*; Springer: Cham, Switzerland, 2017; pp. 117–143.

43. Chemale, F., Jr.; Dussin, I.A.; Alkmim, F.F.; Martins, M.S.; Queiroga, G.; Armstrong, R.; Santos, M.N. Unravelling a Proterozoic basin history through detrital zircon geochronology: The case of the Espinhaço Supergroup, Minas Gerais, Brazil. *Gondwana Res.* **2012**, *22*, 200–206. [CrossRef]

44. Oliveira, R.G.; Andrade, J.B.F. Interpretação Geofísica dos Principais Domínios Tectônicos Brasileiros. In *Metalogênese da Províncias Tectônicas Brasileiras*, 1st ed.; Silva, M.G., Rocha Neto, M.B., Jost, H., Kuyumjian, R.M., Eds.; CPRM—Serviço Geológico do Brasil: Rio de Janeiro, Brazil, 2014; Volume 1, pp. 21–38.

45. Pereira, R.S.; Fuck, R.A. Archean nucleii and the distribution of kimberlite and related rocks in the São Francisco craton, Brazil. *Rev. Bras. Geociências* **2016**, *35* (Suppl. S4), 93–104. [CrossRef]

46. Correa, R.T. *Mapa da Anomalia Magnética do Brasil (Terceira Edição)*; Escala 1:5.000.000; SGB-CPRM—Serviço Geológico do Brasil: Brasília, Brazil, 2019.

47. Pinto, L.G.R.; Ussami, N.; Sá, N.C.D. Aquisição e interpretação de anomalias gravimétricas do Quadrilátero Ferrífero, SE do Cráton São Francisco. *Rev. Bras. Geofísica* **2007**, *25*, 21–30. [CrossRef]

48. Alexandrino, C.H.; Hamza, V.M. Estimates of heat flow and heat production and a thermal model of the São Francisco craton. *Int. J. Earth Sci.* **2008**, *97*, 289–306. [CrossRef]

49. Klein, F.; Bach, W.; McCollom, T.M. Compositional controls on hydrogen generation during serpentinization of ultramafic rocks. *Lithos* **2013**, *178*, 55–69. [CrossRef]

50. Mayhew, L.E.; Ellison, E.T.; McCollom, T.M.; Trainor, T.P.; Templeton, A.S. Hydrogen generation from low-temperature water–rock reactions. *Nat. Geosci.* **2013**, *6*, 478–484. [CrossRef]

51. Miller, H.M.; Mayhew, L.E.; Ellison, E.T.; Kelemen, P.; Kubo, M.; Templeton, A.S. Low temperature hydrogen production during experimental hydration of partially-serpentinized dunite. *Geochim. Cosmochim. Acta* **2017**, *209*, 161–183. [CrossRef]

52. Alexandrino, C.H.; Hamza, V.M. Improved assessment of Deep Crustal Thermal Field based on Joint Inversion of Heat Flow, Elevation and Geoid Anomaly data. *Geophys. Res. Abstr.* **2012**, *13*, EGU2011-10079.

53. Kennedy, B.M.; Kharaka, Y.K.; Evans, W.C.; Ellwood, A.; DePaolo, D.J.; Thordsen, J.; Ambats, G.; Mariner, R.H. Mantle fluids in the San Andreas fault system, California. *Science* **1997**, *278*, 1278–1281. [CrossRef]

54. Donzé, F.V.; Tsopela, A.; Guglielmi, Y.; Henry, P.; Gout, C. Fluid migration in faulted shale rocks: Channeling below active faulting threshold. *Eur. J. Environ. Civ. Eng.* **2020**, 1–15. [CrossRef]

55. Truche, L. Transformations Minéralogiques et Géochimiques Induites par la Présence D'hydrogène dans un site de Stockage de Déchets Radioactifs. Ph.D. Thesis, Université de Toulouse, Université Toulouse III-Paul Sabatier, Toulouse, France, 2009.

56. De Carvalho, O.A.; Guimarães, R.F.; Montgomery, D.R.; Gillespie, A.R.; Trancoso Gomes, R.A.; de Souza Martins, É.; Silva, N.C. Karst depression detection using ASTER, ALOS/PRISM and SRTM-derived digital elevation models in the Bambuí Group, Brazil. *Remote Sens.* **2014**, *6*, 330–351. [CrossRef]

57. Dos Santos, D.M.; Sanchez, E.A.; Santucci, R.M. Morphological and petrographic analysis of newly identified stromatolitic occurrences in the Lagoa do Jacaré Formation, Bambuí Group, State of Minas Gerais, Brazil. *Rev. Bras. Paleontol.* **2018**, *21*, 3. [CrossRef]

58. Galvão, P.; Halihan, T.; Hirata, R. The karst permeability scale effect of Sete Lagoas, MG, Brazil. *J. Hydrol.* **2016**, *532*, 149–162. [CrossRef]

59. Bazarkina, E.F.; Chou, I.M.; Goncharov, A.F.; Akinfiev, N.N. The Behavior of H2 in Aqueous Fluids under High Temperature and Pressure. *Elem. Int. Mag. Mineral. Geochem. Petrol.* **2020**, *16*, 33–38. [CrossRef]
60. Charlou, J.L.; Donval, J.P.; Konn, C.; OndréAs, H.; Fouquet, Y.; Jean-Baptiste, P.; Fourré, E. High production and fluxes of H_2 and CH_4 and evidence of abiotic hydrocarbon synthesis by serpentinization in ultramafic-hosted hydrothermal systems on the Mid-Atlantic Ridge. *GMS* **2010**, *188*, 265–296.
61. Keir, R.S. A note on the fluxes of abiogenic methane and hydrogen from mid-ocean ridges. *Geophys. Res. Lett.* **2010**, *37*. [CrossRef]
62. Cannat, M.; Fontaine, F.; Escartin, J. Serpentinization and associated hydrogen and methane fluxes at slow spreading ridges. In *Diversity of Hydrothermal Systems on Slow Spreading Ocean Ridges*; Rona, P.A., Devey, C.W., Dyment, J., Murton, B.J., Eds.; AGU Geophysical Monograph Series; John Wiley and Sons: Hoboken, NJ, USA, 2010; Volume 188.
63. Myagkiy, A.; Brunet, F.; Popov, C.; Krüger, R.; Guimarães, H.; Sousa, R.S.; Charlet, L.; Moretti, I. H2 dynamics in the soil of a H2-emitting zone (São Francisco Basin, Brazil): Microbial uptake quantification and reactive transport modelling. *Appl. Geochem.* **2020**, *112*, 104474. [CrossRef]

Communication

Earth Tides and H_2 Venting in the Sao Francisco Basin, Brazil

Jacob Simon [1], Patrick Fulton [1], Alain Prinzhofer [2] and Lawrence Cathles [1,*

[1]   Department of Earth and Atmospheric Sciences, Cornell University, Ithaca, NY 14853, USA;
       js2629@cornell.edu (J.S.); pfulton@cornell.edu (P.F.)
[2]   Geo4U Geosciences Integrated Services LTDA (Geo4U), Praia de Botafogo, 501-Botafogo,
       Rio de Janeiro RJ-CEP 22250-040, Brazil; alain.prinzhofer@geo4u.com.br
*   Correspondence: lmc19@cornell.edu

Received: 29 September 2020; Accepted: 14 October 2020; Published: 19 October 2020

Abstract: Hydrogen gas seeping from Proterozoic basins worldwide is a potential non-carbon energy resource, and the vents are consequently receiving research attention. A curious feature of H_2 venting in the Sao Francisco Basin in Brazil is that the venting displays a very regular daily cycle. It has been shown that atmospheric pressure tides could explain this cycle, but solid earth tides might be an alternative explanation. We show here that it is unlikely that solid earth tides are a dominant control because they have two equally strong peaks per day whereas the H_2 venting has only one.

Keywords: diurnal hydrogen gas venting; earth tides

1. Introduction

The quest to reduce dependence of fossil fuels has raised interest in hydrogen as a carbon-free fuel that combusts only to water. Hydrogen seeps are known in a number of Proterozoic basins worldwide [1–5]. Satellite detection of roughly circular depressions with depressed vegetation has become an exploration method, since the H_2 seeps are often associated with these features [3,5]. One H_2 vent in the Sao Francisco Basin in Brazil has been the subject of detailed monitoring since 2018. It shows a remarkably regular diurnal cycle of hydrogen venting wherein H_2 is vented for only about half the day centered on 1:00 p.m. [3]. A recent paper shows that the atmospheric pressure tide at the site could explain this venting [4]. The association with a ~550 m diameter circular depression aligned linearly with others in the area presumably along a fault suggests the venting could be from a deeply penetrating gas pipe. Earth tides are known to induce water flow in and out of wells [6–8]. The purpose of this paper is to evaluate the possibility that earth tides cause the diurnal gas venting at the Sao Francisco vent site studied by Cathles, Prinzhofer, and Donzé et al. [4,5]. Following Donzé et al. [5] we will refer to this site as the H2G site.

2. Methods

To evaluate the possibility that the diurnal venting at the H2G site could be caused by solid earth tides, we computed the solid earth strain at the location studied in Cathles and Prinzhofer [4] using the theoretical tidal loading software SPOTL [9] and compared it to the variations in H_2 concentration at 80 cm depth for sensor 6 in Figure 1b below. The theoretical earth tide strains determined with this software have often been used in the analysis of high-resolution borehole water level time-series measurements, since the tidal amplitude response and phase shift within a well or borehole are controlled by the formation permeability and specific storage [7,8].

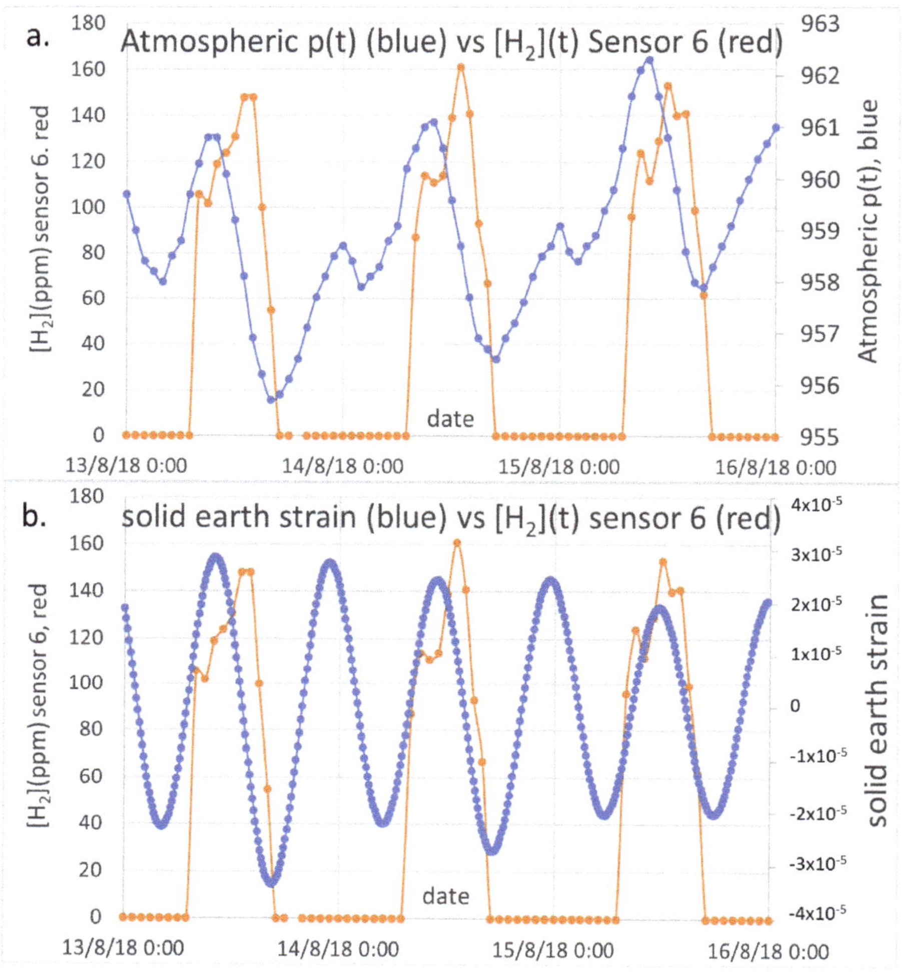

Figure 1. (**a**) Comparison of changes in atmospheric pressure to observe changes in H_2 concentration at 80 cm depth at the location studied by Cathles and Prinzhofer [4] from the 13th to the 16th of August 2018. (**b**) Comparison of changes in solid earth tidal strain at the same site over the same period.

3. Results

As can be seen from Figure 1a and as noted by Cathles and Prinzhofer [4], the changes in H_2 concentration are closely aligned with the maximum rate of decrease in daily atmospheric pressure (the atmospheric tide). Unlike the atmospheric tide, which has one dominant cycle per day, the solid earth tidal strain has two cycles per day (Figure 1b). Although one of the cycles could assist the H_2 venting, no changes in H_2 concentration are associated with the second which is equally strong.

4. Discussion and Conclusions

Because the solid earth tide has two equally strong peaks each day at the H2G site whereas the H_2 venting peaks only one, it appears unlikely that solid earth tides play a dominant role in producing the observed daily changes in H_2 venting in the Sao Francisco Basin of Brazil. It is possible that the solid earth tide contributes to the venting, but the atmospheric pressure tide must be (and could be) the dominant control. Biologic control on venting is, to our knowledge, the only other potential control. It awaits evaluation.

Author Contributions: Conceptualization, J.S., P.F., L.C., methodology, J.S., A.P.; writing-original draft, L.C.; writing-review and editing, all. All authors have read and agreed to the published version of the manuscript.

Funding: This research received no external funding.

Conflicts of Interest: The authors declare no conflict of interest.

References

1. Larin, N.; Zgonnik, V.; Rodina, S.; Deville, E.; Prinzhofer, A.; Larin, V.N. Natural molecular hydrogen seepages associated with surficial, rounded depression on the European craton in Russia. *Nat. Resour. Res.* **2015**, *24*, 369–383. [CrossRef]
2. Zgonnik, V.; Beaumont, V.; Deville, E.; Larin, N.; Pillot, D.; Farrell, K. Evidences for natural hydrogen seepages associated with rounded subsident structures: The Carolina bays (Northern Carolina, USA). *Prog. Earth Planet. Sci.* **2015**, *2*, 31. [CrossRef]
3. Prinzhofer, A.; Moretti, I.; Françolin, J.; Pacheco, C.; D'Agostino, A.; Werly, J.; Rupin, F. Natural hydrogen continuous emission from sedimentary basins: The example of a Brazilian H2-emitting structure. *Int. J. Hydrog. Energy* **2019**, *44*, 5676–5685. [CrossRef]
4. Cathles, L.; Prinzhofer, A. What Pulsating H2 Emissions Suggest about the H2 Resource in the Sao Francisco Basin of Brazil. *Geosciences* **2020**, *10*, 149. [CrossRef]
5. Donzé, F.V.; Truche, L.; Namin, P.S.; Lefeuvre, N.; Bazarkina, E.F. Migration of Natural Hydrogen from Deep-Seated Sources in the São Francisco Basin, Brazil. *Geosciences* **2020**, *10*, 346. [CrossRef]
6. Hsieh, P.A.; Bredehoeft, J.D.; Farr, J.M. Determination of aquifer transmissivity from Earth tide analysis. *Water Resour. Res.* **1987**, *23*, 1824–1832. [CrossRef]
7. Xue, L.; Li, H.-B.; Brodsky, E.E.; Xu, Z.-Q.; Kano, Y.; Wang, H.; Mori, J.J.; Si, J.-L.; Pei, J.-L.; Zhang, W.; et al. Continuous Permeability Measurements Record Healing Inside the Wenchuan Earthquake Fault Zone. *Science* **2013**, *340*, 1555–1559. [CrossRef] [PubMed]
8. Xue, L.; Brodsky, E.E.; Erskine, J.; Fulton, P.M.; Carter, R. A permeability and compliance contrast measured hydrogeologically on the San Andreas Fault. *Geochem. Geophys. Geosyst.* **2016**, *17*, 858–871. [CrossRef]
9. Agnew, D.C. Some Programs for Ocean-Tide Loading, SIO Technical Report, Scripps Institution of Oceanography. Available online: https://igppweb.ucsd.edu/~{}agnew/Spotl/spotlmain.html (accessed on 10 May 2020).

MDPI

St. Alban-Anlage 66

4052 Basel

Switzerland

Tel. +41 61 683 77 34

Fax +41 61 302 89 18

www.mdpi.com

Geosciences Editorial Office

E-mail: geosciences@mdpi.com

www.mdpi.com/journal/geosciences